Other titles in the Collected Studies Series:

DAVID A. KING
Islamic Astronomical Instruments

BERNARD R. GOLDSTEIN
Theory and Observation in Ancient and Medieval Astronomy

DAVID C. LINDBERG
Studies in the History of Medieval Optics

CURTIS WILSON
Astronomy from Kepler to Newton

CHARLES B. SCHMITT
Reappraisals in Renaissance Thought

CHARLES B. SCHMITT
The Aristotelian Tradition and Renaissance Universities

CHARLES B. SCHMITT
Studies in Renaissance Philosophy and Science

JACQUES FONTAINE
Tradition et actualité chez Isidore de Séville

V. I. J. FLINT
Ideas in the Medieval West: Texts and their Contexts

BRUCE S. EASTWOOD
Astronomy and Optics from Pliny to Descartes

GEORGES VAJDA
Etudes de théologie et de philosophie arabo-islamiques
à l'époque classique

GEORGES VAJDA
La transmission du savoir en Islam (VIIe-XVIIIe siècles)

SEBASTIAN BROCK
Syriac Perspectives on Late Antiquity

HAN J.W. DRIJVERS
East of Antioch. Studies in Early Syriac Christianity

WILFERD MADELUNG
Religious Schools and Sects in Medieval Islam

M.J. KISTER
Studies in Jāhiliyya and Early Islam

R.B. SERJEANT
Studies in Arabian History and Civilisation

The Arabs and the Stars

Professor Paul Kunitzsch

Paul Kunitzsch

The Arabs and the Stars

Texts and Traditions on the Fixed Stars, and their Influence in Medieval Europe

VARIORUM REPRINTS
Northampton 1989

British Library CIP data

Kunitzsch, Paul
The Arabs and the stars.
— (Collected studies series; CS307)
1. Astronomy — History
I. Title
520′.9

ISBN 0-86078-255-7

Published in Great Britain by Variorum Reprints
32 Connaught Street
Northampton NN1 3BP

Printed in Great Britain by Galliard (Printers) Ltd
Great Yarmouth Norfolk

VARIORUM REPRINT CS307

CONTENTS

This volume contains xvi + 334 pages.

ACKNOWLEDGEMENTS

Grateful acknowledgement is made to the following for kindly permitting the reproduction of articles originally published by them: The Indian National Science Academy, New Delhi (I); the editor of *Byzantinische Zeitschrift* and C.H. Beck Verlag, Munich (II & XV); Franz Steiner Verlag Wiesbaden GmbH, Stuttgart (III, VII, X, XII, XIII, XIV, XVI, XVII & XXVI); Science History Publications Ltd, Chalfont St Giles (IV, XXI, XXII & XXIII); Munksgaard International Publishers Ltd, Copenhagen (V); the Institute for the History of Arabic Science, University of Aleppo (VI); the Institut für Geschichte der Arabisch-Islamischen Wissenschaften, Frankfurt am Main (VIII & XI); the New York Academy of Sciences (IX); E.J. Brill, Leiden (XVIII, XIX & XX); the Royal Astronomical Society, London (XXIV); Verlag Walter de Gruyter & Co., Berlin (XXV & XXVIII); and Otto Harrassowitz Verlag, Wiesbaden (XXVII).

PUBLISHER'S NOTE

The articles in this volume, as in all others in the Collected Studies Series, have not been given a new, continuous pagination. In order to avoid confusion, and to facilitate their use where these same studies have been referred to elsewhere, the original pagination has been maintained wherever possible.

Each article has been given a Roman number in order of appearance, as listed in the Contents. This number is repeated on each page and quoted in the index entries.

PREFACE

The important role played by 'the Arabs' in the transmission and development of the sciences is well known. This volume presents a selection of 29 articles dedicated mainly to the knowledge of the fixed stars.

To speak of 'the Arabs', in this context, involves two distinct traditions. In pre-Islamic and early-Islamic times, the Arabs — i.e. the inhabitants of the Arabian peninsula, mostly Bedouins — already possessed a good knowledge of the sky. A large number of fixed stars and asterisms were known and used for orientation in the desert (*ihtidā'*), and for calendrical purposes such as determining seasons, agricultural activities and social or religious events. (Many of these 'indigenous Arabic star names' were also mentioned by the writers of classical Arabic poetry.) The astronomy of the Arabs in this early, 'pre-scientific', period was thus a sort of folk-astronomy, as yet uninfluenced by the scientific methods and traditions of other civilizations.

At a later stage, after the spread of Islam through the entire Near and the Middle East, the Arabs became acquainted with the scientific heritage of Greece and with some Indian and Persian traditions. Furthermore, from about the late eighth century A.D. onwards, all the various peoples of the Islamic world took part in the transmission and development of the sciences. Nevertheless, for many centuries, the Arabic language continued to be used as the vehicle for all scientific activity. It seems thus justified to speak of the 'Arabic-Islamic sciences', in which the language was Arabic, but the authors who wrote it belonged to different nations within the Islamic world. In contrast to the 'pre-scientific' astronomical traditions of the early Arabs, from the late eighth/early ninth century onwards there developed a 'scientific' astronomy (and astrology) which was largely based on Greek, partly also on Indian and Persian, material. The most influential source for the 'scientific' knowledge of the fixed stars now became the star catalogue in Ptolemy's *Almagest*, comprising 1,025 stars arranged in the 48 'classical' constellations. (The *Almagest* was translated into Arabic several times from the late eighth through the ninth centuries.)

A third aspect of historical importance is then the transmission of Arabic-Islamic science to medieval Europe, through translations into Latin, in twelfth-century Spain. Through this channel Arabic-Islamic astronomy (and astrology) — quite separately from many other sciences — exerted a fundamental influence on the West and initiated the development of the sciences in late medieval Europe. Testimonies to the Arabic influence of that time remain alive into our own days: technical terms in astronomy such as *zenith*, *nadir*, *azimuth*, and more than two hundred star names (appearing at various stages of the scientific development of the West).

The following articles bring forward evidence for all of these three fields of influence. It is only natural that the *Almagest* is mentioned in nearly every article, being the basic source and reference book for work on the fixed stars in Arabic-Islamic astronomy and, afterwards, in late medieval Europe right down to Copernicus. Items I–V discuss particular star tables and include the edition of some. It may be noted that medieval Byzantium also received and transmitted Arabic-Islamic astronomical knowledge. Item VI provides evidence for the knowledge of the star *Alpha Eridani* in the 'pre-scientific' Arabic period (being invisible in Alexandria in Ptolemy's time, the star was not included in the star catalogue of the *Almagest*). Items VII–X are dedicated to the history of the astrolabe, the most widely used astronomical instrument in the Middle Ages both in the Orient and, subsequently, in Europe. Item XI describes the role played by the most specialized and most influential Arabic-Islamic author on the fixed stars, Abu 'l-Ḥusayn al-Ṣūfī (903–986), and his book on the constellations which left its mark for centuries both in the Orient and in Europe (cf. also item XXIII). My earlier paper, "Ṣūfī Latinus", in *Zeitschrift der Deutschen Morgenländischen Gesellschaft* 115 (1965), 65–74, is superseded by the present article and was therefore not included in this collection. Two biographical notes on al-Ṣūfī were also omitted; they can be consulted in the *Dictionary of Scientific Biography* (ed. C.C. Gillispie), vol. XIII, New York 1976, pp. 149f., and in *Encyclopaedia Iranica*, vol. I, London etc. [1982–] 1985, pp. 148f. (art. 'Abd-al-Raḥmān b. 'Omar Ṣūfī). [An article "al-Ṣūfī and the astrolabe stars" will soon be appearing in the same journal, *Zeitschrift für Geschichte der Arabisch-Islamischen Wissenschaften.*]

The following three articles (XII–XIV) describe the identification of an astrological treatise on the fixed stars ascribed to Hermes in the Latin and Arabic versions, and which ultimately derives from the late Greek tradition on 'The Thirty Bright Stars'. The star list is also edited. Items XV–XVI similarly describe an astrological tradition on stars causing, or indicating, tempests, of which until now only the Greek and the medieval Latin versions have been found; an assumed intermediate Arabic version still awaits discovery amongst the innumerable mass of Arabic manuscripts in collections and libraries all over the world. In addition, item XVII discusses and edits some star lists of astrological interest centered, in this case, around the work of the famous and most influential astrologer Abū Ma'shar (the Latin Albumasar) from ninth-century Baghdad.

It is a great pleasure that I obtained permission for the inclusion of three articles of general interest from the *Encyclopaedia of Islam*, on the Pole (*al-Ḳuṭb*, XVIII), the Milky Way (*al-Madjarra*, XIX) and the Lunar Mansions (*al-Manāzil*, XX). It may be mentioned that a further article, on the Zodiac (*al-Minṭaḳa*), will appear in the same *Encyclopaedia* in due course, and an article on all matters concerning the stars (*al-Nudjūm*) is envisaged for a later volume.

Next there follow three items (XXI–XXIII) illustrating the influence of Arabic-Islamic astronomy in late medieval and Renaissance Europe. A unique example for such influence is found in XXIII: the Renaissance astronomer Peter Apian (1495–1552) incorporated figures of some indigenous Arabic asterisms into one of his star maps, using the book on the constellations of al-Ṣūfī (see above, XI).

Article XXIV was written to warn off historians of astronomy and star names from the book on star names of R.H. Allen (1899, repr. 1963), which, despite being extremely popular right up to the present time, abounds with linguistic mistakes and erroneous associations for the Arabic material and thus has perpetuated misinformation from one generation to the next. For historically and linguistically reliable information on Arabic star names in Western use, the reader should consult my book, *Arabische Sternnamen in Europa*, Wiesbaden: O. Harrassowitz, 1959 (in German), or most recently — written in collaboration with T. Smart — *Short Guide to Modern Star Names and Their Derivations*, Wiesbaden: O. Harrassowitz, 1986 (in English).

It is not the fixed stars, but the planet Mars (and its Arabic name) that is discussed in item XXV, analysing the anecdote on the foundation of the city of Cairo (al-Qāhira).

The next three articles deal with the use of astronomical lore and star names in poetry. In item XXVI the star matter in an anthology of verse compiled by the Spanish-Arabic author Ibn al-Kattānī (died around 1029) is analysed. Item XXVII discusses the use of the Greek name *Zeus* for the planet Jupiter in a poem of the famous poet Abū Nuwās (early ninth-century Baghdad) and describes some other astronomical motifs in his poetry. Lastly, item XXVIII analyses a section from the epic *Vīs u Rāmīn* by the Persian poet Gurgānī (written around 1050), which introduces, in connection with a horoscope, all the classical constellations of the fixed stars.

The final item (XXIX) is an unpublished index of star names used in the writings of the Arab seafarers of the Indian Ocean, Aḥmad ibn Mājid (died around 1500) and Sulaymān al-Mahrī (first half of the 16th century). This index is intended to facilitate the location of discussions and analyses of these names which are scattered through a number of books and articles of mine (not contained in the present selection).

In contrast to the works of several scholarly colleagues which are dedicated to the themes of theoretical astronomy, planetary theory and mathematical astronomy of the Arabic-Islamic civilization, this collection presents a panorama of studies on the matter of the fixed stars. For some major works of mine in this field, the following should be mentioned. On indigenous Arabic star names: *Untersuchungen zur Sternnomenklatur der Araber*, Wiesbaden: O. Harrassowitz, 1961, and *Über eine* anwā'*-Tradition mit bisher unbekannten Sternnamen*, München 1983 (Bayerische Akademie der Wissenschaften, Philos.-Hist. Kl., Sitzungsberichte, Jahrgang 1983, Heft 5); on the Arabic and the subsequent Latin transmission of the *Almagest*: *Der Almagest. Die Syntaxis Mathematica des Claudius Ptolemäus in arabisch-lateinischer Überlieferung*, Wiesbaden: O. Harrassowitz, 1974. The star catalogue of the *Almagest* is now being edited; the first volume containing the two surviving Arabic versions of al-Ḥajjāj ibn Yūsuf ibn Maṭar and of Isḥāq ibn Ḥunayn has recently appeared: Claudius Ptolemäus, *Der Sternkatalog des Almagest. Die arabisch-mittelalterliche Tradition*, I: *Die arabischen Übersetzungen*, Herausgegeben, ins Deutsche übertragen und bearbeitet von Paul Kunitzsch, Wiesbaden: O. Harrassowitz, 1986. The second volume containing the Latin version made from the Arabic by Gerard of Cremona (Toledo, ca. 1150–80) is in the press; the third volume containing a comprehensive concordance of the star coordinates according to seven basic versions, Greek, Arabic and Latin, is ready for publication. The highly important treatise of Ibn al-Ṣalāḥ (died 1154) on the mistakes in the transmission of the star coordinates in the star catalogue of the *Almagest* — the author quotes directly from five versions accessible to him and thereby gives us the only clue for the interpretation of the material in the surviving Arabic texts — has already been edited: Ibn aṣ-Ṣalāḥ, *Zur Kritik der Koordinatenüberlieferung im Sternkatalog des Almagest*, Göttingen 1975 (Abhandlungen der Akademie der Wissenschaften in Göttingen, Philol.-Hist. Kl., 3. Folge, Nr. 94). In addition, there should be mentioned the edition of a collection of the most prominent types of star tables in medieval Europe (including the Arabic originals for some of them): *Typen von Sternverzeichnissen in astronomischen Handschriften des zehnten bis vierzehnten Jahrhunderts*, Wiesbaden: O. Harrassowitz, 1966.

The articles in this collection are reproduced in their original form (adapted in size to the page format here used), but taking the opportunity to correct various misprints and minor errors on the pages themselves. In some instances (items II, V, XXVII and XXVIII) an additional note has been added to an article. The *Index* lists proper names, some titles of works and a choice of topics of major interest, and concludes with a list of the star tables discussed or edited in the present articles.

It is a pleasure to express my sincere thanks to the publishers of these articles for kindly consenting to their inclusion here. Lastly, I feel indebted to my friend and colleague, Prof. D. King of Frankfurt, for initiating the publication of this volume.

PAUL KUNITZSCH

Munich,
June 1989

I

STAR CATALOGUES AND STAR TABLES IN MEDIAEVAL ORIENTAL AND EUROPEAN ASTRONOMY*

In mediaeval times, the knowledge of the fixed stars became manifest mostly in catalogues and tables of these stars. Apart from that, a major point of discussion, in this connexion, was precession which had been fixed at a rate of 1° in 100 years by Ptolemy, a value which was later improved to 1° in 66 or 70 years by Islamic astronomers.[1] In addition to precession, some astronomers also pursued the theory of trepidation which, however, was not accepted by Ptolemy and consequently not by most of the great astronomers in Islamic astronomy.

For convenience, in the following discussion we shall basically distinguish between star catalogues and star tables. Star catalogues will be the complete long lists of the fixed stars following the model of Ptolemy's catalogue in the *Almagest*, whereas star tables will be the smaller lists of fundamental or otherwise important stars, often devised for the astrolabe, and mostly containing from 20 to 70 or so stars.

The model for cataloguing the fixed stars had been set up by Ptolemy. In books vii and viii of the *Almagest* he had listed 1,025 fixed stars with their ecliptical coordinates and magnitudes, arranged in 48 constellations, for the epoch 137 A.D.[2] This catalogue of stars visible to the naked eye became the standard model for describing the fixed stars both in the Islamic Orient and in mediaeval Europe until the invention of the telescope, which led to a new view of the stellar sky and to the formation of new types of star catalogues.

Together with many other scientific works of ancient Greek and Hellenistic times, the *Almagest* was translated into Arabic several times from the late 8th to the late 9th centuries.[3] These Arabic versions of Ptolemy's work, called *al-majasṭī* by the Arabs[4] (and hence *Almagest* by the Europeans[5]), became of greatest influence in the development of scientific astronomy in the Arabic-Islamic civilization. Two

*Lecture delivered in the Colloquium on *History of Oriental Astronomy*, held in November 1985, at New Delhi.

of these translations have survived, in manuscripts form, till our times, the one by al-Ḥajjāj ibn Yūsuf ibn Maṭar, and the other by Isḥāq ibn Ḥunayn (with emendations by Thābit ibn Qurra). A critical edition of the star catalogue from these two Arabic translations is presently under preparation.[6]

All the succeeding star catalogues in the Islamic world imitated Ptolemy's catalogue in every detail—the numbers of the stars and constellations, their distribution, the nomenclature, and the coordinates and locations of the stars. The most famous star catalogues in the Islamic world are those of al-Battānī, al-Ṣūfī, al-Bīrūnī, and Ulugh Bēg, to which can be added Naṣīr al-Dīn al-Ṭūsī's recension of the *Almagest* completed in 1247 which again repeated the *Almagest* catalogue (in Isḥāq ibn Ḥunayn's translation).

The star catalogue of al-Battānī is prepared for the epoch 880 and uses a longitude value of ≪ Ptolemy $+11°10'$ ≫ applying the improved Arabic precession constant of 1° in 66 years. Of Ptolemy's 1,025 stars, al-Battānī has retained only 533, that is slightly more than a half. The catalogue has been edited, incorporated in al-Battānī's complete astronomical work, by Carlo Nallino,[7] who also added a modern Latin translation of his own. In mediaeval times, only the accompanying text had been translated into Latin, but not the star catalogue. However, an Old-Spanish translation made in the time of King Alfonso X of Castile also includes the star catalogue, and Nallino has incorporated in his edition two missing pages in the Arabic text from that Old-Spanish translation. Recently it has been found that Nallino committed a fundamental error in establishing the coordinates of al-Battānī's star catalogue; the reason was that at his time he had no full knowledge of the Arabic transmission of the *Almagest*. Therefore, erroneously, he emended many of the coordinates in the unique Arabic manuscript of al-Battānī's work according to the values found in some versions of the *Almagest* accessible to him. He failed to notice that al-Battānī had not followed those more recent versions. A few years ago, the evaluation of an Arabic critic of the transmission of the coordinates in Ptolemy's star catalogue by the Islamic scholar Ibn al-Ṣalāḥ of the first half of the 12th century[8] has proved that al-Battānī had used the so-called "old Arabic version" of the *Almagest* which has not survived in direct tradition.[9] Thus, it has become evident that most of the coordinate values found in al-Battānī's manuscript, which were emended by the editor Nallino according to more recent versions of the *Almagest*, were in fact correct and derived from the "old version" as cited by that 12th century author. The star catalogue in Nallino's edition, therefore, can no longer be regarded as reliable.

al-Ṣūfī, of Persian origin, is most famous for his *Book on the constellations of the fixed stars*.[10] For each of the 48 Ptolemaic constellations, this work contains first, a general description of the constellation and its individual stars including al-Ṣūfī's criticism, especially with regard to star positions, or magnitudes ; second, a section on the indigenous Arabic star names falling under the respective Greek constellation, and the identification of the indigenous Arabic names with the respective stars of Ptolemy's catalogue ; third, two drawings of the constellation, one as it is seen in the sky, and one as it is seen on a celestial globe ; and fourth, a table of the stars making up the constellation. These 48 tables together form a complete star catalogue. al-Ṣūfī also applies the Arabic precession constant of 1° in 66 years and uses, for his epoch 964, a longitude value of $\ll$ Ptolemy+12°42′ $\gg$. As explained in his own introduction, he did not follow one specific Arabic version of the *Almagest*, but rather drew upon all the versions available to him, because he had realized that the various versions were often in disagreement and so he selected what seemed appropriate to him. al-Ṣūfī's work was of great influence not only in the Islamic Orient but also—as we shall see later—in mediaeval and Renaissance Europe.

The same attitude towards the *Almagest* was displayed by al-Bīrūnī who included in his astronomical *magnum opus*, the *Qānūn al-Mas'ūdī*,[11] the full catalogue of Ptolemy's stars, using for his epoch 1030 a longitude value of $\ll$ Ptolemy+13° $\gg$. He selected values that appeared suitable to him from the various Arabic versions of the *Almagest*. In the terminology and nomenclature of the stars, he abbreviated much of the lengthy formulae of the *Almagest* and introduced many new short or less complicated designations.

Ulugh Bēg, who for a short time (1447-49) was the ruler of Tīmūr's Mongol empire, had assembled a number of outstanding astronomers at his court who constructed for him the observatory at Samarcand and who composed under his supervision an astronomical work, which also contained the full catalogue of Ptolemy's stars. In the introduction to his star catalogue Ulugh Bēg says that he especially relied upon al-Ṣūfī's work, but when he found discrepancies between al-Ṣūfī's indications and the actual situation in the sky he preferred to repeat himself the observation of all those stars with the exception of a small number of stars invisible to him at Samarcand and which he listed in his catalogue by merely adding a precession constant of 1° for 70 years to Ptolemy's longitudes, for the epoch 1437. The Persian text of Ulugh Bēg's catalogue was edited, together with a Latin translation and an ample commentary, by Thomas Hyde.[12] (In the commentary Hyde quoted a great number of Arabic star names from al-Ṣūfī's work many of which were later introduced into modern astronomical use by Giuseppe Piazzi, 1814.[13]) In 1917, E.B. Knobel published in Washington an English version of the catalogue,[14] including recensions of the coordinates from a large number of manuscripts. For the star nomenclature Knobel followed Schjellerup's French translation of al-Ṣūfī's book of 1874. Knobel's philological comments on Ulugh Bēg's Persian nomenclature are often erroneous and incorrect.

The star nomenclature used by Ulugh Bēg is partly Persian, translated from the Arabic, but partly the original Arabic terms have been retained. Recently I have observed that the textual descriptions of the stars in Ulugh Bēg's catalogue are obviously not the work of Ulugh Bēg. It is known that Naṣīr al-Dīn al-Ṭūsī, whose recension of the *Almagest* of 1247 we have mentioned above, completed perhaps in 1250 a Persian translation of al-Ṣūfī's *Book on the constellations of the fixed stars*. There exists, in Istanbul, a manuscript of al-Ṭūsī's translation (*i.e.*, *Aya Sofya* 2595, 104 folios, stemming from Ulugh Bēg's library) dated at that very date (1250),[15] and it has been assumed by some scholars that it might be al-Ṭūsī's autograph. This manuscript has been edited, in facsimile, in Teheran in 1969.[16] The question whether the Istanbul manuscript is indeed al-Ṭūsī's autograph or not, is still in dispute. I own a copy of the facsimile edition, and I must confess that I have strong doubts about its being an autograph of al-Ṭūsī's. (There exists also a printed edition from that same manuscript published in Teheran, 1972,[17] of which I was unable to obtain a copy.) Whatever be the question of the autograph—anyhow, here we have the text of al-Ṭūsī's Persian translation of al-Ṣūfī's work. A detailed comparison between al-Ṭūsī's translation and Ulugh Bēg's catalogue shows that Ulugh Bēg repeats literally the mixed Persian-Arabic descriptions of the stars given by al-Ṭūsī. Thus, it can be taken for granted that Ulugh Bēg did not commit himself to the translation of al-Ṣūfī's Arabic work, or rather his texts of the star descriptions, but instead he merely used and copied the earlier Persian translation of al-Ṭūsī. Ulugh Bēg's own contribution then lies in the new observation of the stars and the fixing of their coordinates according to his observations of 1437.

So much on the most important star catalogues of the Islamic Orient. Let us now turn to mediaeval Europe.

Here also Ptolemy's *Almagest* formed the basic model for star catalogues which was followed down to Copernicus, and till the invention and astronomical application of the telescope.

Mediaeval Europe, as is well known, had almost totally lost its contact with the Greek sources. Works of Greek or Hellenistic origin were only read in Latin translations. Thus, the first Latin translation of the *Almagest* in Europe was made in Sicily, from the Greek, around 1160 as had been assumed earlier, or before 1154 by Hermannus de Carinthia as is at present assumed by Richard Lemay.[18] This translation remained without an echo in Europe. Presently, there are known only two fragmentary manuscripts of it, and one complete manuscript (Vat. lat. 2056, 13th cent.).

Later, perhaps in 1175, in Toledo, Gerard of Cremona translated the *Almagest* from the Arabic into Latin. This translation became the standard version known and used in Western Europe till Renaissance times when Greek texts started again to become available.[19] Around 50 manuscripts of this version are still known today, 35 of which complete with regard to the star catalogue. In 1515, this version was printed in Venice. A critical edition of Gerard's version of the star catalogue is at present under preparation.[20]

An Old-Spanish version of Ptolemy's star catalogue was later included in King Alfonso X's astronomical work *Libros del saber de astronomia*, using the Alfonsine longitude value of ≪ Ptolemy+17°8′ ≫. It has not yet been investigated from which sources this translation was made, but certainly Arabic texts were involved. This translation remained without any visible influence on the subsequent European astronomy. The catalogue is included, in the form of *ruedas* (*i.e.*, the stars of each constellation being arranged in the form of a rosette) in vol. i of Manuel Rico y Sinobas' edition of the *Libros del saber*, Madrid, 1863. The edition, unfortunately, is unreliable in matters of spelling and philology. Around 1341, an Old-Italian translation of the *Libros del saber* was made in Florence which has survived in a unique manuscript (Vat. lat. 8174). Book i from it, including the star catalogue, has been edited by P. Knecht in Zaragoza, 1965.[21]

Ptolemy's star catalogue in Gerard of Cremona's Latin translation was so much favoured in Europe that it was also included in or appended to some other astronomical collections.

One of these are the so-called *Alfonsine Tables*. The Spanish original text of these tables has not survived ; only the theoretical introduction is still extant, and it was included by Rico y Sinobas in vol. iv of his edition of the *Libros del saber* (though not forming part of the *Libros* proper). The form generally known in Western Europe as the *Alfonsine Tables* is a collection of planetary and other tables, in Latin, that were composed, under circumstances still insufficiently known, in Paris in the 1320ies.[22] To this Latin corpus of the *Alfonsine Tables*, of which innumerable manuscripts exist and many printed editions were published between 1483 and 1641,[23] was also added a star catalogue. This catalogue is nothing else than Gerard of Cremona's version of the *Almagest* catalogue, with the longitudes of the stars adapted to the Alfonsine value, *i.e.* ≪ Ptolemy+17°8′ ≫.

Secondly, there exists a mediaeval European tradition of al-Ṣūfī. This so-called *Ṣūfī Latinus* is a compilation of disparate astronomical and astrological elements only some of which are really related to al-Ṣūfī. The fundamental section is a star catalogue accompanied by drawings of the 48 constellations, one drawing of each constellation, variously selected from al-Ṣūfī's two drawings of each of them. The star catalogue again proves to be Gerard of Cremona's Latin translation of the *Almagest* catalogue, with the longitudes of the stars adapted to al-Ṣūfī's value, *i.e.* ≪ Ptolemy+12°42′ ≫. Of the *Ṣūfī Latinus* till now eight manuscripts have come to light.[24]

Thus, the leading specimens of star catalogues in Europe, until Copernicus' time, in the *Almagest* itself, in the *Alfonsine Tables*, and in the so-called *Ṣūfī Latinus* tradition, are all formed by one single text, Gerard of Cremona's Latin translation of the *Almagest* star catalogue. Of these, the catalogue appended to the *Alfonsine Tables* was the most popular. It was often copied alone by itself, without the other pertaining tables, and was inflated by later authors with star and constellation names compiled from various astronomical and astrological sources.[25]

Now regarding star tables. These minor star lists are so frequent both in the Islamic Orient and in mediaeval Europe that it is impossible to go here into details.

In the Islamic Orient nearly every astronomer composed one or the other star table. Of the most frequent types of such tables, I shall name the following.

A very common kind of astronomical works in the Orient are the so-called *Zījes*.[26] These usually consist of a theoretical introduction and a collection of astronomical tables, for chronology, for the planets, sun and moon, *etc.* Mostly, a star table is also included. The star tables in the *zīj*-works mostly are merely the result of computation or compilation, and not the result of independent star observation. Most of them use ecliptical coordinates which are derived from the *Almagest* tradition by adding to the longitudes one of the constants for precession, current in the Islamic Orient.

Very rare, both in the Orient and in mediaeval Europe, are the cases of independent star observation and resulting star tables.

One example in the Islamic world for independent star observation is the observations carried out by the order of the Abbaside caliph al-Ma'mūn. The results of these observations including a table of 24 stars are laid down in the so-called *Zīj al-Mumtaḥan* (or, in Latin translation, *Tabulae Probatae*) for the epoch 829/30. The table has been printed, in translation, by J. Vernet in 1956.[27] This independent star table, afterwards, had the same fate as Ptolemy's catalogue, *i.e.* it was used as a basis for other star tables that were merely compiled and computed from its values. To the star tables derived from the *Mumtaḥan* star table of 829/30 belong, as I have demonstrated in an article published in 1970,[28] the astrolabe star list of al-Farghānī (25 stars, epoch 856/57) ; a table of 30 stars in the *Zīj* of Ḥabash for the epoch 916/17 (I refer to the Berlin MS, Ahlwardt no. 5750) ; and even two star tables translated into Greek, in Byzantine times, found in cod. Vat. gr. 1056 (14th cent.).

Other independent Islamic observations were made by al-Ṣūfī. As for Ulugh Bēg and his presumed observations, these still need closer investigation to make sure whether his star coordinates are really the result of his own observations, or merely computed from earlier existing catalogues, especially from al-Ṣūfī's catalogue.

Since I mentioned Byzantine Greek translations of Arabic star tables, I can add that two other Islamic star tables, that of al-Khāzinī in his *Zīj al-Sanjarī* (epoch 1115) and that of al-Ṭūsī in his *Zīj-i Īlkhānī* (epoch 1232), exist in Greek translation in several manuscripts, included in the so-called *Persian Syntaxis* of Georgios Chrysokokkes from the middle of the 14th century. (They have been edited, in their Arabic and Greek versions, in an article in 1964.[29])

Another frequent type of Islamic star tables are lists of astrolabes tars. Their stars are often accorded, apart from the usual eliptical coordinates, equatorial coordinates which were preferred for practical use on the instrument. The equatorial coordinates mostly show values up to the smallest units of minutes which never could be obtained by observation in those times. The explanation is that these equatorial coordinates were derived through calculation from the well established ecliptical coordinates. The same holds good for similar star tables in mediaeval Europe.

Still another type of star tables consists of mere extracts from complete great star catalogues, as for example, a table of 48 stars by Kūshyār ibn Labbān extracted from the *Almagest* (in the Berlin MS Ahlwardt no. 5751, epoch 932), or a table of 81 stars by al-Kharaqī (in the Berlin MS Ahlwardt no. 5669, epoch 1112) derived from al-Ṣūfī's work on the constellations.

In mediaeval Europe such minor star tables were initially introduced through translations of Arabic tables. Thus, the oldest western star table is a list of 27 astrolabe stars derived from an Arabic model, datable to the late 10th century, in Spain. Its coordinates are *mediatio coeli* (*i.e.*, the degree of the ecliptic culminating together with the respective star), which were derived from Arabic models, and a second value of a non-traditional character which could only be read off from an astrolabe.[30] Very old is also a table of 21 astrolabe stars for the epoch 978 by the Spanish-Arabic astronomer Maslama al-Majrīṭī of which a Latin translation (of uncertain origin)[31] and an expanded Latin version of 45 stars[32] have been transmitted. Succeeding to such translated tables, European astronomers compiled and composed their own star tables using the translated shorter tables or the complete Ptolemaic catalogue as models.

The main types of star tables composed and circulated in mediaeval Europe were lists of astrolabe stars or stars to be used on other instruments such as the quadrant or the celestial globe, and star tables extracted from Ptolemy's catalogue, mostly in that recension of it which was included in the *Alfonsine Tables*. Works of the Arabic *zīj*-type are not frequent in Europe. Of this kind, we could mention the *Toledan Tables* (compiled from material translated from the Arabic)[33] in which several types of star tables were included, all of which translated from the Arabic, *viz.* from various recensions of a star table of the Spanish-Arabic astronomer al-Zarqāllu (or, in Latin, Azarchel ; d. 1100).

The most frequent work of the *zīj*-type circulated in Europe were the so-called *Alfonsine Tables*, in the Latin version composed in Paris in the 1320ies. To these was added Ptolemy's star catalogue in the wording of Gerard of Cremona's Latin translation from the Arabic, as mentioned above.

17 types of European star tables from the tenth to the fourteenth centuries have been edited by myself in a book in 1966,[34] some of which being translations from the Arabic, together with their Arabic originals.

Star tables resulting from independent observation are equally rare in mediaeval Europe as in the Islamic Orient. Most of the tables were formed by merely adding a constant for precession to the Ptolemaic ecliptical coordinates, or by transforming given ecliptical coordinates through calculation into equatorial coordinates. A rare instance of independent observation is found in the table of 40 stars with ecliptical coordinates observed by means of an armillary sphere in Paris in 1246 by John of London. In another paper in "Colloquium 91" a few days ago, I spoke at length about John of London's star table.[35] It has been transmitted in several manuscripts. Many of the new Arabic star names introduced by John of London are still used today in modern international astronomy. Several stars of his table were merged, a few years later, together with a number of stars of the oldest European star table from the end of the 10th century, into a new list of 49 astrolabe stars (with coordinates *mediatio coeli* and declination) which was appended to and transmitted together with the astrolabe treatise commonly ascribed to the Arabic astronomer and astrologer Messahalla.[36] This ascription has been demonstrated to be false ; the entire text is a compilation of the middle or second half of the 13th century from various translated Arabic or other compiled Western astrolabe writings.[37] But Pseudo-Messahalla's treatise became the most famous and most widely spread astrolabe treatise in late mediaeval Europe and, therefore, that offspring from John of London's star table also became one of the star tables with the richest transmission in Europe.

At the end it must be underlined that our present knowledge of mediaeval astronomy, both in the Islāmic Orient and in Europe, is still utterly restricted. The major part of the material still remains unread, unstudied, and unpublished, in the libraries of all the world. It may seem that the main lines of transmission from classical antiquity to the Islamic Orient and further on to mediaeval Europe are now sufficiently clear. Nevertheless, many details still lack documentary evidence not to speak of the minor works and unknown or lost writings which might come to light some day. Therefore, it remains a duty for many generations to come to examine carefully the treasures of the libraries in all the world and to publish as many as ever possible of the original sources in all the languages involved.

References and Notes

[1]*Cf.* Nallino (as in note 7, below), vol. i, p. 292f.

[2]A recent English translation is *Ptolemy's Almagest*, Translated and Annotated by G. J. Toomer, London, 1984.

[3]For a study of the various Arabic translations of the *Almagest* see P. Kunitzsch, *Der Almagest. Die Syntaxis Mathematica des Claudius Ptolemäus in arabisch-lateinischer Überlieferung*, Wiesbaden, 1974.

[4]*See* Kunitzsch, *op. cit.*, pp. 117–125.

[5]*See* Kunitzsch, *op. cit.*, pp. 115–117.

[6]Claudius Ptolemäus, *Der Sternkatalog des Almagest. Die arabisch-mittelalterliche Tradition.* vol. i : *Die arabischen Übersetzungen.* Herausgegeben und bearbeitet von P. Kunitzsch, Wiesbaden, 1986.

[7]*Al-Battānī sive Albatenii opus astronomicum*, Ed. and Transl. by C. A. Nallino, i–iii, Milan, 1899–1907; repr. Hildesheim/New York, 1977.

[8]Ibn aṣ-Ṣalāḥ, *Zur Kritik der Koordinatenüberlieferung im Sternkatalog des Almagest*, Ed. and Transl. by P. Kunitzsch, Göttingen, 1975.

[9]*Cf.* P. Kunitzsch, New Light on al-Battānī's *Zīj*, *Centaurus*, **18**, 270–274, 1974; *idem*, in the work cited in note 8, Appendix II, pp. 97–108.

[10]*Ṣuwaru'l-Kawākib or Uranometry*, Published by the Dairatu'l-Ma'arifi'l-'Osmania, Hyderabad/Dn., 1373/1954; repr. Beirut, 1981; French transl. by H. C. F. C. Schjellerup, *Description des étoiles fixes...par...Abd-al-Rahman al-Sûfi*, St. Pétersbourg, 1874.

[11]*Al-Qānūn al-Mas'ūdī*, Published by the Dairatu'l-Ma'arifi'l-'Osmania, i–iii, Hyderabad/Dn., 1373–75/1954–56.

[12]*Tabulae longitudinis ac latitudinis stellarum fixarum, ex observatione Ulugh Beighi*, Ed. and Transl. by Thomas Hyde, Oxford, 1665.

[13]G. Piazzi, *Praecipuarum stellarum inerrantium positiones...*, Palermo, 1814.

[14]E. B. Knobel, *Ulugh Beg's Catalogue of Stars*, Washington, 1917.

[15]*Cf.* M. Krause, *Stambuler Handschriften islamischer Mathematiker*, Berlin, 1936 (Quellen und Studien zur Geschichte der Mathematik Astronomie und Physik, Abt. B : Studien, Bd. 3, Heft 4), p. 498 (work no. 20).

[16]*Tarjama-i ṣuwar-i kawākib*, be qalam-i Khwāja Naṣīr al-Dīn-i Ṭūsī, Tehran, 1348 Sh. (= 1969).

[17]Ed. by Sayyed Mo'ezz al-Din Mahdavi, Tehran, 1351 Sh. (= 1972).

[18]A paper by R. Lemay, "Hermann de Carinthe, auteur de la traduction "sicilienne" de l'*Almageste* à partir du grec", is presently in preparation.

[19]For a study of Gerard's Latin translation see the work cited in note 3, above.

[20]It will appear later as vol. ii of the work cited in note 6, above.

[21]P. Knecht, *I libri astronomici di Alfonso X in una versione fiorentina del trecento*, Zaragoza, 1965.

[22]*Cf.* the introduction to the recent edition by E. Poulle, *Les Tables Alfonsines avec les canons de Jean de Saxe*, Paris, 1984. (This edition does not include the star catalogue.)

[23]For my studies of the star catalogue, I have used the following editions : Venice 1483; Venice 1492; Venice 1518 (but at the end of the book : 1521); Venice 1524; Paris 1545; Paris 1553; and Madrid 1641.

[24]*Cf.* P. Kunitzsch, Ṣûfī Latinus, *Zeitschrift der Deutschen Morgenländischen Gesellschaft*, **115**, 65–74, 1965. Here, six manuscripts are mentioned. In the meantime, two more manuscripts have been found : Berlin, Hamilton, cod. 556 (78 D 12); and Gotha, cod. M II 141. From the latter, the drawings of the 48 constellations have been edited, in facsimile, by G. Strohmaier, *Die Sterne des Abd ar-Rahman as-Sufi*, Leipzig/Weimar, 1984.

[25]A good example of that inflated type is in Vienna, cod. Lat. 5415 (dated 1444), fol. 217r–251r.

[26]*Cf.* E. S. Kennedy, *A Survey of Islamic Astronomical Tables*, Philadelphia, 1956 (Transactions of the American Philosophical Society, N.S., vol. 46, part 2).

[27]J. Vernet, Las 'Tabulae Probatae', in *Homenaje a J. M. Millás Vallicrosa*, ii, Barcelona, 1956, pp. 501ff., esp. p. 519.

[28]P. Kunitzsch, Die arabische Herkunft von zwei Sternverzeichnissen in cod. Vat. gr. 1056, *Zeitschrift der Deutschen Morgenländischen Gesellschaft*, **120**, 281–287, 1970.

[29]P. Kunitzsch, Das Fixsternverzeichnis in der 'Persischen Syntaxis' des Georgios Chrysokokkes, *Byzantinische Zeitschrift*, **57**, 383–406, 1964.

[30]Edited and discussed as "Type III" in the work cited in note 34, below.

[31]It is "Type I" in the work cited in note 34, below.

[32]"Type II" in the work cited in note 34, below.

[33]*Cf.* G. J. Toomer, A survey of the Toledan Tables, *Osiris*, **15**, 1968.

[34]P. Kunitzsch, *Typen von Sternverzeichnissen in astronomischen Handschriften des zehnten bis vierzehnten Jahrhunderts*, Wiesbaden, 1966.

[35]It will be published in the "Proceedings" of that Colloqium. See also P. Kunitzsch, John of London and his unknown Arabic source, *Journal for the History of Astronomy*, **17**, 51–57, 1986.

[36]It is "Type VIII" in the work cited in note 34, above.

[37]*Cf.* P. Kunitzsch, On the authenticity of the treatise on the composition and use of the astrolabe ascribed to Messahalla, *Archives Internationales d'Histoire des Sciences*, **31**, no. 106, 42–62, 1981; M. Viladrich, On the sources of the Alphonsine Treatise Dealing with the Construction of the Plane Astrolabe, *Journal for the History of Arabic Science*, **6**, 167–171, 1982.

II

DAS FIXSTERNVERZEICHNIS IN DER „PERSISCHEN SYNTAXIS“ DES GEORGIOS CHRYSOKOKKES

In der Kulturgeschichte ist bekannt, welche bedeutende Rolle der islamisch-arabische Kulturkreis bei der Überlieferung antiken Wissens an das moderne Europa einst spielte. Zeugen für diesen Traditionsvorgang wie auch für die eigenen Leistungen des Orients bilden die zahlreichen mittelalterlichen europäischen Werke vieler Disziplinen, die teils aus dem Arabischen übersetzt, teils aus arabischen Vorlagen kompiliert sind. Sie enthalten meist nicht nur das übernommene Wissen, sondern auch Wörter, termini und Namen aus der orientalischen Originalsprache.

Von solchen Werken sind vor allem diejenigen der westeuropäischen lateinischen Überlieferung bekannt, da sie im Traditionskontinuum einen zentralen Platz einnahmen und jahrhundertelang weiterwirkten.

Folgende Abkürzungen werden durchgehend benutzt: I, II und III sind die drei Verzeichnis-Typen; a, b, c usw. sind die entsprechenden Fundstellen; 1, 2, 3 usw. bezeichnen die laufenden Nummern der Sterne innerhalb jedes Typs (II b 12 bedeutet also beispielsweise den 12. Stern des Verzeichnisses von Typ II in der Version von ms Pal. Vat. 312, fol. 93^v). Hierzu kommen die orientalischen Quellen A^1 und A^2 *(az-Zīǧ as-Sanǧarī)* sowie P^1 (mit p^1) und P^2 *(Zīǧ-i Īlḫānī* bzw. *Zīǧ-i Šams-i Munaǧǧim)*.

Für die Namen der Tierkreiszeichen und Sternbilder werden die üblichen astronomischen Abkürzungen gebraucht.

Abgekürzt zitierte Literatur:

BERNARD: E. BERNARD, *The Longitudes, Latitudes, right Ascensions, and Declinations of the chiefest fixt Stars* . . ., Philosophical Transactions, Oxford, XIV (1684).

CCAG: *Catalogus Codicum Astrologorum Graecorum*, I–XII, 1898–1953.

DEVREESSE: R. DEVREESSE, *Introduction à l'étude des manuscrits grecs*, Paris 1954.

GAL: C. BROCKELMANN, *Geschichte der arabischen Literatur*, Leiden, I ²1943, II ²1949, Suppl. I 1937, II 1938, III 1942.

HUNGER: H. HUNGER, *Katalog der griechischen Handschriften der Österreichischen Nationalbibliothek*, I, Wien 1961.

KENNEDY: E. S. KENNEDY, *A Survey of Islamic Astronomical Tables*, Trans. Americ. Philos. Soc., Philadelphia, N. S. 46/2, (1956), p. 123–177.

KNOBEL: E. B. KNOBEL, *The Chronology of Star Catalogues*, Mem. Royal Astron. Soc., London, XLIII (1875–77), p. 1–74.

KRAUSE: M. KRAUSE, *Stambuler Handschriften islamischer Mathematiker*, Quellen u. Studien zur Gesch. d. Math., Astron. u. Phys., Abt. B, Bd. 3, Heft 4, p. 437–532 (Berlin 1936).

KUNITZSCH, *Sternnamen*: P. KUNITZSCH, *Arabische Sternnamen in Europa*, Wiesbaden 1959.

Die wissenschaftliche Literatur der Byzantiner blieb demgegenüber im westlichen Europa lange Zeit unbekannt und hatte an dem großen Traditionsprozeß so gut wie keinen Anteil. So hat ihr auch die Forschung bis heute entsprechend weniger Aufmerksamkeit zugewandt.[1] Auch in der mittelgriechischen Literatur gibt es indessen Texte, die, ähnlich den zahlreichen lateinisch-westeuropäischen Werken, Elemente orientalischer Herkunft aufweisen. In der astronomischen Nomenklatur zeigen sich solche Einflüsse bei Mondstationen,[2] Planeten,[3] Unwettersternen,[4] Fixsternen und natürlich in der Fachterminologie.

Eines dieser auf orientalischen Vorlagen beruhenden mittelgriechischen astronomischen Werke ist die Ἐξήγησις εἰς τὴν σύνταξιν τῶν Περσῶν des Georgios Chrysokokkes. Darin ist ein Fixsternverzeichnis mitenthalten, das im folgenden besprochen werden soll.[5]

Kunitzsch, *Untersuchungen*: P. Kunitzsch, *Untersuchungen zur Sternnomenklatur der Araber*, Wiesbaden 1961.

Mercati: J. Mercati und P. Franchi de' Cavalieri, *Codices Vaticani Graeci*, I, Rom 1923 (Nr. 1–329).

Neugebauer: O. Neugebauer, *Studies in Byzantine Astronomical Terminology*, Trans. Americ. Philos. Soc., N. S. 50/2 (1960).

Sarton: G. Sarton, *Introduction to the History of Science*, Baltimore, I–III, 1927–1948.

Suter: H. Suter, *Die Mathematiker und Astronomen der Araber und ihre Werke*, Leipzig 1900 (Nachträge 1902 und ISIS 18, 1932, p. 166–183).

Usener: H. Usener, *Ad historiam astronomiae symbola*, Bonn 1876 = *Kleine Schriften* III, Leipzig 1914.

[1] Cf. Sarton, Indices und die Übersichten und Resumés bei Usener, Neugebauer, *Studies* p. 3 f., Devreesse p. 241 f., sowie in einem Artikel von D. Pingree, Chicago, über „Palaeologan Astronomy", der demnächst in „Dumbarton Oaks Studies" erscheinen soll und von dem der Autor mir freundlichst einen Durchschlag vorab überließ.

[2] CCAG 5, 3 p. 90 f. (Liste aus ms Vat. gr. 1056, s. XIV, fol. 136; dazu p. 91 f. Liste aus ms Paris B. N. gr. 2419, fol. 27); CCAG 8, 1 p. 217 f (Liste aus ms Paris B. N. gr. 2424, s. XIV ex.); und CCAG 9, 1 p. 138 ff. (Liste aus ms Oxford Cromwell 12, s. XV, p. 402).

[3] CCAG 7 p. 48 f. (ms Berlin Phill. 1577, s. XV, fol. 50, 51^v).

[4] Περὶ ἀστέρων ποιούντων τάραξιν ... Übersetzung von Arsenios Monachos aus der „Geomantie" des „persischen Philosophen" Zanati (Ζανατῆ; aš-Šayḫ Abū ʿAbdallāh az-Zanātī, cf. GAL Suppl. II p. 1037, c. 19, Nr. 5; ausführlicher M. Steinschneider, *Die hebräischen Übersetzungen des Mittelalters*, Berlin 1893, p. 855 f.; siehe auch Devreesse p. 258 n. 3). Dieser Text scheint relativ häufig zu sein: CCAG 4 p. 145 (ed. aus ms Neapol. 19, anno 1495, fol. 398; cf. ebda. p. 118 n. 1); CCAG 4 p. 33 (ms Mutin. 11, s. XV, fol. 94^v; offenbar mitten im Text von Ptolemäus' „Tetrabiblos", fol. 30–126; nicht ediert); CCAG 6 p. 34 (ms Vind. Phil. gr. 179, s. XIV–XV, fol. 110; nicht ediert); CCAG 7 p. 23 (ms München 287, s. XIV ex., fol. 151^v; nicht ediert); CCAG 7 p. 75 (ms Erlangen 93, s. XVI, fol. 3; nicht ediert); wahrscheinlich auch mit enthalten in Vind. Phil. gr. 177, fol. 1^v–211^v (cf. Hunger p. 282); ferner noch Lesarten des gleichen Texts aus ms Petrop. Bibl. Acad. Scient. XX Aa–8, s. XVII ex., fol. 160^v–161^v in CCAG 12 p. 113. Die hier auftretenden 18 Namen (sechs weitere sind an den entsprechenden Stellen ausgefallen) sind etymologisch nicht erklärbar; sie weisen keinerlei Anklang an die bekannten arabischen Gestirnnamen auf. Hier wäre zunächst der arabische Text zu untersuchen.

[5] Auf zwei griechische Sternverzeichnisse mit persischen Sternnamen weist F. Cumont in seiner Anmerkung CCAG 4 p. 145 hin; seine erste Stelle erscheint hier bei mir als

Alles bisher Bekannte über Chrysokokkes hat U. LAMPSIDES zusammengestellt,[6] der auch zahlreiche Handschriften aufführt, die nach Katalogangaben die 'Ἐξήγησις enthalten. Über den Gesamtinhalt und die textgeschichtlichen Zusammenhänge der 'Ἐξήγησις berichtet zuletzt ausführlich D. PINGREE in seinem Aufsatz über „Palaeologan Astronomy".[7] An dieser Stelle soll jetzt nur von dem Fixsternverzeichnis gehandelt werden.

Dieses Sternverzeichnis erschien zum erstenmal gedruckt 1645 in Paris;[8] der Herausgeber BULLIALDUS hatte den Text einer nicht näher bezeichneten griechischen Handschrift entnommen, die sich in der „Bibliotheca Regis Christianissimi" befand.[9] Bereits BULLIALDUS hatte versucht, Näheres über das anzunehmende orientalische Vorbild der Tafeln in Erfahrung zu bringen.[10] E. B. KNOBEL[11] bezeichnet den persischen Autor Cothb-Eddin Schirazi als orientalischen Urheber, was jedoch lediglich ein Mißverständnis ist.[12]

II a; seine zweite ist dort falsch zitiert, es handelt sich um Vind. Phil. gr. 190, fol. 149ᵛ, hier bei mir III h. In dem von D. PINGREE in seinem Artikel „Palaeologan Astronomy" besprochenen ms Vat. gr. 1056 (von dem er mir freundlicherweise seine Abschriften zur Verfügung stellte) sind mehrere kürzere Sternverzeichnisse enthalten, die ebenfalls auf sogar namentlich genannte orientalische Quellen zurückgehen. Es besteht jedoch keinerlei Beziehung zwischen den Verzeichnissen in Vat. gr. 1056 einerseits und den verschiedenen Typen des Chrysokokkischen Verzeichnisses andererseits. Das Fixsternverzeichnis der mittelgriechischen Übersetzung der Toledanischen Tafeln (ms Vat. gr. 212, s. XIV, fol. 84ᵛ; cf. MERCATI p. 271) gehört in den westlichen Traditionszusammenhang und wird von mir an anderer Stelle besprochen werden. Zu nennen wäre ferner u. a. eine Aufzählung von 16 Astrolabsternen (ms Vat. gr. 1058, s. XV, fol. 273ᵛ, lin. 7–20; cf. NEUGEBAUER p. 32. Ich konnte die Photokopien von Prof. O. Neugebauer benutzen), die jedoch nicht in Tabellenform erscheinen und keine Koordinatenangaben enthalten; sie sind lediglich innerhalb der Tierkreiszeichen aufgezählt. Das ptolemäische Gesamtverzeichnis der 1025 Fixsterne in 48 Sternbildern bildet einen eigenen geschlossenen Überlieferungskomplex, der, soweit bisher bekannt, im griechischen Sprachbereich nur in rein griechischer Tradition erscheint; er bleibt daher in unserem Zusammenhang außer Betracht; cf. jedoch Anm. 17.

[6] Byz. Zeitschrift 38 (1938), 312–322; cf. auch SARTON III p. 688 ff.; M. STEINSCHNEIDER, *Die hebräischen Übersetzungen des Mittelalters*, Berlin 1893, p. 629f. (auch 536).

[7] Cf. oben Anm. 1.

[8] Mit Kommentar, p. 225–229, in den „*Tabulae Philolaicae*", angehängt an I. BULLIALDUS, *Astronomia Philolaica*. Nachgedruckt bei M. DELAMBRE, *Histoire de l'Astronomie du Moyen-Age*, Paris 1819, p. 195. Zitiert und besprochen bei L. IDELER, *Untersuchungen über den Ursprung und die Bedeutung der Sternnamen*, Berlin 1809, p. LVI f. sowie bei den einzelnen Sternen; cf. auch I k.

[9] Die Tabelle steht mit den gleichen Abweichungen wie in BULLIALDUS' Druck auf fol. 98ᵛ in ms Paris B. N. gr. 2401 (cf. I i). Also dürfte diese Handschrift (oder eine andere, zur gleichen Klasse gehörende, gegenwärtig jedoch nicht bekannte) als Vorlage für BULLIALDUS gedient haben.

[10] a. a. O. p. 232: Eine arabische Handschrift in der Bibliotheca Patrum Oratorii Iesu zu Paris enthalte Tafeln, die mit denen in der 'Ἐξήγησις identisch sein könnten.

[11] *The Chronology of Star Catalogues*, Mem. Roy. Astron. Soc. (London), XLIII (1875–77) p. 11 Nr. 12.

[12] KNOBEL bezieht sich auf L. A. SEDILLOT, *Prolégomènes des Tables Astronomiques d'Oloug Beg*, Paris 1853, p. 289. Hier sind verschiedene Autoren aufgezählt, am Ende: „COTHB-EDDIN SCHIRAZI, l'anonyme persan auteur des Tables de Chrysococca, etc."

L. A. Sedillot[13] nimmt an, daß diese Tafeln letzten Endes auf Ibn Yūnis[14] zurückgehen.

Die Textuntersuchung des Sternverzeichnisses an insgesamt zweiundzwanzig Fundstellen führt neuerdings zu ganz eigenen Ergebnissen. Es erscheinen drei völlig verschiedene Formen dieses Verzeichnisses, die freilich nach den Angaben der Kataloge und Bibliographien immer in den Chrysokokkes-Text gehören.

Typ I umfaßt 25 Sterne wie bei Bullialdus und ist abgeleitet aus einem arabischen Tafelwerk. Dagegen Typ II mit 67 Positionen stammt aus einem persischen Original. Diese Herkunftsverhältnisse werden durch die anschließende Edition sowohl der griechischen wie der orientalischen Texte verdeutlicht. Typ III schließlich enthält 30 Sterne aus einer alten astrologischen Tradition, die noch unter das dritte Jahrhundert n. Chr. hinabreicht.

Hieraus kann mit aller Vorsicht geschlossen werden, daß wir uns bei den Chrysokokkes-Texten in einem noch nicht genügend geklärten Bereich bewegen. Seine Ἐξήγησις war konzipiert als Einführung und Erklärung „persischer“ astronomischer Texte, die durch Gregorios Chioniades unter der Regierung Alexios' II. Komnenos von Trapezunt (1297–1330) aus Tabrīz,[15] also wirklich aus Persien, nach Byzanz gebracht und auch von ihm ins Griechische übersetzt worden waren.[16]

Die mehrfach vorkommende Jahreszahl 1346 gehört nach allgemeiner Ansicht zur Abfassung der Ἐξήγησις durch Chrysokokkes.[17] Wenn man aber annimmt, daß er es war, der ältere Tafeln auf 1346 umgerechnet hat, so bleibt unerklärlich, warum er dabei in Typ II eine Längendifferenz gegen Ptolemäus von $+18^{0}23'$, dagegen in Typ III eine Differenz von $+18^{0}30'$ für die gleiche Epoche angewendet hat.

Abseits hiervon steht Typ I mit Epoche und Koordinaten von 1115, die unverändert beibehalten sind. I scheint also an der Textbearbeitung von 1346 nicht teilgenommen zu haben.[18]

Wie man sieht, sind also Cothb-Eddin Schirazi und l'anonyme persan auteur einwandfrei als zwei verschiedene Personen zu verstehen,

[13] *Matériaux pour servire à l'histoire comparée des Sciences mathématiques chez les Grecs et les Orientaux*, Paris 1845, p. 80.

[14] In Ägypten; gest. 399 h = 1009; cf. GAL I, 224; Suppl. I, 400 f.

[15] Hierzu cf. D. Pingree in: Bulletin de l'Académie Royale de Belgique (Classe de Lettres), Nov. 1962, p. 323–326.

[16] Cf. die ausführliche Darstellung von D. Pingree in seinem Aufsatz über „Palaeologan Astronomy“.

[17] Für die gleiche Epoche existiert auch ein Gesamtverzeichnis der 1025 ptolemäischen Sterne (Almagest) in ms Vat. gr. 1058, fol. 441–453^{v} (cf. Neugebauer p. 36) sowie auch in ms Escorial I R 14, fol. 128 ff. Die dort benutzten Längenwerte sind unter sich uneinheitlich und stimmen auch zu keiner der verschiedenen Versionen von Typ III.

[18] Ein philologisch-textgeschichtliches Problem aus I (cf. Kunitzsch, *Sternnamen* p. 207, Nr. 183) hat durch die hier ermöglichten Vergleiche von achtzehn handschriftlichen Stellen jetzt seine Erklärung gefunden: siehe unten I k.

Auffällig ist auch, daß mehrere Handschriften[19] das Verzeichnis gleichzeitig in zwei Typen (I und III) enthalten. Hierbei ist zu bemerken, daß es sich bei I und III um in jeder Beziehung völlig verschiedene und untereinander unabhängige Typen handelt.

Es ist daher nicht sicher zu bestimmen, welcher dieser Texte eindeutig primär Bestandteil des Chioniades-Chrysokokkischen Traditionskomplexes war und wieweit hier auch Texte des gleichen Themenbereichs, jedoch anderer Herkunft eingedrungen sind. So wie sich die Text- und Quellengeschichte der drei Verzeichnistypen aus dem augenblicklichen Stand der Erschließung ablesen läßt, spricht die Wahrscheinlichkeit am meisten dafür, daß II zu den von Chioniades aus dem Persischen übersetzten astronomischen Texten gehört, die dann später Chrysokokkes kommentiert hat.[20] Die Datierung der Handschrift I f (vor 1308) erlaubt kaum, Chioniades für den Übersetzer von I zu halten. Eher dürfte es sich bei I um eine ältere Übersetzung[21] handeln, die entweder von Chrysokokkes selbst oder von anderen Kompilatoren dem corpus einverleibt wurde. Eine ähnliche Unsicherheit besteht bei III, wenn auch hier zumindest durch die Gleichheit der Epoche (1346) eine äußere Beziehung zu Chrysokokkes hergestellt ist; ganz sicher gehört III nicht zu den Chioniadischen Übersetzungen und muß von Chrysokokkes oder späteren anderswoher hinzugefügt sein.

Abschließend sei noch darauf hingewiesen, daß sich in unserem Zusammenhang erneut erweist, daß die byzantinische Quellenangabe „persisch“ nicht immer exakt Persien meint, sondern auch generell orientalische Herkunft bezeichnen kann.[22] Das Original von I war arabisch;[23] nur bei II liegt nachweislich ein persischer Text zugrunde. Die Datierung nach persischen Jahren auch in dem rein griechischen Typ III ist

---

[19] Vind. Phil. gr. 190, fol. 149^{v}: III–fol. 222: I; Vind. Phil. gr. 87, fol. 22^{v}: I–fol. 33: III; Paris B. N. gr. 2461, fol. 251: I–fol. 287^{v}: III; Escorial I R 14, fol. 127: III–fol. 127^{v}: I.

[20] In den Auszügen von E. Bernard (1684; cf. I b und III f) wird nur I dem Chrysokokkes zugeschrieben; III dagegen erscheint als „Persischer Kanon“ ohne namentlich genannten Autor.

[21] Spuren älterer Übersetzungen aus orientalischen Quellen in byzantinischen astronomischen Texten siehe z. B. in Vat. gr. 1056; cf. oben Anm. 5.

[22] Cf. Kunitzsch, *Sternnamen* p. 49 f. Interessant ist in dem Zusammenhang der Hinweis auf den mit dieser Bezeichnung bezweckten bewußten Attizismus mittelgriechischer Autoren, bei Sarton III, p. 63. Ein Beispiel vom Ende des 13./Anfang des 14. Jahrhunderts (Theodoros Metochites) auch bei Pingree in seinem Artikel „Palaeologan Astronomy“.

[23] Ob dem Griechen möglicherweise eine persische Version des ursprünglich arabischen Werkes als Vorlage gedient hat, ist aus seiner Übersetzung nicht klar ersichtlich. Ein mögliches Kriterium, der arabische Artikel *al* bei den Sternnamen, wird schon in den arabischen und den daraus abgeleiteten persischen Texten sehr inkonsequent angewendet (teils inkorrekte Weglassung in arabischen Texten, zur Abkürzung; teils dem persischen Gebrauch widersprechende Beibehaltung in persischen Texten), so daß sich hieraus kein sicheres Urteil ableiten läßt.

offenbar stereotyp aus der Überschrift eines anderen Verzeichnisses übernommen, wo sie das Werk des griechischen Übersetzers bzw. Bearbeiters gewesen sein dürfte, dem aus seinem astronomisch-chronologischen Material orientalischer Herkunft verschiedene Ären bekannt waren.

Es folgt nun die Beschreibung und Edition des Verzeichnisses nach den drei Typen.

Typ I

Charakteristik: 25 Fundamentalsterne, angeordnet nach Tierkreiszeichen, mit Ekliptikkoordinaten sowie Angabe der Größe. Angegebene Epoche: der Beginn des Jahres 509 der Hiğra (27. Mai 1115), in einigen Handschriften gleichgesetzt mit dem Beginn des Jahres 1427 der Seleukiden-Ära (Herbst 1115); Längenwert: Ptolemäus $+15^0 0'$. Die Sternnamen sowie die Datierung nach Hiğra-Jahren weisen eindeutig auf orientalische Herkunft.

Die 25 Fundamentalsterne von I stellen einen Auszug aus der längeren Tabelle von 43 Fixsternen dar, die in dem arabischen Tafelwerk *az-Zīğ as-Sanğarī* von ʿAbd ar Raḥmān al-Ḫāzinī[24] enthalten ist. Epoche, Koordinaten (auch die Längen!) und Namen blieben bei der Übernahme unverändert erhalten, lediglich die im Arabischen vorhandenen astrologischen Temperamente sind im Griechischen weggelassen. Der Vergleich jedes griechischen Sterns mit seinem arabischen Original gibt zugleich die Etymologie aller Namen und der verstümmelten arabischen Wörter.

Für den griechischen Text lagen folgende Stellen vor:

a Der Druck von I. BULLIALDUS, Paris 1645, *Tabulae Philolaicae* p. 225 f.; wahrscheinlich nach ms Paris B. N. gr. 2401, fol. 98^v (cf. Anm. 9 und weiter unten sub i). BULLIALDUS erwähnt auch noch einen nicht näher bezeichneten „Codex Basileensis“. Die drei letzten Sterne erscheinen umgestellt, in der Reihenfolge 24, 25, 23.

b Die Längen von 21 dieser Sterne sind abgedruckt bei BERNARD p. 572, col. ϑ: „Ex [Canonibus] Persicis Chrysococcae anno Chr. 1115.“ Es ist nicht ersichtlich, ob hier ein bloßer Nachdruck der Edition von 1645 oder eine eigene Edition vorliegt (bei Stern Nr. 7 hat b abweichend und falsch $10^0 50'$, dagegen a und die Handschriften richtig $10^0 0'$).

---

[24] gest. ca. 582 h = 1186/7; cf. GAL Suppl. I, p. 902, 1 c; SUTER Nr. 293; KRAUSE p. 487; KENNEDY p. 129, Nr. 27 und p. 159–161, Abstract 12, N (aus ms Vat. arab. 761); M. DESTOMBES, *Globes célestes et catalogues d'étoiles orientaux du Moyen-Age*, Tirage à part des „Actes du VIIIe Congrès International d'Histoire des Sciences“ (Florenz 1956), p. 313–324 (p. 10 des Sonderdrucks). Dieser ʿAbd ar-Raḥmān al-Ḫāzinī wird auch in ms Laur. 28, 17, fol. 4 (ed. CCAG 1, p. 89, 2–3: ὁ ᾿Αβδουλαχμάνης ὁ Χαζηνός) als Autor des sicheren und zuverlässigen Tafelwerks Σαντζαρῆ (= *Sanğarī*) genannt; mit Verstümmelungen und Entstellungen wurden diese Angaben hernach von Theodoros Meliteniotes in sein III. Buch übernommen (᾿Αβδουραχμάνης ὀχωζηνὴς [leg. ὁ Χωζηνής], Exzerpt aus Vat. gr. 1059, fol. 350^v, ed. bei USENER 1876 p. 14 = *Kl. Schr.* p. 338; hier gemäß GILDEMEISTER falsch mit aṣ-Ṣūfī [cf. unten Anm. 29] identifiziert).

388

c ms Vind. Phil. gr. 87, s. XIV (2. Hälfte), fol. 22v (cf. HUNGER p. 198). In c sind links am Rand (bei Stern Nr. 1 in der Namenkolumne hinter dem eigentlichen Namen αὔλαξ) neben vielen der aus dem Orient entlehnten Sternnamen die entsprechenden altgriechischen bzw. ptolemäischen Bezeichnungen klein beigeschrieben. (Ich verdanke die Kenntnis der Stelle einem Hinweis von Prof. D. Pingree).

d ms Vat. gr. 1058, s. XV, fol. 363v–364 (cf. NEUGEBAUER p. 35 f.). Die Tabelle ist hier horizontal angeordnet, nicht vertikal, wie sonst üblich. Die Sterne sind gezählt von α bis κε (1 bis 25). (Ich konnte die Fotokopie von Prof. O. Neugebauer benutzen).

e ms Escorial I R 14, fol. 127v (Ich konnte die Fotokopie von Prof. O. Neugebauer benutzen.)

f ms Vat. gr. 211, vor 1308, fol. 153v–154 (Mir lag die eigenhändige Abschrift von Prof. D. Pingree vor.)

g ms Vind. Phil. gr. 190, s. XVI (1. Hälfte), fol. 222 (cf. HUNGER p. 299). g stimmt meist eng mit c überein. (Mir lag die handschriftliche Kollation von Prof. D. Pingree vor.)

h ms Paris B. N. gr. 2461, fol. 251.

i ms Paris B. N. gr. 2401, fol. 98v. Die Tafel weist hier die gleichen Abweichungen sowie die Umstellung von Stern Nr. 23 auf wie der Druck von BULLIALDUS; diese Handschrift kommt daher als dessen Vorlage in Frage.

k Th. HYDE führt im Kommentar zu seiner Ausgabe der Tafeln des Uluġ Bēg[25] bei den betreffenden Sternen jeweils auch die Namen an, unter denen diese Sterne in den „Persischen Tafeln" des Chrysokokkes erscheinen. Er folgt dabei dem Typ I, aus welchem er 22 Namen wiedergibt; ganz offensichtlich lag ihm dabei die Edition von BULLIALDUS vor.[26] Die Namen von Stern Nr. 6 und 8 werden nicht aufgeführt. Stern Nr. 3 erscheint auf p. 20: „in Tab. Persicis Γοργόνιον"; mit diesem Namen kommt er jedoch nicht in I, also auch nicht bei BULLIALDUS vor, sondern vielmehr in III (Nr. 15). HYDE führt dazu noch drei weitere Sterne aus Chrysokokkes auf, die ebenfalls nicht in I vorkommen, sondern nur in III, und gibt dazu auch die Namen, die sie in III haben: p. 36: δ Leo, Ζῶσμα Λέοντος[27] (III, Nr. 26); p. 37: β Leo, Οὐρὰ Λέοντος (III, Nr. 25); und p. 62: α Hya, Ὕδρου αὐχήν (III, Nr. 27). Bei Stern Nr. 5 (= III, Nr. 28) und Nr. 17 (= III, Nr. 9) gibt HYDE sowohl den Namen aus I wie auch denjenigen aus III. Er muß also

[25] *Tabulae longitudinis et latitudinis stellarum fixarum ex observatione Ulugh Beighi* ..., Oxford 1665.

[26] Cf. p. 21 zu Ἀϊούκ und p. 26, wo BULLIALDUS' falsche Schreibweise Ὄμος statt Ὦμος (ἵππου) kritisiert wird.

[27] Die Lesung ζῶσμα ist ein Unikum (cf. auch KUNITZSCH, *Sternnamen* p. 207, Nr. 183); es muß eine Verlesung des Ausdrucks ὀσφύς vorliegen, den bei Stern Nr. 26 einstimmig alle Handschriften von III bieten.

außer dem Druck von BULLIALDUS noch eine andere Quelle benutzt haben, die teilweise oder ganz den Typ III enthielt. Für das Vorhandensein von handschriftlichen Versionen beider Typen in England zeugen die Exzerpte bei BERNARD (Ib und III f).

Κανόνιον τοῦ αὐθημερινοῦ τῶν ἀπλανῶν ἀστέρων · τοῦτο τὸ κανόνιον ἐτέθη εἰς τὴν ἀρχὴν[a] τοῦ $\overline{\varphi\vartheta}$ χρόνου τῶν ἀρράβων · εἰς τοὺς $\overline{\xi\eta}$ οὖν χρόνους τούτων μίαν μοῖραν κινοῦνται.

Lfd. Nr.	Name	Länge	Breite	Größe	Mod. Bez.	Nr. in az-Zīǧ as-Sanǧarī
1.	αὖλαξ[b]	0.15°10′	−53°30[c]	1	ϑ Eri	2
2.	χεὶρ βεβαμμένη	0.22°50	51°40	3	β Cas	4
3.	κεφαλὴ ἵππου	1.14°40	23°0[d]	2	β Per	8
4.	πλευρὰ περσαοῦς	1. 19°50	30°0	2	α Per	9
5.	ὄμμα βοός	1.27°0[e]	− 5°0	1	α Tau	10
6.	πούς διδύμων	2.4°50	−31°30	1	β Ori	11
7.	ἀϊούκ	2. 10°0	22°30	1	α Aur	12
8.	ὦμος διδύμων	2.17°0	−17°0	1	α Ori	13
9.	σοαὶλ ἰαμανή[f]	3[g].2°10	−75°0	1	α Car	14
10.	σιαὴρ ἰαμανή[f]	3.2°40	−39°10	1	α CMa	15
11.	σιαὴρ σιαμή[h]	3.14°10	−16°10	1	α CMi	18
12.	καρδία λέοντος	4.17°30	0°10	1	α Leo	20
13.	μικρὸς κονταράτος	6.11°40[i]	− 2°0	1	α Vir	24
14.	κονταράτος	6.12°0	31°30[k]	1	α Boo	25
15.	πούς κουντουρός	6.23°20	−41°10	1	α Cen	26
16.	πινάκιν κεκλασμένον[l]	6.29°40	44°30	2	α CrB	27
17.	καρδία σκορπίου	7.27°40	− 4°0	2	α Sco	29
18.	γὺψ καθήμενος	9. 2°20	62°0	1[m]	α Lyr	32
19.	γὺψ πετόμενος[n]	9.18°50	29°10	2	α Aql	33
20.	στόμα ἰχθύος μεσ[ημβρινοῦ][o]	10.22°0	−23°0	1[p]	α PsA	38
21.	οὐρὰ ὄρνιθος	10.24°10	60°0	2	α Cyg	39
22.	ὀμφαλὸς ἵππου	11.11°40	19°40	2	α Peg	40
23.[q]	ὦμος ἵππου	11.17°10	31°0	2	β Peg	41
24.	οὐρὰ τοῦ καϊστοῦ	11.19°20	9°40[r]	2	ι Cet	42
25.	πήγασος[s]	11.27°10	12°30	2	γ Peg	43

Anmerkungen zum griechischen Text:

[a] Überschrift: In d am Rande der Einschub ·/. τοῦ χρόνου $\overline{\alpha\upsilon\kappa\zeta}$ εἰς τὸ ἔτος τῶν ῥωμαίων · καὶ εἰς τὴν ἀρχὴν (1427 Seleukiden-Ära, beg. Herbst 1115). [b] Alle HSS: αὔλαξ. [c] acegi: −13°20. [d] ai: 43°0; c: νγ (53°), verbessert in κγ (23°), Pingree liest hier und in g: μγ (43°). [e] So alle Stellen statt zu erwartendem 27°40. [f] acegi: ἰαμανῆ; f: ἰαμάνη. [g] aceghi falsch: 2. [h] c: σιαμῆ. [i] ceghi falsch: 11°2 (hier weicht i von a ab). [k] acegi falsch: 31°1. [l] e: πινακι κεκλασμεν(ον?); f: πινάκις κεκλασμένη; h: πινακ . . . κεκλασμ . . . (Abkürzung); k (p. 16): Πινάκιον κεκλασμένον „Discus confractus“. [m] ai: 2. [n] cfhi: πετώμενος. [o] abehi: μέσον. [p] ai: 2. [q] Dieser Stern ist in a und i hinter Nr. 25 angefügt. [r] ai falsch: 19°40; alle Stellen geben die Breite falsch „nördlich“, während gemäß Ptolemäus „südlich“ zu erwarten wäre, wie es auch richtig in $A^{1,2}$ 42 erscheint. [s] f: πεγασος.

Das Verzeichnis der 43 Sterne aus dem arabischen Tafelwerk *az-Zīǧ as-Sanǧarī* von ʿAbd ar Raḥmān al-Ḫāzinī wird nach zwei Handschriften wiedergegeben:

A¹ ms Vat. arab. 761, fol. 188ᵛ–189[28]: *ǧadwal aḥwāl al-kawākib aṯ-ṯābita fa-mawāḍiʿihā li-awwali sanati tisʿin wa-ḫamsimiʾatin hiǧrīya* „Tabelle der Situationen der Fixsterne und ihrer Positionen für den Beginn des Jahres fünfhundertneun der Hiǧra [= 27. 5. 1115]"; der zweite Teil der Tabelle, fol. 189 oben, hat nochmals eine Überschrift: *ǧadwal aḥwāl al-kawākib aṯ-ṯābita zīdat y̅h̅ daraǧa ʿalā mawāḍiʿihā fī l-maǧisṭī li-sanati 1427 li-l-iskandar* „Tabelle der Situationen der Fixsterne; ihren Positionen im Almagest wurden 15 Grad hinzugefügt für das Jahr 1427 der Ära Alexanders [beg. Herbst 1115]". Das Verzeichnis ist also durch Zufügung einer Konstante von 15^0 aus einer arabischen Version des ptolemäischen Verzeichnisses abgeleitet. Die Namen entsprechen größtenteils den ptolemäischen Definitionen; gelegentlich finden sich die altarabischen vorptolemäischen Bezeichnungen (in der Edition durch einen Stern* gekennzeichnet).

A² ms Brit. Mus. Or. 6669, offensichtlich fol. 155ᵛ–156. 155ᵛ ist stärker beschädigt: es fehlen die Überschrift und die Sterne Nr. 21–23 völlig; von Stern Nr. 20 ist nur der Name zu sehen, von Nr. 18 und 19 nur der Name und die Angabe des Tierkreiszeichens, von Nr. 17 nur die Angaben einschließlich der Länge. Die folgende Seite (fol. 156) zeigt fast alle Angaben vollständig. Am Ende der Tafel steht in großer Schrift: *āḫir az-zīǧ al-muʿtabar as-sanǧarī* „Ende des Tafelwerks ‚az-Zīǧ al-Muʿtabar as-Sanǧarī'", darunter in kleiner Schrift der Schreibervermerk, daß die Abschrift am 20. Ǧumādā al-āḫira 620 h (= 21. Juli 1223) vollendet wurde. Obwohl die Handschrift relativ alt ist, weist sie einige Fehler in den Koordinaten auf, die in A¹ nicht erscheinen; A¹ scheint also einen besseren Text zu bieten.

Die Wiedergabe erfolgt nach A¹ (mit Varianten aus A²). Die Angaben der Temperamente sind in die Edition nicht aufgenommen. Bemerkungen zur Punktierung erfolgen nur in Ausnahmefällen. A¹ und A² geben die Größen getrennt nach der Autorität des Ptolemäus und des Ṣūfī[29] wieder; wo diese untereinander abweichen, notiere ich getrennt P und S.

---

[28] So nach Angaben von Prof. E. S. Kennedy, dessen Fotokopie ich benutzen konnte. Bei C. Nallino, *Opus Astronomicum*, I (Mailand 1903), p. LXVII wird das Sternverzeichnis auf fol. 191ᵛ–192 lokalisiert; auch M. Destombes nennt mir brieflich diese beiden Blätter.

[29] Bei A¹ im Kolumnentitel läßt sich mit einiger Schwierigkeit erkennen: *ʿinda Ibn aṣ-Ṣūfī*; richtig muß es statt „*Ibn aṣ-Ṣūfī*" heißen: *aṣ-Ṣūfī*; es handelt sich um den Verfasser des berühmten Fixsternbuches, gest. 376 h = 986; cf. GAL I, p. 223; Suter Nr. 138; Krause p. 463.

Lfd. Nr.	Name	Länge	Breite	Größe	Mod. Bez.	Nr. in I
1.	*ra'ṣ al-mar'a al-musalsala* „Kopf der angeketteten Frau“	0.2° 50′	26° 0′	2	δ Peg= α And	
2.	*āḫir an-nahr* „Ende des Flusses“	0.15° 10	−13° 30[a]	1	ϑ Eri	1
3.	*baṭn al-ḥūt* * „Bauch des Fisches“	0.18° 50	26° 20	P: 3 S: 2	β And	
4.	*kaff*[b] *al-ḫaḍīb* * „die gefärbte Hand“	0.22° 50	51° 40	P: 3 S: 5	β Cas	2
5.	*an-nāṭiḥ* * „der [mit dem Horn] Stoßende“	0.25° 40	10° 0	3	α Ari	
6.	*ra's al-muṯallaṯ* „Spitze des Dreiecks“	0.26° 0	16° 30	3	α TrB	
7.	*riǧl al-musalsala* „Fuß der Angeketteten“	1.1° 50	28° 0	P: 4 S: 3	γ And	
8.	*ra's al-ġūl* „Kopf der ġūl [= γοργόνιον]“	1.14° 40	23° 0	2	β Per	3
9.	*ǧanb baršāwš* „Seite des Perseus“	1.19° 50	30° 0	2	α Per	4
10.	*ad-dabarān* * *wa-huwa ʿayn aṯ-ṯawr* „ad-dabarān, d. i. Auge des Stiers“	1.27° 40	− 5° 10	1	α Tau	5
11.	*riǧl al-ǧawzā'* * „Fuß der ǧawzā' [= Orion]“	2.4° 50	−31° 30	1	β Ori	6
12.	*al-ʿayyūq* * *wa-huwa al-ʿannāz* „al-ʿayyūq, d. i. der Ziegenhirt[c]“	2.10° 0	22° 30	1	α Aur	7
13.	*yad al-ǧawzā'* * *aw*[d] *mankibuhu* „Hand der ǧawzā' [= Orion] oder auch: seine Schulter“	2.17° 0	−17° 0	1	α Ori	8
14.	*suhayl al-yamānī* * „der südliche suhayl“	3.2° 10	−75° 0[e]	1	α Car	9
15.	*aš-šiʿrā al-yamāniya* * „der südliche šiʿrā [= Sirius]“	3.2° 40	−39° 10	1	α CMa	10
16.	*ḥālib*[f] *at-taw'am at-tālī* „der Samenstrang des nachfolgenden Zwillings“	3.6° 40	− 0° 30	3	δ Gem	
17.	*ra's at-taw'am* „Kopf des Zwillings“	3.8° 20	9° 40	2	α Gem	
18.	*aš-šiʿrā aš-ša'āmiya* * „der nördliche šiʿrā“	3.14° 10	−16° 10	1	α CMi	11
19.	*al-fard* * *wa-huwa ʿunq aš-šuǧāʿ* „der Einzelne, d. i. der Hals der Hydra“	4.15° 0	−20° 30	2	α Hya	
20.	*qalb al-asad* „Herz des Löwen“	4.17° 30	0° 10	1	α Leo	12
21.	*ẓahr al-asad* „Rücken des Löwen“	4.29° 10	13° 40	2	δ Leo	
22.	*aṣ-ṣarfa* * „der [Wetter-]Umschlag“	5.9° 30	11° 50	1	β Leo	

Lfd. Nr.	Name	Länge	Breite	Größe	Bez. Mod.	Nr. in I
23.	*ǧanāḥ al-ġurāb* „Flügel des Raben"	5.18°30	−14°50	3	γ Crv	
24.[g]	*simāk*[h] *al-aʿzal* * „der unbewaffnete simāk"	6.11°40	− 2°0	1	α Vir	13
25.	*simāk*[h]*ar-rāmiḥ* * „der lanzenbewaffnete simāk"	6.12°0	31°30	1	α Boo	14
26.	*riǧl qanṭūris*[i] „Fuß des Kentauren"	6.23°20	−41°10	1	α Cen	15
27.	*al-munīr min al-fakka* * „der Helle von al-fakka"	6.29°40	44°30	2	α CrB	16
28.	*ʿunq al-ḥayya* „Hals der Schlange"	7.9°20	25°20	3	α Ser	
29.	*qalb al-ʿaqrab* „Herz des Skorpions"	7.27°40	− 4°0	2	α Sco	17
30.	*raʾs al-ǧāṯī* „Kopf des Knienden"	8.2°40	37°30	3	α Her	
31.	*raʾs al-ḥawwāʾ* „Kopf des Schlangenträgers"	8.9°50	36°0	3	α Oph	
32.	*an-nasr al-wāqiʿ* * „der stürzende Adler"	9.2°20	62°0[k]	1	α Lyr	18
33.	*an-nasr aṭ-ṭāʾir* * „der fliegende Adler"	9.18°50	29°10	2	α Aql	19
34.	*minqār ad-daǧāǧa* „Schnabel der Henne"	9.19°30	49°20[l]	3	β Cyg	
35.	*ḏanab ad-dulfīn* „Schwanz des Delphins"	10.2°40	29°10	P:3 S:4	ε Del	
36.	*ḏanab al-ǧady* „Schwanz des Steinbocks"	10.11°20	− 2°0	3	δ Cap	
37.	*fam al-faras aw ǧaḥfalatuhu* „Maul des Pferdes, oder auch: seine Lefze"	10.20°20	22°30	P:4 S:3	ε Peg	
38.	*fam al-hūt al-ǧanūbī* „Maul des Südlichen Fisches"	10.22°0	−23°0	1	α PsA	20
39.	*ḏanab ad-daǧāǧa* „Schwanz der Henne"	10.24°10	60°0	2	α Cyg	21
40.	*matn al-faras* „Rücken des Pferdes"	11.11°40	19°40	2	α Peg	22
41.	*mankib al-faras* „Schulter des Pferdes"	11.17°10	31°0	2	β Peg	23
42.	*ḏanab qayṭus* „Schwanz des κῆτος"	11.19°20	− 9°40	P:3 S:4[m]	ι Cet	24
43.	*ǧanāḥ al-faras* „Flügel des Pferdes"	11.27°10	12°30	2	γ Peg	25

Anmerkungen zum arabischen Text:

[a] *yǧ* = 13°, durch Weglassung eines Punktes aus dem richtigen *nǧ* = 53° verdorben. [b] Korrekt: *al-kaff*. [c] Abgeleitet aus der ptolemäischen Darstellung, nach welcher der Fuhrmann zwei Zicklein hält. [d] $A^{1,2}$ haben zwischen dem dritten und vierten Wort nur éin alif; doch ist auf jeden Fall *aw* „oder" zu lesen und nicht *wa* „und" (cf. Nr. 37). [e] A^2: 70°0. [f] A^1: *ǧʾlb*, A^2: *ḥʾkb*; offenbar gemeint ist *ḥālib* als Übersetzung des ptole-

mäischen βουβών. [g] In A^1 und A^2 Beginn der neuen Seite. [h] Korrekt: *as-simāk.* [i] A^1: *qnṭwrs* (*n* unpunktiert); A^2: *qyṭwrs.* [k] A^2: Auf dem Foto schwer lesbar, anscheinend statt *sb* = 62° nur *b* = 2°. [l] A^1: Der Buchstabe vor *ṭ* = 9 ist beschädigt; der Größe und Lage nach muß es *m* = 40 gewesen sein. [m] A^2: Auch bei Ṣūfī 3.

Textvergleich I–$A^{1,2}$

Die Längendifferenz zu Ptolemäus von +15°0′ ist in I aus A übernommen und beibehalten. Vgl. übrigens auch die Anmerkungen von BULLIALDUS selbst sowie von HYDE und IDELER (cf. Anm. 8 und 25).

Nr. 1: $A^{1,2}$ geben die ptolemäische Definition von ϑ Eri. Dieser Stern hatte in der altarabischen Astrothesie den Namen *aẓ-ẓalīm* „der Strauß" (cf. KUNITZSCH, *Sternnamen* p. 155, Nr. 80; *Untersuchungen* p. 119, Nr. 325). Wie bereits IDELER p. 233 f. (und zuvor schon HYDE, Kommentar p. 49) erkannte, hat der griechische Übersetzer das arabische Wort *ẓalīm* in dem durch die arabischen Lexika auch bezeugten Sinn „der [zum erstenmal, oder: an der falschen Stelle] aufgegrabene Boden" verstanden und durch αὖλαξ „Erdfurche" wiedergegeben.

Nr. 2: In der Astrolabsternaufzählung Vat. gr. 1058, fol. 273ᵛ, erscheint χεὶρ κοκκινή.

Nr. 3: Schon IDELER p. 117 hat angenommen, daß es sich hier um β Per handelt. Die entsprechende richtige Breite (+23°0′) erscheint auch wirklich in I cdefh. Die Wiedergabe von *al-ġūl* durch ἵππος beruht vielleicht auf einer Verlesung des arabischen Namens (statt *al-ġwl*: *al-frs* = *al-faras* „Pferd").

Nr. 4: *baršāwš* ist vom griechischen Übersetzer nicht als arabische Wiedergabe von Περσεύς (bzw. dem griechischen Genitiv Περσέως) erkannt worden.

Nr. 6, 8: Zu *al-ǧawzā'* = δίδυμοι und Orion cf. KUNITZSCH, *Untersuchungen* p. 23 ff. und Index (d).

Nr. 7: BULLIALDUS' schon von IDELER p. 92 angezweifelte Erklärung von *al-ʿayyūq* als „corruptum nomen ex Graeco αἴξ Capella" ist nicht haltbar; cf. *Sternnamen* p. 119 f., Nr. 27; *Untersuchungen* p. 20 f. und 46, Nr. 47. Das gleiche gilt für Nr. 10, cf. *Sternnamen* p. 117, Nr. 26, Anm. 3 und *Untersuchungen* p. 21 und 111, Nr. 289b.

Nr. 13: Schon von IDELER p. 171 offenbar richtig interpretiert: Der kleine Lanzenträger, im Gegensatz zu Nr. 14, der „Lanzenträger" schlechthin genannt wird (cf. *Sternnamen* p. 79, Nr. 41, wo auch auf eine hebräische Parallele hingewiesen ist).

Nr. 14: Zu dem griechischen Ausdruck sagt IDELER p. 50 f.: „. . . welches in der späteren griechischen Sprache einen mit einem κοντός oder Wurfspieß Bewaffneten bezeichnet haben muß".

Nr. 15: *qanṭūris* ist vom griechischen Übersetzer nicht als arabische Wiedergabe von κένταυρος erkannt worden; cf. oben Nr. 4.

Nr. 16: Der altarabische Name *al-fakka* wurde schon von arabischen Philologen wie auch in der Volksetymologie mit der Wurzel *fkk* „zerbrechen" in Verbindung gebracht (cf. *Sternnamen* p. 131 f., Nr. 43 mit Anm. 1; *Untersuchungen* p. 21 und 55 f., Nr. 85). Der griechische Übersetzer formulierte seine Version offenbar auf Grund einer entsprechenden Glosse oder in Anlehnung an die persische Form des aus jener Volksetymologie heraus gebildeten Populärnamens (*kāse-i šikaste* „die zerbrochene Schüssel", cf. *Sternnamen* p. 138 Anm. 1, aus aṭ-Ṭūsī); ähnlich urteilte bereits Ideler p. 60. Auch in der Sternbildaufzählung Vat. gr. 1058, fol. 317ᵛ erscheint πινάκϊν κεκλασμένον (sic); cf. ferner II 42.

Nr. 18: Bullialdus in seinem Kommentar und Ideler p. 70 f. bemerken, daß *al-wāqiʿ* „fallend, stürzend" besser durch καθειμένος statt des in den Handschriften gebotenen καθήμενος wiederzugeben wäre; Hyde schreibt καθείμενος (Kommentar p. 19). In der Astrolabsternaufzählung von Vat. gr. 1058, fol. 273ᵛ wie auch in der Sternbilderliste daselbst fol. 317ᵛ erscheint das Sternbild Lyra (dessen Name in den meisten Fällen auch auf α Lyr allein angewendet wird) als (ὁ) σελιάκ; hierzu cf. *Sternnamen* p. 206 f., Nr. 182, mit den Anmerkungen.

Nr. 20: Bei Ideler p. 285 ist μέσον falsch auf στόμα bezogen („der mitten am Maul des Fisches stehende"); das Wort ist natürlich zu ἰχθύς als „Südlicher Fisch" zu ziehen und ist die verkannte Abkürzung für μεσημβρινοῦ „des südlichen".

Nr. 22: ὀμφαλός entspricht wörtlich *surrat al-faras*, siehe II, 1 (δ Peg = α And. Dieser Stern erscheint bei $A^{1,2}$ Nr. 1 als *raʾs al-marʾa al-musalsala*). Die hier, bei Nr. 22 (= $A^{1,2}$ 40) stehenden Koordinaten sowie der arabische Name (*matn* „Rücken", Ptolemäus: ἐπὶ τοῦ μεταφρένου) weisen den Stern einwandfrei als α Peg aus.

Nr. 24: *qayṭus* ist vom griechischen Übersetzer nicht als arabische Wiedergabe von κῆτος (bzw. dem griechischen Genitiv κήτους) erkannt worden (cf. oben Nr. 4 und 15). Ähnlich heißt das Sternbild κῆτος in der Sternbilderliste Vat. gr. 1058, fol. 318 unerkannt ὁ καΐτοῦς.

Typ II

Charakteristik: 67 Positionen – davon 7 Duplikate bzw. Einschübe, also eigentlich 60 Sterne, angeordnet nach Tierkreiszeichen, mit Ekliptikkoordinaten sowie Angabe von Größe und Temperamenten. Angegebene Epoche: der Beginn des Jahres 6854 griechischer Weltära (Herbst 1345), gleichgesetzt mit 715 Yazdağird (beg. 1346). Längendifferenz zu Ptolemäus: $+18^{0}23'$. Zahlreiche arabische Namen sowie die Beziehung auf die persische Yazdağird-Ära weisen auf orientalische Herkunft. Die gleichen

60 Sterne in der gleichen Anordnung und mit den gleichen Angaben, nur mit älterer Epoche und entsprechend kleinerer Längendifferenz zu Ptolemäus, finden sich in dem persischen Tafelwerk *Zīǧ-i Īlḫānī* von Naṣīr ad-Dīn aṭ-Ṭūsī.[30] Man muß also Entlehnung aus dem Persischen annehmen, wobei die spätere Epoche mit der umgerechneten Länge am ehesten das Werk des griechischen Bearbeiters wäre.[31]

Der griechische Text beruht auf drei Stellen, von denen a für die Edition benutzt wurde; aus b und c werden Varianten angegeben. Die Numerierung stammt vom Herausgeber. Die Angaben der Temperamente sind in die Edition nicht aufgenommen.

a ms Angel. 29, anno 1388, fol. 267[r-v] (cf. CCAG 5, 1 p. 54, hier wird noch verwiesen auf III c und h; s. auch Anm. 5): Περσῶν περὶ τῶν ἀπλανῶν καὶ στερεῶν ἀστέρων · ἔτι δὲ περὶ μήκους καὶ πλάτους καὶ κράσεως καὶ μεγέθους αὐτῶν.

b ms Pal. Vat. 312, fol. 93[v]. Enthält nur die zweite Hälfte des Verzeichnisses ab Stern 28.

c ms Marc. 324, s. XIV–XV, fol. 210 (cf. CCAG 2, p. 7). Enthält ebenfalls nur die zweite Hälfte der Tafel ab Stern 28 und ist gegen Ende stellenweise beschädigt und nicht lesbar. (Von b und c standen mir die eigenhändigen Abschriften von Prof. D. Pingree zur Verfügung.)

Lfd. Nr.	Name	Länge	Breite	Größe	Mod. Bez.
[1 a][a]	πτερὰ ἵππου	Ari 0°33	12°35		
1.	στῆθος ἵππου · μέρος	6°13	26°10	2	δ Peg = α And
2.	αὖλαξ ἢ ποταμός	18°33	−13°30[b]	1	ϑ Eri
3.	γυναικὸς πζ[c] ἁλισιδωτοί	22°13	26°20	2	β And
4.	ἀρχὴ τοῦ σιουρετάϊ	25°1[d]	7°20	3	γ Ari
5.	τέλος τοῦ σιουρετάϊ	26°3	8°20	3	β Ari
6.	χεὶρ βεβαμμένη	26°13	51°40	3	β Cas
7.	ναδῆχ· κερατισμὸς δάχ	29°10[e]	10°0	2[f]	α Ari
8.	δὰτ κουρσῆ· ἐπὶ θρόνου καθήμενος	29°10[g]	46°45	3	α Cas
9.	ἀριστερὸς πούς γυναικὸς ἁλισιδωτοί	Tau 5°13	28°0	3	γ And
10.	ἀρχὴ τοῦ θουρέϊα· πλοιάς	15°3	40°30	neb.	χ Per
11.	κεφαλὴ κασιδωτή	18°3	23°14[h]	2	β Per

[30] gest. 672 h = 1274; cf. GAL I 508; 511, Werk 46; Suppl. I 924 f.; 931, Werk 45 (verfaßt 670 h = 1271); SUTER Nr. 368, Werk 4; KRAUSE p. 495, Werk 6; KENNEDY Nr. 6 und Abstract aus Vat. ar. 761, p. 160 ff. Die in der Handschrift gegebene Epoche 1232 deckt sich nicht mit dem meist als Abfassungsjahr des Tafelwerks angegebenen Jahr 670 h = 1271/2; siehe auch C. NALLINO, *Opus Astronomicum*, I (Mailand 1903), p. LXVII: verfaßt für den Īlḫān Ābāqā Ḫān, der ab 664 h = 1265/6 regierte. Der Auszug bei BERNARD p. 571, col. ς [cf. KNOBEL, *Chronology* p. 11 und Nr. 14] gibt die Epoche 1233, was, wie übrigens auch die Koordinaten, zu unserer Handschrift stimmt.

[31] Das ἐποίησα ... in der Einleitung (fol. 266[v], cf. CCAG 5, 1 p. 54) dürfte sich in Verbindung mit der Jahreszahl 1346 sowie im Zusammenhang des Textes der Īḫānischen Tafeln auch am ehesten auf den griechischen Übersetzer bzw. den etwas späteren Bearbeiter beziehen.

Lfd. Nr.	Name	Länge	Breite	Größe	Mod. Bez.
12.	μέση τοῦ θουρέϊα	20°33	4°30	5	Fl. 19 Tau (Ptol. 29)
13.	ἀγκὼν[i] θουρεΐας	23°13	30°0	2	α Per
14.	ὄμμα τοῦ ταύρου· τάπαρ	Gem 1°3	— 5°5	1	α Tau
15.	πούς διδύμων	5°13[k]	—31°30	2[l]	β Ori
16.	ὦμος ὠρίωνος ἡγούμενος[m]	12°23	17°30[n]	2	γ Ori
[16a][o]	ἀκρόποδον ὠρίωνος	8°48	—		
17.	ἀϊούχ· ούάια ἀϊνάμ	13°23	22°30	1	α Aur
18.	ἀρχὴ τῆς ζώνης	13°43	—24°10	2	δ Ori
19.	κεφαλὴ δυνατοῦ	15°23	—13°50	neb.	λ Ori
20.	τὰ μέσα τῆς ζώνης	15°43	—24°50	2	ε Ori
21.	τὸ τέλος τῆς ζώνης	16°33	—25°40	2	ζ Ori
[21a][p]	αἴξ	14°21	+	1	
22.	χεὶρ διδύμων	20°23	—17°0	[q]	α Ori
23.	ῥάχη[r]· ὦμος δεξιὸς διφριλάτου	21°13	20°14[s]	2	β Aur
24.	σοδὴλ ἰαμανή	Cnc 5°33	—25°14[t]	1	α Car
25.[u]	σιαρὰ ἰαμανί	5°40[v]	—39°10	1	α CMa
26.	κεφαλὴ δευτέρα[x] τοῦ μουχαταμὴ ἢ τοῦ ἡγουμένου	11°43	—39°40[w]	2	α Gem
27.	κεφαλὴ δευτέρα[x] τοῦ μουάχαρ	15°3	— 6°10	2	β Gem
28.[y]	σουαρὰ σαμίε· προκύων	17°33	— [q]	1	α CMi
29.	ἡ φάτνη	15°38[z]	0°40	neb.	ε Cnc
[29a][aa]	δευτέρα κεφαλὴ διδύμων	28°43	+		
30.	νοτία κεφαλὴ τοῦ λέοντος	Leo 12°33	9°40	3	ε Leo
[30a][bb]	ἀρκτοῦρος	21°48	0°0		
31.	βόρειον μέρος κεφαλῆ[cc] τοῦ λέοντος	12°43	12°0	3	μ Leo
32.	φαρσιαχά[dd]· ὕδρου τράχηλος	18°23	20°30	2	α Hya
33.	ῥάχις λέοντος	20°33	8°30	2	γ Leo
34.	καρδία λέοντος	21°0[ee]	0°10	1	α Leo
35.	ὀσφὺς λέοντος	Vir 2°33	13°40	2	δ Leo
36.	σιρφά[ff]· οὐρὰ λέοντος	12°53	11°50	1	β Leo
37.	χάετ	18°3[gg]	54°0	2	η UMa
38.	ζάνη ἀλάουα[hh]	Lib 1°33	2°50	3	γ Vir
39.	πτερόν	1°53	—14°50	3	γ Crv
40.	σιμὰκ ἀλάκ · κοντάρατ[ii]· ὁ στάχυς	15°3	— 2°0	1	α Vir
41.	σιμὰκ ῥαμέ	15°23	31°30	1	α Boo
42.	πινακίου ἥμισυ	Sco 3°3	44°35	2	α CrB
43.	χεὶρ νότιος	6°23	0°40	2	α Lib
[43a][kk]	κένταυρος		36°51		
44.	χεὶρ βορεία	10°33	8°50	2	β Lib
45.	μέσος ὀφθαλμὸς τοῦ σκορπίου	24°3	—1°40	3	δ Sco
46.	καρδία σκορπίου· ἀντάρης	Sgr 1°3	—4°0	2	α Sco
47.	κεφαλὴ τοῦ ὀφιούχου	13°13	36°0	3	α Oph
48.	ταβαλὴ ἀλσούλα	10°33[ll]	13°15	neb.	G Sco
49.	ὀφθαλμὸς τοῦ κενταύρου	Cap 3°33	0°0[mm]	3[nn]	ν Sgr
50.	ἀετὸς καθήμενος ὁ λύρας	5°43	62°0	1	α Lyr
51.	πέτασμα ἀετοῦ	22°13	29°10	2	α Aql
[51a][oo]	ὁ ἐν τῷ γόνατι	1°6	0°0		
52.	μυκτὴρ ὄρνιθος	22°53[pp]	49°20	3	β Cyg
53.	οὐρὰ δελφῖνος	Aqr 6°3	29°10	2[qq]	ε Del
54.	στόμα τοῦ σιαγόνος[rr] ἵππου	23°43	32°30[ss]	3	ε Peg

Lfd. Nr.	Name	Länge		Breite	Größe	Mod. Bez.
55.	στόμα τοῦ νοτίου ἰχθύος σιαγόνος[rr]		25°23	−23°0	1	α PsA
56.	ῥίτφη· ὄρνη[tt]		27°43[uu]	60°0	2	α Cyg
57.	ἵππου ὦμος	Psc	20°33	31°0	2	β Peg
58.	οὐρὰ τοῦ κήτους βορείου		22°43	− 9°40	3	ι Cet
59.	οὐρὰ τοῦ κήτους τοῦ νοτίου		24°3	−20°20	3	β Cet
60.	πτερὰ τοῦ ἵππου	Ari	0°33	−12°35	2	γ Peg

Anmerkungen zum griechischen Text:
[a] Der Stern hatte in P¹ die Länge Psc 28°55'; gemäß der späteren Epoche und der dementsprechend um 1°38' größeren Länge muß er bei II im Zeichen Ari aufgeführt werden; gleichzeitig ließ ihn der griechische Bearbeiter von II aber auch an seiner alten Stelle (bei Nr. 60, im Zeichen Psc) stehen, so daß γ Peg in II zweimal erscheint. [b] Der Fehler 13 statt 53 ist häufig und dürfte auf Verlesung des arabisch-persischen Originals beruhen (*yǧ* = 13 statt *nǧ* = 53, nur durch einen Punkt unterschieden). [c] πζ als Abkürzung von περίζωμα? Der Stern steht nach Ptolemäus ὑπερ τὸ περίζωμα. [d] Zu erwarten: 25°3'. [e] Zu erwarten: 29°3'. [f] Ptolemäus und P¹,²: 3. [g] Zu erwarten: 29°13'. [h] Statt 23°0'. [i] ms: ἀγγών. [k] Zu erwarten: 8°13'. [l] Ptolemäus und P¹,²: 1. [m] ms: ἡγουμένου. [n] Himmelsrichtung fehlt. [o] Einschub; Duplikat zu Nr. 15. [p] Einschub; Duplikat zu Nr. 17. [q] Fehlt im ms. [r] Dieses entstellte Wort gehört zu Nr. 33. [s] Statt: 20°0'. [t] Statt: −75°0'. [u] Beginn der neuen Seite in a. [v] Zu erwarten: 6°3'. [w] Statt: −9°40. [x] Das Wort δευτέρα ist zu streichen. [y] Hier setzen b und c ein. [z] Die hierhergehörige Länge (28°43') steht in der folgenden Zeile. [aa] Einschub; Duplikat zu Nr. 26 oder 27, dazu die Länge von Nr. 29. [bb] Einschub; Duplikat zu Nr. 41. [cc] leg. τῆς κεφαλῆς. [dd] bc: φαρσιαχᾶ. [ee] Zu erwarten: 20°53'. [ff] b: σιφρὰ. [gg] Zu erwarten: 18°13'. [hh] b: ζώνη ἀλάουα; c: ζάνη ἀλάουλ. [ii] κοντάρατ gehört zu Nr. 40 (cf. I 14). [kk] Einschub; als Stern nicht identifizierbar. [ll] c: 15°33'. [mm] Ptolemäus und P¹,²: 0°45'. [nn] Ptolemäus: neb. [oo] Einschub; als Stern nicht identifizierbar (cf. III 14). [pp] b: 22°23'; c: 29°13'. [qq] Ptolemäus: 3; P¹,²: 4. [rr] σιαγών: Glosse, mit in den Text geraten. [ss] Ptolemäus: 22°30'. [tt] Anscheinend aus ὄρνις, griechisch-ptolemäischer Name des Sternbilds, in dem der arabische Stern *arridf* lokalisiert ist. [uu] Zu erwarten: 27°33'.

Die komplette Tafel der gleichen sechzig Sterne findet sich in gleicher Anordnung in dem persischen Tafelwerk *Zīǧ-i Īlḫānī* von Naṣīr ad-Dīn aṭ-Ṭūsī. Mir lag vor:

P¹ ms Bodl. Hunt. 143 (eine verso-Seite; leider fehlte die folio-Angabe): *ǧadwal-i asmā-i kawākib-i mašhūr az ṯawābit wa-aṭwāl wa-ʿurūḍ wa-aqdār wa-amziǧa-i īšān be-sana-i ḫ' yazdaǧirdī* (ad marg.: *wa-li-sana 1543 rūmī*) „Tafel der Namen der bekannten Fixsterne und ihrer Längen, Breiten, Größen und Temperamente für das Jahr 601 der Ära Yazdaǧird (Zusatz am Rand: und für das Jahr 1543 „griechischer" [= seleukidischer] Ära) [= 1232]." Die 60 Sterne sind nach Tierkreiszeichen angeordnet und haben Ekliptikkoordinaten; Längendifferenz zu Ptolemäus: +16°45'. In acht Kolumnen folgen von rechts nach links: Nummer des Sterns, Name, Tierkreiszeichen und Länge, Breite und Himmelshälfte, Größe, Temperament.

p¹ Die gleiche Handschrift enthält auf der folgenden recto-Seite eine Tafel von 18 Sternen[32] mit Ekliptikkordinaten (*ǧadwal-i mawāḍiʿ-i*

[32] In P¹ = 2. 6. 11. 14. 15. 17. 22. 25. 28. 34. 40. 41. 46. 47. 50. 51. 56. 57.

ba'ḍī az ṯawābit be-ḥasab-i arṣād-i muḫtalif „Tabelle der Positionen einiger Fixsterne nach verschiedenen Beobachtungen"); in vier Rubriken folgen hier von rechts nach links die Angaben gemäß Ibn al-A'lam,[33] die Angaben der Ḥākimitischen Tafeln von Ibn Yūnis,[34] „Ptolemäus", sowie aṭ-Ṭūsīs eigene Beobachtungsergebnisse.[35] Die Längen der Rubrik „Ptolemäus" sind identisch mit denen in P¹; man hat also die Kolumne so zu verstehen, daß darin Werte geboten werden, die nur durch Addition einer Konstante aus den alten Werten des Ptolemäus errechnet sind (im Gegensatz zu den „eigenen Beobachtungen" der vierten Kolumne); es handelt sich also nicht um die ursprünglichen Werte von Ptolemäus selbst.

P² Eine Ableitung hieraus findet sich in dem persischen Tafelwerk *Zīǧ-i Šams-i Munaǧǧim* von Muḥammed ibn 'Alī Ḫwāǧa Šams al-Munaǧǧim al-Wābiknawī[36] in ms Aya Sofya 2694, fol. 137ᵛ: *ǧadwal-i taqwīm-i kawākib-i ṯābita dar awwal-i dawr-i aṣl ya'nī awwal qfḥ malikī ke ḫld yazdaǧirdī būd* „Tabellenaufstellung der Fixsterne für den Beginn der

[33] gest. 375 h = 985. Nicht in GAL; cf. SUTER Nr. 137; KENNEDY Nr. 70; KNOBEL, *Chronology* p. 8 und Nr. 10; die Auszüge bei BERNARD p. 571, col. δ, stimmen im wesentlichen mit p¹ überein.

[34] Siehe oben Anm. 14; cf. ferner SUTER Nr. 178; KENNEDY Nr. 14; KNOBEL, *Chronology* p. 8 f. und Nr. 11; die Auszüge bei BERNARD p. 571, col. ε stimmen im wesentlichen mit p¹ überein.

[35] *mā ḥaṣala lanā* „was wir ermittelten"; cf. KENNEDY p. 162, N.

[36] Um 716–736 h = 1316–1336; nicht in GAL und bei SUTER; cf. KRAUSE p. 518 f., Nr. 18; KENNEDY Nr. 35. Das Tafelwerk *Zīǧ-i Šams-i Munaǧǧim* schreibt BROCKELMANN, GAL Suppl. II p. 297, 1 b, Werk 3, irrtümlich dem Muḥammad 'Alī ibn Mubārakšāh Šams ad-Dīn Mīrak al-Buḫārī zu; das Incipit bei KRAUSE sichert jedoch Šams al-Munaǧǧim al-Wābiknawī als Autor. USENER nennt als ältesten bekannten griechischen Text über „persische" Astronomie und somit als Vorläufer des Theodoros Meliteniotes den cod. Laur. 28, 17 (anno 1323). Von diesem Text nimmt BROCKELMANN a. a. O. an, daß er eine griechische Übersetzung des *Zīǧ-i Šams-i Munaǧǧim* enthalte. Im Laur. 28, 17 ist jedoch, wie aus dem Wortlaut der Einleitung hervorgeht (cf. SUTER p. 219; USENER a. a. O.; CCAG 1, p. 3; sowie den Handschriftenkatalog zu dieser Stelle) von der Niederschrift eines angehörten Kollegs die Rede, nicht aber von der Übersetzung eines bestimmten Werkes. Nach den Forschungen von D. PINGREE (mitgeteilt in seinem Artikel über „Palaeologan Astronomy") ist in Laur. 28, 17 neben Tafeln diverser Herkunft (darunter jedoch kein Fixsternverzeichnis) der *Zīǧ* eines Šams ad-Dīn al-Buḫārī enthalten (Σαμψπουχαρης, geb. 11. Juni 1254 in Buḫārā, der von dem Übersetzer Gregorios Chioniades als sein Lehrer bezeichnet wird. Nicht zu verwechseln mit dem obengenannten Šams ad-Dīn Mīrak al-Buḫārī bei BROCKELMANN!), dessen Text einen Kommentar zu einem *Zīǧ al-'Alā'ī* (von Farīd ad-Dīn Abū l-Ḥasan 'Alī ibn 'Abd al-Karīm aš-Širwānī al-Fahhād ? – zu diesem cf. SUTER p. 218, Anm. 72; KRAUSE p. 519, in Nr. 1; KENNEDY Nr. 23, 84, auch 42; nicht in GAL) darstellt. Das GAL Suppl. II p. 296 f., Werk 29 genannte Tafelwerk *az-Zīǧ as-Sulṭānī* (von Quṭb ad-Dīn aš-Šīrāzī, einem Schüler von Naṣīr ad-Dīn aṭ-Ṭūsī; cf. SUTER Nr. 387; KRAUSE p. 507 f.; KENNEDY Nr. 25; SARTON II, p. 1017 ff.; oder von dem obengenannten Šams ad-Dīn Mīrak al-Buḫārī, cf. GAL Suppl. II, p. 297, 1 b; SUTER Nr. 397) in ms Teheran, Majlis 2–184, enthält ein Verzeichnis von 56 Fixsternen für das Jahr 217 Ǧalālī [= 664 Yazdaǧird = 1295], das offensichtlich in Anlehnung an P¹ zusammengestellt wurde: Die Längen entsprechen bei 38 Sternen Ptolemäus + 17°42′, d. h. *Ilḫānī* + 57′, was bei Zugrunde-

Grundperiode [?], d. h. den Beginn [des Jahres] 188 der Ära des Sultans Maliksāh, was [dem Jahr] 634 Yazdağird [= 1266] entsprach“. P^2 hat, ähnlich wie die griechische Ableitung oben, ebenfalls geringfügige Veränderungen gegenüber P^1 erfahren und umfaßt 69 Positionen, von denen 13 neu eingefügt sind, während vier Sterne von P^1 fehlen. Der Aufbau und die Nomenklatur sind gleich P^1. Oben am Rande rechts findet sich der Zusatz: *taqwīm-i ṯawābit darīn ğadwal az ğadwal-i zīğ-i īlḫānī sī wa-yak daqīqa ziyāde ast* „die Aufstellung der Fixsterne in dieser Tabelle ist gegegenüber der Tabelle in den Īlḫānischen Tafeln um einunddreißig Minuten vermehrt“. Hiermit ist zugleich die Längendifferenz zu Ptolemäus festgestellt (Ptolemäus $+17^0 16' =$ *Īlḫānī* $+0^0 31'$, was auch der Zeitdifferenz von 34 Jahren gegenüber P^1 entspricht), wie auch der Zusammenhang von P^2 mit P^1 unzweifelhaft zum Ausdruck gebracht. Prof. D. Pingree macht mich brieflich darauf aufmerksam, daß II abc in Handschriften auftreten, in denen die dem „Perser Achmet“ zugeschriebene Kompilation arabischer Astrologie enthalten ist, deren Übersetzung dem elften Jahrhundert angehört. Die einwandfreie Herleitung von II aus P^1 bringt diese Tabelle jedoch in unmittelbare Beziehung zu aṭ-Ṭūsī und seinen Īlḫānischen Tafeln.

(Von P^1 und P^2 konnte ich die Fotokopien von Prof. E. S. Kennedy benutzen). Ich folge bei der Wiedergabe der Reihenfolge in P^1; Zusätze aus P^2 stehen an den entsprechenden Stellen in []. P^2 ist sehr nachlässig geschrieben; mehrfach treten bei den Koordinaten Verschiebungen um eine Zeile auf; sie werden nicht eigens in den Apparat aufgenommen. Fragen der Punktierung werden nur in wichtigen Fällen im Apparat besprochen. Die astrologischen Temperamente sind auch hier in die Edition nicht aufgenommen. Die Numerierung stammt vom Herausgeber; die Nummern stimmen mit denen im griechischen Text II überein. Auch hier sind altarabische vorptolemäische Bezeichnungen mit einem Stern* gekennzeichnet, wie oben in $A^{1,2}$.

Lfd. Nr.	Name	Länge	Breite	Größe	Mod. Bez.
1.	*surrat al-faras* „Nabel des Pferdes“	Ari 4°35′	26°0′	2	δ Peg = α And
2.	*āḫir an-nahr* „Ende des Flusses“	16°55	−53°30	1	ϑ Eri
3.	*ğanb al-musalsala*[a] „Seite der Angeketteten“	20°35	26°20	2	β And
4.	*muqaddam aš-šaraṭayn** „der vordere der beiden šaraṭ“	23°25	7°20	3	γ Ari

legung einer Präzession von 1° in 66 Jahren genau der Differenz von 63 Jahren zu P^1 entspräche. Freilich fehlen hier 10 Sterne aus P^1, statt dessen finden sich 6 neue, die nicht in P^1 stehen. Ein unmittelbarer Zusammenhang zwischen der Liste in *az-Zīğ as-Sulṭānī* und den Chrysokokkischen Listen besteht nicht.

400

Lfd. Nr.	Name	Länge	Breite	Größe	Mod. Bez.
5.	*mu'aḫḫar aš-šaraṭayn** „der hintere der beiden šaraṭ"[b]	24°25	8°20	3	β Ari
6.	*kaff al-ḫaḍīb** „die gefärbte Hand"[c]	24°35	51°40	3	β Cas
7.	*an-nāṭiḥ** „der [mit dem Horn] Stoßende"[d]	27°25	10°0	3	α Ari
8.[e]	*ḏāt al-kursī* „die Frau auf dem Thron"	27°35	76°45[f]	3	α Cas
9.	*riǧl al-musalsala* „Fuß der Angeketteten"	Tau 3°35	28°0	3	γ And
10.	*mi'ṣam aṯ-ṯurayyā** „Handgelenk der ṯurayyā"	13°25	40°30	neb.	χ Per
11.	*ra's al-ġūl* „Kopf der ġūl"	16°25	23°0	2	β Per
12.	*wasaṭ aṯ-ṯurayyā* * „Mitte der Plejaden"	18°55	4°30	5	Fl.19Tau (Ptol. 29)
13.	*mirfaq aṯ-ṯurayyā** „Ellenbogen der ṯurayyā"[g]	21°35	30°0[h]	2	α Per
14.	*'ayn aṯ-ṯawr* „Auge des Stieres"	29°25	— 5°5[i]	1	α Tau
15.	*riǧl al-ǧawzā'** „Fuß der ǧawzā' [=Orion]"[k]	Gem 6°35	—31°30	1	β Ori
16.[e]	*katif-i yusrā*[l] „linke Schulter [des Orion]"	10°45	—17°30	2	γ Ori
17.	*'ayyūq**[m]	11°45	22°30	1	α Aur
18.	*muqaddam al-minṭaqa* „der vordere am Gürtel [des Orion]"	12°5	—24°10	2	δ Ori
[18a]	*ka'b ḏī l-'inān* „Ferse des Zügelhalters [=ἡνίοχος]"	12°56	24°10	—[n]	β Tau = γ Aur
19.	*ra's al-ǧabbār* „Kopf des Riesen [= Orion]"	13°45	—13°50	neb.	λ Ori
20.	*wasaṭ al-mintaqa*[o] „Mitte des Gürtels"	14°5	—24°50	2	ε Ori
21.	*mu'aḫḫar al-minṭaqa* „der hintere am Gürtel"	14°55	—25°40	2	ζ Ori
[21a]	*al-ǧudayy** „das Böckchen"	17°26	66°0	3	α UMi
22.	*yad al-ǧawzā'** „Hand der ǧawzā' [=Orion]"	18°45	—17°0	1	α Ori
23.	*mankib ḏī l-'inān* „Schulter des Zügelhalters"	19°35	20°0	2	β Aur
24.	*suhayl**[p]	Cnc 3°55	—75°0	1	α Car
25.	*ši'rā-i yamāniya**[q] „der südlichere ši'rā"	4°25	—39°10	1	α CMa
26.	*ra's at-taw'am al-muqaddam* „Kopf des vorderen Zwillings"	10°5	9°40	2	α Gem
27.	*ra's at-taw'am al-mu'aḫḫar* „Kopf des hinteren Zwillings"	13°25	6°10	2	β Gem
28.	*ši'rā-i ša'āmiya**[r] „der nördlichere ši'rā"	15°55	—16°10	1	α CMi
29.	*ma'laf*[s] „die Krippe"	27°5	0°40	neb.	ε Cnc
[29a]	. . . *aš-šuǧā'* „. . . der Hydra"[t]	Leo 1°16	15°0	—[n]	σ Hya

Lfd. Nr.	Name	Länge	Breite	Größe	Mod. Bez.
[29b]	*minḫar al-asad* „Nase des Löwen“	5°36[u]	10°0	–[n]	κ Leo
[29c]	*farš as-safīna* „Deck [Ptol.: κατάστρωμα] des Schiffes“	8°26	58°20	–[n]	ζ Pup
30.	*ra'ʾs al-asad al-ǧanūbī* „der südliche am Kopf des Löwen“	10°55	9°30	3	ε Leo
31.	*ra'ʾs al-asad aš-šamālī*,„ der nördliche am Kopf des Löwen“	11°5	12°0	3	μ Leo
32.	*fard*-i šuǧāʿ*[v] „der Einzelne in der Hydra“	16°45	–20°30	2	α Hya
33.	*mankib al-asad* „Schulter des Löwen“[w]	18°55	8°30	2	γ Leo
34.	*qalb al-asad al-malikī* „Herz des Löwen, βασιλικός“[x]	19°15	0°10	1	α Leo
35.	*ẓahr al-asad* „Rücken des Löwen“[y]	Vir 0°55	13°40	2	δ Leo
36.	*aṣ-ṣarfa** „der [Wetter-]Umschlag“	11°15	11°50	1	β Leo
[36a]	*ẓahr al-asad* „Rücken des Löwen“[z]	12°6	30°0	–[n]	15,cCom. (Ptol.: 6 ext.)
37.	*al-qāʾid** „der Anführer“	16°35	54°0	2	η UMa
38.	*zāwiyat al-ʿawwāʾ** „der Winkel [an der Figur] von al-ʿawwāʾ“	29°55	2°50	3	γ Vir
39.	*ǧanāḥ al-ġurāb al-ayman* „der rechte Flügel des Raben“	Lib 0°15	–14°50	3	γ Crv
40.	*as-simāk al-aʿzal** „der unbewaffnete simāk“	13°25	– 2°0	1	α Vir
41.	*as-simāk ar-rāmiḥ** „der lanzenbewaffnete simāk“	13°45	31°30	1	α Boo
42.	*nayyir al-fakka** „der helle von al-fakka“[aa]	Sco 1°25	44°30	2	α CrB
43.	*al-kiffa*[bb] *al-ǧanūbīya* „die südliche Waagschale“	4°45	0°40	2	α Lib
44.	*al-kiffa*[bb] *aš-šamālīya* „die nördliche Waagschale“	8°55	8°50	2	β Lib
[44a]	*ʿunq al-ḥayya* „Hals der Schlange“	11°36	25°20	–[cc]	α Ser
45.	*mā bayna ʿaynay*[dd] *al-ʿaqrab* „Was zwischen den Augen des Skorpions ist“	22°25	– 1°40	3	δ Sco
[45a]	*riǧl qanṭūris*[ee] „Fuß des Kentauren“	25°36	–41°10	1	α Cen
46.	*qalb al-ʿaqrab* „Herz des Skorpions“	29°25	– 4°0	2	α Sco
47.	*raʾs al-ḥawwāʾ* „Kopf des Schlangenträgers“	Sgr 11°35	36°0	3	α Oph
48.	*tālī aš-šawla* „der dem erhobenen Schwanz des Skorpions Folgende“	17°55	–13°15[ff]	neb.	G Sco

Lfd. Nr.	Name	Länge	Breite	Größe	Mod. Bez.
[48a]	*zuǧǧ an-nuššāba* „Spitze des Pfeils“[gg]	22°46[hh]	6°20	–[n]	γ Sgr
49.	*ʿayn ar-rāmī* „Auge des Schützen“	Cap 1°50[ii]	0°45	neb.	$\nu^{1,2}$ Sgr
50.	*an-nasr al-wāqiʿ** „der stürzende Adler“	4°5	62°0	1	α Lyr
[50a]	*ʿurqūb ar-rāmī* „Achillessehne des Schützen“	4°56	23°0	–[n]	β Sgr
51.	*an-nasr aṭ-ṭāʾir** „der fliegende Adler“	20°35	29°10	2	α Aql
52.	*minqār ad-daǧāǧa* „Schnabel der Henne“	21°15	49°20	3	β Cyg
53.[e]	*ḏanab ad-dulfīn* „Schwanz des Delphins“	Aqr 4°25	29°10	4	ε Del
[53a]	*šawkat ḏanab al-ǧady* „Spitze am Schwanz des Steinbocks“	22°56	– 4°20	–[cc]	[kk]
54.	*fam al-faras* „Maul des Pferdes“	22°5[ll]	22°30	3	ε Peg
55.	*fam al-ḥūt al-ǧanūbī* „Maul des Südlichen Fisches“	23°45[mm]	–23°0	1	α PsA
56.	*ar-ridf** „der [auf dem gleichen Pferd, hinter dem Reiter] Mitreitende“	25°55[nn]	60°0	2	α Cyg
[56a]	*rukba min ad-daǧāǧa*[oo] „Knie der Henne“	29°26	63°45	–[n]	$\omega^{1,2}$ Cyg (Ptol.: 17 Cyg)
[56b]	*matn al-faras* „Rücken des Pferdes“	Psc 14°56	19°40	–[n]	α Peg[pp]
57.	*mankib al-faras* „Schulter des Pferdes“	18°55	31°1[qq]	2	β Peg
58.	*ḏanab qayṭus aš-šamālī* „Nördlicher Teil vom Schwanz des κῆτος“	21°5	– 9°40	3	ι Cet
59.[e]	*ḏanab qayṭus al-ǧanūbī* „Südlicher Teil vom Schwanz des κῆτος“	22°25	–20°20	3	β Cet
60.	*ǧanāḥ al-faras* „Flügel des Pferdes“	28°55	12°30	2	γ Peg

Anmerkungen zum Text:

[a] P²: *ǧanb al-marʾa al-musalsala* „Seite der angeketteten Frau“. [b] P²: *kaff al-ḫaḍīb*, am Rande korrigiert in: *muʾaḫḫar*. [c] P²: *an-nāṭiḥ min al-ḥamal* „der [mit dem Horn] Stoßende, aus dem Widder“, am Rande korrigiert in: *kaff al-ḫaḍīb*; P¹: korrekt wäre: *al-kaff al-ḫ*. [d] P²: *riǧl al-musalsala*, am Rande korrigiert in: *nāṭiḥ*. [e] Der Stern fehlt in P². [f] P¹: Schreibfehler *ʿw mh* (76°45′) statt *mw mh* (46°45′). [g] P² hat statt der altarabischen die ptolemäische Bezeichnung *ǧanb al-ḥāmil* „Seite des Trägers [scil. des Medusenhauptes]“; „Träger des Medusenhauptes“ ist eine häufig gebrauchte arabische Paraphrase für den griechischen mythologischen Sternbildnamen Perseus; cf. auch unten Textvergleich zu Nr. 11. [h] P²: Von hier bis einschließlich Stern 21 reichen die zeilenweisen Verschiebungen von Längen- und Breitengraden. [i] p¹ und P² haben die richtige Breite von –5°10′. [k] P²: *riǧl al-ǧawzāʾ al-yusrā* „linker Fuß der ǧawzāʾ“. [l] Der Name

erscheint in persischer Form; ebenso in P^1 die Namen von Nr. 17, 25, 28, 29 und 32; auch Nr. 6 hat nicht die korrekte arabische Form. [m] P^2: *al-ʿayyūq*. [n] P^2: Himmelshälfte und Größe nicht angegeben. [o] Wohl besser zu lesen: *awsaṭ al-minṭaqa* „der mittlere am Gürtel". [p] P^2: *as-suhayl*, mit falsch gebrauchtem arabischem Artikel. [q] P^2: *aš-šiʿrā al-y.*, arabische Form. [r] P^2: *aš-šiʿrā aš-š.*, arabische Form. [s] P^2 *al-maʿlaf*, arabische Form. [t] Das erste Wort des Namens ist auf der sehr schwachen Fotografie nicht lesbar. Zu erwarten wäre *minḫar* . . . „Nase . . .", was als ptolemäische Bezeichnung zu dem durch die Koordinaten als σ Hya ausgewiesenen Stern paßt (cf. KUNITZSCH, *Sternnamen* p. 183, Nr. 137). [u] Ptolemäus hat die Länge Cnc 18°30'; vermehrt um die Konstante 17°16' von P^2 wäre also zu erwarten Leo 5°36'; auf der schwachen Fotografie ist jedoch *kw* (26') und nicht *lw* (36') zu erkennen. [v] P^2: *al-fard min aš-šuǧāʿ*, arabische Form gleicher Bedeutung. Erscheint in P^2, da vom Schreiber am richtigen Platz vergessen, nachgetragen am Ende der Tabelle hinter Nr. 60. [w] Ptolemäus: ἐν τῷ τραχήλῳ. [x] Entstellung aus βασιλίσκος, in den arabischen Ptolemäusversionen. [y] Bei Ptolemäus: ἐπὶ τῆς ὀσφύος. P^2: *az-zubra* „die Mähne", der altarabische Name der 11. Mondstation (gebildet von δϑ Leo). [z] Zu erwarten wäre der Name *aḍ-ḍafīra* = πλόκαμος (cf. KUNITZSCH, *Sternnamen* p. 111 f., Nr. 18). Der persische Schreiber hat anscheinend das arabische Wort der Vorlage nicht verstanden. Der Name *ẓahr al-asad* gehört traditionsgemäß zu Nr. 35. [aa] P^2: *an-nayyir min al-fakka*, gleicher Bedeutung. [bb] P^2: *al-fakka*. [cc] P^2: Größenangabe fehlt. [dd] ms: *ʿayn*. [ee] Die schwache Fotografie läßt bei dem Wort *qanṭūris* den Ansatz des arabischen Artikels *al* erkennen, der freilich im arabischen Gebrauch bei diesem Fremdwort nicht zu stehen pflegt, wie etwa auch in Nr. 58 und 59: *qayṭus* (= κῆτος). [ff] P^2: *d yh* (4°15') statt *yǧ yh* (13°15'). [gg] Das erste Wort (*zuǧǧ*) ist auf der Fotografie nicht klar zu erkennen; ein derartiger Name ist jedoch bekannt (cf. *Sternnamen* p. 129, Nr. 38). [hh] Ptolemäus hat Sgr 4°30'; es liegt also eine unregelmäßige Differenz vor von + 18°16' (statt + 17°16'). [ii] P^1 falsch *n* (50') statt *nh* (55'). P^2 hat richtig Cap 2°26. [kk] Name und Breite gehören zu c^1 Cap (Ptolemäus: 28 Cap), die Länge paßt auf keinen Stern des ganzen Sternbildes. [ll] P^2: 24°16', gehört zu Nr. 55. [mm] P^2: unleserlich, ausgehend mit 26', also zu Nr. 56 gehörig. [nn] P^2: 29°16' (statt 26°26'). [oo] Die Lesung der ersten beiden Wörter ist wegen der undeutlichen Fotografie nicht sicher. Die gelesene Bezeichnung paßt zu dem durch die Koordinaten definierten Stern. [pp] Die Stellung innerhalb der Liste sowie die Breite entsprechen genau α Peg; die Länge ist dann um 1° zu groß (Ptolemäus: Aqr 26°40'; + 17°16' ergäbe Psc 13°56', das Foto von P^2 zeigt jedoch klar *yd* = 14° statt zu erwartendem *yǧ* = 13°). Das erste Wort des Namens ist auf dem Foto nicht klar zu lesen; statt *matn* „Rücken" könnte der Schriftzug auch *ʿunq* „Hals" bedeuten; zu dieser Definition gehören nach Ptolemäus ζξ Peg, doch würden die Koordinaten in P^2 zu keinem dieser beiden passen. [qq] sic in P^1; p^1 und P^2 haben richtig 31°0' (cf. im Griechischen einmal ähnlich I, 14).

Textvergleich II – $P^{1,2}$

Die Längendifferenz zu Ptolemäus beträgt bei II: +18°23' und bei P^1: +16°45; von P^1 zu II besteht also eine Längendifferenz von +1°38'; die Zeitdifferenz der Epochen beträgt 114 Jahre.

Nr. 1: στῆϑος und μέρος entsprechen als Wiedergabe nicht genau dem arabischen *surra* „Nabel" (auch Ptolemäus: ἐπὶ τοῦ ὀμφαλοῦ).

Nr. 2: αὖλαξ, cf. oben zu I, 1.

Nr. 3: Bei Ptolemäus ὑπὲρ τὸ περίζωμα, entsprechend in den arabischen Versionen: *fawqa l-miʾzar*. Hierauf scheint sich die Abkürzung in II a zu beziehen. Daneben wird schon bei Ṣūfī *ǧanb al-musalsala* als Eigenname von β And angeführt (cf. KUNITZSCH, *Sternnamen*

p. 67, Nr. 7 und p. 184, Nr. 140); auch dieser Name beruht auf der ptolemäischen Darstellung des Sternbilds. ἁλισιδωτοί (cf. das Sternbilderverzeichnis Vat. gr. 1058, fol. 317ᵛ, lin.-8: ἁλισειδωτή = Andromeda) ist Korruption aus ἁλυσίδετος „angekettet" (Hesychios) und soll die arabische Paraphrase *al-musalsala* „die Angekettete" (= Andromeda) wiedergeben.

Nr. 4, 5: In den Mondstationenlisten (an den ersten drei von den vier in Anm. 2 genannten Stellen) als Station 1: Σουρουτάιν, Σουρατάιν, Στουρτάην.

Nr. 6: cf. oben zu I, 2.

Nr. 9: cf. oben zu Nr. 3.

Nr. 10, 12, 13: Es handelt sich hier um eine altarabische Ausgestaltung der *ṯurayyā*-Figur (cf. Kunitzsch, *Untersuchungen* p. 114 f., Nr. 306). Zu den Plejaden im griechischen wie im heutigen astronomischen Sinne gehört hier nur Nr. 12. Als 3. Mondstation in den Listen: Σουρεντιά, Σουραία.

Nr. 11: Ptolemäus' γοργόνιον ist bei den Arabern durchgehend *ra's al-ġūl*. κασιδωτή läßt sich hieraus nicht ableiten. Nach einer Vermutung von Prof. H. Hunger ist vielleicht κεφαλὴ λασιῶτις „das zottige Haupt" (scil. der Gorgo) zu lesen. Im Sternbilderverzeichnis Vat. gr. 1058, fol. 317ᵛ heißt Perseus: ὁ τὴν κεφαλὴν κρατῶν τοῦ δράκοντος (offenbar entsprechend arab. *ḥāmil ra's al-ġūl*, deferens caput algol).

Nr. 14: Die von Ptolemäus abweichende Breitenangabe von $-5^0 5'$ (statt $-5^0 10'$) erscheint sowohl in $P^{1,2}$ wie in IIa. τάπαρ ist aus dem altarabischen Namen von α Tau, *ad-dabarān*, abzuleiten, hier in persischer Form ohne den arabischen Artikel gebraucht (cf. auch als 4. Mondstation in den Listen Τεπεράν, Οὐταρπαράν).

Nr. 15: cf. oben zu I, 6 und 8.

Nr. 16: Der Stern ist allgemein als „linke Schulter des Orion" bekannt; diese ist zugleich im Gefüge des ptolemäischen Sternbilds Orion „die vorangehende" (gegenüber der rechten Schulter). Ein entsprechendes Adjektiv seiner Vorlage hat der Übersetzer von IIa irrig auf ὠρίων statt auf ὦμος bezogen, was bei der arabisch-persischen Wortkonstruktion möglich ist. $P^{1,2}$ enthalten dieses Adjektiv *(mutaqaddim)* nicht.

Nr. 17: IIa entspricht hier der Version, die auch in $A^{1,2}$ erscheint (cf. oben I, 7).

Nr. 19: *al-ǧabbār* „der Riese, der Gewaltige", die arabische Paraphrase der griechischen mythologischen Sternbildfigur Orion.

Nr. 22: Nach Ptolemäus eigentlich ἐπὶ τοῦ δεξιοῦ ὤμου, und so auch in den arabischen Versionen (*mankib* „Schulter"). Daneben erscheint früh die Bezeichnung „Hand der ǧawzā'" (cf. Kunitzsch, *Untersuchungen* p. 77, Nr. 158 und p. 116, Nr. 317a).

Nr. 24: II a entspricht hier wieder einer Version, die auch in A[1,2] erscheint (cf. oben I, 9).

Nr. 26: Für Parallelen zur griechischen Wiedergabe von arab. *muqaddam* cf. die Mondstationenlisten, Station 26: Μουκάδιμ, Μουκατάμ, Μουκάν.

Nr. 27: Für Parallelen zur griechischen Wiedergabe von arab. *mu'aḫḫar* cf. die Mondstationenlisten, Station 27: Μουάχαρ, Μουαχάρ.

Nr. 32: φαρσιαχά sicher Korruption aus *fard-i šuǧā'* (P[1]). Zu ὕδρου τράχηλος, der ptolemäischen Definition dieses Sterns, cf. oben I, A[1] 19.

Nr. 33: Nach Ptolemäus ἐν τῷ τραχήλῳ. *mankib* „Schulter“ in P[1,2] ist auf jeden Fall eine Korruption, das Wort erscheint sonst nie in Beziehung auf Sterne des Löwen. Nach arabischer Tradition gehört der Stern zu der Gruppe *ǧabhat al-asad* „Stirn des Löwen“ (ζγηα Leo), die die 10. Mondstation bildet (cf. Kunitzsch, *Untersuchungen* p. 24 und p. 61, Nr. 103a; auch *Sternnamen* p. 115, Nr. 23). ῥάχις scheint als Übersetzung eher zu *ẓahr* (P[1]: 35, P[2]: 36a) zu passen; die Koordinaten wie auch die Reihenfolge in der Liste weisen den Stern jedoch eindeutig als γ Leo aus.

Nr. 35: Nach Ptolemäus ἐπὶ τῆς ὀσφύος, wofür die arabischen Versionen *matn* oder *qaṭan* haben. Daneben erscheint bei Ṣūfī als Eigenname *ẓahr al-asad* (cf. *Sternnamen* p. 159, Nr. 90).

Nr. 36: Als 12. Mondstation in den Listen: Ζάρφε, Ζερφά, Σαλφά.

Nr. 38: Zu dem arabischen Namen cf. *Sternnamen* p. 220 f., Nr. 203 und 205; *Untersuchungen* p. 45, Nr. 44. *al-'awwā'* als 13. Mondstation in den Listen: Ἄουε, Ἀββά, Ναάλ.

Nr. 40: Zu κοντάρατ etc. cf. oben zu I, 13. Als 14. Mondstation in den Listen: Σιμάκ, Σαμειάν.

Nr. 42: ἥμισυ beruht vielleicht auf Lesefehler *niṣf* „Hälfte“ statt *nayyir*.

Nr. 43, 44: χείρ Mißverständnis *kaff* „Hand“ statt *kiffa* „Waagschale“, also unabhängig von der altarabischen Tradition; für diese cf. *Untersuchungen* p. 117, Nr. 318.

Nr. 46: Als 18. Mondstation in den Listen: Κάλπ.

Nr. 48: *aš-šawla* als 19. Mondstation in den Listen: Σέουλε, Σουλλά, Ἀγάουλα.

Nr. 49: κένταυρος als Übersetzung des eindeutigen arabischen *ar-rāmī* „Schütze“ ist schwer erklärlich. Freilich ist zu erinnern an die sehr ähnliche figürliche Darstellung der beiden Sternbilder Schütze und Kentaur.

Nr. 50: Zu καθήμενος cf. oben zu I, 18.

Nr. 58, 59: Falsche Beziehung der Beiwörter „nördlich“ und „südlich“; ähnlich oben Nr. 16.

406

Nr. 60 und 1a: In IIa steht jedesmal der Plural πτερά, im Arabischen dagegen immer in Übereinstimmung mit Ptolemäus der Singular *ǧanāḥ*.

Typ III

Charakteristik: 30 Sterne mit Ekliptikalkoordinaten sowie Angabe der Größe und der astrologischen Temperamente. Als Epoche wird, außer in a, an allen Stellen angegeben das Jahr 6854 griechischer Weltära (beg. Herbst 1345) = 715 Yazdaǧird (beg. 1345). Dieses Datum ist auch die Epoche von II.[37] Die Sterne sind nicht nach den üblichen Prinzipien[38] angeordnet, sondern nach den astrologischen Temperamenten. Terminologie und Nomenklatur sind rein griechisch und zeigen – bis auf die Epoche – keine Spuren islamisch-orientalischen Einflusses.[39]

Wir haben es hier mit dem letzten Ausläufer einer alten griechischen astrologischen Texttradition zu tun, der nach weit über tausendjähriger Wanderung Aufnahme in das Chrysokokkische Corpus fand. Es handelt sich um die dreißig „λαμπροὶ ἀστέρες" gewisser alter Astrologentexte, die nach F. Boll[40] über Ptolemäus letztlich auf babylonische Texte zurückgehen; den Ausgangspunkt für die Beziehung dieser Fixsterne auf bestimmte Planeten[41] bildeten Farbvergleiche. Über diese Gleichsetzung gewannen dann die betreffenden Fixsterne auch ihrerseits astrologische Aussagekraft. Im Gegensatz zu Boll nimmt W. Gundel[42] hermetischen und letzten Endes ägyptischen Ursprung an.

Die dreißig λαμπροὶ ἀστέρες treten für uns zum erstenmal in dem Kapitel Ἀποτελέσματα τῆς τῶν ἀπλανῶν ἀστέρων ἐποχῆς im Werk eines in Rom schreibenden anonymen Astrologen ägyptischer Herkunft im Jahre 379 n. Chr. auf,[43] der die noch ältere Lehre des Antiochus von Athen wiedergibt[44] (ed. CCAG 5, 1 p. 196 ff.). In gleicher Anordnung

---

[37] Cf. auch Anm. 17.

[38] Bekannt ist die Anordnung nach Tierkreiszeichen (Ekliptikkoordinaten), nach Rektaszensionen oder Mediatio coeli, nach Größen oder nach der Reihenfolge innerhalb der 48 Sternbilder im Almagest.

[39] Die Benutzung einer persischen Datierung in der Epoche, die gleichförmig in den Versionen c–i erscheint, dürfte hier nicht original sein, sondern, wie diese Epoche an sich, als ganzes stereotyp aus einem anderen Text bzw. einer anderen Tabelle hierher in die Überschrift übernommen sein.

[40] *Antike Beobachtungen farbiger Sterne*, Abh. Bayer. Ak. Wiss., Bd. 30, 1. Abh., München 1916, p. 71 ff.

[41] terminus: κρᾶσις, „Temperament", arab. *mizāǧ*. In den griechischen Texten sind sie durch die bekannten Planetensymbole bezeichnet, die aus Verkürzungen der Anfangsbuchstaben der Planetennamen hervorgegangen sind (cf. W. Gundel, *Sternglaube, Sternreligion, Sternorakel*, Leipzig 1933, p. 107), während sie in den arabisch-persischen durch den letzten Buchstaben der Planetennamen bezeichnet werden.

[42] *Neue astrologische Texte des Hermes Trismegistos*, Abh. Bayer. Ak. Wiss., N. F. 12, München 1936, p. 124 f. Cf. auch A.-J. Festugière, *La révélation d'Hermès Trismégiste*, vol. I, Paris ²1950, p. 120, 171 ff., der für die 30 Sterne Gundel und für den Pseudo-Ptolemäus Boll folgt.

[43] CCAG 5, 1 p. 194

[44] CCAG 8, 3 p. 108 Anm. 3.

erscheinen sie ca. 505 mit dem Längenwert Ptolemäus $+3^{0}40'$ in einer Bearbeitung dieses Textes bei Rhetorius Aegyptius[45] (Περὶ τῆς τῶν ἀπλανῶν ἀστέρων δυνάμεως, ed. CCAG 8, 4 p. 174 ff.); später in einer Bearbeitung von Theophilus Edessenus (am Baghdader Kalifenhof, gest. 785)[46] mit dem Längenwert Ptolemäus $+6^{0}20'$ (ed. CCAG 5, 1 p. 212 ff.); und ferner noch in einer Version von 884 mit der Länge Ptolemäus $+7^{0}30'$ („Excerpta Parisina“, ed. CCAG 5, 1 p. 217 ff.), die einen Iulianos als ursprünglichen Autor nennt.[47].

Die Reihenfolge ist in allen diesen Texten die gleiche; bestimmte Gruppen sind nach gemeinsamen Temperamenten zusammengefaßt.[48] Später sind die Sterne dann aus dem Text herausgesammelt und, mit den gleichen Angaben, zu einer übersichtlichen Tabelle zusammengestellt worden.

Dem alten Text am nächsten steht die Version a; hier ist noch die ursprüngliche Reihenfolge beibehalten (Stern 29 und 30 zwischen Nr. 17 und 18 der „jüngeren“ Tabelle); für die angegebene Epoche 6832 griechischer Weltära (beg. Herbst 1324) ist der Längenwert Ptolemäus $+11^{0}55'$ benutzt, d. h. es ist noch der ptolemäische Präzessionswert von 1^{0} in 100 Jahren beibehalten; die Versionen d–i enthalten Umrechnungen auf der Basis der durch die arabische Astronomie eingeführten verbesserten Präzessionswerte (1^{0} in 66 bzw. 68 Jahren). Version b spiegelt den Übergang zum eigentlichen Tabellentyp III: die Koordinatenwerte sind noch in der ursprünglichen Reihenfolge aufgeführt wie in den alten Texten und in a; jedoch sind die Namen von Stern 29 und 30 hier bereits an das Ende der Tafel geraten;[49] der Längenwert ist bei 20 Sternen einheitlich um $4^{0}46'$ geringer als bei a (also Ptolemäus $+7^{0}10$, so daß sich eine

---

[45] Cf. CCAG 8, 4 p. 115 ff. und 174 Anm. 1.

[46] CCAG 5, 1 p. 195.

[47] Cf. die nähere Erörterung über diesen Iulianos CCAG 5, 1 p. 195.

[48] Bei Rhetorius Aegyptius in seinen Exzerpten aus dem babylonischen Astrologen Teukros werden innerhalb der zwölf Tierkreiszeichen 32 Sterne nach aufsteigenden Längen aufgeführt (ed. CCAG 7, p. 194 ff., aus ms Berlin Phill. 1577, s. XV, fol. 139 ff.; hier fehlen von den „dreißig hellen Sternen“ α CMi und δ Leo; statt dessen sind eingefügt Plejaden, ε Cnc, β Vir, ε Vir). Etwa die gleichen Sterne, in derselben Anordnung, aber in Tabellenform, bietet ms Vind. Phil. gr. 108, s. XVI, fol. 283 (cf. Hunger, p. 215, 35); die Tabelle ist in der Überschrift dem Ptolemäus zugeschrieben, jedoch kann es sich aus mehreren Gründen keinesfalls um einen reinen Ptolemäus-Text handeln (cf. Boll, *Antike Beobachtungen* p. 74 f.); ediert von Boll als „Pseudo-Ptolemäus“, *Antike Beobachtungen* p. 77 und 82 (hier finden sich folgende Abweichungen: Von den „dreißig hellen Sternen“ fehlen ε Ori, β Aur und δ Leo; eingeschoben sind, wie auch bei Rhetorius, Plejaden, β Vir und ε Vir; α Cyg ist zweimal vertreten, einmal fälschlich bei Cap 8°, das zweitemal richtig bei Aqr 19°. Die Liste des Pseudo-Ptolemäus umfaßt daher insgesamt 31 Positionen). Auch diese beiden Verzeichnisse gehören eindeutig in den Überlieferungskomplex der „dreißig hellen Sterne“; durch sie wird zugleich Bolls oben zitierte These erhärtet, die diese Liste über Ptolemäus aus Babylonien herleitet.

[49] Der Schreiber hat nachträglich die beiden Namen durch einen Bogen am Rande der Tabelle an ihre „richtige“ Stelle oben zwischen Position 17 und 18 eingewiesen.

Epoche um 850 ergäbe, während die Überschrift jedoch, wie in allen anderen Versionen, 1346 nennt). Damit erweist sich natürlich b als älter gegenüber a; daß a durch einfache Hinzufügung der Konstante von $4^0 45'$ aus b abgeleitet ist, wird an éinem Falle ganz klar.[50] In c sind Namen und Breiten angeordnet wie allgemein in Typ III; dagegen die Längen haben noch die Reihenfolge wie in b (ab Position 8 bis zum Schluß stimmen b und c genau überein). c ist eine schlechte Kopie, in der die Kolumnen verständnislos nebeneinandergeschrieben wurden, so daß in den 30 Zeilen die Namen und die zugehörigen Werte nicht übereinstimmen. Die Versionen d–g bieten zu der feststehenden Epoche 6854 griechischer Weltära = 715 Yazdağird = 1346 den Längenwert Ptolemäus + $18^0 30'$, und die Versionen hi zu der gleichen Epoche den Längenwert Ptolemäus + $19^0 30'$.

Die Edition folgt der Anordnung des jüngeren Tabellentyps (Version d–i); es werden die verschiedenen Längenwerte aller Versionen geboten; die Temperamente sind in die Edition nicht aufgenommen.

Die Stellen von Typ III[51]

a ms Paris B. N. Suppl. gr. 38, fol. 182v–183 (den Hinweis auf die Stelle verdanke ich Mme. M.-L. Concasty). Überschrift: κανόνιον ἀστέρων ἀπλανῶν; links oben neben der Tabelle die Epoche: κατὰ τὸ ϛωλβ (6832 griechischer Weltära, beg. Herbst 1324); Längenwert: Ptolemäus + $11^0 55'$. Abweichungen vom allgemeinen Typus: der Name von Stern 8 ist ausgefallen, alle folgenden Namen sind dadurch um eine Zeile heraufgerückt, die letzte Zeile der Seite (Position 17) enthält infolgedessen keinen Namen. Die Sterne 29 und 30 stehen gemäß der ursprünglichen Tradition zwischen Position 17 und 18. Am Ende sind als Unikum vier zusätzliche Positionen angefügt, die in sämtlichen anderen Versionen nicht enthalten sind.

b ms Vat. gr. 214, s. XIV–XV, fol. 4v (cf. Mercati I p. 277). Überschrift: ἀστέρες ἀπλανεῖς; Epoche nicht genannt. Längenwert (in 20 Fällen): Ptolemäus + $7^0 10$. Sämtliche Koordinaten in der Reihenfolge wie in a; die Namen von Stern 29 und 30 jedoch nicht zwischen Position 17 und 18, sondern wie im allgemeinen Typus am Ende.

c ms Vind. Phil. gr. 108, s. XV bzw. XVI, fol. 30 (cf. Hunger p. 213; CCAG 6, p. 2). Überschrift und Epoche wie der allgemeine Typus. Längenwerte ab Stern 8 in Reihenfolge und Wert gleichlautend mit b; Namen und Breiten jedoch wie im allgemeinen Typus (also Stern 29 und 30 am Ende).

d ms Vat. gr. 209, s. XIV, fol. 23 (cf. Mercati I p. 258). Überschrift und Epoche wie im allgemeinen Typus; Längenwert: Ptolemäus + $18^0 30'$.

[50] Der Längenwert von ϑ Eri (b, Position 26; infolge der Verschiebung um zwei Zeilen neben δ Leo stehend) ist richtig Ari (ms falsch: Cap) $7°20'$; der gleiche Wert – nicht um $4°45'$ weitergerechnet auf Ari $12°5'$ – erscheint in a, Position 26.

[51] Siehe auch oben I k.

e ms Escorial I R 14, fol. 127 (Ich konnte die Fotokopie von Prof. O. Neugebauer benutzen). Überschrift und Epoche wie im allgemeinen Typus. Stimmt eng mit d überein. Längenwert: Ptolemäus + 18°30'.

f BERNARD p. 572, col. ι: „Ex Canonibus Persarum Pembrochianis an. Chr. 1346. Melixae 338".[52] Hier sind die Längen von 16 der dreißig Sterne abgedruckt; sie stimmen in elf bzw. dreizehn Fällen mit deg überein; zwei Werte stimmen zu b.

g ms Paris B. N. gr. 2461, fol. 287v. Überschrift und Epoche wie im allgemeinen Typus; Längenwert: Ptolemäus + 18°30'.

h ms Vind. Phil. gr. 190, s. XV bzw. XVI, fol. 149v (cf. HUNGER p. 299; CCAG 6, p. 51). Überschrift und Epoche wie im allgemeinen Typus; Längenwert: Ptolemäus + 19°30'.

i ms Vind. Phil. gr. 87, s. XIV (2. Hälfte), fol. 33 (cf. HUNGER p. 198; nicht in CCAG 6). Überschrift und Epoche wie im allgemeinen Typus; Längenwert: Ptolemäus + 19°30'.

---

[52] Diese Gleichsetzung stimmt nicht. 1346 wäre 268 Ära Malikšāh; 338 Malikšāh entspräche 1416.

Nachtrag

S. 383, Fn. 4, Z. 7-8: streiche "offenbar ... 30-126".

S. 397 unten: Von der Sterntafel von aṭ-Ṭūsī gibt es eine russische Übersetzung von M.M. Roshanskoj und B.A. Rosenfeld, in: *Istoriko-Astronomičeskije Issledovanija*, 8 (Moskau 1962), 174-176, mit Anmerkungen auf S. 190-192.

Κανόνιον τῶν ἀπλανῶν ἀστέρων· ἐκτεθειμένον ἐν ἔτει ϛωνδ ἀπὸ κτίσεως κόσμου· περσῶν δὲ ψιε.[a]

Lfd. Nr.	Name	Längen: a Ptol.+11°55	Längen: bc Ptol.+7°10	Längen: (c)	Längen: defg Ptol.+18°30	Längen: hi Ptol.+19°30	Breite	Größe	Mod. Bez.
1.	στάχυς	Lib 8°35	Lib 4°20	Lib 5°59	Lib 15°10	Lib 15°30[b]	— 2°0	1[c]	α Vir
2.	λύρα	Sgr 29°15	Ari 24°9	Cap 25°48	Cap 5°50[d]	Cap 6°50	62°0	1	α Lyr
3.	ἰχθὺς μέγας	Aqr 18°55	Aqr 14°10	Aqr 15°58	Aqr 25°30	Aqr 26°30	—23°0[e]	1[c]	α PsA
4.	ὄρνις	Aqr 21°5	Aqr 16°20	Aqr 18°16	Aqr 27°40[f]	Aqr 28°40	60°0	2	α Cyg
5.	βόρειος στέφανος	Lib 26°35	Lib 21°0	Sco 22°48	Sco 3°10	Sco 3°10[g]	44°30[h]	2	α CrB
6.	καρδία λέοντος	Leo 14°25	Leo 9°31	Leo 11°19	Leo 21°0[i]	Leo 22°0	0°10[k]	1[c]	α Leo
7.	ἀρκτοῦρος[l]	Lib 8°55	Lib 4°10	Lib 5°58	Lib 15°30	Lib 16°30	31°30	1[c]	α Boo
8.	ἀετός	Cap 15°45	Cap 15°0		Cap 22°20	Cap 23°20	29°10[m]	2	α Aql
9.	ἀντάρης	Sco 24°35	Sco[n] 19°20		Sgr 1°10	Sgr 2°10	— 4°0	2	α Sco
10.	κύων	Gem 29°35	Gem[o] 24°50		Cnc 6°10[p]	Cnc 7°10	—39°10[q]	1	α CMa
11.	ἀριστερὸς πούς ὠρίωνος[r]	Gem 2°45	Tau[s] 27°0		Gem 8°20	Gem 9°20	—31°30	1	β Ori
12.	μέσος τῆς ζώνης ὠρίωνος	Gem 9°15	Gem 4°30		Gem 15°50	Gem 16°50	—24°50[t]	2	ε Ori
13.	ὦμος δεξιὸς ἡνιόχου	Gem 14°45	Gem 20°0		Gem 21°10[u]	Gem 22°10[u]	20°0	2	β Aur
14.	γόνυ τοῦ τοξότου[v]	Sgr 28°55	Cap[w] 24°10		Cap 5°30	Cap 6°30	—18°0	2	α Sgr
15.	γοργόνιον	Tau 11°35	Tau 2°50		Tau 18°10	Tau 19°10	23°0[x]	2	β Per
16.	αἴξ	Gem 6°55	Gem 2°10		Gem 13°30[y]	Gem 14°30	22°30	1	α Aur
17.	κεφαλὴ τοῦ ἑπομένου διδύμου	Cnc 8°35	Cnc 3°50		Cnc 15°10	Cnc 16°10	6°15	2	β Gem
18.	ὦμος ἡγούμενος[z] ὠρίωνος	20.[aa] Gem 5°55	20.[bb] Gem 1°10		Gem 12°30	Gem 13°30	—17°30	2	γ Ori
19.	προκύων	21. Cnc 11°25[cc]	21. Cnc 6°40[cc]		Cnc 18°00[cc]	Cnc 18°40[dd]	—16°10	1	α CMi
20.	ὦμος δεξιὸς ὠρίωνος	22. Gem 13°55	22. Gem 9°10		Gem 20°30	Gem 21°30	—17°0	1	α Ori
21.	κοινὸς ἵππου καὶ ἀνδρομέδας	23. Psc 29°45	23. Psc 25°0		Ari 6°20	Ari 7°20	26°0	2	δ Peg = α And
22.	ὦμος ἵππου δεξιός[ee]	24. Psc 14°5	24. Psc 9°20		Psc 20°40	Psc 21°40	31°0[ff]	2	β Peg

Lfd. Nr.	Name	Längen a Ptol.+11°55	Längen bc (c) Ptol.+7°10	Längen defg Ptol.+18°30	Längen hi Ptol.+19°30	Breite	Größe	Mod. Bez.
23.	ποὺς δεξιὸς κενταύρου	25. Sco 20°15	25. Sco 15°30	Sco 26°50	Sco[gg] 27°50	—41°10	1	α Cen
24.	ὁ ἔσχατος τοῦ ποταμοῦ	26. Cap[hh] 7°20[ii]	26. Cap[hh] 7°20	Cap[hh] 18°40	Ari 19°40	—53°30[kk]	1	ϑ Eri
25.	οὐρὰ λέοντος	27. Vir 6°25	27. Vir 1°40	Vir 13°0	Vir 14°0	11°50	1	β Leo
26.	ὀσφὺς λέοντος ἑπομένη	28. Leo 27°5	28. Leo 21°20	Vir 2°40	Vir 3°40[ll]	13°40	2	δ Leo
27.	ὕδρου αὐχήν	29. Leo 11°55	29. Leo 7°10	Leo 18°30	Leo 19°30	—20°30	2	α Hya
28.	ὁ λαμπρὸς τῶν ὑάδων	30. Tau 24°35	30. Tau 19°50	Gem 1°30[mm]	Gem 1°30[nn]	— 5°10	1	α Tau
29.	βόρειος χηλὴ τοῦ σκορπίου	18.[oo] Sco 4°5	18. Lib 29°20	Sco 10°40	Sco 11°40	8°50[pp]	2	β Lib
30.	κεφαλὴ τοῦ ἡγουμένου διδύμου	19.[oo] Cnc 5°15	19. Cnc 0°30	Cnc 11°50	Cnc 12°50	9°30[qq]	2	α Gem
—	πλειάς	31.[rr] Tau 14°25				3°30	–	Fl. 23 Tau (Ptol. 30)
—	νεφέλη τοῦ καρκίνου ἢ φάτνη	32. Cnc 22°15				0°20	–	ε Cnc
—	κέντρον τοῦ σκορπίου	33. Sgr 13°5				—13°15	–	G Sco
—	ἀκὶς τοῦ τοξότου	34. Sgr 21°25[ss]				— 6°20	–	γ Sgr

[a] ab: Abweichungen in der Überschrift und Epoche, siehe die Beschreibung der Stellen. [b] Zu erwarten: 16°10′. [c] b: 2. [d] f: 5°0′. [e] b: Angabe fehlt. [f] f: 10°20′ (cf. b !). [g] Zu erwarten: 4°10′. [h] b: 44°0′. [i] f: 9°31′ (= b !). [k] h: 6°10′. [l] a: Der Name ist ausgelassen, alle folgenden Namen bis einschl. 17 um eine Zeile vorgerückt, Position 17 ohne Namen. [m] b: 28°50′. [n] c: Sgr. [o] c: Cnc. [p] f: 18°0′ (Wiederholung des Wertes von Nr. 19). [q] c: —31°10′. [r] b: ὄρνιθος. [s] c: Gem. [t] a: Breite nördlich. [u] Zu erwarten: 20′. [v] a: γωρυτὸς τοξότου. [w] b: Ari. [x] h: 22°0′. [y] f: 13°3′. [z] a: ἀριστερός. [aa] a folgt von hier an in den Koordinaten der ursprünglichen Reihenfolge der alten Tradition, mit allen Angaben. [bb] b folgt der gleichen Reihenfolge wie a, doch sind die Namen von 29 und 30 nicht zwischen 17 und 18 eingefügt, sondern folgen am Ende. c hat nur die Längen von Nr. 29 und 30 zwischen 17 und 18 eingeschoben, jedoch die Namen, Tierkreiszeichen und alle anderen Angaben in der Reihenfolge des allgemeinen Typus (d–i) beibehalten. [cc] Ausgehend von Ptolemäus = Gem 29°30′. [dd] Ausgehend von Ptolemäus = Gem 29°10′. [ee] h: δεξιοῦ. [ff] g: Breite südlich. [gg] h: Lib; i: Sco (korrigiert aus Lib). [hh] Statt Ari. [ii] Zu erwarten: 12°5′. [kk] a: —43°30′; d: —50°30′; hi: —53°20. [ll] h: 10°40′. [mm] defg: sic statt zu erwartendem 1°10′. [nn] sic statt zu erwartendem 2°10′. [oo] In abc Umstellung, siehe oben Anm. [aa] und [bb]. [pp] c: —8°10; deg: 8°10′. [qq] cdeghi: Breite südlich. [rr] 31–34 Zusatz in a. [ss] Ausgehend von Ptolemäus = Sgr 9°30′.

III

Die arabische Herkunft von zwei Sternverzeichnissen in cod. Vat. gr. 1056

Unlängst habe ich an anderer Stelle die verschiedenen Typen des Sternverzeichnisses der „Persischen Syntaxis" des Georgios Chrysokokkes besprochen[1]. Bei dieser Gelegenheit wies ich auf weitere Sternverzeichnisse arabisch-islamischer Herkunft in griechischen Handschriften hin[2]. Für zwei davon ergab sich inzwischen die Möglichkeit einer näheren Fixierung der Herkunft. Es handelt sich um zwei Verzeichnisse aus cod. Vat. gr. 1056 (14. Jhd.)[3].

Das erste dieser beiden Verzeichnisse umfaßt 29 Sterne nebst Länge, Breite, Größe und planetären Temperamenten. Die Koordinaten fallen dadurch auf, daß sie auch kleinste Einerwerte von Minuten angeben, wodurch sie sich von vornherein von den ptolemäischen unterscheiden. Im Sternkatalog des Almagest sind die Minutenwerte der Gradangaben stets in folgenden Brüchen von Graden ausgedrückt: ½ (= 30′), ⅓ (= 20′), ¼ (= 15′) oder 1/6 (= 10′), zuweilen in Kombinationen hiervon; Werte wie 4′ oder 18′ jedoch kennt Ptolemäus nicht. Solche Werte bilden dagegen ein Charakteristikum arabischer Sternverzeichnisse, und zwar auch dort nur solcher, die nicht durch Addition einer Konstante aus dem Almagest abgeleitet sind, sondern solcher, die auf eigenen, vom Almagest unabhängigen Beobachtungen beruhen. Hierin wäre also ein erster

[1] Byzantinische Zeitschrift 57, 1964, 382—411.

[2] a.a.O., Anm. 5 auf S. 384. Cf. inzwischen auch D. Pingree, *Gregory Chioniades and Palaeologan Astronomy*, Dumbarton Oaks Papers 18, 1964, S. 138f. (dessen globale Angaben werden jetzt hierdurch in Einzelheiten modifiziert).

[3] Das eine auf fol. 30v mit Fortsetzung auf 33r, das zweite auf fol. 31r; cf. *CCAG* 5, 3 S. 11. Mir liegt der Text beider Verzeichnisse in einer eigenhändigen Abschrift von Prof. Pingree vor, die er mir seinerzeit freundlicherweise zur Verfügung stellte. Aus Gründen der Authentizität verzichte ich daher hier auf eine vollständige Edition und beschränke mich auf die Darlegung des Zusammenhangs. Immerhin reicht das vorliegende Material aus, um die vorgetragenen Zusammenhänge mit absoluter Sicherheit daraus abzuleiten. An der gleichen Fundstelle stehen ferner noch drei andere, kleinere Verzeichnisse (zwei von Abū Maʿšar, eines von Simeon Seth; zu ersteren cf. jetzt übrigens meinen Artikel in ZDMG 120/1, 1970, S. 103—125).

unbestreitbarer Hinweis auf arabisch-islamische Herkunft der beiden Tafeln zu erblicken. In die gleiche Richtung deutet die Beobachtung, daß die Längen keine einheitliche, konstante Zunahme gegenüber den Werten des Almagest aufweisen. Was die Nomenklatur angeht, so sind die Sterne zwar alle mit griechischen Ausdrücken bezeichnet, doch geben diese häufig nicht die eigentlichen griechischen bzw. ptolemäischen Bezeichnungen wieder, sondern wörtliche Übersetzungen arabischer Namen[4]. Laut Überschrift ist das Verzeichnis berechnet für den Beginn des Jahres 6664 griechischer Weltära (beg. Herbst 1155) und abgeleitet aus drei Sternverzeichnissen (ἀπὸ τριῶν κανονίων).

Näher informiert uns die Überschrift des zweiten Verzeichnisses. Dieses enthält 33 Sterne mit den gleichen Angaben wie das erste und ist berechnet für den Beginn des Jahres 6669 griechischer Weltära (beg. Herbst 1160). Hier wird ebenfalls gesagt, es sei aus drei Sternverzeichnissen kompiliert, und diese werden auch namentlich genannt:
ἀπό τε τοῦ κανονίου τοῦ Χέκεμ καὶ ἀπὸ τοῦ κανονίου τοῦ Κουσιὰρ καὶ ἀπὸ τῆς ζητηθείσης παρ' ἡμῶν καὶ ἀποκομισθείσης ἡμῶν εἰδήσεως ἀπὸ Αἰγύπτου.

Zugrunde liegen also, nach Angaben des Übersetzers bzw. Kompilators, die Tafel des Χέκεμ, wohinter mit großer Wahrscheinlichkeit das Tafelwerk *az-Zīǧ al-ḥākimī* des ägyptischen Astronomen Ibn Yūnus[5] zu verstehen ist; zweitens die Tafel des arabisch schreibenden Astronomen persischer Herkunft Kūšyār ibn Labbān[6]; und drittens eigene Beobachtungen (des Übersetzers oder Kompilators?) in Ägypten. Die arabische Herkunft ist damit also auch explicite im griechischen Text ausgesprochen.

Als weiteres erhebt sich danach die Frage, ob und wie weit die in der griechischen Überschrift behauptete Herkunft aus arabischen Quellen wirklich zutrifft.

Die wertvollen Detailforschungen von M. Destombes über arabische Sternkoordinaten und Sternkataloge bahnen uns den Weg zu einer sehr alten arabischen Sterntafel, die nun tatsächlich mit Sicherheit als letzter Ausgangspunkt auch für die Herkunft der beiden ins Griechische übersetzten Verzeichnisse im cod. Vat. gr. 1056 anzusetzen ist. Und zwar hat

[4] Z. B. χεὶρ Διδύμων (aus arab. *yad al-ǧawzā'*), statt original griech. ὁ ὦμος τοῦ 'Ωρίωνος, für den Stern α Orionis.

[5] Gestorben 1009; cf. Brockelmann, *Gesch. d. arab. Literatur* I 224; Suppl. I 400; E. S. Kennedy, *A Survey of Islamic Astronomical Tables*, Trans. Amer. Philos. Soc., N.S., 46/2 (1956), Nr. 14.

[6] Um 1000; cf. Brockelmann *a.a.O.* I 222; Suppl. I 397; Kennedy *a.a.O.* Nr. 9 und S. 156f.

M. Destombes als erster das Sternverzeichnis aus dem Astrolabtraktat von al-Farġānī aufgefunden und ediert[7].

Dieses Astrolabsternverzeichnis von al-Farġānī enthält nahezu völlig die gleichen Sterne wie die beiden Verzeichnisse in cod. Vat. gr. 1056 und weist dabei, was besonders wichtig ist, in der überwiegenden Zahl der Fälle in den Breiten die gleichen nichtptolemäischen Grad- und Minutenwerte auf wie jene. Ein Zusammenhang erscheint also fraglos als gegeben. In den Längen ist, ähnlich wie bei den griechischen Verzeichnissen, keine konstante Präzession gegenüber Ptolemäus festzustellen. Bereits Destombes weist darauf hin, daß die Koordinaten der Farġānī-Tafel „correspondent aux valeurs attribuées par Ibn Yunus, dans le chapitre 5 de sa Grande Table Hakemite, écrite en 1007 A.D., à la „Table vérifiée" d'Al-Mamoun rédigée en 829".

Glücklicherweise existiert auch von der „Table vérifiée" bereits eine, wenn auch nicht fehlerfreie, Übersetzung[8]. Die „Table vérifiée" geht zurück auf astronomische Beobachtungen, die auf Geheiß des Abbassidenkalifen al-Ma'mūn (reg. 813—833) in Bagdad und Damaskus stattfanden. Ihre Ergebnisse wurden in dem Tafelwerk *az-Zīǧ al-mumtaḥan* („Table vérifiée, *Tabulae Probatae*") zusammengetragen, das unter dem Namen des um 830 gestorbenen Astronomen Yaḥyā ibn Abī Manṣūr[9] läuft. Selbstverständlich gehörte zu dessen verschiedenen Tafeln auch ein Fixsternverzeichnis; in der von Vernet aus ms Escorial 927 entnommenen Fassung enthält es 24 Sterne. Die Längenwerte ergeben keine konstanten Differenzen zu Ptolemäus, womit erwiesen ist, daß es sich hier in der Tat um selbständige Beobachtungen der arabischen Astronomen handelt, und nicht — wie sonst üblich — um mechanisch mit Hilfe einer Präzessionskonstante aus dem Almagest abgeleitete Werte. Das gleiche gilt von den Breiten. Dabei zeigt sich für uns das Überraschende, daß die

[7] 25 Sterne mit Ekliptikal- und Äquatorialkoordinaten; aus ms Brit. Mus. or. 5479, fol. 56v ediert in *Note sur le catalogue d'étoiles du Calife Al-Mamoun*, in: "Actes du VIIIe Congrès Internat. d'Histoire des Sciences (Florence 3—9 Septembre 1956)", S. 309—312; von Destombes auf 845 bzw. 846 datiert. Cf. dazu auch kurz Destombes in seiner wichtigen Arbeit *Un astrolabe carolingien* ... in: Archives Internationales d'Histoire des Sciences, no. 58—59, 1962, S. 13, Anm. 13. Das gleiche Verzeichnis fand ich inzwischen noch, zum Teil mit besserem Text, in drei Berliner arabischen Handschriften: or. quart. 99 (= Ahlwardt 5792), fol. 3rv; Landberg 56 (= Ahlwardt 5790), fol. 30v—32r; und Landberg 58 (= Ahlwardt 5791), fol. 23rv.

[8] J. Vernet: *Las „Tabulae Probatae"*, in: *Homenaje a J. Millás Vallicrosa*, II (Barcelona 1956) S. 501ff., speziell 519, aus der arabischen Handschrift Escorial 927, fol. 95v. Einige Korrekturen hierzu unten in meinen Anmerkungen zur synoptischen Übersicht aller dieser Tafeln.

[9] Brockelmann *a.a.O.* I 221; Suppl. I 393; Kennedy *a.a.O.* Nr. 51 und S. 145ff. (speziell sub N).

Breiten identisch sind mit denen in den zwei griechischen Verzeichnissen von cod. Vat. gr. 1056[10]. Damit wäre als letzter Ausgangspunkt das Sternverzeichnis der *Tabulae Probatae* aus Bagdad ermittelt.

Von hier aus sind nun auch sehr leicht die Längenwerte in den verschiedenen daraus abgeleiteten Sterntafeln zu erklären. Die Epoche der *Tabulae Probatae* ist das Jahr 214 h = 829/30. al-Farġānī hat diese Tafel in seinen Astrolabtraktat übernommen und dabei für die Epoche 845 (bzw. 846, nach Destombes) den Längen eine Konstante von 0° 15′ hinzugefügt. Das laut Überschrift für 1155 berechnete Verzeichnis in Vat. gr. 1056, fol. 30ᵛ hat die Werte der *Tabulae Probatae* konstant um 5° 0′ vergrößert, und das unmittelbar darauf folgende zweite Verzeichnis auf fol. 31ʳ, laut Überschrift berechnet für 1160, hat wiederum die Werte des vorangehenden, ersten Verzeichnisses um die Konstante von 0° 5′ vermehrt. Original waren auch in dieser Tradition nur die Beobachtungen der Astronomen des Kalifen al-Ma'mūn, die späteren Ableitungen dagegen kamen ebenso mechanisch zustande wie sonst die Ableitungen aus dem Almagest.

Der letzte Ausgangspunkt der gesamten Tradition ist damit sichergestellt. Ungeklärt bleibt lediglich noch, aus was für einem Text der griechische Kompilator bzw. Übersetzer seine beiden Verzeichnisse entnommen haben könnte. Hierzu ist vorläufig keine abschließende Stellungnahme möglich. Die genannte, von Vernet benutzte Escorial-Handschrift enthält Einschübe mehrerer Autoren (Kūšyār ibn Labbān, Ibn al-Aʿlam und Abū l-Wafā'[11]), die sämtlich jünger sind als die *Tabulae Probatae*. Kūšyār wird auch im Vat. gr. 1056 genannt[12], dazu der *Zīǧ al-ḥākimī* des 1009 gestorbenen Ibn Yūnus. Die Epoche der griechischen Verzeichnisse liegt jedoch um ca. einhundertfünfzig Jahre später. Es wäre also möglich, daß der griechische Übersetzer bzw. Kompilator eine arabische Hand-

[10] In einigen Fällen hat das zweite griechische Verzeichnis anscheinend den „richtigeren", zu den *Tabulae Probatae* stimmenden Wert. Auf eine detaillierte Gegenüberstellung muß hier verzichtet werden, da mir die griechischen Verzeichnisse nur in handschriftlicher Kopie und nicht in authentischer Fotokopie vorliegen. Auch die Lesung der arabischen Werte bei Vernet ist nicht völlig zuverlässig (allein dreimal nimmt er bei den Breiten n = 50 fälschlich für y = 10); die vier Fotokopien der Farġānī-Tafel ermöglichen eine Reihe von Emendationen.

[11] Brockelmann *a.a.O.* Suppl. I 393, 2a.

[12] Die Berliner arabische Handschrift or. quart. 101 (= Ahlwardt 5751) enthält auf p. 158—159 ein Verzeichnis von 48 Sternen von Kūšyār für den Beginn des Jahres 301 Yazdaǧird = April 932. Das Verzeichnis ist nach der Reihenfolge der 48 ptolemäischen Sternbilder aufgebaut und zeigt eine schematische Berechnung der Längen durch Addition von 12° 0′ zum Wert des Almagest; es gehört somit nicht in die Reihe der hier behandelten Verzeichnisse aus der Tradition der *Tabulae Probatae*.

schrift gemischten Charakters benutzte, ähnlich der Escorial-Handschrift, in der zu der alten Grundtafel aus den *Tabulae Probatae* auch spätere Zusätze von anderen arabischen Astronomen, darunter Kūšyār ibn Labbān und Ibn Yūnus, hinzugetreten waren. Ein Text von Ibn Yūnus könnte dabei die Basis gebildet haben. Die Beurteilung solcher Überlieferungsprobleme ist für uns jedoch sehr schwierig, da uns trotz der erdrückenden Unzahl arabischer Manuskripte von bestimmten Texten oft eben doch zu wenige oder gar keine authentischen Exemplare erhalten sind.

Ich beschließe diese Notiz mit einer synoptischen Übersicht über die Sterne der aus den *Tabulae Probatae* hergeleiteten Traditionen, soweit oben dargestellt[13]. Zum Ausgangspunkt nehme ich dabei das erste der beiden griechischen Verzeichnisse. Kritische Anmerkungen bleiben auf ein Minimum beschränkt, da mir kein authentischer griechischer Text vorliegt.

Stern	Vat. gr. 1056		*Tabulae Probatae*	Farġānī
	30v/33r λ = *Prob.* + 5° 0′	31r λ = 30v + 0° 5′	λ = eigene Beobachtung	λ = *Prob.* + 0° 15′
θ Eri	1	1	1	25
β Ari	2	—	3	—
β Cas	3	2	2	1
16, p¹ Per[a]	4	3	—	—
β Per	5	4	4	2
χh Per[b]	6	5	—	—
α Tau	7	7	6	4
α Per[c]	8[d]	6	5	3
β Ori	9	8	7	5
α Aur	10	9	10	8
ε Ori[e]	11[f]	10[g]	8[h]	6[i]

[13] Ebenfalls aus den *Tabulae Probatae* abgeleitet ist ein Sternverzeichnis des arabischen Astronomen Ḥabaš (cf. Brockelmann *a.a.O.* I 221, Suppl. I 393; Kennedy *a.a.O.* Nr. 15 und S. 151ff., speziell sub N) in ms Wetzstein I, 90 = Ahlwardt 5750, fol. 62rv (Deutsche Staatsbibliothek, [Ost-]Berlin). Das Verzeichnis ist bisher nicht ediert oder übersetzt. Als Epoche ist in Ziffern angegeben 304 h = 916/17. Der zugehörige Längenwert ist gegenüber den *Tabulae Probatae* konstant um 1° 17′ vermehrt, die Breiten sind mit den dortigen identisch. Das Verzeichnis von Ḥabaš enthält die gleichen 25 Sterne, in derselben Reihenfolge und mit denselben Benennungen, wie das Astrolabsternverzeichnis von al-Farġānī; am Schluß sind 5 weitere Sterne hinzugefügt, einer davon (α Cygni) ist bereits in der Grundtafel enthalten und erscheint somit bei Ḥabaš zweimal (mit verschiedenen Koordinaten!).

Stern	Vat. gr. 1056 30v/33r λ = *Prob.* + 5° 0′	Vat. gr. 1056 31r λ = 30v + 0° 5′	*Tabulae Probatae* λ = eigene Beobachtung	Farġānī λ = *Prob.* + 0° 15′
λ Ori[b]	12	11	—	—
α Ori	13	12	9	7
α CMa	14	13	11[k]	9
α CMi	15	14	12[l]	10
α Leo	16	18	13	11
α Vir	17	19	14	12
α Boo	18	20	15	13
α Cen	19	21	24	23
α CrB	20	22	16	14
α Sco	21	23	17	15
α Oph	22	24	18	16[m]
α Lyr	23	28	19	17
β Sgr	24	—	—	22
α Aql	25	29	20	18
α PsA	26	30	21	21[n]
α Cyg	27	31	22	19
ω Cyg[b]	28	32	—	—
β Peg	29	33[o]	23	20[p]

Anmerkungen:

[a] Bei Ptolemäus im Almagest: der 3. externe Stern des Perseus.

[b] Diese drei Sterne gehören nach Ptolemäus zu den „nebligen". Sie sind hier nachträglich eingeschoben.

[c] Der Stern trägt in allen hier besprochenen arabischen Quellen, d. h. in den *Tabulae Probatae*, bei al-Farġānī und bei Ḥabaš die Benennung *al-kaff al-ǧaḏmā'*; analog dazu griech. fol. 30v (am Rande): ἡ χεὶρ τῆς ἀνδρομέδης, fol. 31r: ὁ ἐπὶ τῇ χειρὶ τῆς ᾿Ανδρομέδας. Der arabische Name bedeutet „die verstümmelte Hand" (scil. der *ṯurayyā*, der Plejaden; cf. P. Kunitzsch, *Untersuchungen zur Sternnomenklatur der Araber*, Wiesbaden 1961, Nr. 137, dazu auch Nr. 306. Der griechische Übersetzer behalf sich damit, diese „Hand" der Andromeda zuzuteilen) und bezeichnet nach aṣ-Ṣūfī die sechs Sterne λαγδνμ Ceti, im Kopf des ptolemäischen Sternbildes Walfisch. Durch die Koordinaten ist unser Stern in der Tabelle astronomisch jedoch einwandfrei als α Persei ausgewiesen. Diese Abweichung im Gebrauch des Namens *al-kaff al-ǧaḏmā'* hat auch bereits aṣ-Ṣūfī bei „Autoren von Tafelwerken" (*aṣḥāb az-zīǧāt*) gekannt und getadelt: *Kitāb ṣuwar al-kawākib*, ed. Hyderabad 1954, S. 260 [= ed. Schjellerup, St. Pétersbourg 1874, S. 202]. (Korrekt gehört α Persei innerhalb dieses arabischen Bildzusammenhangs zu der anderen, „ausgestreckten" Hand der *ṯurayyā* und trägt darin die Bezeichnung *al-mirfaq*

„Ellenbogen", cf. Kunitzsch, *Untersuchungen* Nr. 163; auch id., *Arabische Sternnamen in Europa*, Wiesbaden 1959, S. 176, Nr. 121).

[d] Im ms am Rand nachgetragen.

[e] Auch Destombes richtig: 27 Orion (nach Ptolemäus); dagegen Vernet abweichend: δ Ori.

[f] ὀμφαλὸς Διδύμων; daß dies (und nicht etwa ὀφθαλμὸς Διδύμων) die richtige Lesung ist, ergibt sich aus dem arabischen Original (vgl. unter h und i).

[g] Auch hier ist die richtige Lesung ὀμφαλός (und nicht ὀφθαλμός), vgl. unter f.

[h] Vernet liest den arabischen Namen irrtümlich *Al-Našra*; die richtige Lesung ist jedoch *as-surra* „der Nabel" (cf. unter i).

[i] Alle vier Farġānī-Handschriften haben die richtige, zum Teil sogar vokalisierte Lesung *surrat al-ǧawzā'* „Nabel der ǧawzā'". Es handelt sich hier um eine recht seltene Bezeichnung altarabischer, nichtptolemäischer Herkunft, die in meinem Verzeichnis der Sternnamen echtarabischen Ursprungs (*Untersuchungen zur Sternnomenklatur der Araber*, Wiesbaden 1961) nachzutragen wäre. Die Benennung *surrat al-ǧawzā'* findet sich auch auf folgenden zwei arabischen Astrolabien: Astrolab von Ḥāmid al-Ḫiḏr al-Ḫoǧandī, 374 h = 984/85 (Gunther Nr. 111; M. Destombes, *Astrolabe carolingien* S. 16 mit Anm. 23 und 24; ders., *Diffusion des Instruments ...*, Cahiers d'Histoire Mondiale X, 1, 1966, S. 40; diese Angaben verdanke ich brieflichen Hinweisen von Herrn Destombes); und Astrolab von 'Abd al-Karīm al-Miṣrī, 638 h = 1240/41 (Gunther Nr. 104; M. Destombes, *Les chiffres coufiques ...*, Physis 2, 1960, S. 205, Anm.).

[k] Vernet hat den gewöhnlichen arabischen Namen *al-yamāniya* verlesen in *Al-Baḥāniyya*, das er nicht zu identifizieren weiß. Die hierher gehörige Identifizierung α CMa setzt er irrtümlich zu dem folgenden Stern.

[l] Vernet setzt die Identifizierung des vorangehenden Sterns, α CMa, irrtümlich hierher.

[m] Drei Handschriften (London, Ahlwardt 5790 und 5791, 23v) schreiben falsch *ra's al-ǧawzā'* („Kopf des Orion"); dagegen Ahlwardt 5792 und 5791, 23r haben das richtige *ra's al-ḥawwā'* „Kopf des Schlangenträgers".

[n] In dieser Tafel haben vier südliche Sterne (Nr. 21, 22, 23, 25) den stereotypen Längenwert Ptolemäus + 11° 12′ sowie die Breiten des Almagest. Nach M. Destombes, *Note* ... S. 310 sind sie vielleicht nicht aus den originären Beobachtungen der *Tabulae Probatae* hervorgegangen, sondern aus anderer Quelle übernommen.

[o] Diese Liste ist um sechs zusätzliche, meist ebenfalls „neblige" Sterne erweitert: 15 (ε Cnc), 16 (μ Leo), 17 (ε Leo), 25 (G Sco), 26 (μ Sgr) und 27 (ν Sgr).

[p] Stern Nr. 24 dieser Liste ist α Car, *suhayl*, tief im Süden, der in den anderen Versionen nicht vorkommt. Sein Längenwert beträgt Gem 29° 55′, d. h. eine Zunahme gegenüber dem Almagest um 12° 45′ (zum Vergleich: Kūšyār benutzt für 933 eine Zunahme von 12° 0′; aṣ-Ṣūfī für 964 eine Zunahme von 12° 42′; und Maslama für 978 eine Zunahme von 12° 40′).

IV

TWO STAR TABLES FROM MUSLIM SPAIN

The Arabic MS. Istanbul, Aya Sofya 2671 (dated 621 H = AD 1224), is a most interesting document for the history of astronomy as it contains, among others, the Arabic version of Ptolemy's *Planisphaerium* (ff. 76^{v}–97^{r}) which has not survived in the original Greek, and of which only the Latin translation made from the Arabic by Hermannus Secundus in Toulouse, AD 1143, is at present accessible.[1]

Apart from this, the MS. also contains two other texts of astronomical interest each of which is equipped with a star table. These two star tables will be discussed and edited in this present paper.

I

On ff. 1^{r}–75^{r} our MS. shows a version in eighty chapters of the treatise by Azarquiel (d. AD 1100) on the universal astrolabe invented by him, generally known as *al-ṣafīḥa al-zarqālīya* (*Azafea*).[2] The treatise is accompanied, on f. 12vb, by a table of twenty-nine stars with ecliptical coordinates intended to be marked on the instrument. In a note on f.10^{r},14–10^{v},1 the author explicitly states that he has computed the coordinates of these stars for his present time, 459 H = AD 1066/7, by adding to their longitudes a constant increment of *c.* 2°37′ against the values in al-Battānī's table.[3] As for al-Battānī (d. AD 929) himself, he had, in the star catalogue of his *Zīj*, augmented the longitudes of his 533 stars, for the epoch AD 880, by 11°10′ by comparison with the values of Ptolemy's star catalogue in the *Almagest*.[4] This would imply that Azarquiel's star table, for the epoch AD 1066/7, should have been augmented by comparison with Ptolemy's values by 13°47′.

A quick glance at Azarquiel's table (see below), however, shows that this is not the case, since the minutes of the longitudes in this table, without exception, consist of full *tens*, and no single *seven* turns up. Furthermore, a more detailed inspection shows that the longitudes in our table are augmented by comparison with Ptolemy's value not by 13°47′ as should be expected according to the author's own statement, but by 12°40′. This value, apart from being quite different from the 13°47′ explicitly noted by the author, shows still another defect: it would hardly fit for an epoch AD 1066/7, but points to a much earlier date.[5]

These obvious discrepancies require a more detailed investigation, and fortunately we have at our disposal some additional source material, which will help us to elucidate some of the problems posed by that star table.

The Arabic MS. Paris, B.N. ar. 4821, contains some scattered astronomical writings of the famous Spanish-Muslim astronomer Maslama al-Majrīṭī (d. *c.* AD 1007), *viz* (i) his notes to Ptolemy's *Planisphaerium* (ff. 69^{r}–75^{v});[6] (ii) a special chapter added by Maslama to the *Planisphaerium* (ff. 76^{r}–79^{r});[7] and (iii) some "Chapters indispensable for everyone who wishes to construct an astrolabe" (ff. 79r–81v).[8] Item (iii) of these concludes (on f. 81v) with a table of twenty-one stars.

The Paris star table is entitled "Table of the places of the fixed stars as observed by Maslama ibn Aḥmad at the end of the year 367 [H = AD 978], according to al-Battānī's method;[9] these are the stars that are put on the astrolabe".[10] It offers for all the twenty-one stars ecliptical longitude and latitude together with *mediatio coeli* and declination. The precession value adopted for the longitudes, for the said epoch of AD 978, is Ptolemy + 12°40′. (Of this table, a Latin version also exists. It is found in a MS. of the *Toledan tables*; see the edition in Kunitzsch [1], 15ff., 'Typ I'.)

We then find that the first twenty-one stars in Azarquiel's table in MS. Istanbul are identical to the twenty-one stars of Maslama's table of AD 978, as far as the identities of the stars and the precession values are concerned. The only difference—apart from the equatorial coordinates, which are not given in MS. Istanbul—is that the Istanbul table apparently groups the stars approximately according to their magnitudes, whereas in the Paris table they are arranged strictly according to (increasing) longitudes (from 42°20′ to 347°0′).[11] In the Latin version of the Paris table, however, the sequence of the stars is totally identical with the Istanbul table (though the coordinates in the Latin table are in great disorder).

Beyond these two, we have a third parallel to the Istanbul table, in a Latin table of forty-five stars with equatorial coordinates, in MS. Paris, B.N. lat. 7293 A, f. 16r, appended to the treatise on the use of the astrolabe by Johannes Hispalensis (but not originally forming part of it), edited in Kunitzsch [1], 19ff., as 'Typ II'. The first twenty stars of this table are identical with 'Typ I' (also in its sequence of the stars!), and 'Typ I A' (just α Leo from these two has been omitted there). Most important, the declinations in 'Type II' are identical with those in 'Typ I' and 'Typ I A'. 'Typ II' therefore is closely related to the basic type represented by 'Typ I A' and 'Typ I'. Further, we see that the stars no. 22–29 in MS. Istanbul that exceed the twenty-one stars found in the basic type of Maslama, correspond exactly—in their sequence—to the next following stars no. 21–28 in 'Typ II', of the same family.[12]

We can thus say that Azarquiel's treatise in MS. Istanbul, contrary to the author's statement, does not contain a star table computed for 459 H = AD 1066/7 with a precession value of Ptolemy + 13°47′, but instead an augmented version of a star table first established by Maslama al-Majrīṭī for 367 H = AD 978. The Istanbul table is closely related to Maslama's table by several elements: the precession value of Ptolemy + 12°40′ (equally found in 'Typ I A' and in 'Typ I'), the existence of the same stars, no. 1–21 (even in the same sequence, with regard to 'Type I'), and the additional stars, no. 22–29 (found, in the same sequence, in 'Type II'). The question cannot be answered here why, and how, this star table of the Maslama-type, and with Maslama's coordinates, was added to a recension of Azarquiel's treatise on the *Azafea*, which was composed almost a century later.

Now follows the annotated edition of the star table from MS. Istanbul, Aya Sofya 2671, f. 12vb:

Number	Name	Longitude	Latitude	Modern designation
1.	*ra's al-ghūl* The Demon's Head (Algol)	Gem 12°20′	N 23°0′	β Per
2.	*al-dabarān* Aldebaran	Tau 25°20′	S 5°10′	α Tau
3.	*al-'ayyūq* Capella (Alhajoth)	Gem 4°40′	N 22°30′	α Aur
4.	*al-'abūr* Sirius (Alhabor)	Cnc 0°20′	S 39°10′	α CMa
5.	*al-ghumayṣā'* Procyon (Gomeisa)	Cnc 12°0′	S 16°10′	α CMi
6.	*al-rāmiḥ* The [Simāk] Armed with a Lance (Alramech)	Lib 9°40′	N 31°50′	α Boo
7.	*al-a'zal* The Unarmed [Simāk] (Spica, Azimech)	Lib 8°20′	S 2°0′	α Vir
8.	*al-wāqi'* The Falling [Eagle] (Vega)	Cap 0°0′	N 42°0′	α Lyr
9.	*al-ṭā'ir* The Flying [Eagle] (Altair)	Cap 16°30′	N 29°10′	α Aql
10.	*al-ridf* The One Sitting Behind the Rider [on the Same Animal] (Arided, Deneb)	Aqr 21°50′	N 60°0′	α Cyg
11.	*dhanab al-'aqrab* The Scorpion's Tail (Antares)	Sco 26°20′	S 4°0′	α Sco
12.	*qalb al-asad* The Lion's Heart (Regulus)	Leo 15°40′	N 0°10′	α Leo
13.	*mankib al-jawzā'* Orion's Shoulder (Betelgeuze)	Gem 14°40′	S 16°0′	α Ori
14.	*rijl al-jawzā'* Orion's Foot (Rigel)	Gem 2°30′	S 31°50′	β Ori
15.	*mankib al-faras* The Horse's Shoulder (Scheat)	Psc 14°50′	N 31°0′	β Peg
16.	*muqaddam al-dhirā'ayn* The Preceding One of the Two dhirā'	Leo 10°20′	S 16°0′	
17.	*munīr al-fakka* The Bright One of al-fakka (Alphecca)	Lib 27°20′	N 44°50′	α CrB
18.	*al-kaff al-khaḍīb* The Stained Hand (Caph)	Psc 17°0′	N 29°0′	[β Cas]
19.	*ṭaraf banāt na'sh* The End of the Daughters of the Bier (Benetnasch)	Sco 12°0′	N 55°0′	η UMa
20.	*dhanab al-asad* The Lion's Tail (Denebola)	Vir 6°40′	N 11°30′	β Leo
21.	*ra's al-ḥāwī* The Head of the Snake-Charmer (Ras Alhague)	Sgr 8°40′	N 36°0′	α Oph
22.	*al-dulfīn* The Dolphin	Cap —	N —	[ε Del]
23.	*'unuq al-ḥayya* The Neck of the Snake	Sco 12°50′	N 16°30′	[Ser]
24.	*al-rukba* The Knee	Leo 16°0′	N 8°50′	
25.	*janāḥ al-ghurāb* The Wing of the Raven	Gem —	S —	[γ Crv]
26.	*dhanab al-jady* The Tail of the Young Goat (Deneb Algedi)	Aqr —	S —	[δ Cap]
27.	*dhanab qayṭūs* The Whale's Tail (Deneb Kaitos)	Psc 19°50′	S 9°20′	ι Cet
28.	*baṭn qayṭūs* The Whale's Belly (Baten Kaitos)	Ari 6°0′	S 23°0′	ζ Cet
29.	*al-rijl* The Foot	—	—	

NOTES TO THE STAR TABLE

MS., title of the column of the names: *al-kawākib al-thābita allatī taqaʿu fī l-asṭurlāb*, "The fixed stars that are on the astrolabe"; title of the column of the longitudes: *al-ṭūl wa-huwa mawḍiʿ al-kawkab min al-burūj*, "Longitude, which is the place of the star in the zodiacal signs"; title of the column of the latitudes: *al-ʿarḍ wa-huwa buʿduhū min minṭaqat al-burūj*, "Latitude, which is its distance from the zodiac".

1. The zodiacal sign, *Gem*, is wrong and should correctly be *Tau*; *cf.* 'Typ I A', no. 1: 42°20′ (= Tau 12°20′).
2. Identical with 'Typ I A', no. 2.
3. Long.: MS. mistake *d* (= 4°), for *z* (= 7°), easily explicable in the Arabic script; the correct value in 'Typ I A', no. 4, is 67°40′ (= Gem 7°40′).
4. Identical with 'Typ I A', no. 6.
5. Identical with 'Typ I A', no. 7. Long. departs from a Ptolemaic value Gem 29°20′ which is not found in Heiberg's edition of the *Almagest* (there, Gem 29°10′, with variant 29°30′). al-Battānī has 100°20′ (departing from Gem 29°10′).
6. Identical with 'Typ I A', no. 13. Lat.: MS. mistake *n* (= 50′), for *l* (= 30′), a frequent miswriting in the Arabic script.
7. Identical with 'Typ I A', no. 12. Long.: the degrees, here and in 'Typ I A', identically are too small against Ptolemy by 1°. al-Battānī correctly has 187°50′ (departing from Ptolemy's Vir 26°40′). The change between 9° (*ṭ*) and 8° (*ḥ*) cannot be explained as a mistake in the Arabic script.
8. Identical with 'Typ I A', no. 16. Lat.: MS. wrongly has *mb* (= 42°), miscopied for *ṣb* (= 62°, in the Maghrebin numbering system). *Cf.* below, Table 2, no. 11.
9. Identical with 'Typ I A', no. 17.
10. Identical with 'Typ I A', no. 19.
11. Identical with 'Typ I A', no. 15. Long.: MS. *kw* (= 26°), instead of the required *kh* (= 25°). Name: *dhanab* ('Tail') in the MS. is an error for the correct *qalb* ('Heart').
12. Identical with 'Typ I A', no. 9. Long.: Maslama's Leo 15°40′ exceeds his current increment over Ptolemy by 30′ (Ptolemy: Leo 2°30′. al-Battānī has 134°0′, apparently departing from a Ptolemaic value 2°50′, instead of 2°30′). Azarquiel, in a treatise preserved only in a Hebrew translation, relates that Maslama had observed α Leo in 369 H (= AD 979/80) and found its longitude at 135°40′. It can be doubted whether this really was an independent observation, or whether Azarquiel had just extracted this value from Maslama's table of 367 H (the year, then, being corrupted to 369 H in the Hebrew translation).[13] However that may be, it attracts our special attention that Maslama's table here shows a unique deviation from its usual increment over Ptolemy. (For another instance of a change in the minutes of a longitude from expected 10′ to 40′, see no. 20 below.)
13. Identical with 'Typ I A', no. 5. Lat.: Both 'Typ I A' and MS. Istanbul have *yw* (= 16°) instead of the correct *yz* (= 17°), an easy mistake in the Arabic script.
14. Identical with 'Typ I A', no. 3. Lat.: Both 'Typ I A' and MS. Istanbul have *n* (= 50′) for the correct *l* (= 30′); *cf.* above, no. 6.
15. Identical with 'Typ I A', no. 20.
16. Name and coordinates are identical with 'Typ I A', no. 8. There is a constant mistake about this star in most of the sources. While the Arabic name is that of α Gem, the coordinates point to a place between θ and α Hya (or, roughly, to the bright star α Hya itself). *Cf.* the note to no. 8 in Kunitzsch [1], 18, and Catalá (as in ref. 8 above), 47.
17. Identical with 'Typ I A', no. 14. Lat.: *n* (= 50′) for the correct *l* (= 30′); *cf.* above, no. 6 and 14.
18. Identical with 'Typ I A', no. 21. The name unequivocally is that of β Cas, the coordinates however (identically here and in 'Typ I A') point to a place east of β Peg where no conspicuous star is to be found. *Cf.* the note to no. 21 in Kunitzsch [1], 18.
19. Name: 'Typ I A', no. 11, has the more usual *qā'id banāt naʿsh* ('The Leader of the Daughters of the Bier') instead. Long.: MS. *Sco* instead of *Vir* (explicable in the Arabic writing, confounding *ʿadhrā'* and *ʿaqrab*); both sources, MS. Istanbul and 'Typ I A', equally have dropped the minutes, putting zero instead of *l* (= 30′. The correct value, departing from Ptolemy's Leo 29°50′, ought to be Vir 12°30′. al-Battānī correctly has 161°0′). Lat.: Both sources *nh* (= 55°) instead of the correct *nd* (= 54°).
20. Name: 'Typ I A', no. 10, has the more complete *ṭaraf dhanab al-asad* ('The End of the Lion's Tail'). Long.: identical in MS. Istanbul and 'Typ I A'; correctly, departing from Ptolemy's Leo 24°30′, it should be Vir 7°10′; perhaps 6° is due to an Arabic miswriting (*w* = 6, for *z* = 7); to explain the divergence of *m* (= 40′) for *y* (= 10′) would be

more difficult (but *cf.* no. 12 above). Lat.: MS. Istanbul wrongly *l* (= 30′) for *n* (= 50′); 'Typ I A' has correctly 11°50′ (*cf.* no. 6, 14, 17 above).

21. In 'Typ I A', no. 19, the coordinates of the star are omitted from the table. 'Typ I', no. 21, has 255°30′ (= Sgr 15°30′) which evidently is a mistake of the copyist (departing from Ptolemy's Sco 24°50′, 'Typ I' should have Sgr 7°30′, or 247°30′ respectively). The value in MS. Istanbul, Sgr 8°40′, here, in a unique instance, shows that difference to al-Battānī of +2° and approximately 37′ (or, rounded up, 2°40′) that was indicated by the author in his introductory notice (al-Battānī has 246°0′, or Sgr 6°0′ respectively, correctly adding his constant 11°10′ to Ptolemy's Sco 24°50′).

Here ends the section that is identically found in 'Typ I A' (and 'Typ I'). The following stars correspond to the continuation of 'Typ I A' in 'Typ II' where, however, equatorial coordinates only are given. It can be observed that the coordinates in MS. Istanbul in this next following section are remarkably disturbed as compared to the first section, which has a counterpart in the ecliptical coordinates of 'Typ I A' (and 'Typ I').

22. Corresponds to 'Typ II', no. 21: *delphin*. The coordinates are omitted from the MS. According to al-Ṣūfī it is ϵ Del that is marked on the astrolabe. For this star, however, in the longitude the sign *Aqr* would be required, and not *Cap* which is given in the MS.
23. Corresponds to 'Typ II', no. 22: *heynelheye*. This would be the Arabic *'ayn al-ḥayya*, 'The Eye of the Snake'. There is, however, no star placed on the eye of this constellation; *'ayn*, 'eye', is a frequent Arabic miswriting for *'unuq*, 'neck'. The name, *'unuq al-ḥayya*, belongs to α (or, rarely, β) Ser. The coordinates, however, do not fit either of the two; they come nearest to those of μ Ser (Ptolemy: Lib 28°50′, which would yield Sco 11°50′ in our table; lat.: 16°30′, equal to our table); this faint star certainly would not have been entered into a table of fundamental stars. Obviously the author of this section of the table has committed an error when he picked up the coordinates for α Ser from a copy of the *Almagest*, falling two lines below this star, in his source.
24. This name can designate the star θ UMa, not rarely found on astrolabes and in tables of astrolabe stars (*cf.* Kunitzsch [1], the indexes). Here, however, the coordinates point to a place slightly east of γ Leo where no conspicuous star is to be found. The corresponding coordinates in 'Typ II', no. 24, point to ϵ Per, a star usually not counted among the astrolabe stars.
25. The name refers to γ Crv, which is used rather frequently as an astrolabe star (*cf.* Kunitzsch [1], the indexes). For this star, however, in the longitude, *Vir* would be required instead of *Gem*. 'Typ II', no. 25, has *yenahalfaraz* which looks like a transcription of the Arabic *janāḥ al-faras*, 'The Horse's Wing', which, therefore, could refer to γ Peg; in Kunitzsch [1], 21, I have therefore explained this star as γ Peg (with certain modifications in the coordinates). In the light of the evidence of the Arabic text in MS. Istanbul, I should now prefer to understand 'Typ II', no. 25, as γ Crv, for which the declination nicely fits; RA (in Vir), then, has to be read 256°30′ (instead of 56°30′. Note that RA is counted, in 'Typ II', from the first degree of Cap.).
26. According to al-Ṣūfī, δ Cap is the astrolabe star. The corresponding coordinates in 'Typ II', no. 26, would roughly fit this star. In our table, the sign of *Aqr* also would suit δ Cap.
27. Obviously, ι Cet is intended though the coordinates do not exactly correspond to this star (Ptolemy, Psc 4°20′, would yield Psc 17°0′ in our table; lat. had to be S 9°40′). The same star also is intended in 'Typ II', no. 27.
28. The coordinates only very roughly correspond to those of ζ Cet (Ptolemy's Psc 25°0′ would result, in our table, in Ari 7°40′; lat. should be S 20°0′). The corresponding star in 'Typ II', no. 28, equally is ζ Cet.
29. The corresponding star in 'Typ II' would be no. 23, *errigel*, which, as can be inferred from parallel sources, could be μ UMa (*cf.* the indexes in Kunitzsch [1]).

II

On ff. 133v–150r,[14] the Istanbul MS. Aya Sofya 2671 contains an anonymous treatise on the construction of the planisphaeric astrolabe accompanied, on f. 148v, by a table of thirty-seven stars with ecliptical coordinates computed for 459 H = AD 1066/7. This being the epoch of Azarquiel's star table in his treatise on the *Azafea*, D. King argues[15] that this anonymous treatise could well be ascribed to this author. We would then possess, apart from his treatise on the *Azafea* (existing in three different recensions of 100, 80, and 61 chapters respectively), also a treatise on the 'normal' astrolabe by Azarquiel of which nothing had been known until now.

The star table on f. 148ᵛ is computed, according to the author's notice on f. 148ʳ, for the epoch 459 H = AD 1066/7, adding the necessary value for the precession[16] which, however, is not explicitly named. The inspection of the table shows that the value is Ptolemy + 14°7′.

Of this star table, first established by Azarquiel, further copies exist, which will be used here for a comparative study.

The Arabic MS. Paris, B.N. ar. 4824, contains the hundred-chapter recension of Azarquiel's treatise on the *Azafea*. It is accompanied, on f. 4ᵛ, by a table of thirty-seven stars that is identical with our Istanbul table apart from the fact that the longitudes in MS. Paris are in most of the cases greater by 13′ or 15′ than in MS. Istanbul. The Paris star table has been edited in Kunitzsch [1], 73ff. (especially 83ff.), as 'Typ XII A'.

Of this star table there exists a Latin translation found in many MSS. as part of the *Toledan tables*. The Latin table is fully identical with the two Arabic specimens of the table; the longitudes, in the Latin, are Ptolemy + 14°7′, which is the value also used in MS. Istanbul. One MS. of the Latin gives as the epoch of the table the year 400 H (= AD 1009/10)[17] which, however, would be much too early. Compared to the date given in MS. Istanbul (459 H), 400 in that Latin MS. could well be a mistake for 460 which would then come close to the Arabic recension.[18] The same table afterwards had been adopted by several mediaeval Latin writers on the astrolabe, and incorporated in their treatises (*e.g.* Raymund of Marseille; Arialdus; Wilhelm of Anglia. See Kunitzsch [1], 76f., MSS. s–x). This Latin version of the table is edited in Kunitzsch [1], 73ff., as 'Typ XII'.

Another copy of the hundred-chapter recension of Azarquiel's treatise on the *Azafea*, in Arabic, has recently been found by D. King in MS. Cairo, Dār al-Kutub, *mīqāt* 647.[19] This contains again the same star table,[20] with longitudes identical with those in the Paris MS., and one doubtful star (no. 38) added to its end.

Contrary to the first table discussed above, this second table does not pose major problems. It is recognizably identical in all the sources except for the longitudes, which are Ptolemy + 14°7′ in the Latin and in MS. Istanbul, and increased by 13′ or 15′ in MSS. Paris and Cairo.

There now follows the edition of the table, from MS. Istanbul, Aya Sofya 2671, f. 148ᵛ, with comparative notes from the other sources.[21] In this, the following sigla will be used for the different MSS.; I = MS. Istanbul; P = MS. Paris; C = MS. Cairo; and L = the Latin version.

Number	Name	Longitude	Latitude	Modern designation
1.	*al-dabarān* Aldebaran	Tau 26°47′	S 5°10′	α Tau
2.	*rijl al-jabbār* Orion's Foot (Rigel)	Gem 3°57′	S 31°30′	β Ori
3.	*al-ʿayyūq* Capella (Alhajoth)	Gem 9°7′	N 22°30′	α Aur

Number	Name	Longitude	Latitude	Modern designation
4.	*mankib al-jabbār* Orion's Shoulder (Betelgeuze)	Gem 16°7′	S 17°0′	α Ori
5.	*al-shiʿrā al-ʿabūr* Sirius (Alhabor)	Cnc 1°47′	S 39°10′	α CMa
6.	*al-ghumayṣāʾ* Procyon (Gomeisa)	Cnc 13°17′	S 16°10′	α CMi
7.	*qalb al-asad* The Lion's Heart (Regulus)	Leo 16°37′	N 0°10′	α Leo
8.	*al-ṣarfa* The Changing [of the Weather] (Denebola)	Vir 8°37′	N 11°50′	β Leo
9.	*al-simāk al-aʿzal* The Unarmed Simāk (Spica, Azimech)	Lib 10°47′	N 2°0′	α Vir
10.	*al-simāk al-rāmiḥ* The Simāk Armed with a Lance (Arcturus, Alramech)	Lib 11°7′	N 31°30′	α Boo
11.	*al-nasr al-wāqiʿ* The Falling Eagle (Vega)	Cap 1°27′	N 42°0′	α Lyr
12.	*fam al-ḥūt al-janūbī* The Mouth of the Southern Fish (Fomalhaut)	Aqr 21°7′	S 23°0′	α PsA
13.	*raʾs al-marʾa* The Head of the Woman	Ari 1°37′	N 26°0′	α And (= δ Peg)
14.	*raʾs al-ghūl* The Head of the Demon (Algol)	Tau 13°47′	N 23°0′	β Per
15.	*janb ḥāmil raʾs al-ghūl* The Flank of the Carrier of the Demon's Head (Algenib)	Tau 19°0′	N 30°0′	α Per
16.	*mankib al-jabbār al-aysar* Orion's Left Shoulder (Bellatrix)	Gem 4°27′	S 17°30′	γ Ori
17.	*katif mumsik al-ʿinān al-yumnā* The Right Shoulder of the Rein-Holder (Menkalinan)	Gem 16°37′	N 20°0′	β Aur
18.	*raʾs al-tawʾam al-muqaddam* The Head of the Preceding Twin	Cnc 7°27′	N 9°40′	α Gem
19.	*raʾs al-tawʾam al-muʾakhkhar* The Head of the Following Twin	Cnc 10°47′	N 6°15′	β Gem
20.	*al-munīr alladhī fī ʿunuq al-asad* The Bright One on the Lion's Neck	Leo 16°17′	N 8°30′	γ Leo
21.	*al-fard min ʿunuq al-shujāʿ* The Isolated One on the Hydra's Neck (Alphard)	Leo 14°7′	S 20°30′	α Hya
22.	*muqaddam al-zubānayn* The Preceding One of the Two Claws (of the Scorpion)	Sco 2°7′	N 0°40′	α Lib
23.	*al-muʾakhkhar minhumā* The Following One of Them	Sco 6°17′	N 8°30′	β Lib
24.	*qalb al-ʿaqrab* The Scorpion's Heart (Antares)	Sco 26°47′	S 3°0′	α Sco
25.	*rukbat al-rāmī al-yusrā* The Left Knee of the Archer	Cap 1°7′	S 18°0′	α Sgr
26.	*ʿurqūb al-rāmī al-aysar* The Left Ankle of the Archer	Cap 1°52′	S 23°0′	β Sgr
27.	*al-nasr al-ṭāʾir* The Flying Eagle (Altair)	Cap 17°0′	N 29°10′	α Aql
28.	*dhanab al-dajāja* The Tail of the Hen (Deneb)	Aqr 23°17′	N 60°0′	α Cyg
29.	*al-aʿlā min al-fargh al-awwal wa-huwa mankib al-faras* The Upper One from the First Spout, *i.e.* the Horse's Shoulder (Scheat)	Psc 16°17′	N 31°0′	β Peg

Number	Name	Longitude	Latitude	Modern designation
30.	*warik al-mar'a wa-huwa qalb al-ḥūt* The Woman's Hip, *i.e.* the Heart of the Fish	Ari 18°0′	N 26°20′	β And
31.	*al-shamālī min al-naṭḥ* The Northern One of the [Ram's] Horns	Ari 21°47′	N 8°20′	β Ari
32.	*rijl al-mar'a al-yusrā wa-huwa ʿanāq al-arḍ* The Woman's Left Foot, *i.e.* the Caracal (Alamak)	Tau 1°0′	N 28°0′	γ And
33.	*al-ʿurqūb al-ayman min mumsik al-ʿinān* The Left Ankle of the Rein-Holder	Gem 9°47′	N 0°0′	γ Aur (= β Tau)
34.	*munīr al-hanʿa* The Bright One of al-hanʿa (Alhena)	Gem 28°47′	S 10°30′	ξ Gem
35.	*ra's al-ḥāwī* The Snake-Charmer's Head (Ras Alhague)	Sgr 8°37′	N 36°0′	α Oph
36.	*jaḥfalat al-faras* The Horse's Lip (Enif)	Aqr 19°20′	N 22°30′	ε Peg
37.	*ṭaraf dhanab qayṭūs* The End of the Whale's Tail (Deneb Kaitos)	Psc 19°47′	S 20°20′	β Cet

NOTES TO THE STAR TABLE

4. In I, the name is slightly corrupted (*mankib al-jw'r*); here the copyist has confounded the two Arabic designations of Orion, *viz*, *al-jabbār* and *al-jawzā'*. In concordance with P and C, we therefore adopt the reading *mankib al-jabbār*.
8. Lat.: The isolated number 50′ (= *n*), in I, looks quite like the letters *z* in other stars designating isolated sevens. In other cases, this isolated *n* was turned into an *l* (thus designating 30, instead of 50; *cf.* no. 23).
9. Lat.: I wrongly has N, PCL correctly S.
11. Lat.: All the Arabic sources, IPC, wrongly write *mb* (= 42°), miscopied for *ṣb* (= 62°, in the Maghrebin numbering system); *cf.* Table 1 above, no. 8. L has correctly 62°.
12. Lat.: All the texts 23°, following the tradition of MS. D of the *Almagest*; Heiberg, in his edition, had adopted 20°20′, from MSS. ABC.
13. Long.: *lz* (37′), slight graphical error for *nz* (57′), in I.
15. Name: C wrongly *mankib*, 'The Shoulder', easy graphical error for the correct *janb*, 'The Flank'. Long.: 19°0′ is rounded up for the correct 18°57′ (*cf.* also below, no. 30 and 32). L, no. 20a, has the correct 18°57′. PC departed from 19°0′, as they both have 19°15′.
16. Long.: Departs from a Ptolemaic value of Tau 20°20′, which is not mentioned in Heiberg's edition (there, Tau 24°0′ only). But *cf.* the note to no. 16, in Kunitzsch [1], 80f.
17. Name: In I, the adjective *al-yumnā* is mistakenly placed in front of the whole expression, instead of the end where it correctly should be. (In P, the adjective is distorted into *alladhī*; in C, it is omitted altogether.) Instead of the correct *katif*, 'The Shoulder' (*sic.* in I and C), P mistakenly writes *kaff*, 'The Hand'. From a similarly distorted source, the name was translated in L as '*palma*' (instead of the correct *humerus*). The 'hand' does not occur in Ptolemy's description of this constellation, so that undoubtedly *katif*, 'The Shoulder', is the correct reading.[22] Long.: IL have 37′ which is the common Arabic miswriting *lz* (37) for *nz* (57). P, departing from this modified value, has 16°50′ which, in turn, is again miswritten as 16°30′ in C.
20. Name: Instead of *al-munīr* (in I and C), P has *al-nayyir*, of the same meaning.
23. Lat.: All the sources 30′, from the frequent Arabic miswriting *l* (= 30), for *n* (= 50).
24. Lat.: 3°, resulting from an Arabic miswriting *j* (= 3), for the correct *d* (= 4).
26. Long.: Cap 1°52′, departing from a Ptolemaic value of Sgr 17°45′ as found in the Arabic version of the *Almagest* in MS. London, Brit. Libr. Add. 7475.[23]
27. Long.: Departing from Ptolemy's 3°50′, 17°57′ is to be expected. By some reason, however, the minutes had been dropped in the Azarquiel tradition. Thus, while IL have 17°0′, PC have 17°15′, departing from the same wrong value.
29. In the MS., the long name of the star covers two lines. Thus, neglectfully, the scribe has written part of the coordinates in the upper line, and the other part in the lower line.
30. Name: For I's *qalb al-ḥūt*, 'The Heart of the Fish', P has *dhanab al-ḥūt*, 'The Tail of the Fish'. In C, the star is called *baṭn al-ḥūt*, 'The Belly of the Fish', alone. Of these three, *qalb* and *baṭn* would be equally correct, while *dhanab* is not admitted with this star.[24]

L, *cor piscis*, is in concordance with I. Long.: 18°0′ (in I and L) is rounded up for the correct 17°57′ (*cf.* also no. 15 and 32). PC have 18°15′, departing from this modified number.

31. Name: IC have the correct *al-shamālī*, 'The Northern One', which is corrupted, in P, to *al-thānī*, 'The Second One'. L, *septentrionale*, follows the correct version.
32. Long.: 1°0′, rounded up for the correct 0°57′ (*cf.* also no. 15 and 30 above). L, by some mistake of transmission, has 1°2′. PC departed from the modified value and have 1°15′.
33. Lat.: PC have correctly 5°0′, while the scribe of I neglectfully has put down two zeros, 0°0′. L has distorted 5°0′ into 8°0′.
34. The name refers to the bright one of the two stars constituting *al-han'a*, the 6th lunar mansion, which would be γ Gem.[25] The coordinates however are those of ξ Gem, the smaller star in this constellation. Name: Instead of *al-han'a* (thus in ICL), P has the graphical mistake *al-haq'a*.[26]
35. Long.: All sources depend on the same mistake in this tradition, which had put *lz* (37′) for *nz* (57′), a sort of a graphical error in the Arabic which frequently occurs (*cf.* no. 13 and 17 above). (Thus, the correct reading of 'Typ I A', no. 35, must be Sgr 8°52′, instead of 8°12′, in Kunitzsch [1].) Name: For *al-ḥāwī*, P has *al-ḥawwā'*, of the same meaning.
36. Long.: IL both follow a tradition which mistakenly had changed the value of the minutes (20′, instead of 27′). PC however have 19°40′, thus departing from the correct 19°27′.
38. At the end of the table, C adds an extra star not belonging to the table in all the other sources: *al-ḍifdi'*, 'The Frog', Aqr 23°46′, S 22°0′. This obviously is a 'ghost star', as the name refers to either α PsA, or β Cet[27] (both of which exist in this table separately, *cf.* no. 12 and 37), while the coordinates point to a region east of α PsA empty of conspicuous stars. Perhaps, indeed α PsA was here intended, a second time in this version of the table, compiled from some independent source, but then the coordinates are in bad order.

REFERENCES

1. Edited by J. L. Heiberg in *Cl. Ptolemaei opera*, ii (Leipzig, 1907), 227–59; German translation from the Latin, by J. Drecker, in *Isis*, ix (1927), 255–78. For the Arabic, *cf.* M. Krause, *Stambuler Handschriften islamischer Mathematiker* (Berlin, 1936; Quellen und Studien zur Geschichte der Mathematik Astronomie und Physik, Abt.B, Bd.3, Heft 4), 443; F. Sezgin, *Geschichte des arabischen Schrifttums*, v (Leiden, 1974), 170 (no. 1).
2. *Cf.* Krause, *op. cit.* (ref. 1), 482 (no. 255, 2); D. King, "On the early history of the universal astrolabe in Islamic astronomy", *Journal for the history of Arabic science*, iii (1979), 244–57, p. 253, footnote 21.
3. "For the sake of precession since the epoch of his [= al-Battānī's] observation", *bi-qadr mā aqbala al-falak min waqt qiyāsihī.*
4. *Cf.* the edition by C. Nallino, *Al-Battānī sive Albatenii Opus astronomicum*, trans. and comm. in vol. ii (Milan, 1907), 144ff. and 269ff.
5. *Cf.* P. Kunitzsch, *Typen von Sternverzeichnissen in astronomischen Handschriften des zehnten bis vierzehnten Jahrhunderts* (Wiesbaden, 1966), 15 with footnote 1, and 132f. (sub d). (This book will henceforth be cited as Kunitzsch [1].)
6. These were also translated into Latin and are found in some older editions of the *Planisphaerium*, *cf.* Heiberg, *op. cit.* (ref. 1), pp. clxxx–clxxxiii.
7. Also translated into Latin and edited in some older editions, *cf.* Heiberg, *l.c.*, p. clxxxiii; see further ref. 8.
8. This section was not translated into Latin (except for the star table; see below). Sections ii and iii have been edited, and translated into Spanish, by J. Vernet and M. A. Catalá, in *Al-Andalus*, xxx (1965), 15–45 (with additional remarks on the star table by M. A. Catalá, *ibid.*, 46f.). The star table (in the Arabic version) was separately edited by M. Destombes, in *Archives internationales d'histoire des sciences*, xv (1962), 26. Another edition of this star table, both in its Arabic text and in a Latin version, is in Kunitzsch [1], 15ff., as 'Typ I' and 'Typ I A'.
9. With "al-Battānī's method", Maslama seems to refer to the continuous counting of the longitudes from 0° to 360°, as against the usual system (following the *Almagest*) of attributing the longitudes to one of the 30 degrees within one of the twelve zodiacal signs (*e.g.*, for α Leo: 135°40′, instead of Leo 15°40′).
10. See Kunitzsch [1], 17.
11. Sequence of the Istanbul numbers in the Paris table (Kunitzsch [1], 'Typ I A'): 1, 2, 14, 3, 13, 4, 5, 16, 12, 20, 19, 7, 6, 17, 11, 8, 9, 21, 10, 15, 18.
12. The only exception is Istanbul no. 29 (*al-rijl*) which, in 'Typ II', is placed before Istanbul no. 24 (*al-rukba*). The 'Typ II'-numbers are less by one against the Istanbul numbers because α Leo is missing in 'Typ II'.

13. *Cf.* Vernet and Catalá, *op. cit.* (ref. 8 above), 16f.; Destombes, *op. cit.* (ref. 8), 20f.
14. The subsequent matters on ff. 150v–151r apparently do not belong to this treatise which clearly is declared to end at the bottom of f. 150r. See also Krause, *op. cit.* (ref. 1), 525, no. 15.
15. See ref. 2, above.
16. "According to what results from the calculation of the movement of precession", *'alā mā yakhruju min ḥisāb ḥarakat al-iqbāl.*
17. MS. m in Kunitzsch [1], 'Typ XII'.
18. The Arabic sources now fix the date of 'Typ XII', with longitude = Ptolemy+14°7′, to 459 H = AD 1066/7 which brings the discussion in Kunitzsch [1], 73, to a conclusion.
19. See D. King, as cited in ref. 2 above. The treatise in MS. Cairo, *miqāt* 647 contains, in the preface, the same dedication to the *amīr* al-Mu'tamid ibn 'Abbād that is cited by D. King, *op. cit.*, from a sixty-chapter version of the same treatise which also survives in several manuscripts.
20. I owe to the kindness of Prof. David King copies of the relevant pages from this MS. Unfortunately, the folio numbers are not visible in the copies so that I cannot here indicate the exact page where the star table is placed in the MS.
21. I am greatly indebted to my friend and colleague, Mme B. Atsiz of the University of Munich, for kindly putting her photographs of the relevant pages from the Istanbul MS. at my disposal.
22. *Cf.* P. Kunitzsch, *Der Almagest. Die Syntaxis Mathematica des Claudius Ptolemäus in arabisch-lateinischer Überlieferung* (Wiesbaden, 1974), 245f., esp. no. 155.
23. *Cf.* C. H. F. Peters and E. B. Knobel, *Ptolemy's catalogue of stars* (Washington, 1915), 167, no. 592.
24. *Cf.* P. Kunitzsch, *Untersuchungen zur Sternnomenklatur der Araber* (Wiesbaden, 1961), 50, no. 64a, and 91, no. 217 (henceforth quoted as Kunitzsch [2]). The word *warik*, 'Hip', originally does not occur in the Arabic versions of the *Almagest*; see Kunitzsch, *Almagest* (ref. 22 above), 260, no. 247.
25. *Cf.* Kunitzsch [2], 64, no. 114.
26. This would be the 5th lunar mansion, consisting of $\lambda\phi^{1,2}$ Ori; *cf.* Kunitzsch [2], 64, no. 115a.
27. *Cf.* Kunitzsch [2], 52f., no. 75 a/b and 76 a/b.

V

New Light on al-Battānī's Zīj

The greatest and most important textbook on astronomy in antiquity and the middle ages was undoubtedly Ptolemy's *Almagest* (written about A.D. 138). It is well known that this work found its way into the Islamic culture of the Near and Middle East through a series of several translations which seem to have started as early as about A.D. 250, under the Sassanide ruler Shāhpūr I., and the last of which is the version of Isḥāq ibn Ḥunayn in a revision made by Thābit ibn Qurra.

In the ninth century A.D. two Arab astronomers, viz. al-Farġānī and al-Battānī, composed, each of them, a concise treatise on the fundaments of the science of astronomy, being largely based on the so far existing Oriental versions of the *Almagest*. al-Battānī's treatise, generally called *al-Zīj al-Ṣābī*, added to his introductory text quite a number of astronomical tables, among which two star catalogues. (For the relevance and rôle of al-Battānī in the history of science we may refer to the respective article, by Willy Hartner, recently published in the first volume of the *Dictionary of Scientific Biography*, New York 1970.)

The first one of al-Battānī's two star catalogues in his *Zīj* is an excerpt of 533 stars from the 1025 stars being cataloguized in books VII and VIII of Ptolemy's *Almagest*. It was computed for the epoch 1st of March 880, adding constantly a value of 11°10′ to the longitudes of Ptolemy's catalogue.

al-Battānī's *Zīj* was printed, together with a modern Latin translation and rich commentaries, in the most valuable and monumental edition of Carlo Alfonso Nallino, in three volumes, Milano 1899–1907.

In editing the star catalogue and establishing the correct readings of the star coordinates, Nallino had at his disposal, at that time, only the older editions of Ptolemy's catalogue by N. Halma (Paris 1813–16) and F. Baily (London 1843), further the Venise print of 1515 of the Latin translation by Gerard of Cremona from the Arabic, and Schjellerup's French translation of al-Ṣūfī's *Book on the Constellations of the Fixed Stars* (St. Pétersbourg 1874).

The method adopted by Nallino in editing the star catalogue was to adapt and harmonise the values given in the unique Battānī-manuscript with the values established in those other corresponding editions.

From these sources, however, it was not possible to gain a true picture of the full score of the Arabic tradition of the *Almagest*.

Recent research[1] brings evidence for the survival of two main Arabic versions of the *Almagest*: one by al-Ḥajjāj ibn Yūsuf ibn Maṭar and Sarjūn ibn Hilīyā al-Rūmī, completed in the year 212 hijra = A.D. 827–28, and the other one by Isḥāq ibn Ḥunayn and improved by Thābit ibn Qurra, about A.D. 880–90. Gerard's Latin translation was based on the Ḥajjāj-version for books I–IX, and on the Isḥāq-version for the rest (books X–XIII); the star catalogue, though being placed in the Ḥajjāj-portion of Gerard's translation, nevertheless was translated after a special copy of the Isḥāq-text. Nallino, on his side, wrongly regarded Gerard's translation identical with al-Ḥajjāj's version in all places.

Let us mention additionally that al-Ṣūfī in his *Book of the Constellations* quotes the *Almagest* nearly always in Isḥāq's wording whereas the star coordinates are given partly based on al-Ḥajjāj, partly on Isḥāq, and partly even on a third version.

Now, recently I have studied a special Arabic treatise by Aḥmad ibn Muḥammad ibn al-Sarī Ibn al-Ṣalāḥ (died A.D. 1154) on the mistakes in the tradition of the star coordinates in Ptolemy's star catalogue, existing in four manuscripts at Istanbul, Manisa and Oxford.[2]

This short treatise proved to be one of the most valuable sources for the history of the Arabic tradition of the *Almagest*. The author, Ibn al-Ṣalāḥ, had at his disposal, in the middle of the 12th century in Baghdad, five different versions of the *Almagest*, viz. a Syriac one, and four Arabic versions: the so-called "old version", made under the caliph al-Ma'mūn by one al-Ḥasan ibn Quraysh (about whom we have no further information), and the versions by al-Ḥajjāj and Isḥāq (the last one in two recensions). We are lucky enough to find in Ibn al-Ṣalāḥ's treatise full discussions of 88 out of Ptolemy's 1025 stars. For all of them he quotes from his sources the respective values given in them, and judges which one should be preferred as the "correct" one. Here we have a unique document not only for the situation of the *Almagest*-text in the Islamic world in general, but we are also presented with numerous quotations from the two oldest extant versions, the Syriac and the "old" Arabic one, of which, until now, no details had been known.

Beyond producing this invaluable evidence from the internal situation of the *Almagest*-text in the Islamic culture, Ibn al-Ṣalāḥ's quotations from these different five versions provide us with a key for a new, and better, understanding of al-Battānī's star catalogue.

Let me give you an example in full detail so as to let you understand what I mean.

I chose the 10th star of the constellation of Virgo (modern δ Virginis), being the 6th star of Virgo in al-Battānī's catalogue (which, as I mentioned before, contains approximately half of Ptolemy's stars only), and being the 47th among Ibn al-Ṣalāḥ's 88 stars. The value of the longitude of this star is in the manuscript *qʿb l* = 172°30′, which is, in *Almagest* terms, Virgo 22°30′. From this we have to reduce al-Battānī's constant for the precession, namely 11°10′, and so we find that al-Battānī departed from a value of Virgo 11°20′ for the longitude. Nallino, however, comparing the afore mentioned parallel sources, in which the value for the longitude of this star was 14°20′ (instead of 11°20′), thought it necessary, in the light of this evidence, to regard the manuscript's value *qʿb l* = 172°30′ a scribal error and to emend this text to *qʿh l* = 175°30′, so that dropping al-Battānī's constant for the precession (which is 11°10′) he would arrive at the very value of Virgo 14°20′ which he had observed in his parallel sources.

Turning, then, to Ibn al-Ṣalāḥ's treatise we find him quoting a value of Virgo 14°20′ for the versions of Isḥāq (the original one, and the one which was improved by Thābit ibn Qurra), against 11°20′ for the Syriac and the old Arabic versions. Here we have a proof that Virgo 11°20′ was really a value being transmitted in certain versions, namely in the Syriac and the old Arabic version of the *Almagest*. If we cast a look on what the existing Arabic manuscripts of the *Almagest* are showing, we find them all (al-Ḥajjāj as well as Isḥāq) giving Virgo 14°20′. Ibn al-Ṣalāḥ declares 14°20′ to be the correct value, and 11°20′ to be a mistake which arose from an error in reading the Greek text of the *Almagest*, where someone wrongly read (or wrote) IA = 11 instead of IΔ = 14.

In the light of this evidence we are entitled to change Nallino's edited emended text from 175°30′ back to the 172°30′ which are shown in the manuscript.

Extending these studies to all the stars of al-Battānī's which are among those discussed by Ibn al-Ṣalāḥ, and which are suitable for such a comparison, namely 29 stars, including 30 values, I arrived at the following results:

In 5 longitudes and in 2 latitudes the value which was established by Nallino has to be abolished in favour to the value which is actually in the manuscript and which is identical with the respective values as being quoted by Ibn al-Ṣalāḥ from the Syriac and the old Arabic versions of the *Almagest*.

Another 15 values which were adopted by Nallino in accordance with the manuscript, prove to be identical with Ibn al-Ṣalāḥ's quotations from the Syriac and the old Arabic versions and should be retained in this form.

Here we have a share of 22 out of 30 cases, which is near to 70%, in which the coordinates used by al-Battānī are proved to be taken from a version of the *Almagest* as being tradited in the Syriac and in the old Arabic translations.

It may seem justified to reconsider the whole of al-Battānī's star catalogue in the light of this new evidence, in order either to prove Nallino's text, or change it for the manuscript's data, in accordance with the respective variants observed in the existing text traditions of the *Almagest*.

Ibn al-Ṣalāḥ could help us with his quotations only in this restricted number of stars. It is obvious, however, that all the different values which he quotes from the five different translations ultimately result from mistakes in reading the Greek text. That makes it seem worth while to extend the examination of Nallino's text to the rest of al-Battānī's stars relating them to the variants of the Greek text in Heiberg's edition, or to similar sorts of misreadings.

We have heard that in Ibn al-Ṣalāḥ's quotations the Syriac and the "old" Arabic versions are nearly always showing identical values. Therefore, regarding al-Battānī, it appears difficult to distinguish whether he picked up his values from the Syriac, or from the old Arabic versions. In one case Nallino could show a word in a genuine Syriac form in al-Battānī's text (the name of a wind: *zahfurus*, from Greek ζέφυρος). But this does not really mean that al-Battānī used the *Almagest* throughout, or portions of it, in the Syriac translation.

In the presence of these recent observations we are happy enough to be able to show that the coordinates in al-Battānī's star catalogue apparently are taken from one of these two versions of the *Almagest* and that their tradition in the manuscript, in their way, is "correct".

NOTES

1. Cf. Paul Kunitzsch, *Der Almagest. Die Syntaxis Mathematica des Claudius Ptolemäus in arabisch-lateinischer Überlieferung*, Wiesbaden (Harrassowitz) 1974.
2. An edition, accompanied by a translation, of the treatise is presently prepared by P. Kunitzsch.
   [This edition has now appeared as: Ibn aṣ-Ṣalāḥ, *Zur Kritik der Koordinatenüberlieferung im Sternkatalog des* Almagest, Göttingen 1975 (Abhandlungen der Akademie der Wissenschaften in Göttingen, Philol.-Hist. Kl., 3. Folge, Nr. 94).]

VI

On the Mediaeval Arabic Knowledge of the Star Alpha Eridani

The only first magnitude star (out of about fifteen to twenty) which is not included in the standard catalogues of fixed stars of classical antiquity and mediaeval times is *alpha Eridani.* It was not until the discoveries of the European seafarers in the 15th and 16th centuries that this bright star became known to western astronomers. Johann Bayer introduced it into his famous celestial atlas, *Uranometria*, of 1603, and assigned it the Greek letter *alpha*, while he gave to Ptolemy's "bright and last star" in the constellation of ποτ αμός, *Eridanus*, the letter *theta.*

These facts have been known to the historians of astronomy for a long time, and have been widely discussed by the editors and commentators of Ptolemy's star catalogue, such as Baily,[1] Ideler,[2] Knobel,[3] Nallino,[4] etc.

The reason for Ptolemy's omitting this star from his catalogue is obvious. It was due to the limits of visibility of southern stars in the region of Alexandria where Ptolemy is reported to have executed his astronomical observations. The geographical latitude of Alexandria is roughly $31^{0}20'$, which limits the visibility of stars in the southern hemisphere to a line of declination of $-58^{0}40'$. The position of *alpha Eridani*, in Ptolemy's time (around A.D. 150), and taking into account the value of precession, was at a declination of roughly $-66\frac{1}{2}^{0}$. This makes it clear that *alpha Eridani* remained invisible, at that time, north of the geographical latitude of $23\frac{1}{2}^{0}$ which corresponds to a line running between Medina and Mecca, and through Mascat in Oman, approximately.

The southernmost stars registered by Ptolemy were some stars of his constellation of *Centaurus*, now commonly known as the "Southern Cross". Assuming a medium declination for them of -60^{0},[5] they were visible at

* Institute of Semitic Languages, University of Munich, West Germany.

1. F. Baily, "The Catalogues of Ptolemy, Ulugh Beigh, Tycho Brahé, Halley, Hevelius", *Memoirs of the Royal Astronomical Society*, 13 (London, 1843).

2. L. Ideler, *Untersuchungen über den Ursprung und die Bedeutung der Sternnamen*, (Berlin, 1809), pp. 231, 234.

3. E. B. Knobel, "The Chronology of Star Catalogues", *Memoirs of the Royal Astronomical Society*, vol. XLIII, (London, 1875-1877), p. 64, n. 3; C. H. F. Peters and E. B. Knobel, *Ptolemy's Catalogue of Stars*, (Washington, 1915), p. 110 ad no. 805.

4. *Al-Battānī sive Albatenii opus astronomicum*, ed. and trans. C. A. Nallino, I-III, (Milan, 1899-1907). See vol. II, p. 170.

5. In 1950. See: *Norton's Star Atlas and Reference Handbook*, (15th ed., reprint, London, 1966, and 16th ed., London, 1973), maps 10 and 16.

Ptolemy's time up to the geographical latitude of 39⁰, that is Athens.

The stellar astronomy of the Arabic-Islamic culture relied heavily on Ptolemy. Together with his *Almagest*, his star catalogue was translated into Arabic and served as the standard catalogue for the Islamic astronomers, from al-Battānī[6] through al-Ṣūfī[7] and al-Bīrūnī[8] to Ulugh Beg,[9] just to mention the most important names. This canonized catalogue was also adopted, through Latin translation, in mediaeval Europe, where it was used either in its original text, the *Almagest*, itself,[10] or, derived from it, in the *Alfonsine Tables*[11] and other similar works, until the introduction of modern astronomy. In this tradition, the constellation of *Eridanus* was generally known to have its southern end at the star designated by Bayer with the Greek letter *theta*.

Turning then to the Arabs, it is known that they had a certain knowledge of the stellar sky already a long time before their acquaintance with Greek astronomy. The bedouins are famous for having used the stars for orientation in their migrations in the desert. Many star names also found their way into the classical Arabic poetry which was developed to its climax already in pre-Islamic times. Later on, Arabic philologists and lexicographers, in their efforts to collect the genuine ancient Arabic terminologies and vocabulary, composed special books in which they collected all the star names they could find in those old traditions. And it was the astronomer al-Ṣūfī who then made an attempt, in his book on the constellations composed in A. D. 964,[12] to identify the respective celestial objects according to the scientific Ptolemaic tradition. In a monograph on the indigenous Arabic star names, I arrived at a total number of 329 names which are mentioned in those old traditions.[13] But there may be still more, as some may have escaped my attention.

In view of this huge number of star names, one would of course expect to find among them also the bright first magnitude star *alpha Eridani*, which was clearly visible in the Arabian peninsula, south of the latitude of 23 ½⁰ at Ptolemy's

6. Edited by Nallino, see footnote 4 above.

7. *Kitāb ṣuwar al-kawākib or Uranometry* (ed. Hyderabad, 1954) (this ed. is quoted here). Also: H.C.F.C. Schjellerup, *Description des étoiles fixes par Abd-al-Rahman Al-Sûfi*, (French trans. and partial ed. of the Arabic text), St. Petersburg 1874.

8. *Al-Qānūn al-Masʿūdī*, (ed. Hyderabad, 1954-1956). See vol. III, pp. 1012-1126.

9. Th. Hyde, *Tabulae longitudinis et latitudinis stellarum fixarum ex observatione Ulugh Beighi..*, (Oxford, 1665); 2nd ed., by Dr. G. Sharpe, *Syntagma Dissertationum*, (Oxford, 1767); E. B. Knobel, *Ulug Beg's Catalogue of Stars* (Washington, 1917).

10. Translation by Gerard of Cremona from the Arabic, A. D. 1175; existing in many manuscripts, printed Venice 1515.

11. Existing in numerous manuscripts and several printed editions: Venice 1483, 1492, 1518 (at the end 1521), 1524, Paris 1545 and 1553, Madrid 1641.

12. See above, footnote 7.

13. P. Kunitzsch, *Untersuchungen zur Sternnomenklatur der Araber* (Wiesbaden, 1961).

time, or south of 26¼° (that is a line through Khaibar and Bahrain, approximately) in A.D. 700. This, however, seems not to be the case. Al-Ṣūfī, in his identifications of these names, was limited to Ptolemy's catalogue in which, as we have seen, *alpha Eridani* was not included. Beyond that, al-Ṣūfī was living and working in Iraq and, occasionally, at Shiraz in Iran. So his own visibility of the southern sky was limited to a declination of —56°, or at most —60½°. Whereas, on the other hand, Arab tribes were living as far south as the Yemen at a geographical latitude of 13° which allowed them a visibility up to —77° in the southern sky. So, a belt of 17° to 20° could not actually be controlled by al-Ṣūfī. This led to a number of errors and doubtful cases among his identifications of certain traditions relating to southern star names.[14]

An example is the pair of stars called by the Arabs *al-ẓalīmān* (الظليمان), "the two ostriches". This name also occurs in the respective collections of the philologists Ibn Qutayba (d. A. D. 884 or 89)[15] and al-Marzūqī (who declares that he follows, in this section, the philologist Abū Ḥanīfa al-Dīnawarī, d. A.D. 895).[16] The two citations are nearly identical, and explain that *al-ẓalīmān* are two bright stars above another pair of stars consisting of *alpha* + *beta Gruis*,[17] and that they are separated from each other, when both reach the same height above the horizon, by 100 *dhirāᶜ*.

The value of 100 *dhirāᶜ* given in this definition is strongly misleading, and apparently a fault in the textual transmission. One *dhirāᶜ* with al-Ṣūfī, and also, approximately, in the definitions of the philologists, equals 2°20′. 100 *dhirāᶜ* would then mean a distance of 233° between those two stars, which is of course impossible.

Al-Ṣūfī identified the "two ostriches" as *alpha Piscis Austrini* and *theta Eridani*. The distance between these two is about 60°.

In 1974, I spent some time at Malindi, Kenya, in order to study and control the indigenous Arabic traditions on certain star names and al-Ṣūfī's identifications. The place is situated just south of the equator, so that I had the opportunity of observing the sky down to the southern pole. My observations confirmed that the descriptions of the Arabic philologists were mostly correct and adequate to identify the objects mentioned in their texts.[18]

With regard to the pair of stars called *al-ẓalīmān* I found that they comply with the philologists' definition as to being at the same altitude and above the

14. Cf. P. Kunitzsch, in *Der Islam* 51 (1974), 52f., with footnote 19.
15. Ibn Qutayba, *Kitāb al-anwā'*, (ed. Hyderabad, 1956), p. 73.
16. Abū ᶜAlī al-Marzūqī, *Kitāb al-azmina wa al-amkina*, (ed. Hyderabad, vols. I-II, 1332 H.). See vol. II, p. 383.
17. I.e. *al-yamāmatān*; cf. P. Kunitzsch, *Untersuchungen* (as in footnote 13 above), p. 117, no. 319.
18 Cf. my report "Die arabischen Sternbilder des Südhimmels" (II), in *Der Islam* 52 (1975), 263-277.

pair *alpha* + *beta Gruis* when these are setting. In this situation, there appears a pair of stars of equal brightness above *alpha* + *beta Gruis*, equally high in the sky. These two stars are *alpha Piscis Austrini* (which was also fixed by al-Ṣūfī), and *alpha Eridani* (not *theta*, as stated by al-Ṣūfī).[19]

Al-Ṣūfī, who had no knowledge of the actual view of the southern sky, and was dependent entirely on his written sources, and perhaps a celestial globe, saw no better way than to identify the pair of the "two ostriches" with the two Ptolemaic stars *alpha Piscis Austrini* and *theta Eridani*. Actually, however, *theta Eridani* is not only perhaps too far distant from *alpha Piscis Austrini* to be included together with this in an asterism, but moreover it is apparently much less bright than *alpha Piscis Austrini* (*theta* is of magnitude $3\overset{m}{.}4$, *alpha* $0\overset{m}{.}6$). To form a pair of equal brightness, in that position, as required by the texts of the Arabic philologists, besides *alpha Piscis Austrini* the only suitable component can be *alpha Eridani*.[20]

So, from a critical examination of the texts, combined with actual observation of the sky, it was found that al-Ṣūfī committed a mistake in his identification of the old Arabic asterism of the "two ostriches", and that the bright first magnitude star *alpha Eridani* was not unknown and not unnamed with the old Arabic star gazers. They included it, together with *alpha Piscis Austrini*, in the name of the "two ostriches", *al-ẓalīmān* (in the dual).

Apart from the passages cited above, I have found some additional evidence, again in the compilation of al-Marzūqī already mentioned. In another place, and apparently again quoting the same Abū Ḥanīfa, he gives a list of thirteen bright stars, i.e. first magnitude stars (in Arabic: *darārī*).[21] Here there occurs an otherwise unknown name which can be read *al-maḥnath*, or *al-muḥnith* (المحنث) .

The word appears in the texts usually in connection with two stars called *ḥaḍāri* and *al-wazn* (حضار و الوزن).[22] Their identification was disputed even among the pre Islamic Arabs, and so the philologists said these two are *muḥlifān* or *muḥnithān*, i.e. "disputed, and causing a man to perjure himself with regard to their identity". Al-Ṣūfī wavered in their identification between *alpha* + *beta Centauri*, or *alpha* + *beta Columbae*.[23] According to my findings, only the second of these two pairs can be correct.[24] Nevertheless, from these texts it can be inferred that al-Marzūqī's *al-muḥnith* (which seems to be the better reading) could designate one of the two first magnitude stars *alpha* and *beta Centauri*.

19. Cf. P. Kunitzsch in *Der Islam* 52 (1975), 271 f.
20. The distance between these two is roughly 45°.
21. Al-Marzūqī (as in footnote 16, above) vol. II, p. 370.
22. Cf. P. Kunitzsch, *Untersuchungen*, p. 65, no. 118, and p. 116, no. 315; also p. 81 f., nos. 174 and 175.
23. *Kitāb ṣuwar al-kawākib*, pp. 289, 302, 333.
24. Cf. P. Kunitzsch in *Der Islam* 51 (1974), 43 f.

There are, however, later testimonies showing that *al-muḥnith* should be identified with *alpha Eridani*. These are contained in a younger class o texts giving rich material on stellar nomenclature.

Around A.D. 1500, two Arabic navigators of the Indian Ocean, Aḥmad ibn Mājid and Sulaymān al-Mahrī, composed a number of texts, both in prose and in verse, on the art of sailing in the Indian Ocean.[25] They had of course a good knowledge of the sky, and of the southern hemisphere especially, because, as in navigation until today, they used the stars for keeping their course and fixing their position.

Both these authors knew *alpha Eridani* and made constant use of it. Their name for it is either *al-salbār* (السلبار), a non-Arabic word of uncertain pronunciation and origin,[26] or the Arabic *al-maḥnath* or *al-muḥannith* (as they pronounce it). In view of their use of the latter, the name in al-Marzūqī (quoted before) seems also to refer to *alpha Eridani.*

Thus it is proved that *alpha Eridani* was known to the Arabs in different epochs: in their ancient indigenous stellar traditions, and again in their nautical traditions of the 15th and 16th centuries. To their scientific astronomers, however, who strictly followed Ptolemy and his catalogue of stars, and who were living too far north to observe this region of the sky themselves, *alpha Eridani* and several other objects of the southern sky remained unknown or unidentified.

25. There is a facsimile edition of two Paris manuscripts by G. Ferrand, *Instructions nautiques et routiers arabes et portugais*, vol.I-II,(Paris 1921-23 and 1925). Recently I. Khoury has published five text volumes in print:*Al-ᶜUlūm al-baḥrīya ᶜinda al-ᶜarab* vols. I, 1, 2 and 3 (Sulaymān al-Mahrī), Damascus, 1970 and 1972, and vol. II,1 (Aḥmad ibn Mājid), Damascus, 1971, and another of Ibn Mājid's works in *Bulletin d'Etudes Orientales* (Damascus, t.XXIV, 1971), pp. 249-386. Ibn Mājid's *Kitāb al-fawā'id* was translated into English by G. R. Tibbetts, *Arab Navigation in the Indian Ocean Before the Coming of the Portuguese*, (London, 1971).

26. Cf. P. Kunitzsch,*Arabische Sternnamen in Europa*,(Wiesbaden 1959), p.100, footnote 1; Kunitzsch, *Untersuchungen*, p. 104, no. 160. Another pronunciation of the name is *al-sillibār*, which is metrically supported by a verse, in the metre *ṭawīl*, of Aḥmad ibn Mājid himself, see ed. Khoury, vol. II,1, p. 120 and P. Kunitzsch in *Der Islam* 51 (1974), 47, with footnote 8.

VII

OBSERVATIONS ON THE ARABIC RECEPTION OF THE ASTROLABE

The astrolabe may well be called the most famous and widespread astronomical instrument of the middle ages, both in the orient and the occident. This is testified by the great number of instruments displayed in museums and kept in private collections all over the world, as well as by the huge number of texts on the "fabrication" and/or "use" of the astrolabe that are found in so many libraries, in different languages. While the descriptions of surviving instruments are rather large in number, the existing texts have been much less well studied. The broad outline of the history of the astrolabe is, by now, generally known: the instrument seems to have originated gradually in antiquity and the hellenistic period, to have penetrated into Syriac circles of erudition, to have passed on to the Arabic-Islamic civilization, and finally to have reached Byzantium and the mediaeval Latin west. It was a living device, both in the orient and in the occident, well into the sixteenth century and later. Many details, however, of the singular stages of reception and transmission still remain uncertain and need more detailed investigation.

Our purpose in this paper is to study some problems regarding the reception of the astrolabe by the Arabs. As for the early history of the astrolabe, before the Arabs, this has been the object of a series of studies of learned authorities which, however, have not yet led to definite final conclusions[1].

The general stream of transmission of sciences went from Greek sources, through, among others, Syriac stages, to the Arabs. The same, obviously, was the case with the knowledge of the astrolabe. Of earlier complete Greek texts on the astrolabe we possess only one, by John Philoponus (6th century A.D.) [G2][2], apart from scattered notices in other authors that are difficult to ascertain.

Of the Syriac stage, we fortunately possess the treatise by Severus Sebokht (7th century A.D.) [S1].

* This article is the revised version of a paper read at the "Second International Symposium for the History of Arabic Science", held in Aleppo, April 5–12, 1979. For technical reasons, the full text of the paper is printed here, and the forthcoming *Proceedings* of the Symposium will only contain a summary. During the year 1979, I obtained copies of some more Arabic treatises (A17–A20; cf. also footnote 21) the material of which is added to this paper in some additional notes in order to supply as much material as available.

[1] O. Neugebauer, "The Early History of the Astrolabe", in *Isis* 40, no. 121 (1949), 240–256; J. Vogt and M. Schramm, "Synesios vor dem Planisphaerium", in *Das Altertum und jedes neue Gute. Für Wolfgang Schadewaldt zum 15. März 1970*, Stuttgart, 1970, pp. 265–311; A. K. Tagi-Zade, "Iz istorij izobretenija astroljabij", in *Voprosy Istorij Estestvoznanija i Techniki* 49 (1974), 42–45 (in Russian); O. Neugebauer, *A History of Ancient Mathematical Astronomy*, 3 vols., Berlin etc., 1975, pp. 868–879.

[2] For the system of quotation of sources, see the 'List of texts consulted' at the end of this paper.

As for the Arabs, the oldest texts hitherto known date from the ninth century A.D. No earlier documents have come to our knowledge until now though it must be assumed that there have existed earlier stages of transmission.

No attempt has yet been made to investigate the intimate relations between the early Arabic texts and their Greek and Syriac predecessors, nor has research into the foundation of the Arabic technical terminology of the astrolabe, its parts and its operations, yet properly begun.

In two successive teaching periods at the University of Munich, I have undertaken, with the assistance of the late Dr. Hermelink, and Mr. Reich, both of Munich, a first attempt to start a close comparison of the different texts. For this purpose we had at our disposal 36 different texts, in Greek, Syriac, Arabic, and Latin, most of them copies of manuscripts collected by Mr. Reich.

In our work it proved difficult to compare the Arabic texts directly with the Greek ones. This was partly due to the fact that the internal relations among the Greek texts are far from being clearly established. It goes without saying that those texts that were translated from the Arabic or Persian differ greatly from those of pure Greek tradition. Among these, however, the relations and dependences need further investigation. So, a general agreement in the concepts of the astrolabe, its parts and its operations, can be found between the Arabic and the Greek texts, as is naturally to be expected. But the comparison of the texts does not show striking details of immediate relationship.

For a comparative study of the terminology of the astrolabe, one of the existing two major types of treatises is of especial interest: *viz.* that describing the "use" of the instrument, and comprising, in most cases, an introductory chapter naming the constituent parts of the astrolabe, followed by a series of chapters describing individual operations that can be done by means of the instrument. In the Arabic literature, this type of treatise shows a remarkable uniformity of tradition, from the ninth century down to the late manuscripts of the eighteenth century, examples of which will be given shortly. It should be mentioned, however, that in this type of treatise the terminology occasionally differs slightly, in such a way that the practical chapters use a simplified form of certain technical terms which were given in a fuller form in the explanatory introduction. This can obviously be explained as the result of the historical development: in addition to the literal tradition in books which suffered no major changes, the practical use of the instrument had produced its own technical language which in certain cases was slightly simpler than that of the basic literal tradition. The second type of treatise, on the "fabrication" of the astrolabe, provides the theoretical basis for the construction of the instrument, theory of projection etc., and is therefore of less interest for the study of terminology. But several terms are also found in this group, and these may be counted among those simplified terms of practical use which I have just mentioned.

Nau's edition and translation of Severus Sebokht's Syriac treatise on the astrolabe [= S1] were made from a single manuscript, at Berlin, dated A.D. 1556[3].

[3] The Syriac dating was read by the editor as 1267 of the Greeks = A.D. 956 [S1, p. 58]. A better reading, however, is 1867 of the Greeks = A.D. 1556, see *Die Handschriftenverzeichnisse*

Though the text is poor, and the translation often rather superficial, the material is sufficient for the sort of comparisons envisaged here. The treatise contains a great many Greek terms, and thus shows a close relationship to Greek sources.

When comparing the terminology of the Arabic treatises on the "use" of the astrolabe with the Syriac terminology we were struck by several cases of close relationship between the Arabic and the Syriac. It must be kept in mind that there is a gap of almost two hundred years between the compositions of the Syriac and the earliest known Arabic treatises. Adding these new findings to other known facts, the Arabic terminology of the astrolabe can for the time being be divided, historically, into three major classes:

1) terms which are related to models found in the Syriac text;
2) terms which are formed after models found equally in Greek and in Syriac texts;
3) terms which have no recognizable ancient or Syriac model, and therefore seem to have been introduced by the Arabs themselves, at a stage as yet unknown to us.

The most striking example for class one is the term for the circles parallel to the horizon, on the face of the astrolabe. These were called παράλληλοι κύκλοι and the like in the Greek texts [G2, G3, G4, G5; also G7; but G8 has πόλοι instead]. The current Arabic term for them is *al-muqanṭarāt*, a term belonging to class three, the etymology of which it has not yet been possible to explain adequately, and which was ultimately retained by the Latin translators in the west and subsequently penetrated the European languages (*almucantarat*, etc.). Al-Farġānī [A2], who also speaks of *al-muqanṭarāt*, wrote in one instance: *dawā'ir al-irtifā' wa-hiya ad-dawā'ir al-muwāziya li-saṭḥ al-ufq* ("the circles of altitude which are the circles parallel to the plane of the horizon"), thus almost literally following the Greek of Philoponus [G2] who said: οἱ . . . κύκλοι παράλληλοί εἰσι τῷ ὁρίζοντι. But, then, 'Alī ibn 'Īsā [A4] explains the *al-muqanṭarāt* saying: *hiya l-ḫuṭūṭ al-muqawwasa*[4] *al-mutaḍāyiqa* etc. ("they are the curved lines that are close to each other"). Aṣ-Ṣūfī [A5] explains the *al-muqanṭarāt* as *ad-dawā'ir al-mutaḍāyiqa* ("the circles that are close to each other"), while the *Mafātīḥ al-'ulūm* [A6] show a wording very close to that of 'Alī ibn 'Īsā: *hiya al-ḫuṭūṭ al-muqawwasa*[5] *al-mutaḍāyiqa* ("they are the curved lines that are close to each other"). Abū ṣ-Ṣalt [A10], in explaining the *al-muqanṭarāt*, says: *hiya ad-dawā'ir wa-qiṭa' ad-dawā'ir al-mutatāliya al-mutaḍāyiqa* ("they are the circles, and portions of circles, which follow each

der Königl. Bibliothek zu Berlin, 23. Band: E. Sachau, *Verzeichniß der syrischen Handschriften*, ii, Berlin, 1899, p. 607b (in no. 186 = coll. Petermann no. 26). Also, after the publication of S1, a second manuscript containing the same treatise became known. It was purchased by the Bibliothèque Nationale, Paris, and was assigned the number 346. This manuscript, dated 1620 of the Greeks = A.D. 1309, and so almost 250 years older than the Berlin manuscript, proved to belong to the same branch of transmission, offering the same mistakes, omissions, etc.; cf. F. Nau, in *Revue de l'Orient Chrétien* 15 (1910), 226 and 229.

4 The printed text has *al-maqsūma* which needs to be emended to *al-muqawwasa*, in the light of the parallels in the other Arabic texts.

5 Variant readings: *al-maqsūma*, and *al-mutaqawwisa*.

other and are close to each other"). The two Arabic anonymous manuscripts of A.D. 1698 and 1702 [A12, A13] write: *wa-hiya al-ḫuṭūṭ al-mutawāliya al-muqawwasa* ("they are the curved lines which are closely following each other"), while that one of A.D. 1737 [A14] has: *wa-hiya al-qisīy al-mutawāliya al-mutaḍāyiqa* ("they are the arcs that follow each other closely and are close to each other"). And the Vienna anonymous [A15] says: *hiya al-ḫuṭūṭ al-mutawāliya al-mutaḍāyiqa* ("they are the lines that are closely following each other and are close to each other")[5a].

Knowing that the current Arabic rendering of Greek παράλληλος is *muwāzin* ("parallel to something"), or *mutawāzin* ("parallel to each other"), one wonders how, in describing the parallels to the horizon, the astrolabe texts use terms like *mutaḍāyiqa*, *mutawāliya*, *mutatāliya*. The reason for this is obvious from the Syriac text. Because in Severus Sebokht's treatise, in describing the parallels to the horizon, the Greek term παράλληλοι is retained, and afterwards explained in Syriac as *hānaw qarrīḇay la-ḥḏāḏē aw d-ḇāṯar ḥḏāḏē* ("that is those which are close to each other or behind each other")[6]. The same wording is repeated also in other places where the parallels are mentioned[7].

It is obvious that the Arabic use of words meaning "close to each other", "closely following each other", in describing the term παράλληλοι, depends on the Syriac model, although due to the lack of documents from the period between Severus Sebokht and the Arabic treatises of the ninth century, we are not able to determine when, where, and by whom, the contacts between Syriac and early Arabic astrolabists were established.

Two other instances may be added in which the Arabic texts show resemblances to certain phrases in the Syriac version which do not occur in the Greek, or the early Greek texts.

In describing the three circles, *viz.* the two tropics and the equator, on the face of the astrolabe, Philoponus says that when one moves the spider around, the beginning of the sign of Cancer will be seen describing that circle (i.e., the tropic of the summer), . . . τὸν κύκλον τοῦτον γράφουσαν [G2, p. 133f.]. For the equator, and the winter tropic, he says that the beginnings of Aries, Libra, and Capricornus, respectively, will run through the two corresponding circles (διέρχονται). This detail of description is not mentioned in the other Greek texts [G4, G5]. Now some of the Arabic texts, while treating this item, do not simply say that the beginnings of the corresponding signs "run through" those circles, but they use expressions of much greater force: ʿAlī ibn ʿĪsā [A4]: . . . *lazimathā* [sic] *luzūman mustawiyan* ("it [i.e., the beginning of Cancer] clings to it [i.e., the circle of the summer tropic] in an equal and constant manner"). Aṣ-Ṣūfī [A5] says of the summer tropic: *wa-hiya lāzima abadan li-ra's as-saraṭān* ("it constantly clings to the

[5a] Ibn as-Samḥ [A18] writes: . . . *dawā'ir baʿḍuhā ġayr tāmma wa-baʿḍuhā tāmma mutaqāriba baʿḍahā* [sic] *min baʿḍ yuqālu lahā l-muqanṭarāt*, ". . . circles, some of which are incomplete and some complete, being near each other, which are called *al-muqanṭarāt*". Kūšyār ibn Labbān [A17] and Ibn an-Naṭṭāḥ [A19] introduce the *al-muqanṭarāt* without producing a detailed explanation.

[6] S1, p. 79, 11–12.

[7] E.g., S1, p. 239, -3: *pāralēlū awkēṯ qarrīḇay la-ḥḏāḏē* ("παράλληλοι, that is the ones [plural] close to each other").

beginning of Cancer"). And Abū ṣ-Ṣalt [A10] speaks of the summer tropic: *wa-hiya allatī yadūru ra's as-saraṭān mumāssan lahā ġayra mufāriqin* ("it is that [circle] which the beginning of Cancer, in its rotation, is constantly touching and never leaving"); of the equator and of the winter tropic he equally says, they are those circles through which the beginnings of Aries, and Capricornus respectively, are revolving, "never separating from them" (*ġayra mufāriqin lahā*)[7a]. Turning, then, to Severus Sebokht, we find that he also used terms of such intensity in describing these three circles: the circle "along which the beginning of Capricornus walks continually" (*da-ʿlaw mhallek̲ āmīnā'it̲ rēšeh d-g̲ad̲yā*; S1, p. 83, ult. – 84, 1), i.e. the winter tropic; the circle "along which all the time are wandering the beginnings of Aries and Libra" (*da-ʿlaw b-k̲ull-zb̲an rēšeh ʿāb̲ar d-emmrā w-haw tūb̲ d-massaṭā*; S1, p. 84, 2–3), i.e. the equator; and the circle "along which always is moving the beginning of Cancer" (*da-ʿlaw rād̲ē kull-šnā rēšeh d-sarṭānā*; S1, p. 84, 5), i.e. the summer tropic. Here, the quoted Arabic sources show closer resemblance to the Syriac than to the Greek.

A third case is the description of the pivot in the center of the astrolabe. In Greek texts, only the late Nikephorus Gregoras [G5] and the somewhat related MS. Vat. gr. 178 [G4] speak of it, here called ἧλος ("a nail, or tack"), saying that a hole is to be made in the center of the instrument so that the "nail" may hold together the spider and the plates (ὡς ἂν εἷς ἧλος συνέχῃ τήν τε ἀράχνην καὶ τὰ τύμπανα[8]). Philoponus [G2] does not mention these technical devices[9]. Of the known early sources it is Severus Sebokht who describes the "hole" (*neqb̲ā*; S1, p. 77, 8) and the "nail" (*sekkt̲ā*; S1, p. 77, 11), saying that it is the duty of the latter "to hold together all the parts of the astrolabe in order that they may not be dispersed" (*āḥdā hāy . . . dlā netpalhad̲*; S1, p. 77, 13–14). This somewhat expanded description seems to be echoed by some Arabic texts, cf. Abū ṣ-Ṣalt [A10]: *al-miḥwar huwa al-mismār an-nāfid̲ fī l-quṭb al-munaẓẓim bi-ǧamīʿ aṣ-ṣafā'iḥ* ("*al-miḥwar* is the nail inserted in the hole [*al-quṭb*] and which holds all the plates well together"); Ibn aṣ-Ṣaffār [A7]: *wa-fīhā al-quṭb al-mumsik li-ṣ-ṣafā'iḥ* ("and in it [i.e. *al-miḥwar*, the hole] is the pivot [*al-quṭb*] which retains the plates"[10]); and *Mafātīḥ al-ʿulūm* [A6]: *al-quṭb huwa al-watad al-ǧāmiʿ li-ṣ-ṣafā'iḥ wa-l-ʿankabūt* ("*al-quṭb* is the pivot holding together the plates and the spider")[10a]. Perhaps, the same topic of "holding together" was also extended, in the Arabic texts, to the description of *al-faras*, a horse-shaped transversal pin which was inserted in the pivot in order to prevent it from falling out (cf. A6: *al-faras huwa qiṭʿa šabīha bi-ṣūrat faras yušaddu bihā l-ʿankabūt ʿalā ṣ-ṣafā'iḥ*; A7: *al-faras huwa ad-dāḫil fī l-quṭb al-mumsik lahū*; A9:

[7a] Ibn an-Naṭṭāḥ [A19] has a more striking description for this detail, saying *wa-lā yamurru [yaqaʿu] ʿalayhā ġayr . . .*, "and no other [degree of the great circle] rotates along [falls upon] this [respective] circle", while Ibn as-Samḥ [A18] and Kūšyār ibn Labbān [A17] do not quote this detail.

[8] G5, p. 202, 13–14; G4, 133ᵛ.

[9] In G3 the "nail" is mentioned under the designation of ὀμφαλός, "navel" (fol. 1ʳ and 17ʳ).

[10] In the Latin translation [L6]: *(foramen . . .) in quo est axis qui tenet tabulas.*

[10a] Ibn as-Samḥ [A18]: *yaṣluḥu hād̲ā l-quṭb an yuḍbaṭa bihī al-ʿiḍāda wa-l-umm wa-ṣ-ṣafā'iḥ wa-l-ʿankabūt . . .*, "this pivot serves to fix the rule, the mother, the plates, and the spider . . .". Kūšyār ibn Labbān [A17] and Ibn an-Naṭṭāḥ [A19] have no detailed description for these devices.

[quṭb] yadḫuluhū faras yataʿallaqu bihī yumsiku al-quṭb māsikan lahū; A10: *al-faras hiya aš-šaẓīya allatī tadḫulu fī l-miḥwar fa-tašuddu ṣ-ṣafā'iḥ wa-tulṣiqu suṭūḥahā baʿḍ* [sic] *bi-baʿḍ ilṣāqan muḥkaman*; etc.).

For class two of the Arabic terminology of the astrolabe, i.e. those terms for which models are found both in the Syriac and in early Greek texts so that no decision can be made as to from which of the two the Arabic might be derived, I just mention three examples:

The Arabic term for "zenith" always comprises the element ". . . of the head[s]" (*samt ar-ra's*, or *ar-ru'ūs*; also in Latin, *cenit capitis*, or *capitum*). The same was the case in Syriac[11] and in the early Greek texts[12].

The "spider" of the astrolabe, in Arabic *al-ʿankabūt*, was equally called ἀράχνη in the Greek texts and by Severus Sebokht who, however, did not translate the word but moreover retained it in its Greek spelling[13].

The small excrescence on the outer rim of the circle of the zodiac, on the spider, at the beginning of Capricornus, serving as a pointer, is generally called in Arabic *al-murī* ("the pointer")[14], and hence, in Latin, *almuri* etc. In some instances, however, some texts have retained the more complete designation *murī al-ağzā'* ("pointer of the degrees")[15], and this might be a reflection of the earlier Greek μοιρογνωμόνιον[16], and/or the corresponding Syriac *mḥawwyānā d-mūrās*[17,18].

The third class of the Arabic terminology of the astrolabe comprises such typical terms as *an-naẓīr* ("the point opposite to a given point", the Latin *nadir*, which is still used in modern times)[19]; *al-umm* ("the mother", that is the basic plate of the

[11] Cf. S1, p. 75, 4–5: *hāy d-sīmā trīṣā'iṯ lʿel men qenṭrōn d-arʿā awkēṯ lʿel men qarqapṯā d-ḵoll-ḥaḏ mennan* ("that is the point directly over the centre of the earth, that is, over the head of everyone of us"), and also p. 78, -2 – 79, 1: *sīmīyōn d-meṣʿaṯ šmayyā awkēṯ qenṭrōn d-qāymā men koll-pros lʿel men rēšan* ("the point [σημεῖον] of the middle of the sky, that is, the point [κέντρον] standing directly over our head").

[12] G2, p. 130, 11; 131, -4; etc.: τὸ κατὰ κορυφὴν σημεῖον; the same in G3, 5ʳ etc.

[13] But invariably in the form *arkānā*, instead of *araḵnā*; see, e.g., S1, p. 85, 11; 241, 14; 248, 13; etc.

[14] al-Bīrūnī explicitly says that it is called by the mere word *murī* without the addition of any attribute (. . . *tusammā muriyan bi-l-iṭlāq min ġayri ṣifa bi-šay'*) [A9, Arabic text opposite to p. 195, line 1].

[15] So, e.g., A1; A5, p. 3, 16; A17 [mis-spelt *murī al-'ḫr*]; A18; appearingly also represented by the Latin *almuri alagee*, L7, p. 327, -4.

[16] G2, p. 141, 7 (τὸ τῆς ἀράχνης μοιρογωμόνιον), etc.

[17] S1, p. 241, 14–15 (*mḥawwyānā d-mūrās da-ḇ-arkānā hānaw rēšeh d-ḡadyā*, "the pointer of degrees on the spider which is the beginning of Capricornus"); p. 248, 12–13 (*mḥawwyānā d-mūrās haw da-ḇ-arkānā*, "the pointer of degrees which is on the spider").

[18] It should be added that the cited texts, Greek and Syriac respectively, use the same terms of μοιρογνωμόνιον and *mḥawwyānā d-mūrās* also for the pointed ends of the alidade which, however, in the Arabic texts were not accorded a term of their own, but only called *ra's*, or *ṭaraf (muḥaddad)* ("[sharpened] end"), with the exception of al-Bīrūnī [A9] who cites for them the designation *muriyā al-ʿiḍāda* ("the two pointers of the alidade"). (The same also in Ibn as-Samḥ [A18], fol. 32ᵛ, line 3.)

[19] With Philoponus [G2, p. 138, 22; etc., and similarly in G3, 18ʳ] we find the expression ἡ κατὰ διάμετρον τοῦ ἡλίου μοῖρα, which almost literally is repeated by Severus Sebokht: *mūrā d-zōdyōn diyāmeṭrōn d-sarṭānā* ("the degree of the zodiacal sign diametrically [opposite] to Cancer", S1, p. 241, 3–4; cf. also p. 240, 3; 248, -3). There is no immediate linguistic relation between

instrument, the Latin *mater*); *al-muqanṭarāt* (see above); *al-ʿiḍāda* ("the alidade", the turnable index, or ruler on the back side of the astrolabe, also called *regula* by the Latin translators), etc. These terms seem to have been introduced by the Arabic astrolabists, as neither the Syriac nor the old Greek treatises call these devices by names that might have served as models for the forming of the respective Arabic designations.

This contribution is meant to be a first attempt to analyse the material referring to the Arabic reception of the astrolabe, and to study the formation of the Arabic terminology of this instrument and its dependence on earlier Syriac and Greek models. It must be stressed that these investigations and their results are far from being the last word in this field. Our research, in Munich, was based on a rather limited number of texts. It is known that many more texts exist, most of them in more than one manuscript, and some of them in very many, scattered through the libraries of the world. The task is left to the scholarly world to prepare critical editions of every single text, before proceeding to their analysis and to the establishment of a comparative history of the terminology of the astrolabe[20] in those civilizations which had their part in the chain of transmission: Greek, Syriac, Arabic, and mediaeval Latin.

LIST OF TEXTS CONSULTED

I. Greek

1. Synesius of Cyrene (around A.D. 400), Letter to Paeonius. Edited in Migne, *P. G.* 66, 1577–1588; N. Terzaghi, *Synesii Cyrenensis opuscula*, Rome, 1944, pp. 132–142. A new German translation, accompanied by a study, was recently published by J. Vogt and M. Schramm, "Synesius vor dem Planisphaerium", in: *Das Altertum und jedes neue Gute. Für Wolfgang Schadewaldt zum 15. März 1970*, Stuttgart, 1970, pp. 265–311.
2. John Philoponus (Alexandria, 6th century A.D.), *De usu astrolabii eiusque constructione*, ed. H. Hase, in *Rheinisches Museum für Philologie* 6 (1839), 127–156; German translation by J. Drecker, in *Isis* 11 (1928), 15–44.
3. MS. Vat. gr. 212, 1r–25v: Περὶ κατασκευῆς ἀστρολάβων ὀργάνων (anonymous; terminology partly similar to Philoponus [no. 2]).
4. MS. Vat. gr. 178, 132r–135v: Ἀστρολάβου κατασκευή (anonymous; terminology largely similar to Nikephorus Gregoras [no. 5]).
5. Nikephorus Gregoras (Byzantium, 14th century A.D.), Περὶ κατασκευῆς καὶ γενέσεως ἀστρολάβου, ed. A. Delatte, *Anecdota Atheniensia et alia*, ii, Liège and Paris, 1939, pp. 195–208.
6. Makarius Monachus, *scholion* to Nikephorus Gregoras' treatise, ed. H. Hase, *l.c.*, p. 157f. (chapter 1 only); also A. Delatte, *l.c.*, pp. 208–210 (two chapters).

these terms and the Arabic *an-naẓir* (literally, "the corresponding one"). It is significant that the Greek text of G8, which presumably was translated from the Persian, renders *naẓir (-i daraǧa-i šams)* as κατ' ἐναντίον τῆς μοίρας τοῦ ἡλίου, totally different from the above cited old Greek texts, and etymologically well corresponding to the Arabic *naẓir*.

[20] A selected number of astrolabe terms, in Latin and Arabic, have lately been studied by P. Kunitzsch, *Mittelalterliche astronomisch-astrologische Glossare mit arabischen Fachausdrücken*, Munich, 1977 (Sitzungsberichte der Bayerischen Akademie der Wissenschaften, Philos.-Hist. Kl., Jahrgang 1977, Heft 5), see especially i, nos. 18–34 (with commentary on pp. 18–24), and ii, nos. 10–14 (commentary on pp. 40–42).

7. Greek text, anonymous, *inc.* Εἰ βούλει γιγνώσκειν τὴν τοῦ ἀστρολάβου μέθοδον. Edited in part by H. Hase, *l.c.*, pp. 158–160, and in full by A. Delatte, *l.c.*, pp. 254–262. Mr. Reich has collated this text in more than 30 manuscripts and regards it as a translation from the Arabic.
8. MS. Vat. gr. 210, 3rff.: Ποίημα Σιάμψ τοῦ Περσοῦ περὶ τῆς διδασκαλίας τοῦ ἀστρολάβου. Greek text, translation from the Persian (probably 14th century A.D.). Partly edited by A. Delatte, *l.c.*, pp. 263–271. Fully collated by Mr. Reich. *Inc.* Περὶ τῶν τεμμαχίων τοῦ ἀστρολάβου ιζ ὄντων. Τὸ πρῶτον καλεῖται βραχίων ...

II. Syriac

1. Severus Sebokht, *Le traité sur l'astrolabe plan* (North Syria, written before A.D. 660), ed. with French translation by F. Nau, in *Journal Asiatique*, 1899, pp. 58–86 and 238–273 (text), and 87–101 and 274–303 (translation). An English translation, made from Nau's French version, is given in R. T. Gunther, *The Astrolabes of the World*, Oxford, 1932, i, pp. 82–103. A new critical edition, accompanied by a German translation, is presently prepared by Mr. E. Reich of Munich.

III. Arabic

1. Muḥammad ibn Mūsā al-Ḫwārizmī (d. ca. A.D. 835), treatise on the use of the astrolabe. MS. Berlin, Ahlwardt 5793, 81vff.; German translation by J. Frank, *Die Verwendung des Astrolabs nach al Chwârizmî*, Erlangen, 1922 (Abhandlungen zur Geschichte der Naturwissenschaften und der Medizin, Heft iii).
2. Al-Farġānī, On the fabrication of the astrolabe (written 225 Yazdağird = A.D. 857/8). MS. Berlin, Ahlwardt 5790, 2vff.
3. Al-Ya'qūbī (around A.D. 875), *Historiae*, ed. M. T. Houtsma, i–ii, Leiden, 1883. In vol. i, pp. 154–161, quotations from *Kitāb ḏāt al-ḥalaq*, *Kitāb ḏāt aṣ-ṣafā'iḥ*, and *Kitāb al-qānūn*, allegedly by Ptolemy, containing a number of technical terms pertaining to our domain. German translation by M. Klamroth, in *Zeitschrift der Deutschen Morgenländischen Gesellschaft* 42 (1888), 18–27.
4. 'Alī ibn 'Īsā (physician under the caliph al-Mu'tamid, A.D. 870–892). *Kitāb al-'amal bi-l-aṣṭurlāb*, ed. L. Cheikho, in *al-Mašriq* 16 (1913), 29–46. German translation by C. Schoy, in *Isis* 9, no. 30 (1926), 239–254.
5. 'Abd ar-Raḥmān ibn 'Umar aṣ-Ṣūfī (d.986), *Kitāb al-'amal bi-l-asṭurlāb*, Hyderabad/Dn., 1962.
6. Muḥammad ibn Aḥmad al-Ḫwārizmī (around A.D. 980), *Kitāb mafātīḥ al-'ulūm*, ed. G. van Vloten, Leiden, 1895. An early encyclopaedia of sciences; on the astrolabe: pp. 232–235; summary, in German, by E. Wiedemann in a paper of 1912, reprinted in his *Aufsätze zur arabischen Wissenschaftsgeschichte*, i–ii, Hildesheim and New York, 1970; see vol. i, pp. 551–553.
7. Ibn aṣ-Ṣaffār (Spain, d. A.D. 1035), *Kitāb al-'amal bi-l-asṭurlāb*, ed. J. M. Millás Vallicrosa, in *Revista del Instituto Egipcio de Estudios Islamicos en Madrid* 3 (1955), 35–49 (Spanish) and 47–76 (Arabic)[21]. The literal Latin translation (see below, group iv, no. 6) gives the author as Maslama al-Mağrīṭī (d. A.D. 1007), Ibn aṣ-Ṣaffār's teacher.
8. 'Abdallāh ibn Muḥammad ibn Sa'd at-Tuğībī, *Risāla fī l-asṭurlāb*. MS. Berlin, Ahlwardt 5805, 68v–81r. This is an abbreviated version of Ibn aṣ-Ṣaffār's treatise (no. 7, above).
9. Al-Bīrūnī, *Kitāb at-tafhīm li-awā'il ṣinā'at at-tanǧīm* (written A.D. 1029). Ed. with English translation by R. R. Wright, London, 1934. Section on the astrolabe: pp. 194–210 (§§ 324–346).

[21] Millás' edition was made from MS. Escurial 964. Another MS. of the same treatise mentioned in the bibliographies, is MS. London, British Library, Add. 9600, 262v–280v. Here, the title runs *Kitāb fī l-'amal bi-l-asṭurlāb waḍa' abū aš-šayḫ al-'ālim Abū l-Qāsim aṣ-Ṣaffār* [262v]; the text, however, is much different from the one edited by Millás. After a first quick inspection I would say that the London MS. contains a later redaction of Ibn aṣ-Ṣaffār's original treatise (which would be the one edited by Millás, and which is also represented in the Latin translation ascribed to Maslama, cf. below, L6), with several additions, alterations, omissions, and some changes in the order of chapters.

10. Abū ṣ-Ṣalt Umayya ibn ʿAbd al-ʿAzīz, of Denia (Spain, d. 1134), *Risālat al-ʿamal bi-l-aṣṭurlāb.* MS. Berlin, Ahlwardt 5798, 44ʳ–71ᵛ.
11. Abū Ǧaʿfar Muḥammad ibn Ayyūb aṭ-Ṭabarī, *Kitāb maʿrifat al-asṭurlāb* (written A.D. 1234). MS. Munich 347 (Persian).
12. Treatise on the astrolabe, anonymous. MS. Berlin, Ahlwardt 5810, 61ᵛ–69ʳ (MS. dated A.D. 1698).
13. Treatise on the astrolabe, anonymous. MS. Berlin, Ahlwardt 5811ᵃ, 1ᵛ–11ᵛ (MS. dated A.D. 1702). Nearly identical with no. 12 above.
14. Treatise on the astrolabe, anonymous. MS. Berlin, Ahlwardt 5812, 1ᵛ–12ᵛ (MS. of around A.D. 1737). Closely related to nos. 12 and 13 above.
15. *Tuḥfat al-lubāb bi-šarḥ ʿamal al-aṣṭurlāb* (in 10 *faṣl* and a *ḫātima*), anonymous. MS. Vienna 1364.
16. *Muḫtaṣar aṣ-ṣafīḥa az-zarqālīya* (in 50 *bāb*), anonymous. MS. Florence, Laur. or. 416, 1ᵛff.
17. Kūšyār ibn Labbān (end of the 10th century A.D.), *Risāla fī ʿilm al-asṭurlāb.* MS. London, British Library, Add. 7490, 211ᵛ–221ᵛ (apparently a very late copy, incomplete from ch. 10 in section ii, full of mis-spellings).
18. Ibn as-Samḥ (Spain, d. A.D. 1035), *Kitāb al-ʿamal bi-l-asṭurlāb.* MS. London, British Library, Add. 9602, 25ᵛ–55ᵛ.
19. Muḥammad ibn ʿAlī ibn Yaḥyā Ibn an-Naṭṭāḥ (presumably in Spain; after Abū l-Qāsim Ibn aṣ-Ṣaffār [see above A7], whom he quotes, foll. 15ʳ, ult.; 15ᵛ, 9; 16ʳ, 4), *Kitāb fī l-asṭurlāb.* MS. London, British Library, Add. 9602, 1ᵛ–24ᵛ.
20. ʿAbdallāh ibn Faḫr ad-Dīn al-Aʿraǧ al-Mawṣilī (d. A.D. 1771/2, or 1774/5), *Sawāniḥ al-qarīḥa fī šarḥ aṣ-ṣafīḥa*, ed. Ṣ. M. Muḥammad, Baghdad, n.d. (ca. 1979), being a commentary to the treatise on the astrolabe *aṣ-Ṣafīḥa fī l-aʿmāl al-asṭurlābīya* by Bahāʾ ad-Dīn al-ʿĀmilī (d. A.D. 1620/1, or 1621/2).

IV. Latin

1. *Sententiae astrolabii*, 10th century A.D., written after Arabic models, but obviously no direct translation. Ed. J. M. Millás Vallicrosa, *Assaig d'història de les idees físiques i matemàtiques a la Catalunya medieval*, i, Barcelona, 1931, pp. 275–288 (in 387 numbered lines). Inc. *Quicumque uult scire certas horas noctium* . . .
2. Gerbertus (around A.D. 1000), *De utilitatibus astrolabii*, ed. N. Bubnov, *Gerberti postea Silvestri II papae Opera Mathematica*, Berlin, 1899, pp. 114–147. Inc. *Quicumque astronomicae discere peritiam disciplinae* . . . The attribution to Gerbertus is dubious; the same treatise is also attributed to Hermannus Contractus (see next no.).
3. Hermannus Contractus (d. A.D. 1054), *De mensura astrolabii*, ed. Migne, *P. L.* 143, 381–390 (Inc. *Herimannus Christi pauperum* . . .; *In metienda igitur* . . .), and *De utilitatibus astrolabii*, ed. *ib.* 389–412 (Inc. *Quicumque astronomicae peritiam disciplinae* . . ., identical with no. 2 above). *De mensura* was also edited, with a commentary, by J. Drecker, in *Isis* 16, no. 49 (1931), 200–219.
4. Pseudo-Māšāʾallāh: *Nos iuxta rectam imaginationem* . . ., ed. by J. M. Millás Vallicrosa, *Las traducciones orientales en los manuscritos de la Biblioteca Catedral de Toledo*, Madrid, 1942, p. 313 (– 316, 8). Certainly not by Māšāʾallāh, and probably not from a treatise on the astrolabe (F. J. Carmody, *Arabic Astronomical and Astrological Sciences in Latin Translation*, Berkeley and Los Angeles, 1956, p. 118f., no. 3, attributes it to Ṯābit ibn Qurra; cf. also E. Poulle, in *Studi medievali*, 3ª Serie, v, 2, 1964, p. 871, footnote, under "*Astrologie speculationis* . . .").
5. Pseudo-Māšāʾallāh: *Astrologie speculationis exercitium habere uolentibus* . . ., ed. by J. M. Millás Vallicrosa, *Las traducciones, l.c.*, pp. 316, 14–321. Most certainly not by Māšāʾallāh (Carmody, *Sciences, l.c.*, p. 169f., no. 2b, attributes it to John of Seville).
6. On the use of the astrolabe, Inc. *Primum horum armilla* . . ., attributed to Maslama al-Maǧrīṭī (d. A.D. 1007), Latin translation by John of Seville. Ed. by J. M. Millás Vallicrosa, *Las traducciones, l.c.*, pp. 261–284. The corresponding Arabic text is attributed to Maslama's disciple, Ibn aṣ-Ṣaffār (see above, group iii, no. 7).
7. John of Seville (12th century A.D.), On the construction of the astrolabe, Inc. *Cum uolueris facere astrolabium accipe auricalcum optimum* . . . Partly edited by J. M. Millás Vallicrosa, *Las traducciones, l.c.*, pp. 322–327.

8. Pseudo-Māšā'allāh, *De compositione astrolabii*, Inc. *Scito quod astrolabium est nomen grecum* . . ., ed. R. T. Gunther, *Chaucer and Messahalla on the Astrolabe*, Oxford, 1929 [Early Science in Oxford, v], pp. 195–216.
9. Idem, *De utilitate astrolabii*, Inc. *Nomina instrumentorum sunt hec* . . ., ed. R. T. Gunther, *l.c.*, pp. 217–231. In my opinion, nos. 8 and 9 are western compilations closely related to the Latin version of the treatise by Maslama (above no. 6), or his disciple Ibn aṣ-Ṣaffār (group iii, no. 7), as has been shown in a preceding article in this journal, vol. 31, no. 106 (1981), 42–62.
10. *Liber de compositione astrolabii uniuersalis qui dicitur ptolomei quod uix credo*, Inc. *Quoniam in mundi spera est motus perfectus* . . . MS. Vienna 5311, 33ra–34vb (cf. Carmody, *l.c.*, p. 19, no. 17). Ptolemy's authorship is impossible; a number of Arabic technical terms are borrowed from *Sententiae*, above no. 1. Thus, it would appear to be a western compilation. (Cf. J. D. North, ed., *Richard of Wallingford*, Oxford, 1976, vol. iii, pp. 162–164, who discusses a similar text from MS. Digby 40.)
11. Treatise on the astrolabe by Abraham ibn 'Ezra (Spain, 12th century A.D.), Inc. *Genera astrolabiorum duo sunt: dextrum et sinistrum* . . ., ed. J. M. Millás Vallicrosa, in *Al-Andalus* 5 (1940), 1–29. (Another treatise on the astrolabe by the same author, in Hebrew, is: *Sefer kelī neḥōšet*, ed. by H. Edelmann, Königsberg/Pr., 1845.)

These sources are quoted in the present paper as G1, G2, etc.; S1; A1, A2, etc.; L1, L2, etc., for the Greek, Syriac, Arabic, and Latin texts respectively.

VIII

REMARKS REGARDING THE TERMINOLOGY OF THE ASTROLABE[1]

The astrolabe has been handed down to the mediaeval West by the Arabs, in Spain, from the late 10th century onwards. This historical development is proved, among others, by two easily recognizable facts: by the many Arabic star names which can be read both on the instruments themselves and in most of the treatises written on this subject in Europe from the late 10th century onwards, and second by the great number of technical terms of Arabic origin which are applied to different parts or features of the instrument and which are continuously used in those treatises.

The study of astrolabe star names, therefore, has proved to be useful for research work into the astrolabe and its history in the West. The names read on an instrument can be of help for its dating once their dependence from a certain written source has been established. Examples for this are given in a book of mine on astrolabe star tables, published in 1966[2].

Similarly useful is the study of the terminology of the astrolabe. Also here, chronological observations can be made and the relations among different texts can be established by comparing the technical terms used in them.

In the margin, I should mention that three astronomical terms of Arabic origin which are used in astronomy all over the world until now, came into the West through texts on the astrolabe: I mean the three terms *zenith*, *azimuth*, and *nadir*. All these three

[1] The text was written for the annual meeting of the International Society of the Astrolabe, July 1982 in Oxford. Since for technical reasons I was unable to go to Oxford, I present the paper here.

[2] P. Kunitzsch, *Typen von Sternverzeichnissen in astronomischen Handschriften des zehnten bis vierzehnten Jahrhunderts*, Wiesbaden 1966. See pp. 25, 40, 52 and 68.

were originally brought into the West in treatises on the astrolabe and its use, *zenith* referring to the highest point above the horizon, the center of the so-called *almucantarat*, that is the parallels to the horizon. In Arabic it was *samt al-ra'ṣ*, "the direction of the head", signifying the point in the sky directly over one's head. Latin translators borrowed it as *cemt/zemt capitis*, and finally *cemt/zemt* was misread and miswritten, in Latin, as *cenit/zenit*. Further, *nadir* in Arabic was *naẓīr*, "the corresponding one", and mostly employed in phrases like "the degree corresponding to the degree of the sun", that is "opposite to the degree of the sun" at a distance of 180°; today it is used only for the point in the sky opposite to the *zenith*. Finally, *azimuth* in modern astronomy (taken as a word in the singular) refers to an arc of the horizon between the South point and the point at which a vertical circle passing through the *zenith* and through a given star cuts the horizon. In Arabic, *azimuth* was the plural, *al-sumūt*, of the same word *samt* which we saw shortly before in the term *samt al-ra's, zenith*. It designated, in the singular *samt* – apart from the special term *zenith* which, in Arabic, was always complemented by a subsequent genitive: *samt* of the head, or of the heads, or of a place or town – any of the vertical great circles, in the horizontal system, passing through the *zenith* and cutting the horizon at right angles, and also the arc on the horizon between the East or West point and the place where the respective *samt*, or vertical circle, cuts the horizon. Thus, altogether there were three uses of the word *samt* in Arabic: composed with "of the head" etc., it designated the *zenith*; as a normal noun, with the plural *sumūt*, it designated the vertical circles in the horizontal system; third, mostly in the singular and complemented by a genitive like "of a given star", "of the sun, or moon", etc., and also "of a place or town", it referred to that coordinate in the horizontal system which nowadays is counted from the South point while the mediaeval astrolabists counted it from the East or West points. Mediaeval authors misused the Arabic plural *azimuth* as equivalent to the singular *samt*, in the sense of that coordinate, and thus it came that in modern astronomy the transliteration of the Arabic plural, *azimuth*, is used in the sense of a singular for that coordinate in the horizontal system.

It may be added that these three terms did not enter the West together at a time. Moreover, *nadir* belongs to the oldest borrowings of the late 10th century, in northeastern Spain, while *zenith* and *azimuth* did not come before the middle of the 12th century, through the great translation movement in Spain.

Here, we meet with an observation of fundamental methodical importance. The more than sixty Arabic technical terms of the astrolabe that where found and studied so far, could not be introduced into western use by any mediaeval western author, at any time, in any place. The means of transmission of such oriental matters were direct translations from the Arabic into western idioms. Of these, among the 42 western treatises dating from the late 10th to the 16th centuries that I have lately studied[3], there exist only three major groups: the "Sententie astrolabii" and some contemporaneous texts from the end of the 10th century, in northeastern Spain; the two Latin translations, above all the one by Johannes Hispalensis, and in the second line that one by Plato Tiburtinus (both around the middle of the 12th century), of the Arabic treatise on the use of the astrolabe by the Spanish-Arabic astronomer of the 11th century, Ibn al-Ṣaffār, which sometimes was wrongly referred to his master Maslama of Madrid; and third, the old Castilian writings on the astrolabe in King Alfonso X's "Libros del Saber" from the 13th century. These three texts, or groups of texts, introduced certain and distinguished numbers of Arabic technical terms of the astrolabe into the West. All the other authors normally followed these, or else other secondary sources that were, in their turn, following the translations.

Thus, it can be clearly observed that the terminology of the "Sententie astrolabii" and their accompanying texts was adopted by Gerbert of Aurillac, Fulbert of Chartres, Hermannus Contractus and Walcher of Malvern. Even the translators Johannes Hispalensis and Plato Tiburtinus, as well as Adelard of Bath who knew Arabic himself, knew this oldest group of translations and occasionally used some of their terminology.

The greatest and farthest reaching was the influence of Johannes Hispalensis' translation of Ibn al-Ṣaffār's treatise, from the middle of the 12th century. Its influence is visible in such authors, or texts, as Raymond of Marseille, and Pseudo-Messahalla (*De operatione*

[3] P. Kunitzsch, *Glossar der arabischen Fachausdrücke in der mittelalterlichen europäischen Astrolabliteratur*, Göttingen 1983 (Nachrichten der Akademie der Wissenschaften in Göttingen, I. Philol.-Hist. Kl., Jahrg. 1982, Nr. 11). On the texts see p. 476–508.

astrolabii), and through this most popular last mentioned text further in Arialdus, the Bruxelles version of Messahalla, Pierre de Maricourt, Robertus Anglicus, Andalò di Negro, and others.

The Alphonsine texts, on the other hand, were of no visible influence with regard to terminology though also they, in their relatively late epoch, had created some new Arabic terms in western transliteration.

It is worth while stressing that three of the fundamental astronomical terms, namely the three afore mentioned terms for the *zenith*, for the vertical circles, and for the horizontal coordinate called *azimuth*, were not transmitted to the West through the first group of translations of the late 10th century. Moreover they became known in the West only in the 12th century, through the two Latin translations of Ibn al-Ṣaffār by Johannes Hispalensis and Plato Tiburtinus (of unknown date, sometime between A. D. 1133 and 1153 or 1133 and 1145, respectively). Afterwards they were also mentioned by Raymund of Marseille (whose dating, by E. Poulle, as "antérieur à 1141"[4] therefore cannot be maintained without reluctance); further in Maslama's notes to Ptolemy's "Planisphaerium" perhaps translated together with the text itself by Hermannus Secundus in A. D. 1143; by Rodolfus Brugensis, this last mentioned translator's pupil, in A. D. 1144; by Adelard of Bath whose treatise on the astrolabe is regarded to be a late work, between A. D. 1142–1146, by Haskins; in a treatise on the universal astrolabe ascribed to Robert of Chester and dated to A. D. 1147; and in works dependent on these, like the most widely spread Pseudo-Messahalla (*De constructione* and *De operatione astrolabii*)[5].

I shall not touch here the older history of the astrolabe terms, before the ninth century, in the Orient[6]. Also there, many problems are still unsolved. Let me just say that until now it was not possible

[4] E. Poulle, "Le traité d'astrolabe de Raymond de Marseille", in *Studi medievali*, 3a Serie, V,2 (1964), 866ff.; esp. p. 866f. and 868.

[5] On Pseudo-Messahalla cf. P. Kunitzsch, "On the authenticity of the treatise on the composition and use of the astrolabe ascribed to Messahalla", in *Archives Internationales d'Histoire des Sciences* 31 (1981), 42–62; M. Viladrich, "On the Sources of the Alphonsine Treatise Dealing with the Construction of the Plane Astrolabe", in *Journal for the History of Arabic Science* 6 (1982) [app. Dec. 1983], 167–171.

[6] Cf. P. Kunitzsch, "Remarks on the Arabic reception of the astrolabe", in *Archives Internationales d'Histoire des Sciences* 31 (1981), 243–252.

to find out which preceding Greek, or Syriac, term it was that the Arabs converted into their curious word *samt* which literally means "way, path, or direction", since those vertical circles to the horizon are not mentioned in the older Greek astrolabe texts. I found the term only in one Greek manuscript, dating from the 14th century, in a text which unfortunately is suspicious of being of a very late date, and perhaps itself under Arabic influence.

One last example: In the "Sententie astrolabii" of the late 10th century, in one instance the Arabic expression *vacadib* is used, with the appended translation: *id est uirgula*, "that is the stick, or bridge". Here, the translator had not only transliterated the Arabic noun itself, but also, connected to it, the Arabic word for "and", according to the phrase he was forming. *Vacadib*, thus, literally means "and the stick, or bridge". It is not surprising that in later times compilers of astrolabe treatises who were not in command of Arabic themselves, picked up this complete term *vacadib* unaware that only *cadib* precisely is the noun, and *va* means "and", and used it in the sense of that very part of the spider of the astrolabe. It is, however, absolutely improbable that a new translator, translating another text from the Arabic around 170 years later, would commit the same mistake again and write down again the same wrong form *vacadib*. Equally it cannot be expected that this second translator would borrow the first translator's wrong form and use it in his translation, instead of using a form found by himself. Apart from this – it is a fact that the younger translators, of the 12th century, used Arabic forms much lesser than the "archaic" translators of the first period. Altogether, on the grounds of these reflexions, we cannot accept the theory that a certain treatise on the universal astrolabe, found in three manuscripts, said to be by Ptolemy and to have been translated from Arabic into Latin by Robert of Chester in A. D. 1147 in London, really was thus translated. Moreover, judging from the Arabic terminology used in it, we must assume that this most interesting treatise is a pure western fabrication, or compilation[7].

Let me conclude with a general remark: Like the star names, also the technical terms of Arabic origin are of great importance for the historical study of the astrolabe. It is, however, a methodical duty, not just to pick out single words and to explain them at best

[7] Cf. Kunitzsch (as in note 3 above), p. 489–491 ('RC').

knowledge from some possible Arabic word, in an isolated way. They have, moreover, to be traced back, step by step, from one text to its next predecessor and so on, until the file is complete and the last origin reached. Only this method guarantees the correct interpretation of the term itself, and its correct understanding in every new context, in the process of its historical development.

IX

Al-Khwārizmī as a Source for the *Sententie astrolabii*

APPARENTLY THE oldest extant medieval Western text on the astrolabe, and one of the most influential, is the treatise known as *Sententie astrolabii*, compiled in Northeastern Spain in the late 10th century. The text is found in a number of manuscripts dating to the 11th century and one (MS. Ripoll 225) sometimes attributed to the 10th.[1] The *Sententie* have been edited, from six manuscripts (MS. Ripoll; three 11th c. MSS.; two 12th c. MSS.), by J. Millás Vallicrosa in 1931.[2]

The text is divided into three sections: (i) an introduction (*Inc. Quicumque uult scire certas horas noctium et dierum. . .* , lines 1–22); (ii) a description of the instrument and its parts (*Inc. In primis almucantarat, id sunt circuli. . .* , lines 23–110); and (iii) a description of the "use" (title: *Incipiunt capitula orologii regis ptolomei*; index of chapters, *inc. Quomodo scias altitudinem solis*; Ch. 1, *inc. Iste titulus est quomodo debes in primis laborare. . .* , lines 111–387).

From the large number of Arabic technical terms used, it is obvious that the *Sententie* are either a direct translation from the Arabic, or at least were composed according to Arabic sources. Millás understood the *Sententie* to be a direct translation from the Arabic, made probably by Lupitus of Barcelona in the last few decades of the 10th century in Northeastern Spain.[3]

In the course of studies of different astrolabe treatises conducted during the last few years, I found that part of Section iii of the *Sententie*, on the use of the instrument, is a literal translation from the Arabic treatise on the use of the astrolabe by Muḥammad ibn Mūsā al-Khwārizmī (first half 9th century).[4]

Al-Khwārizmī's treatise, so far, is known from a single manuscript, *viz.* Berlin, Ahlwardt 5793 (= Landberg 56), fol. 81^{v}–91^{v}.[5] It is still unedited.

In some previous publications, I have briefly mentioned the newly found relationship between the "use" of the *Sententie* and al-Khwārizmī's treatise,[6] and now publish the identical sections of the two texts in full. This presentation of the oldest medieval Latin translated text in the field of astronomy together with its Arabic original is a small token of gratitude to Edward S. Kennedy for making the study of Arabic-Islamic mathematics and astronomy his main life's work.

In the following, the two texts will be edited in two columns, side by side. First, under A, are given the sections for which the Latin is the literal translation of al-Khwārizmī's Arabic text; second, under B, will follow the sections in which the Latin is apparently based on the Arabic but does not follow it literally. The Latin is taken from Millás' edition with the respective line numbers, but since this edition is not always fully reliable, occasional better readings will be added from MS Bern, Burgerbibliothek 196 (11th century), fol. 3^{v}ff. The Arabic text is quoted from the Berlin manuscript (see above).

A. *SECTIONS TRANSLATED LITERALLY FROM AL-KHWĀRIZMĪ*

Sententie	*al-Khwārizmī*
[127] Dum [Bern: Cum] queris scire altitudinem solis turna *dahar* [= Arab. *ẓahr*] id est costam astrolapsum [Bern: astrolapsus] [128] contra te, et pendeat de manu tua dextera et humerum tuum sinistrum [129] pone contra solem, et ipsi xcta *alhotoi* [Bern: *alhutut*, = Arab. *al-khuṭūṭ*[7]] id est linee uel uersus stent contra [130] oculum [Millás and Bern: oculos[8]] solis et ipsam *hahidada* [= Arab. *al-ʿiḍāda*[9]] tantum exalta et demerge quousque radius [131] solis transeat amba foramina, et considera ubi stat *almeri* [= Arab. *al-murī*[10]] id est sumitas [132] accuta *alhidade* [= *al-ʿiḍāda*] in ipsis xc ordinibus qui sunt in *daar* [= *ẓahr*] ipsius astrolapsus. [133] Et quod ibi statim inuenies ipsa est altitudo solis in ipsa hora.	Fa-idhā aradta akhdh al-irtifāʿ fa-qlib al-asṭurlāb bi-ẓahrihā thumma ʿalliqhā bi-yamīnika wa-ḥādhi [MS: ḥādhī] al-shams bi-mankibika al-aysar wa-jʿal al-khuṭūṭ al-tisʿīn allatī [MS: alladhī] fī ẓahr al-asṭurlāb naḥwa ʿayn al-shams thumma lā tazālu tarfaʿu al-ʿiḍāda ḥattā tarā al-shams qad dakhalat min thaqbayhā jamīʿan thumma nẓur ayna waqaʿa al-murī alladhī fī l-ʿiḍāda wa-huwa ṭarafuhā al-muḥaddad min al-ajzāʾ [MS: ajzāʾ] al-tisʿīn allatī fī ẓahrihā fa-mā kāna fa-huwa irtifāʿ al-shams fī tilka al-sāʿa (fol. 81^{v},-2 / 82^{r},6).
[150] et ipsum gradum simul cum casa sua [Bern: signo suo] pone super ipsum *almvcantarat* [= Arab. *al-muqanṭarāt*[11]] [151] id [Bern: in] quo inuenies numerum de altitudine solis uel sibi similem quem [152] in [Bern *om.*] antea in aprehensione altitudinis solis reperisti, in dorso astrolapsus. Si [153] est ante meridiem, in parte *almaserec* [= Arab. *al-mashriq*[12]] pone, et si post meridiem in parte [154] *almagrib* [= Arab. *al-maghrib*[13]].	Thumma ḍaʿ darajat al-shams min burjihā ʿalā mithli dhālika l-irtifāʿ min al-muqanṭarāt min nāḥiyat al-mashriq in kāna qiyāsuka qabla niṣf al-nahār wa-in kāna qiyāsuka baʿda niṣf al-nahār fa-min nāḥiyat al-maghrib (fol. 82^{r},9-12).

[161] pendeat de dextera tua, et contemplare per ipsa foramina [162] *aizdade* [= *al-ʿiḍāda*] ipsa [Bern: ipsam] stellam cuius altitudinem queris scire.

Fa-ʿalliq al-aṣṭurlāb bi-yamīnika wa-staqbil bi-thaqbay al-ʿiḍāda al-kawkab alladhī turīdu al-qiyās bihī (fol. 83^{r},-3/ult.).

[164] et cum [165] uideas stellam per ambo foramina intuere ubi stat *almery* [Bern: *al-meri*, = Arab. *al-murī*] *ahildade* [= Arab. *al-ʿiḍāda*; Bern *omis.*] in ipso [166] numero nonagenario [Bern: xc], et quod [Bern: quot] sunt in ipso loco uel quotus numerus est ipse [Bern: quotus est numerus ipsa est] [167] altitudo ipsius stelle.

ḥattā tarā al-kawkab bi-iḥdā ʿaynayka min al-thaqbayn jamīʿan thumma nẓur ilā murī al-ajzāʾ mūrī al-ʿiḍāda [*sic* MS., leg. ilā murī al-ʿiḍāda] ayna waqaʿa min al-khuṭūṭ fa-dhālika irtifāʿ al-kawkab alladhī qistahū (fol. 83^{r}, ult. – 83^{v},3).

[173] Postea uero turna ipsum astrolapsum et capud ipsius stelle [174] cuius altitudinem aprehendisti, pone super ipsum *almucanitarat*, in quo [175] inuenies similem numerum . . .

Thumma qlib al-aṣṭurlāb wa-ḍaʿ raʾs al-kawkab al-muḥaddad ʿalā mithli dhālika l-irtifāʿ (fol. 83^{v},3–4).

[179] de hinc considera ubi sit kasa [Bern: signum] solis, et super quam [180] horam posita sit.

Thumma nẓur darajat al-shams ayna waqaʿat min al-sāʿāt (fol.83^{v},7–8).

[226] Cum uolueris scire arcum diei pones ordinem uel gradum cum casa [Bern: signo] solis [227] in primo *almucantarat* de parte *almaserech* [*cf.* note 12], et postea considerabis ubi [228] uel in quo numero stat *almeri* [= Arab. *al-murī*, here different from that cited above; *cf.* Kunitzsch, 1982, 538, no. 32a] in ordinibus ccclx, et gradum solis circumuolue [229] de parte *almaserech*, usque ad partem *almagrib* [*cf.* note 12], quousque perueniat [230] ad primum *almucantarat*, de parte *almagrib*, et scias ubi uel in quo numero [231] est *almeri*, et quantum se de primo loco suo promouebat, hoc est quantum [232] est de loco unde se mouebat usque ad ipsum locum ubi tunc stat et [233] quantum est, ipse est arcus diei.

Idhā aradta an taʿrifa qaws al-nahār fī kull yawm fa-ḍaʿ darajat al-shams ʿalā muqanṭarat al-mashriq thumma nẓur murī [MS: mūrī] al-ajzāʾ ayna waqaʿa fa-taʿallam ʿalayhi thumma adir darajat al-shams ḥattā taḍaʿahā ʿalā muqanṭarat al-maghrib thumma nẓur murī [MS: mūrī] al-ajzāʾ ayna waqaʿa thumma ʿuddi mā bayna mawḍiʿ murī [MS: mūrī] al-ajzāʾ al-awwal wa- . . . [the verb that follows here is illegible because of a blur in the MS; at the end the two letters *-ny* are visible] min al-ʿadad alladhī ʿalā l-ḥujra fa-mā kāna fa-huwa qaws al-nahār (fol.84^{r},5–10).

[236] Quando desideras scire arcum noctis, sic facies cum *nadair* [= Arab. *naẓīr*[14]] gradus solis.

Fa-in aradta qaws al-layl bi-l-aṣṭurlāb fa-ʿmal bi-naẓīr darajat al-shams (fol. 84^{r},12–13).

[251] Quando queris scire rectas horas [horas: *om.* Millás] diei partire arcum diei per xv [Bern: xxx] ordines, quot [252] inuenies ipse sunt hore recte diei.

Wa-in aradta sā'āt al-nahār fa-qsim qaws al-nahār 'alā 15 fa-mā kāna fa-huwa sā'āt dhālika l-yawm (fol. 84^{r},13–15).

[254] Et si queris minuere [Bern: minue] horas rectas [255] diei quote sunt, de xxiiii, quot remanent, ipse sunt hore noctis recte.

Fa-nquṣ sā'āt al-nahār min 24 fa-mā baqiya fa-huwa sā'āt al-layl (fol. 84^{r}, 15–16).

[307][15] Quando queris scire stellas que sunt in astrolapsu [Millás: astrolapsum] in qua casa [Bern: quo signo] sint [308] uel in quo ordine eiusdem case [Bern: signi], mitte caput ipsius case [Bern: stelle] in lineam medii [309] celi, et uide que casa [Bern: quod signum] sit super ipsam lineam medii celi, uel quis ordo in ipsa [301] casa [Bern: ipso signo] et in ipso ordine est ipsa stella.

Idhā aradta an ta'rifa mawāḍi' al-kawākib al-thābita fī al-asṭurlāb fa-ḍa' ra's al-kawkab al-muḥaddad 'alā khaṭṭ wasaṭ al-samā' thumma nẓur ayy burj wa-ayy daraja wāfaqa khaṭṭ wasaṭ al-samā' fa-huwa mawḍi' al-kawkab min al-ṭūl (fol. 85^{r},10–13).

B. *SECTIONS CLOSELY RELATED TO AL-KHWĀRIZMĪ'S TEXT*

Sententie / *al-Khwārizmī*

[124] quomodo debes in primis laborare per astrolapsum, et cur [125] tibi necessarius est.

Awwalu mā yaḥtāju ilayhi al-nāẓir fī 'amal al-asṭurlāb (fol. 81^{v},4–5).

[137] Postea si quesieris [Bern: queris] scire quota sit ora [Bern: hora] diei, certissime scias et teneas firmiter [138] in animo tuo locum et numerum, in quo aprehendisti altitudinem solis [139] ut diximus, et considera casam [Bern: signum] solis in qua [Bern: quo] stet, et in quo loco uel gradu eiusdem [140] case [Bern: signi] maneat.

Idhā aradta an ta'rifa al-ṭāli' wa-mā maḍā min al-nahār min sā'a wa-kasr sā'a fa-khudh al-irtifā' 'alā mā waṣaftu laka wa-mithl dhālika mā qad 'arafta mawḍi' al-shams fī burjihā wa-darajatihā (fol. 82^{r},6–9).

[174] pone super ipsum *almucanitarat*, in quo [175] inuenies similem numerum quilibet sit in numero illi quem in aprehensione [176] altitudinis stelle inuenisti in dorso astrolapsus, ita ut si ipsam stellam apprehendisti [177] in parte *almaserec* pones caput ipsius super *almucantarat*, in [178] parte *almaserec*. Et si in parte *almagrib*, eam aprehendisti, pones caput [179] illius in parte *almagrib*.

. . . min al-mashriq in kāna l-kawkab lam yazal 'an wasaṭ al-samā' fa-'alā mithli dhālika l-irtifā' min nāḥiyat al-maghrib thumma nẓur muqanṭarat al-mashriq ayy burj qaṭa'at wa-ayy daraja fa-huwa al-ṭāli' wa-mā qaṭa'a khaṭṭ wasaṭ al-samā' fa-huwa darajat wasaṭ al-samā' (fol. 83^{v},4–7).

[180] Ea considerata scias quia ipsa est hora quam queris. . . . [182] et hoc diligenter obserua ut per *nadair*, id est [183] oppositam kasa [Bern: oppositum signum] solis, semper apprehendas horas diei, et per kasam [Bern: signum] solis, [184] semper horas noctis.	Fa-huwa mā maḍā min al-layl min sāʿa (fol. 83^v,8–9) wa-ʿmal bi-kusūr al-sāʿāt bi-darajat al-shams bi-l-layl ka-mā taʿmalu bi-naẓīr darajat al-shams bi-l-nahār (fol. 83^v,9–10).
[273] Quando queris turnare [Bern: mutare] horas rectas [Bern: naturales] ad horas tortas [Bern: artificales], multiplica tempus [274] id est xv quod est ordo unius hore recte, per horas rectas tantas quantas uis [275] tornare [Bern: mutare] tortas [Bern: artificales].	Idhā aradta an taʿrifa al-sāʿāt al-muʿwajja min al-sāʿāt al-mustawiya fa-khudh al-sāʿāt al-mustawiya fa-ḍribhā fī 15 (fol. 87^v,7–9).
[292] Quando queris tornare [Bern: mutare] horas tortas [Bern: artificales] ad horas rectas [Bern: naturales] . . . [293] pone *nadair* [294] solis, et uide ubi stat *almeri*, et pone ibi signum, postea circum uolue ipsum [295] *nadair* solis ab ultima linea usque ad primum *almucantarat* prime hore et [296] uide ubi stat *almeri*, et pone ibi signum et ipsos ordines quos ambulabat [297] *almeri*, partire per ordinem rectarum [Bern: naturalium] horarum, id est per xv [Bern: per x et u], et uidebis quot [298] inde colligis horas rectas [Bern: naturales].	Wa-in aradta an tuḥawwila al-sāʿāt al-mustawiya ilā l-muʿwajja aw al-muʿwajja ilā l-mustawiya fa-ḍaʿ naẓīr darajat al-shams ʿalā ayy sāʿa shiʾta thumma nẓur murī [MS: mūrī] al-ajzāʾ ayna waqaʿa fa-ḥfaẓhu thumma adir darajat naẓīr darajat al-shams ḥattā taḍaʿahā ʿalā muqanṭarat al-maghrib thumma nẓur ayna waqaʿa murī al-ajzāʾ fa-taʾkhudhu ʿadad mā baynahumā min al-ḥujra fa-mā kāna fa-huwa mā dāra min al-falak mundhu ṭalaʿat al-shams ilā tilka l-sāʿa fa-qsimhā ʿalā 15 fa-mā kāna fa-sāʿāt mustawiya (fol. 87^v,1–7).
[300] pone *nadair* [*sic leg.*] solis [301] super primum *almucantarat*, et uide ubi stat *almeri*, et ibi pone signum. . . . [303] et [*sic leg.*] uide ubi stat *nadair* [304] solis, et quot hore sunt a prima hora usque ad ipsam horam in qua stat *nadair* [305] solis, tot hore sunt torte [torte: Bern *om.*].	Thumma ḍaʿ naẓīr darajat al-shams ʿalā muqanṭarat al-maghrib thumma nẓur murī [MS: mūrī] al-ajzāʾ ayna waqaʿa fa-ḥfaẓhu (fol. 87^v,9–11) . . . thumma nẓur naẓīr darajat al-shams ʿalā kam waqaʿat min al-sāʿāt fa-hiya al-sāʿāt al-muʿwajja (fol. 87^v,13–14).

As noted above, the text of the *Sententie* comprises three sections: (i) an introduction, (ii) a description of the astrolabe and its parts, and (iii) the "use" of the astrolabe. Of these, the introduction shows no distinct signs of being a translation. As it seems, it is styled in the own words of the (Latin) author, or compiler, of the whole treatise. Sections ii and iii together represent, though in a modified manner, what is usually contained in the Arabic treatises on the use of the astrolabe. These normally open with an introductory chapter on the different parts of the instrument and their names, which is followed by the more or less numerous chapters on the various tasks that can be solved by means of the astrolabe, mostly starting with a chapter on measuring the altitude of the sun. In the *Sententie*, the standard introductory chapter of the Arabic treatises on the parts of the instrument and their names is replaced by the lengthy section ii (lines 23–110).

This section is of a curious character. On the one hand, it shows signs of Arabic influence: it contains 25 different Arabic technical terms of the astrolabe, many of which are continuously repeated throughout the section, as well as the Arabic names of four zodiacal signs.[16] In some instances, even syntactical elements of the Arabic are incorporated in the Latin phrases.[17] On the other hand, however, the text does not convey the impression of being a translation. Also, the sequence in describing the singular items of the astrolabe in the Arabic treatises is strikingly different from the *Sententie*; the Arabic treatises never begin with the *almucantarat*, nor do they apply to these such significance as the *Sententie*.[18]

My impression, therefore, is that section ii of the *Sententie* is not a direct translation from the Arabic, but that it was written in Latin by someone who had an astrolabe in front of him (and perhaps also Arabic text material on the astrolabe), and who had at his side a person with sufficient knowledge of Arabic to help him in identifying all the Arabic names of the individual parts of the instrument and their descriptions, and in putting them down in Latin in a reasonable way.

Different again is the case of section iii, on the use of the instrument. Here, as I have just shown, we really have a translation from the Arabic, *viz.* from a treatise on the use of the astrolabe by al-Khwārizmī. But of a total of 276 lines (*i.e.* lines 111–387), only roughly a seventh could be proved to be a literal translation from al-Khwārizmī's text as we have it in the Berlin manuscript. If we include the sections also that show a close relationship to al-Khwārizmī's text, we arrive at a maximum of roughly a quarter of the text lines of section iii of the *Sententie* that have an Arabic counterpart in the treatise of al-Khwārizmī. It is evident that the Latin author, or compiler, has added a great many explanatory phrases. This can be observed in the portions that are literal translations from the Arabic (*cf.* the lengthy quotations above, where often parallel terms are offered as translations for certain Arabic words or terms, usually introduced by *uel*, "or"). Thus on the whole, section iii, though basically being a translation from the Arabic (of al-Khwārizmī), seems to have been largely inflated by the Latin author who added to the basic stock more or less long portions of his own additions and explanations.

Unfortunately we cannot exactly control the relations between *Sententie*, section iii, and al-Khwārizmī in all places. The main reason is that al-Khwārizmī's text as preserved in the Berlin manuscript is obviously not in a good state of transmission. Not only can we not be certain where it really ends (*cf.* the remarks in note 5, above), but also the succession of chapters sometimes differs from that of the *Sententie* (see the chapters on the "hore recte" and "hore torte" [*Sententie* lines 273–305] which, in the Arabic, come five pages after the "places of the fixed stars" [*Sententie* lines 307–10]). On the other hand, in a long paragraph of *Sententie* without an Arabic counterpart (lines 256–71), there is, unexpectedly, a phrase of pure Arabic construction: (horas) . . . quia non rectas damus sed indirectas, *que*[19] *dicunt eas ezemenie,*[20] id est temporales (266–67); the words *que dicunt eas ezemenie* literally repeat an Arabic relative clause of the type *allatī yaqūlūna lahā zamāniya*, "which they call 'temporal' [*scil.* hours]"; in genuine Latin this would have read "quas dicunt ezemenie".

It would, therefore, be of great value to find more manuscripts of al-Khwārizmī's treatise both in order to obtain a reliable picture of this text in itself, and to establish with greater completeness and exactness the relations between it and the *Sententie*. Both texts deserve more intensive studies since each is the oldest extant sample of an astrolabe treatise, al-Khwārizmī in the Arabic-Islamic tradition,[21] and the *Sententie* in the medieval Latin Western tradition.

In the West, the *Sententie* were of remarkable far-reaching influence. Together with them, and certainly often depending on them, there originated a number of anonymous treatises on the astrolabe.[22] In addition to these, also some early astrolabe treatises and other related writings that are associated with specific authors (Gerbert, Fulbert of Chartres, Herman the Lame, Walcher of Malvern) are derived from the *Sententie*, either directly or through one of the authors mentioned who, in turn, depended immediately on the *Sententie.*[23] Also in the later development of the Western literature on the astrolabe one can observe direct or indirect influence of the *Sententie*, even so late as such authors as J. Stoeffler (1513).[24] A special item is, further, a list of 27 astrolabe stars appended to some of the oldest treatises related to the *Sententie* (though not contained in these themselves). It was copied for centuries in the Western astrolabe literature, [25] and its stars, in the same nomenclature, were used on several medieval astrolabes.[26]

Also, the Arabic terminology of the astrolabe was first introduced into the West through the *Sententie* (especially Section ii), and it continued to be cited until the latest Latin writings in the field. In practical use, however, most of these Arabic terms were replaced, over the centuries, by Latin terms.[27]

NOTES

1. *Cf.* the recent discussion and citations in Kunitzsch, 1982, 477 with note 3. In the same year that Millás' study and edition were published, there appeared, independently, the dissertation listed in the bibliography as van de Vyver.
2. Millás, 1931, 275–88, in 387 numbered lines. (Our subsequent quotations will be according to these line numbers.)
3. *Cf.* Millás, 1931, 177ff. and 275, and 1960, 79ff. (ch. v). On Lupitus *cf.* also Lattin, and Lindgren (179f.).
4. On al-Khwārizmī, *cf.* the articles by G.J. Toomer in *DSB*, vii: 358–65, and by J. Vernet in *EI*$_2$, iv: 1070f.; see also Sezgin, vi: 140ff.
5. There is a study and German translation of the treatise in Frank. It is not quite evident at which point the treatise really ends. Probably, the chapters from fol. 88^r, bottom (= Frank, 13, no. 32) until 91^v, middle (= Frank, 16, no. 40) do not belong to the same text. The last three chapters in Frank (16f., nos. 41–43) follow, in the Arabic MS, several pages later and certainly are not part of al-Khwārizmī's treatise. It appears noteworthy that in one of the doubtful chapters (fol. 89^v,7 = Frank, 14, no. 36,b) it is said: "In another copy I found: . . . ," which seems to point to the fact that the copyist (but not the *Verfasser* [author], as Frank has it) has collected the sections ascribed to al-Khwārizmī, and the related material, from several source manuscripts all of which not complete and reliable.
6. See Kunitzsch, 1981, 55, note 59, and Kunitzsch, 1982, 478, note 6. The information in Kunitzsch, 1981 seems to have been overlooked in Viladrich and Martí, since these two authors write (98) that the source of MS Ripoll 225 is unknown and nothing can be reported on that point.
7. On this term, and a criticism raised against the *Sententie* by E. Poulle because of its use, *cf.* Kunitzsch, 1982, 525, no. 15. For *dahar* see *ibid.*, 562, no. 60.
8. *Cf.* again Kunitzsch, 1982, 525, no. 15, also with regard to Poulle's proposed emendation that failed.
9. For this term, *cf.* Kunitzsch, 1982, 527, no. 19.
10. For this term, *cf.* Kunitzsch, 1982, 539, no. 32b.
11. *ibid.*, 535, no. 31.
12. *ibid.*, 532, no. 27.
13. *ibid.*, 531, no. 24.
14. *ibid.*, 542, no. 36.
15. 306 was omitted by Millás in his numbering of the lines.
16. These are *halmei* [Bern: *alhamel*] = *al-ḥamal*, aries (lines 79 and 83); *aldre* [Bern: *alhadre*] = *al-ʿadhrāʾ*, virgo (l. 79); *alzubule*, in the Latin genitive feminine = *al-sunbula*, again virgo (l. 83); *almizen* = *al-mīzān*, libra, and *alhaut* = *al-ḥūt*, pisces (both in l. 84).
17. Line 55–56: (linea equalitatis) . . . et ipsa est que currens *almaserec rikile almagrib* [better in MS. Bern: *almaric ile almagrib*, Arab. *al-mashriq ilā al-maghrib*, "(from) the east to the west"], id est de oriente usque ad occidentem; 76–77: in quo habentur . . . nomina casarum [Bern: signorum] scripta *vacadib* [= Arab. *wa-qaḍīb*, "and a rod"] id est uirgula que . . . ; 84–85: Et maior circulus incohatur *avvilalmizeni iligair alhaut* [better in MS. Bern: *a(vv)il almizen ile ahir alhaut*, Arab. *awwal al-mīzān ilā ākhir al-ḥūt*, "(from) the beginning of libra until the end of pisces"], id est de capite libre usque ad postremitatem piscis.
18. . . . in quibus precipue ferme tota consistit ars qui quasi fundamentum sunt. Super quod tota construitur scientia (209–210).
19. MS. Bern (fol. 5va,8): quia. In this case, MS Bern's reading is inferior to Millás' printed text. The scribe obviously has confounded the word which, in this place, correctly should be "que",

with "quia" twice occurring in the near neighborhood, just before and behind this relative clause, in lines 266 and 268.
20. For this term, *cf.* Kunitzsch 1982, 561, no. 58.
21. It apparently antedates the treatise by al-Farghānī (on the construction of the astrolabe) which was written near the year 225 Yazdajird = A.D. 856/7 (*cf.* the Berlin MSS. of al-Farghānī, Ahlwardt 5791 [= Landberg 58], fol. 14^v and 20^r, and Ahlwardt 5790 [= Landberg 56], fol. 19^{v}f. and 27^r, in the calculation of the places of the fixed stars on the astrolabe).
22. *I.e.*, all those subsequent anonymous treatises that were edited by Millás, 1931, 288ff.; *cf.* also Kunitzsch, 1982, 478–481.
23. See Kunitzsch, 1982, 481–3.
24. *Cf. ibid.*, 467 and 508.
25. *Cf.* Kunitzsch, 1966, "Type iii" (23ff.), and continuations in types viii-xi *ibid.*
26. *Cf. ibid.*, 25 (astrolabe Gunther, no. 166).
27. See Kunitzsch, 1982, *passim*. Occasionally, these Arabic terms and their Latin equivalents were collected in real glossaries, *cf. ibid.*, 469; see also Kunitzsch, 1977.

BIBLIOGRAPHY

1970-1980 *DSB Dictionary of scientific biography*, I-XVI, New York: Scribner's Sons.
1960-present *EI₂ Encyclopaedia of Islam*, new edition, Iff., Leiden: E.J. Brill.

Frank, J.
1922 Die Verwendung des Astrolabs nach al Chwârizmî. *Abhandlungen zur Geschichte der Naturwissenschaften und der Medizin*, Heft iii.

Kunitzsch, P.
1966 *Typen von Sternverzeichnissen in astronomischen Handschriften des zehnten bis vierzehnten Jahrhunderts*. Wiesbaden: O. Harrassowitz.
1977 *Mittelalterliche astronomisch-astrologische Glossare mit arabischen Fachausdrücken*. München: Bayerische Akademie der Wissenschaften, Philos.-Hist. Klasse, Sitzungsberichte, Heft 5.
1981 On the authenticity of the treatise on the composition and use of the astrolabe ascribed to Messahalla. *Arch. Int. d'Hist. Sci.* 31: 42–62.
1982 *Glossar der arabischen Fachausdrücke in der mittelalterlichen europäischen Astrolabliteratur.* Göttingen: Nachrichten der Akademie der Wissenschaften in Göttingen, Philolog.-Hist. Klasse, Nr. 11: 455–571.

Lattin, H.P.
1932 Lupitus Barchinonensis. *Speculum* 7:58–64.

Lindgren, U.
1971 Die Spanische Mark zwischen Orient und Occident. . . , *Spanische Forschungen der Görres-Gesellschaft*, (Münster) I. Reihe, Bd. 26: 151–200.

Millás Vallicrosa, J.M.
1931 *Assaig d'història de les idees físiques i matemàtiques a la Catalunya medieval*, vol. i. Barcelona: Estudis Universitaris Catalans. Sèrie Monogràfica i.
1960 *Nuevos estudios sobre historia de la ciencia española*, Barcelona: Consejo Superior de Investigaciones Científicas. Instituto "Luis Veves" de Filosofía. Sección de Historia de la Filosofía Española y Asociación para la Historia de la Ciencia Española. Estudios, No. 7.

Sezgin, F.
1967-present *Geschichte des arabischen Schrifttums*. iff., Leiden: E.J. Brill.

Viladrich, M. and
R. Martí
1981 En torno a los tratados Hispánicos sobre construcción de astrolabio hasta el siglo XIII. In *Textos y estudios sobre astronomía española en el siglo XIII*, edited by J. Vernet. Barcelona: Facultad de Filosofía y Letras, Universidad Autónoma de Barcelona, 79–99.

van de Vyver, A.
1931 Les premières traductions latines (X[e] - XIe s.) de traités arabes sur l'astrolabe. *1er Congrès International de Géographie Historique*, t.ii, Bruxelles, 266–90.

X

ON THE AUTHENTICITY OF THE TREATISE ON THE COMPOSITION AND USE OF THE ASTROLABE ASCRIBED TO MESSAHALLA

The Latin treatise on the construction[1] and use[2] of the astrolabe circulating under the name of Messahalla (= Māšā'allāh)[3] can be regarded as the most popular and widespread treatise on the subject in the middle ages. The number of manuscripts in which this text has survived amounts to almost 200[4]. It was repeatedly printed[5], and its fame was still more increased since Chaucer took Messahalla's work as a basis for the elaboration of his own treatise on the astrolabe, in English. In spite of its historical importance, no critical edition of the text has previously been made. The best available, though not fully reliable, is at present that made by R. T. Gunther in 1929[6]. Gunther's version was used by Millás in 1931, and we shall take it as a basis of our present study[7].

Though Māšā'allāh was credited with a treatise on the astrolabe by the well-known Arabic bibliographer Ibn an-Nadīm, in his *Fihrist* of 987 A.D.[8], no specimen of this has been found in the original Arabic until today. On the other hand, quite a number of astrological and some astronomical writings of Māšā'allāh, both in Arabic, and in Latin and Byzantine translations, have survived and are extant in many libraries, in manuscripts and printed books. Of the *Astrolabe*, however,

[1] Incipit: *Scito quod astrolabium est nomen grecum cuius interpretatio est acceptio stellarum* ...

[2] Incipit: *Nomina instrumentorum sunt hec. Primum est armilla suspensoria* ...

[3] A famous astrologer and astronomer of the early period at the caliphs' court in Baghdad, d. around 815 A.D. Cf. D. Pingree, art. 'Māshā'allāh', in: *Dictionary of Scientific Biography*, vol. 9, New York, 1974, pp. 159–162; M. Ullmann, *Die Natur- und Geheimwissenschaften im Islam*, Leiden, 1972, pp. 303–306; Sezgin, vol. vi, pp. 127–129; Carmody, p. 24f. (work no. 1).

[4] Information by Mr. M. Masi of Chicago, in a letter of July 1, 1973. Cf. also Poulle [2], p. 36.

[5] Included in G. Reisch, *Margarita Philosophica*, frequently printed from 1503 to 1583. Cf. E. Zinner, *Geschichte und Bibliographie der astronomischen Literatur in Deutschland zur Zeit der Renaissance*, Leipzig, 1941 (2nd ed., Stuttgart, 1964), nos. 829, 849, 850, 909, 910, 967, 1036, 1079, 1608, and 3105. Messahalla's treatise appears in the prints from 1512 onwards.

[6] Included in his book *Chaucer and Messahalla on the Astrolabe*, Oxford, 1929 (Early Science in Oxford, v), pp. 195–231. This edition was based, for part i (the 'composition' of the astrolabe), on MS. Cambridge, University Library, Ii.3.3, compared with MSS. Ashmole 1796 and Ashmole 1522, and one of Reisch's old prints. For part ii (the 'use' of the astrolabe), Gunther closely followed the earlier edition by W. W. Skeat (see below, footnote 28), introducing, however, several mistakes.

[7] A critical edition is now prepared by M. Masi (its completion is announced for 1981, cf. *Speculum* 54 [1979], p. 212b, under the name Messahala). However, in two letters of July 1 and September 12, 1973, Mr. Masi has informed me that his edition is not planned to be a comprehensive edition of Messahalla, but that his aim is to edit certain MSS. of Messahalla's text that might have served as a source for Geoffrey Chaucer when composing his own treatise on the astrolabe.

[8] Ed. G. Flügel, vol. i, Leipzig, 1871, p. 273, ult.: *Kitāb ṣanʿat al-asṭurlābāt wa-l-ʿamal bihā* ("Book on the fabrication and use of the astrolabes").

only that vulgate Latin text seems to exist which is mostly – but not always – ascribed to *Messahalla* (and variants), "Messahalla" being a mediaeval Latin transcription for the name of the famous Arabic astrologer and astronomer Māšā'allāh.

During the last few decades, however, suspicions have arisen as to whether the ascription to Messahalla of this astrolabe treatise is correct.

Millás has pointed to certain anachronisms in part i ("construction")[9] which testify to the existence of later additions and modifications to the text allegedly composed by Māšā'allāh. And I myself brought forward a number of cases where the author is called Maslama – i.e., Maslama al-Mağrīṭī – instead of Māšā'allāh[10] which, together with some other reflexions[11], caused me to suggest that the author of the treatise might rather be Maslama, or some of his disciples, but not Māšā'allāh.

After re-reading this and a great number of related texts, both in Arabic and in Latin, and the research papers involved, the material, and our understanding, is now sufficient to reconsider the problem of the authenticity of the authorship of this treatise.

In accordance with the historical situation, the two parts of the treatise will be investigated in this paper in two separate sections, part i for the 'composition', and part ii for the 'use' of the astrolabe.

Part i

De compositione astrolabii

The edition of this part of the text is in Gunther[12], pp. 195–216, Incipit: *Scito quod astrolabium est nomen grecum cuius interpretatio est acceptio stellarum ...*

In Gunther's edition, *De compositione* consists of a *Prohemium* and 22 chapters, and contains two star tables inserted between chapters 18 and 19.

The authenticity of this complex, and its ascription, as a whole, to Messahalla, have previously been doubted by several scholars[13]. Our analysis of the text follows the same line and will demonstrate, presenting more detailed arguments, that these doubts were well founded.

De compositione as edited by Gunther obviously falls into four or five distinct sections: the *Prohemium*; section ch. 1–6; section ch. 7–16; section ch. 17–22; and, perhaps as a separate item, the two star tables.

For the *Prohemium*, I am able to present, here, an Arabic original. Thus, this section now can be identified as a true translation from the Arabic. The Arabic

[9] See Millás [1], p. 154; [3], p. 106f.; [4], p. 72f.

[10] See Kunitzsch [1], p. 51 footnote 1; cf. also Poulle [1], p. 85 (*Mascelamah*).

[11] E.g., the use of several different types of star catalogues which originally belonged to various other traditions. Cf. Kunitzsch [1],]types v, vii, viii, ix, x, and xi; see *ibid.*, pp. 7 and 51.

[12] Cf. above, footnote 6.

[13] Millás [1], pp. 153, 192; *id.*, [3], pp. 106ff.; H. Michel, *Traité de l'astrolabe*, 2nd ed., Paris, 1976, p. 8f.; *id.*, in *Homenaie a Millás Vallicrosa*, Barcelona, 1956, ii, p. 45f.; Poulle [1], p. 85f.

text is found in the introduction to a treatise on the composition of the astrolabe, in Istanbul, MS. Aya Sofya 2671 (dated 621 H = 1244 A.D.), foll. 133v–150r[14]. The treatise is anonymous; but on good reasons[15] D. King argues that it could well be by the famous Spanish-Muslim astronomer Azarquiel (az-Zarqāllu, d. 1100 A.D.). The identities are as follows: *Scito quod astrolabium* ... (Gunther, p. 195, l. 2) corresponds to *I'lam anna ism al-asṭurlāb* ... (Aya Sofya 2671, fol. 133v, l. 6). The two texts continue, verbatim identical, until *in eodem climate* (l. 9) = *fī ḏālika l-iqlīm* (l. 10). The following Latin words *et inicium eorum est ex circulo emisperii eiusdem climatis* (l. 10) are not in the Arabic; perhaps they are an explanatory addition by the Latin translator, or by the compiler of *De compositione*. *cui lineate sunt* (l. 10–11) again corresponds to *allaḏī ḫuṭṭat lahū* in the Arabic (l. 10). Now follows, in the Arabic, a lengthy portion on different parts of the astrolabe which is absent from the Latin. This seems quite reasonable, because this portion interrupts the discussion of the seven climates which clearly is the main topic of this introduction. Obviously, the Arabic text here is somehow disturbed. The intercalation, in the Arabic, ends on fol. 134r, l. 14. Here, the discussion of the seven climates is resumed, and from here the two texts, the Arabic and the Latin, continue identically until the end of the Latin *Prohemium*: *Invenimus quoque antiquos* ... (l. 11) = *wa-qad waǧadnā l-qudamā'* (134r, l. 14), until ... *et brevior 8 horarum* (l. 24) = *wa-aqṣaruhū 8 sā'āt* (134v, l. 6). Thus, it can be definitely said that the *Prohemium* of *De compositione* is a true and literal translation from an Arabic treatise on the composition of the astrolabe, probably by Azarquiel. It remains, however, uncertain, when, where, and by whom the Latin translation was made, and whether the *Prohemium* existed, in Latin, as an independent unit or whether it was part of a complete treatise. It was selected by the compiler of *De compositione* and put at the beginning of his compilation, as a suitable introduction. Millás was of the opinion that *De compositione* bears close relations to some of the oldest Latin treatises on the astrolabe, from the end of the 10th century A.D., *viz.* h'' (*Philosophi qui sua sapientia* ..., edited in Millás [1], pp. 293ff.), and h' (*Philosophi quorum sagaci studio* ..., edited *ib.*, pp. 296ff.), and that all these texts derive, though with several modifications, from an authentic treatise by Messahalla[16]. If this observation is correct, however, its interpretation should be contrary to that given by Millás. It would mean that the late compiler of *De compositione* had made use, among others, of those oldest Latin texts in compiling his work. But Messahalla

[14] Cf. Krause, p. 525, no. 15. My attention was drawn to this treatise by two articles on matters relating to the astrolabe by D. King, one of which in the meantime has appeared: "On the Early History of the Universal Astrolabe in Islamic Astronomy and the Origin of the Term *Shakkāzīya* in Medieval Scientific Arabic", in *Journal for the History of Arabic Science* 3 (Aleppo, 1979), 244–257; see especially p. 253 footnote 1. I am greatly indebted to my friend and colleague, Mme B. Atsiz of the University of Munich, for supplying me with photographs of the relevant passages in the Istanbul MS.

[15] The treatise contains, on fol. 148v, a table of 37 astrolabe stars computed, according to a notice on the preceding folio, for 459 H = 1066/67 A.D. which is the date of Azarquiel's star table. A separate article on this and a second star table contained in this MS., by P. Kunitzsch, has been published lately in *Journal for the History of Astronomy* 11 (1980), 192–201.

[16] Cf. Millás [1], pp. 153, 192; [3], pp. 106ff.

is not involved altogether, not in those oldest Latin texts[17], and not in *De compositione*.

The *Prohemium* is the only portion from the Arabic text in MS. Aya Sofya 2671 which, in Latin translation, has been incorporated in the compilation *De compositione*. None of the following technical chapters in *De compositione* shows any further relation to that Arabic text.

The next section in *De compositione* comprises chapters 1–6. At the end of ch. 6 we read (ed. Gunther, p. 202, l. 10–11): *Huc usque intermisimus vero de diversis tractatibus, sed nunc redeamus ad librum*. This clearly shows two things: that these chapters did not originally form part of the present book, and that they were compiled "from several treatises". These chapters treat at great length the fabrication of the *mater* (ch. 1), the back of the astrolabe (ch. 2), the shadow quadrant (ch. 3), the *allidada* (ch. 4), the marking of the hours on the *allidada* (ch. 5), and the central axis of the instrument (ch. 6). I do not know one Arabic, or Latin, treatise on the astrolabe in which the fabrication of these parts of the instrument is described at such a length. Also, to my knowledge, Arabic astrolabes do not usually carry hour marks on the *allidada*. So, altogether, it would seem that this section of *De compositione* may well be compiled "from several treatises", but the single topics have been considerably inflated. The use of Arabic terms[18], and the repeated phrase *si Deus voluerit*[19], need not mean that the sources for the compilation were direct translations from the Arabic, for these terms, and this phrase, were well-known at the time of the translations, so that everybody in the Latin west concerned with the matter of astronomy who had well read the "classical" text-books knew them and could freely use them himself. Perhaps further studies in the Latin texts on the astrolabe from the 10th century to the 13th will reveal some relationship between *De compositione* and the partial sources of this section.

Next follows what would be, according to the notice at the end of ch. 6, the "book" itself, i.e., the text proper on the composition of the astrolabe. This section comprises ten chapters (i.e., chapters 7–16) and describes the construction of the essential lines, curves, etc., to be put on the astrolabe and some of its parts. I have not found immediate similarities between this and the hitherto known

[17] Millás [1], p. 198 footnote 7 points out that the etymology of the Latin word *astrolabium* in h''' (*Cum hominum habitaciones ...*, ed. in Millás [1], pp. 308ff.; the etymology is on p. 309, l. 22–23) and in *Astrologie speculationis exercitium ...* [cf. footnote 35 below; the etymology is in Millás' ed., Millás [2], p. 316, l. 21–25] are equal (igual explicació). It must, however, be stressed that the etymology in these two texts is pure western speculation based on the similarity of the element *-labium* (in *astrolabium*) to the Latin word *labium*, 'lip', which must be taken as a hint that these texts were free Latin compositions, or compilations, and not direct translations from the Arabic. The etymology given in *De compositione*, however, is a correct one and was first given in our Arabic treatise in MS. Aya Sofya 2671, and then translated into Latin. It is, by the way, the only instance in the Arabic literature on the astrolabe where this word is correctly explained; cf. a forthcoming paper by D. King, "The Origin of the Astrolabe According to the Medieval Arabic Sources" (here, item no. 9; cf. also items no. 13 and 27). I am greatly indebted to Prof. D. King for most kindly sending me the typescript of his article before publication.

[18] They are: *almucanthant, cenit capitum, almuri, allidada, architob, alpheram.*

[19] Translated from the well-known Arabic: *in šā'a llāh.*

Arabic astrolabe treatises. It remains, therefore, uncertain whether this section, the "book" proper of *De compositione*, is a direct Latin version of some Arabic text, or whether it again is a compilation, or a modified compilation, from one, or several, existing Latin texts. The Arabic terms found in this section[20] are part of the general knowledge of astronomers of that period and do not testify to the nature of the text as a translation. Further, in ch. 7 when discussing the value for the inclination of the ecliptic, *Albategni*[21] and *Almeon*[22] are quoted. As Millás has pointed out on several occasions[23], this proves that the section, at least, is a later addition, since Māšā'allāh died earlier (ca. 810–815 A.D.). From the technical point of view, H. Michel calls the Latin text of *De compositione* "fortement remaniée". He continues: "L'auteur du manuscrit latin y a introduit des méthodes et des tracés dont Mashallah aurait été absolument incapable"[24]. Altogether, we can, therefore, say that also for the central section, the "book" proper, nothing speaks in favour of Māšā'allāh's authorship. Moreover, at the end of this section (after ch. 16, Gunther, p. 211), some manuscripts have the notice: *Finit opus astrolabii secundum Marcellania* (thus, in Gunther's MS. Ashmole 1796). Here, we come to another argument against Māšā'allāh's authorship, because the name given in the MS., is unequivocally a transcription of the Arabic Maslama, not of Māšā'allāh. Other MSS., quoted by other authors, similarly give this name in the Maslama-form[25]. But also the ascription of this central section to Maslama is erroneous because, as will be discussed below (see especially footnote 39), no such treatise on the astrolabe was ever written by Maslama. Thus, no specified author, either Arabic or Latin, can be named for this section of *De compositione*.

There remains the last section, chapters 17–22. After the concluding phrase at the end of ch. 16, and from their contents, these chapters can be clearly judged to

[20] *alancabut, azimuth, alliladath* (better *alilacat*, for Arabic *al-ʿilāqa*, the suspension apparatus), *cenit capitis, nadayr*.

[21] This is al-Battānī, the famous Arabic astronomer (d. 929 A.D.), whose *Astronomy* was translated into Latin by Plato of Tivoli (Barcelona, ca. 1134–1145 A.D.).

[22] To this name, Gunther (p. 202 footnote 10) gives the puzzling explanation "The son of Albumazar". (Albumazar would be Abū Maʿšar, famous Arabic astrologer, d. 886 A.D.). In reality, *Almeon* is a Latin transcription of the name of the Abbasid caliph al-Ma'mūn (reigned 813–833 A.D.), upon whose order a number of astronomical observations were made around 829 A.D. in order to check and improve some of the parameters transmitted from Ptolemy and other classical authorities. The results of these observations were put down in an astronomical work called *az-Zīğ al-mumtaḥan* (known as *Tabule probate* in the Latin west). This observation was also recorded by al-Farġānī (middle of the 9th century A.D.), in his 30 chapters on astronomy which were twice translated into Latin, by Johannes Hispalensis (1135 A.D.) and again by Gerard of Cremona. In Johannes Hispalensis' translation, al-Ma'mūn's name was given as *Almeon*; hence, this name, in this form, spread in the Latin west (Gerard wrote *Maimon*, here, and in his translation of the *Almagest*). Cf. R. Campani, in *Rivista degli Studi Orientali* 3 (1910), 229f.

[23] See footnote 9, above.

[24] Michel, *Traité* (as in footnote 13, above), p. 8.

[25] Cf. Kunitzsch [1], p. 51 footnote 1: MS. Erfurt, Amplon. 4° 355 (14th c.), fol. 56ᵛ: *macellama* (also Amplon. 4° 363, 94: *Macelama*, according to Schum). See also Poulle [1], p. 85: *Mascelamah*. Gunther, p. 133f., was led far astray in his explanation of *Marcellania* in MS. Ashmole 1796.

be another set of additions to the whole complex[26]. They contain a theoretical dissertation on the projection of the sphere on a plane (ch. 17 and 18)[26a]. Further chapters deal with items that had already been described in the preceding section (ch. 19 and 20, on the *azimuth*; cf. earlier, ch. 15; and ch. 22, on the fixed stars, cf. earlier ch. 10). As E. Poulle has shown, some of the Paris MSS. of *De compositione* end with ch. 16 and do not contain this last section, which might prove that *De compositione* as edited by Gunther was not compiled by one author, at one time, but moreover that in different stages additions were made to the complex, at successive times.

As for the star tables, those in MS. Cambridge, Univ. Lib. Ii. 3. 3 are inserted between chapters 18 and 19. The first one (not printed in Gunther's edition, but given in facsimile, fol. 70v; printed by Skeat [as in footnote 28], pp. xxxvii–xxxix; = type viii, in Kunitzsch [1]) contains 49 stars and is compiled from two basic tables, *viz.* the 27-star table circulated in the Latin west from the oldest Latin astrolabe treatises in Northeastern Spain, late 10th century A.D. (= Kunitzsch [1], type iii); and a star table composed by John of London in Paris, 1246 A.D. (= Kunitzsch [1], type vi). The second table (in Gunther, facsimile, fol. 71r; printed by Skeat, *loc. cit.*, pp. xl f.; = Kunitzsch [1], type vii) is an abridged version made directly from John of London's table of 1246 A.D. Subsequently, Poulle gave the opinion that the additional section of *De compositione* (ch. 17–22) together with the two star tables which so closely are related to John of London's table, probably were all added to the complex by John of London himself[27]. I should not subscribe to this judgment without hesitation, because it is not only these two star tables of types viii and vii that are found inserted in *De compositione*. In many manuscripts we find other star tables, as, e.g., type v (in Kunitzsch [1]; this is a derivation from an Arabic table by Azarquiel), type xi (which, to my opinion, is a modification made from type viii, though its date, 1223, or 1233 A.D., then must be wrong), and types ix and x (both derivations from type viii). I doubt, therefore, whether the addition of the star tables is owed to one person, and whether this person was the same who added the whole section ch. 17–22 to the complex. The star tables, sowith, can be regarded either as part of the additional section ch. 17–22, or as an item separately added to the whole complex.

De compositione thus proves to be a compilation of different text elements, made by different persons at different times. Its latest parts cannot have been added before the second half of the 13th century A.D. (see the star tables, most of which are derived from John of London's table which was composed in Paris, 1246 A.D.). Some of the oldest known manuscripts date from this period: Cambridge, University Library, Ii. 3. 3, the basic MS. for Skeat's and Gunther's editions, is dated 1276 A.D.; further MSS. of the late 13th century are Darmstadt, Landes-

[26] *Sic* also Poulle [1], p. 86.

[26a] These two chapters have been edited from three Paris MSS., and compared to Gunther's edition of 1929, by R. B. Thomson in Appendix 1 to his book *Jordanus de Nemore and the Mathematics of Astrolabes: De plana spera*, Toronto, 1978, pp. 179–185.

[27] Poulle [1], p. 86.

bibliothek, 2661, foll. 146–168, and Munich, Staatsbibliothek, Clm 353, foll. 41–54. Names and dates for the composition, or compilation, of the different sections of the compilation are not known. The ascription of the basic part of the whole work to Messahalla is wrong, perhaps a misinterpretation of the name of Maslama who is named, at the end of the section in ch. 16, as a source. But neither can this be correct since Maslama is not known to have composed an astrolabe treatise of this sort.

Part ii

De operatione uel utilitate astrolabii

The edition of this part of the text is in Gunther[28], pp. 217–231, Incipit: *Nomina instrumentorum sunt hec. Primum est armilla suspensoria ...*

Further, the following texts are involved in the present investigation:

a) Arabic

* Abū l-Qāsim Aḥmad ibn ʿAbdallāh Ibn aṣ-Ṣaffār (Spain, d. 1035 A.D., a disciple of Maslama)[29]: *Kitāb al-ʿamal bi-l-asṭurlāb* ("Book on the use of the astrolabe"), edited by J. M. Millás, in: *Revista del Instituto Egipcio de Estudios Islamicos* (Madrid), 3 (1955), 35–49 (introduction, in Spanish), and Arabic text pp. 47–76 (Arabic pagination), from MS. Escurial 964. Modern translation into Catalan, in Millás [1], pp. 29–48. (The chapters are not numbered, my own numbering amounts to 42 chapters, *bāb*)[30]. My quotations of Ibn aṣ-Ṣaffār will generally refer to this text, unless otherwise stated.
* idem, text by the same title, but in different wording and distribution of chapters, apparently a recension and amplification of the printed version, in MS. London, British Library Add. 9600, fol. 262ᵛ–280ᵛ. (The chapters are not numbered in the MS., my own numbering amounts to 31 chapters, *bāb*)[31].

[28] See above, footnote 6. This part was previously edited by W. W. Skeat, *A Treatise on The Astrolabe ... by Geoffrey Chaucer, A.D. 1391*, (1st ed., London, 1872; several reprints, the latest is: Oxford University Press, 1968), pp. 88–105, from MS. Cambridge, University Library, Ii.3.3 (dated 1276 A.D.), with some comparisons added from MS. Cambridge, Univ. Lib., Hh.6.8. Gunther, in his print of 1929, closely followed Skeat's edition, adding some comparisons from one of Reisch's old prints, and introducing several mistakes.

[29] Cf. Sezgin, vol. vi, p. 250f.

[30] As Steinschneider, p. 582, reports, some of the versions mention the years 413–416 and 431 H = 1023–1026 and 1031 A.D.

[31] Further MSS. equally show modifications and amplifications against the version edited by Millás; cf. J. Samsó, in: *Al-Andalus* 31 (1966), p. 385 no. 1, and p. 387 no. 2; id., in *Acta del II Coloquio Hispano-Tunecino de Estudios Históricos*, Madrid, 1973, p. 175 footnote 24. Apart from these, there exists another recension of Ibn aṣ-Ṣaffār's treatise, by ʿAbdallāh ibn Muḥammad ibn Saʿd at-Tuǧībī, of unknown date; cf. MS. Berlin, Ahlwardt 5805, fol. 68ᵛ–81ʳ; see further Samsó, l.c., *Al-Andalus* p. 388 no. 3, and *Coloquio* p. 175 footnote 25. See also G. ʿAwwād, "al-Asṭurlāb wa-mā ullifa fīhi ...", in: *Sumer* 13 (1957), 154–178, nos. 29, 83, and 139 (in Arabic).

* Muḥammad ibn ʿAlī ibn Yaḥyā Ibn an-Naṭṭāḥ (apparently in Spain, after Ibn aṣ-Ṣaffār whom he quotes[32]): Book (*kitāb*) on the astrolabe, without specified title, in MS. London, British Library Add. 9602, fol. 1v–24v. This text also bears recognizable relationship to Ibn aṣ-Ṣaffār's treatise. (The chapters are not numbered in the MS., my own numbering amounts to 51 chapters, *bāb*).

b) Latin

* Johannes Hispalensis, translation of a treatise on the use of the planispheric astrolabe. In the MS. acribed to Maslama al-Maǧrīṭī, in a later hand of s. xvii[33]. It is, however, *totally identical* with Ibn aṣ-Ṣaffār's Arabic treatise as edited by Millás (above). Edited by Millás [2], pp. 261–284, from MSS. Madrid, Biblioteca Nacional 10053, and Toledo, Biblioteca Catedral 98–27[34]. 40 chapters. Incipit: *Primum horum armilla per quam suspenditur astrolabium* ..., Explicit: ... *de annus* [sic] *Xristi et ex annus* [sic] *arabum, si Deus uoluerit*. The translation is very literal[35].

* Plato Tiburtinus, translation of Ibn aṣ-Ṣaffār's Arabic treatise. Not edited. I used MS. Oxford, Digby 51, fol. 28ra–35rb. Chapters not numbered, my own numbering arrives to 41 chapters. Incipit of prologue: *Translacio platonis tiburtini de opere astrolabii. Suo serenissimo amico johanni dauid in quatuor mateseos disciplinis peritissimo* ..., Incipit of text, ch. 1: *Quoniam interpretationes nominum instrumentorum astrolabij fidelissima descreptione* [sic] ..., Explicit: ... *et hec est figura, sicut in dorso astrolapsus iuxta centrum ponitur. Finit liber operis astrolapsus ab abucazin filio asafar editus, et a platone tiburtino translatus in ciuitate barchinona.* The

[32] fol. 15r, ult.; 15v,9; 16r,4. Sezgin, vol. vi, p. 249 (work no. 1 of Ibn as-Samḥ) erroneously refers these quotations to Ibn as-Samḥ, while, correctly, they belong to Ibn aṣ-Ṣaffār.

[33] In other MSS., as e.g. Oxford, Merton 259, the attribution to Maslama is made in the same hand that copied the whole text.

[34] Further MSS. are registered by Carmody, p. 142f., work no. 1a. See further Poulle [1], p. 102; Thorndike-Kibre, 1118; Sezgin, vol. vi, p. 227. MS. Oxford, Merton 259 (s. xiii), was examined for me *in situ* by Dr. R. Lorch who kindly informs me that the MS., fol. 48–53, contains this text as edited by Millás.

[35] Millás [2], pp. 313–321, has further edited a Latin treatise on the construction of the astrolabe, which he claims to be by Māšā'allāh, and translated by Johannes Hispalensis. The Incipit is: *Nos iuxta rectam imaginagionem* [sic] ...; this text is registered by Carmody, p. 118f., work no. 3, as a work by Ṯābit ibn Qurra (see ed. in F. J. Carmody, *The Astronomical Works of Thabit b. Qurra*, Berkeley and Los Angeles, 1960, pp. 141–143, and notices, there, pp. 82f., 118f., 201). The Explicit as given by Carmody occurs in Millás' edition, on p. 316 (until line 8). Then follow five lines with definitions of *latitudo*, *altitudo*, and *declinatio*, and in line 14 Millás' text continues: *Astrologie speculationis exercitium habere volentibus* ... This, according to Carmody, p. 169, work no. 2b, is an astrolabe treatise by Johannes Hispalensis himself; its Explicit in Carmody is identical with that in Millás [2], p. 321. It seems, therefore, that the text edited by Millás is of disputed attribution and, most probably, does not belong to Māšā'allāh, but we shall not here enter on more details with regard to this text. As for Johannes Hispalensis, his best known astrolabe treatise is the one opening: *Cum uolueris facere astrolabium accipe auricalcum optimum* ..., cf. Carmody, p. 169, work no. 2a, partially edited by Millás [2], pp. 322–327. *Astrologie speculationis* ... is also extant in MS. Oxford, Merton 259, fol. 89v, as Dr. R. Lorch informs me, who has kindly examined the MS. for me *in situ*.

text is identical with Ibn aṣ-Ṣaffār's as edited by Millás (above). The translation is not very literal, but rather diffuse and circumstantial[36].

Our investigation will proceed in two successive steps. First we shall have to disentangle the internal relations among the aforementioned texts, then we shall proceed to establish the relationship of the treatise allegedly by Messahalla to those texts.

One of the most famous and most influential astronomers of Islamic Spain was Maslama al-Maǧrīṭī (d. around 1007 A.D.). His name spread in the Latin world because two major astronomical works were translated into Latin from recensions that he had made: the astronomical tables of al-Ḫwārizmī[37], and Ptolemy's *Planisphaerium*[38]. He is, however, not reported by his biographers to have composed a treatise on the astrolabe[39].

Now, we are confronted with a treatise on the use of the astrolabe of which two identical versions exist, in Arabic and in Latin. The arabic version is clearly attributed to Ibn aṣ-Ṣaffār, while the Latin one (translated by Johannes Hispalensis) is ascribed to his teacher Maslama, and if only in later hands. Millás, who first accepted the two versions as two independent, though identical texts[40], later arrived at the conclusion that there is no evidence for an astrolabe treatise by

[36] Further MSS. are registered by Carmody, p. 143, work no. 2a. See further Thorndike-Kibre, 312, 1282, 1543; Sezgin, vol. vi, p. 250f. Millás [1], p. 28, quotes MS. Digby 51[5] which is our MS.

[37] For editions of the Latin version, see below, footnote 69.

[38] Translated by Hermannus Secundus, in 1143 A.D. Edited by J. L. Heiberg, *Ptolemaei Opera*, vol. ii, Leipzig, 1907: pp. xiif., clxxx–clxxxix, 225–259 (German translation, by J. Drecker, in *Isis* 9, no. 30 [1926], pp. 255–278). It should be stressed that Maslama was probably not the Arabic translator of the *Planisphaerium*. Most probably this was translated into Arabic in the eastern regions, like all the other Greek works on astronomy and astrology. To this text, Maslama added his comments and an additional chapter (see the next following footnote). An Arabic version of the *Planisphaerium* is still extant in Istanbul, MS. Aya Sofya 2671 (dated 621 H = 1224 A.D.), fol. 76^v–97^r; cf. Krause, p. 443; Sezgin, vol. v, p. 170.

[39] There was discovered, some years ago, a longer portion of his writings in MS. Paris, B.N. ar. 4821. This comprises: i) his notes to the *Planisphaerium* (fol. 69^v–75^r. They were also translated into Latin, and are contained in some older prints; cf. Heiberg, l.c., pp. clxxx–clxxxiii); ii) a special chapter added by him to the *Planisphaerium* (fol. 76^r–79^r = ed. Vernet-Catalá, pp. 22–26, 3; also translated into Latin, see Heiberg, l.c., p. clxxxiii); and iii) some theoretical chapters on the astrolabe (fol. 79^r–81^v, including a star table for the epoch end of 367 H = 978 A.D.). The section fol. 76^r–81^v was edited and translated by J. Vernet and M. A. Catalá, in *Al-Andalus* 30 (1965), pp. 15–45 (with additional remarks to the star catalogue, by M. A. Catalá, on pp. 46f.). The star catalogue was also edited by M. Destombes [1], p. 24, and by Kunitzsch [1], type i (both in the Arabic and in a Latin version). The portion on the astrolabe can hardly be regarded a 'treatise'; it only consists of a few chapters, and Maslama himself entitles this portion as "chapters indispensable for everyone who wishes to construct an astrolabe" (*abwāb lā yastaǧnī man yarūmu ʿamal al-asṭurlāb ʿanhā*). Samsó, l.c. (*Coloquio*, as in footnote 31 above), p. 174, calls this piece of text "un des premiers traités d'astrolabe rédigés dans l'Espagne Musulmane". Destombes [1], p. 23f., finds a relationship between Maslama's additions to the *Planisphaerium* (in this MS.) and the Latin text *b''* (Incipit: *Philosophi qui sua sapientia* ...; edited by Millás [1], pp. 293–295), which, however, needs a closer study of these texts. "S'il en était bien ainsi, nous aurions une confirmation de ce que la source du traité de Lupitus est bien l'ouvrage de Maslama."

[40] Millás [1], pp. 27, 29.

Maslama[41] and that both these texts correctly belong to Ibn aṣ-Ṣaffār[42]. This opinion has been accepted, since[43], and I fully subscribe to it.

We thus have a treatise on the use of the astrolabe by Maslama's disciple, Ibn aṣ-Ṣaffār, of which several Arabic recensions and two different Latin translations exist.

Next, we know that Millás was the first to compare Pseudo-Messahalla's *De operatione* to Ibn aṣ-Ṣaffār's treatise. He found a large conformity between the two and, thus, concluded that Maslama (and his school) learnt about the astrolabe through Māšā'allāh's (Arabic) treatise and used this as a base for their own works in this field. That means that, to Millás, Ibn aṣ-Ṣaffār's treatise was a close imitation of Māšā'allāh's Arabic text. In order to lay a safe ground for his hypothesis, Millás made a thorough comparison, chapter by chapter, between Pseudo-Messahalla's *De operatione* (in Gunther's edition of 1929) and Ibn aṣ-Ṣaffār's treatise[44], and found the two fully congruent[45].

For the convenience of readers, I here insert a synopsis of the congruent chapters of this treatise in the six sources I have examined[46].

Millás concluded that the knowledge of the astrolabe had come to Muslim Spain through Māšā'allāh's treatise. According to him, the works of Maslama and his disciples, and their Latin translations, ultimately descended from Māšā'allāh's treatise. Thus, he judged our Pseudo-Messahalla's *De operatione* (as edited by Gunther) to be a genuine work of Māšā'allāh. This theory was maintained by Millás from 1931 until his recent works[47], and was followed, afterwards, by his successors[48]. Apart from the works of Maslama's school, Millás extended the relationship to Māšā'allāh also to those oldest extant Latin texts on the astrolabe that were translated and composed near the end of the 10th century in Northeastern Spain, and which he has edited in his fundamental book *Assaig* (= Millás [1])[49].

[41] But see footnote 39 above.

[42] In his edition of Ibn aṣ-Ṣaffār (see above), in the introduction, pp. 38–42; reprinted in Millás [4], pp. 65–69.

[43] Vernet-Catalá, l.c. (as in footnote 39 above), p. 18.

[44] Along with his Catalan translation of Ibn aṣ-Ṣaffār's treatise, in Millás [1], pp. 29–48.

[45] I started the same comparison (but with a different aim), and found three instances where Millás' comparison was not complete: Millás [1], p. 33f. (coneixement exacte de les fraccions = ch. 10 in the Arabic), corresponds to ch. 7 of Pseudo-Messahalla (Millás, p. 34 footnote 1, says: "Amplifica quelcom el cap. vi"); Millás [1], p. 39f. (hores diurnes per les línies ... = ch. 27 in the Arabic), not compared by Millás, is identical with ch. 11 of Pseudo-Messahalla; and Millás [1], p. 47 (coneixement dels Orients = ch. 41 in the Arabic), not compared by Millás, is identical with ch. 40 of Pseudo-Messahalla.

[46] I have also examined the astrolabe treatise of another of Maslama's disciples, Ibn as-Samḥ (in Granada, d. 1035 A.D.; cf. Sezgin, vol. vi, p. 249), in MS. London, British Library Add. 9602, fol. 25v–55v, in 130 chapters (the text is incomplete at the end, part of ch. 129 and all of ch. 130 are missing). But I found Ibn as-Samḥ's treatise an independent work of a different composition, not following the structure of the Ibn aṣ-Ṣaffār text family. Thus, we need not in our present study consider Ibn as-Samḥ's treatise in detail.

[47] Millás [3], pp. 106–110; introduction to the edition of Ibn aṣ-Ṣaffār (see above), p. 45f. = [4], pp. 72–75.

[48] Vernet-Catalá, l.c. (as in footnote 39 above), p. 19; Vernet, p. 107.

[49] Concentrated survey in Millás [1], pp. 210f. = Millás [4], pp. 94–96.

Ibn aṣ-Ṣaffār Arabic, ed. Millás	Johannes Hispalensis, Latin, ed. Millás [2]	Ps.-Messahalla, *De operatione*, ed. Gunther	Plato Tiburtinus, Latin, MS. Digby 51	Ibn aṣ-Ṣaffār Arabic, MS. London Add. 9600	Ibn an-Naṭṭāḥ Arabic, MS. London Add. 9602
1	1	Introduction	1	(1)	1
2	2	} 1	3	4	2
3	3		4	5	3
4	4	2	5	3	4
5	5	} 3	6	6	10
6	6		7	7	6
7	7	4a/b	8	8	~ 8.9
8	8	5	10	9	12
9	9	6	11		
10	10	7	12		~ 7
11	11	6 contin.	13	11	
12	12	8	14		17.18
13	13	9	15	12	19.20
14	14	10	16	13	23
15	15	13	17		30c
16	16	14	18	22	32
17	17	15	19		~31
18	19	30	20	28	38
19	18	—	21		25
20	20	16	22	16	
21	21, part	17.18, parts	23	25	34
22	—	19 →	→	} 26 →	} 35
23	—	—			36
24	22	20	24	14	27.28a
25	23	21.22	25	23	28b.29.30b
26	24	24	26	31	
27	25	11	27	29	
28	26	26	28	24	
29	27	27, part	29	27	
30	28	42	30	19	47
31	29	43	31	20	48
32	30	44	32	21	
33	31	—	33		
34	32	—	34		
35	33	45.46	35	*	51
36	34	—	36	*	
37	35	28	37	17	42
38	36	29a	38	18	
39	37	32.36	39		44.45
40	38	37	9	*	
41	39	40	40		
[42]	—	—	—		
—	40	—	41		

* the headings of these chapters are mentioned, but the text is omitted as not pertaining to the calculation of times (fol. 280^{v}).

I fully agree with Millás' findings as to the close relationship, or nearly identity, between Pseudo-Messahalla's *De operatione* and the texts of the Ibn aṣ-Ṣaffār family, but my conclusions and explanations are quite contrary to his.

Examining these texts closely, we have to compare Ibn aṣ-Ṣaffār (Arabic), his Latin translation by Johannes Hispalensis, and the Pseudo-Messahalla. We then find that Johannes Hispalensis' translation verbatim reflects Ibn aṣ-Ṣaffār's Arabic version as edited by Millás. In comparing, further, the two Latin texts, we find that the Pseudo-Messahalla is closely related to Johannes Hispalensis' translation and *proves to be a derivation thereof*. This turns Millás' theory into its contrary, because this means that Pseudo-Messahalla's *De operatione* – instead of stemming from the same model that was used by Ibn aṣ-Ṣaffār and Johannes Hispalensis, as Millás assumed – was rather a western recast of Johannes Hispalensis' translation. The source of the whole tradition is hereby restored to Ibn aṣ-Ṣaffār, that is: Maslama's school, and Pseudo-Messahalla's *De operatione* looses any connection with the name of Messahalla, or Māšā'allāh.

A close comparison of Pseudo-Messahalla and Johannes Hispalensis' translation presently has to content itself with Gunther's text, leaving aside the possibility that among the almost 200 more MSS. there could be some which, in certain details, come still closer to Johannes Hispalensis' wording.

It is not possible to present here a total juxtaposition of all the identical phrases in which Pseudo-Messahalla follows Johannes Hispalensis' translation, but I shall give some examples.

The introductory chapter is largely identical in both texts. This refers even to the Arabic terms that were retained in Johannes Hispalensis' translation, and which are used equally in the Pseudo-Messahalla. I here list these terms in the sequence of the text[50]: *alorwat* JH / *alhabor* Ps-M., *almucantharath* / *almucantherath*, *sunt alraz* and *zenitum capitum* / *cenit capitum*, *azimut* / *azimuth*, *allancabuth* / *alhanthabuth*, *almuri* / *almuri*, *almahuar* / *almenath*[51], (*equus*)[52] / *alphaeraz*, *allidadah* / (*regula*)[53].

In the following chapters (with the exception of 4, 9, 11, 15, 19, 43, and 46), Pseudo-Messahalla not only generally follows the sequence of chapters and matters in Johannes Hispalensis, but he even repeats full phrases or sequences of words, especially in chapters 1, 2, 3, 8, 10, 13, 18, 28, 30, 36, 42, and 45. But also in the other chapters, Johannes Hispalensis' intentions are well represented.

50 The first term in Pseudo-Messahalla, *alhabucia* (= Arabic *al-ḥalqa*), is absent from Johannes Hispalensis' edited text, but might exist in other MSS. of this translation. The Latin equivalent of it is nearly identical in both texts: *armilla per quam suspenditur astrolabium* (JH) / *armilla suspensoria* (Ps-M.).

51 Arabic *al-miḥwar*. Another spelling of the word, in the Pseudo-Messahalla texts, is *almehan*; see the explanations by P. Kunitzsch, *Mittelalterliche astronomisch-astrologische Glossare mit arabischen Fachausdrücken*, München, 1977 (Bayer. Akademie der Wissenschaften, Philos.-Hist. Kl., Sitzungsberichte, Jahrg. 1977, Heft 5), p. 18f. (no. 18), especially footnote 32.

52 As with the first term (see footnote 50 above), the Arabic form here is absent from the MS. edited by Millás; the Latin forms are identical in both texts.

53 Here, Pseudo-Messahalla has only retained the Latin form, *regula*, and omitted the Arabic word. But *allidada* occurs later on; see ch. 11, 12, etc.

The conclusion from these resemblances and relations can only be that the Pseudo-Messahalla is a pure western fabrication, made on the basis of Johannes Hispalensis' translation[54], but modified in several ways: simplified and concentrated in certain instances, and augmented by some additional chapters[55]. The (western) author of this fabrication, or compilation, is not known.

It is easy, however, to explain how the name of Messahalla (Māšā'allāh) could have been added to this popularized text. Since Maslama was a name well known to the Latin west, and since this name was connected, through Ptolemy's *Planisphaerium*, with the astrolabe and its theories, some time his name was related to Johannes Hispalensis' Latin translation, instead of the name of the true author, Ibn aṣ-Ṣaffār, Maslama's pupil (a reflexion of this can be seen in the addition of Maslama's name to some MSS. of Johannes Hispalensis' translation[56]). This name, Maslama, was afterwards also appended to some MSS. of the western recasting of Johannes Hispalensis' translation. There, in this much popularized text, some scribes confounded Maslama's name (in its latinized form) with the latinized forms of Māšā'allāh's name, which was much more famous in the west than Maslama's because of his many astrological texts. From then, the new text was circulated more and more under Māšā'allāh's name, as we find it, e.g., in the MSS. edited by

54 To assume, as Millás does, that *De operatione* is an independent translation from a hypothetical Māšā'allāh text, different from Johannes Hispalensis' translation from Ibn aṣ-Ṣaffār's text (which itself, according to Millás, derived from Māšā'allāh), is impossible. Because a second translator would never arrive at the same wording, the same transliterations of Arabic words, and other identities that are found between the *De operatione* and Johannes Hispalensis, not to speak of the difference in time between the two assumed derivations from the hypothetical Māšā'allāh text, Ibn aṣ-Ṣaffār (early 11th century) and the *De operatione* (of unknown date, but most probably after Johannes Hispalensis, that is late 12th, or 13th centuries), which would even mean that the two departed from different manuscripts of Māšā'allāh's hypothetical text and which would most probably have been a cause for further divergencies between the two.

55 Additional chapters in Pseudo-Messahalla are: ch. 12, 19 (but extant in the Arabic MS. Brit. Libr. Add. 9600, and with Ibn an-Naṭṭāḥ), 23, 25, 31, 33, 34, 35, 38, 39, 41, and 47. Of the "geometrical chapters" in the *De operatione*, which Poulle considers later additions, not belonging to the original text translated by Johannes Hispalensis (cf. Poulle [1], pp. 86f., and [2], p. 35), the two chapters 45 and 46 fully correspond to ch. 33 in Johannes Hispalensis' version, and, thus, to ch. 35 in Ibn aṣ-Ṣaffār's Arabic text. It is only for ch. 47 of *De operatione* that its later addition to the corpus must be assumed. It may be added here that ch. 40 in Johannes Hispalensis (equally extant with Plato Tiburtinus as ch. 41), on the beginnings of the Christian years and months, which is absent from the two Arabic texts of Ibn aṣ-Ṣaffār which we are studying here, and from Ibn an-Naṭṭāḥ, have been found by J. Samsó in one of the Tunisian MSS. of Ibn aṣ-Ṣaffār's treatise, cf. *Al-Andalus* 31 (1966), p. 388, ch. liii. Pseudo-Messahalla has not retained this chapter from Johannes Hispalensis' translation.

56 E.g., in the Madrid MS. edited by Millás (cf. above); see further the cases related by Steinschneider, pp. 582, 583 (footnote 331b). See also footnote 33 above, for MS. Merton 259. In the *Speculum astronomiae*, perhaps by Albertus Magnus, Johannes Hispalensis' text is equally ascribed to Maslama (thus all the MS. readings as cited by the editors) although the editors, following some older prints, thought it right to put in the text Messahalla's name instead; cf. the partial edition by F. Cumont, in *Catalogus Codicum Astrologorum Graecorum*, vol. v, 1, Bruxelles, 1904, p. 90, l. 15–16, and *app. crit.*; and Alberto Magno, *Speculum astronomiae*, ed. S. Caroti, M. Pereira, S. Zamponi, P. Zambelli, Pisa, 1977, p. 12, l. 71–72, with *app. crit.* (not correctly explained in Kunitzsch [1], p. 51 footnote 1).

Gunther. But Maslama's name was not totally suppressed, because in some MSS. he is still mentioned as the author[57].

Thus, contrary to Millás, we arrive at the conclusion that Messahalla (Māšā'allāh) was not discernibly involved in this whole group of traditions on the astrolabe. The astrolabe was discussed in Maslama's time (second half of the 10th century A.D.) in Muslim Spain – what sources were used remains uncertain, but there was a large amount of (eastern) Arabic text material on the subject that could have reached Maslama, like the *Planisphaerium*, al-Ḫwārizmī's tables, al-Battānī's work, and the *Almagest*[58]. These discussions of the astrolabe gave rise, on the one hand, to the studies of the astrolabe in western Noıtheastern Spain towards the end of the 10th century which led to the translations and compilations of treatises centered around the *Sententiae astrolabii*, and edited in Millás [1][59]. On the other hand, they produced Ibn aṣ-Ṣaffār's treatise, which was afterwards twice translated into Latin (by Johannes Hispalensis, and Plato Tiburtinus). And one of these two translations, *viz.* the one by Johannes Hispalensis, furthermore gave rise to the treatise *De operatione*, which was later falsely attributed to Messahalla and which we are discussing here[60].

The conclusive element which has enabled us to interpret the close parallels and identities between *De operatione* and Ibn aṣ-Ṣaffār / Johannes Hispalensis correctly, is the great number of literal reproductions of entire phrases and sequences of words in *De operatione* from Johannes Hispalensis' Latin version. This element had not been sufficiently considered by Millás, so that he arrived at an explanation contrary to the historical development. Our analysis of the text material, taking this element into account as well as other observations of the interrelations of

[57] Thus, in Gunther's ed., p. 211 (*De compositione.*) Cf. Kunitzsch [1], p. 51 footnote 1, and Steinschneider (as in footnote 56 above).

[58] Destombes considers MS. Paris, B.N. ar. 4821, which contains some of Maslama's writings (cf. footnote 39, above) together with a number of relevant treatises by eastern Arabic astronomers anterior to Maslama, a specimen of texts that had possibly reached Muslim Spain and might have served as sources for Maslama's studies. See Destombes [1], p. 19, and [2], p. 40. For Maslama's knowledge of the *Almagest*, cf. Vernet-Catalá (as in footnote 39), p. 17.

[59] For parts of *Sententiae astrolabii* I have recently found a possible Arabic source in a treatise on the use of the astrolabe ascribed, in the MS., to al-Ḫwārizmī! In detail, the following lines of *Capitula orologii regis Ptolomei* (ed. Millás [1], pp. 280–288, = lines 111–387 of *Sententiae astrolabii*) appear to be literally translated from that Arabic source: 124f., 127–135, 150–154, 161–163, 164–167, 173f., 179f., 226–234, 236f., 251f., 254f., 306–310, while lines 174–178, 182–184, 273–275, 293–298, and 300–305 look like a free rendering of the Arabic. The Arabic text of al-Ḫwārizmī (first half of the 9th century A.D., in Baghdad; cf. Sezgin, vol. vi, pp. 140ff., and the literature mentioned there) is in MS. Berlin, Ahlwardt 5793 (= Ldbg. 56), fol. 81^{v}, –3ff. (until 88^{v}, or perhaps 91^{v}). As it looks, this Arabic MS. of ca. 1500 A.D. is not complete and not in good order; thus, it might be doubted whether the attribution of the treatise on fol. 81^{v}ff. to al-Ḫwārizmī is correct. However that may be, it can be discerned that the above mentioned passages of *Capitula orologii regis Ptolomei* indisputably correspond to the respective sections of that Arabic text. It would seem that the old translator had used a better, and more complete, copy of the Arabic text than that one in the Berlin MS.

[60] Destombes [2], p. 49, calls the Pseudo-Messahalla treatise (which was Chaucer's source in composing his treatise of the astrolabe), "une adaptation de celui [*viz.* traité de l'astrolabe] de Maslama", which comes quite near to the historical truth. But he does not present the evidence for this judgment (see, however, footnote 39, above, at the end).

the texts involved, now leads to a better understanding of the filiation of these texts. The line of tradition starts with Ibn aṣ-Ṣaffār's Arabic treatise, continues to its Latin translation by Johannes Hispalensis, and finally arrives at the *De operatione*, a western recasting of Johannes Hispalensis' Latin version, falsely circulated under Messahalla's name.

*

The preceding study has shown that nothing of the popularized treatise on the composition and use of the astrolabe generally ascribed to Messahalla can be traced back to this author. Part i, *De compositione*, is a western Latin compilation, perhaps from the middle or second half of the 13th century A. D., consisting of four or five distinct sections of different origin, none of which bears a recognizable relation to Messahalla. Not even Maslama, who is quoted as source for the main part of *De compositione*, seems to be involved in the history of the treatise, as far as his works are known. Part ii, *De operatione*, has proved to be a purely western compilation and fabrication made on the basis of Johannes Hispalensis' Latin translation of the Arabic treatise on the use of the astrolabe by the Spanish-Muslim astronomer Ibn aṣ-Ṣaffār, one of Maslama's disciples. Thus, not only does each of the two parts show its internal elements of compilation, but even the two parts as a whole are of different origin. The question cannot yet be answered whether the composition of the whole was the work of one man, or whether the two parts were conceived independently and merely put together later by readers or scribes. What would be most needed for a further study of the text is a critical edition comprehending all the surviving manuscripts[60a].

APPENDIX

Rodolfus Brugensis

Some authorities have registered a third Latin translation of Māšā'allāh's (or Maslama's) treatise on the composition and use of the astrolabe, by Rodolfus Brugensis[61]. In connection with this present investigation into the authenticity of

[60a] In further studies of mediaeval astrolabe texts, I lately found a reference to the end of ch. 4 of *De compositione* (p. 192f. in Gunther's edition), in the astrolabe treatise by Pierre de Maricourt (in MS. Vienna 5311, fol. 29va, citing *Mesealma*). As Pierre's treatise is datable around 1263 A. D. (cf. Poulle [1], p. 91f.), this year of 1263 can therefore be regarded a *terminus ante quem* for the composition of Pseudo-Messahalla's *De compositione* (and, perhaps, *De operatione* also).

[61] Suter, no. 176 (for Maslama); in an addition of 1902, on p. 167, Suter withdraws this assertion on the authority of Curtze, but obviously without having examined the respective manuscipts himself. Rodolfus' text is equally declared a translation of a treatise by Maslama, by Millás [4], p. 114, and Vernet, p. 115 (but Millás [1], p. 26, and [4], p. 69, blames Suter for attributing to Rodolfus the Latin translation of our present treatise by Maslama, or rather by Ibn aṣ-Ṣaffār). See further Poulle [1], p. 84 and 101 (citing, in the second place, MS. Paris, B.N., lat. 16652, fol. 24–28), and [2], p. 34; Carmody, p. 143, work no. 3 of Maslama's ('Tract', 'description of an astronomical instrument'. Here he notes MS. Naples C 8 50, fol. 80, following Haskins; but MS. 'Cotton p. 104' is wrongly inserted among the MSS. of John of Seville's translation, work no. 1, on top of p. 143); Thorndike-Kibre, 285 (*Cum celestium* ..., Incipit of prologue. Here wrongly related to Bubnov 114–115, where the treatise *Quicumque astronomicae discere peritiam*

Messahalla's treatise, I have therefore examined the London MS. of Rodolfus Brugensis' work[62]. I have arrived at the conclusion that this is not a translation – neither from Māšā'allāh nor from any other Arabic author – but a work by Rodolfus himself.

That Rodolfus' work is not related to our Pseudo-Messahalla tradition is evident from a mere comparison of the two texts. Of the nine columns which the text comprises in the London MS., eight are dedicated to the description of the construction, and one only to the use of the astrolabe.

In the prologue Rodolfus clearly announces that he will describe the instrument as well as he can – which doubtless indicates that the following text is not a translation – and dedicates the work to Johannes David, one of the well-known scientific translators in mid-twelfth century Spain.

The prologue runs as follows (fol. 35rb):

Cum celestium sperarum diuersam positionem stellarum diuersos ortus diuersosque occasus mundo inferiori ministrare manifestum sit, huiusque uarietatis descriptio ut in plano representetur sit possibile prout ptolomeo eiusque sequaci mezlem[63] *qui dictus est Aoukakechita*[64] *uisum est, pro posse suo huius instrumenti formulam dilectissimo domino suo Johanni David Rodolfus brugensis Hermanni secundi discipulus describit.*[65]

The text itself begins: *Primum igitur huius instrumenti est postica quam quidam matrem dicunt eo quod infra limbum eius cetere tabule uti filij in utero matris contineantur...* The construction of the different parts of the astrolabe is described at great length. The author shows his thorough acquaintance with the current literature on the astrolabe, which in all cases was translated from the Arabic, or based on such translations. Thus, he uses several Arabic technical terms[66], and in some cases adds explanations of them – which would not have been required if this had been a translation[67].

disciplinae ... is edited which is counted among the dubious works of Gerbert and bears no relation to Rodolfus' text), and 1119 (*Primum igitur* ..., Incipit of Rodolfus' text).

62 It is in London, British Library, MS. Cotton. Vespasian A.ii. (12–13th c.), fol. 35rb–37va. (Suter, no. 176, and Carmody, p. 143, wrongly locate the MS. at Oxford, as 'Cotton p. 104'). Rodolfus' work is followed, in the MS. (fol. 37va–40vb), by another astrolabe treatise, Inc. *Genera astrolabiorum duo sunt* ..., probably by Abraham ibn 'Ezra, edited by J. Millás Vallicrosa, in: *Al-Andalus* 5 (1940), 1–29, from this MS., together with MS. Arundel 377.

63 *mezlem* = Maslama.

64 *Aoukakechita*, a distortion of *Almacherita* = al-Mağrīṭī. Maslama al-Mağrīṭī was of course well known to Rodolfus, because his teacher Hermannus Secundus had translated Ptolemy's *Planisphaerium* into Latin, from an Arabic version, together with Maslama's notes and additions; cf. also below, footnote 68. The reference, therefore, is only to Ptolemy's theory of projection of the sphere on a plane as transmitted in an Arabic version commented upon by Maslama. But it does not mean that Rodolfus here translates a text on the astrolabe by Maslama, or describes an astrolabe designed by Maslama.

65 Also printed by Haskins, p. 56, from the Naples MS.

66 E.g., the well known *alhaidada, (al)hancabuth, almuri, almucantarath*, etc.

67 Thus, for the zenith: *uerticalium circulorum centrum* ..., *a quibusdam uero nationibus centaroz* [= Arab. *samt ar-ru'ūs*] *dictum* (fol. 36vb). Mostly, however, he writes *zumpruz* for the zenith, from the same Arabic term; cf. fol. 37ra: ... *de uerticalibus circulis quos quidam azimuth* [Arab. = *as-sumūt*], *quidam uero cenit capitum* [= Arab. *samt ar-ru'ūs*], *alij zumpruz appellant.* It is obvious that Rodolfus has collected these different terms from different Latin translations from the Arabic; but he is wrong in equating the three terms, of which only the first one (*azimuth*)

For more detailed information on the obliquity of the ecliptic, Rodolfus refers his readers to Ptolemy's *Planisphaerium*[68]. In discussing the positions of the fixed stars on the rete of the astrolabe, he mentions, as an authority, *Alcaurismus* [= al-Ḫwārizmī][69] *ceterique philosophi*. He then gives (fol. 36va) a star table, which is edited and discussed below.

For the calculation of latitudes of places, he gives an example (fol. 36vb) for Béziers, *biteris in gotia iuxta mare mediterraneum*, for the date April 24, 1144 A.D. Here we have a date for the composition of Rodolfus' treatise. As the quotation shows, this took place not quite a year after the translation of Ptolemy's *Planisphaerium* by Rodolfus' teacher, Hermannus Secundus.

On fol. 37[rb], when discussing the seasonal and the equal hours, Rodolfus remarks that in *Arin*[70] which is assumed to be situated in the centre of the world, the sun always revolves along the equator so that days and nights there consistently comprise twelve hours throughout the year.

The 'composition' ends at the bottom of fol. 37[ra]. It is accompanied by five drawings of certain details of the instrument[71]. Then, on 37[rb]–37[va], follow some much abbreviated rules for the use of the astrolabe. The text ends, on fol. 37[va], with the words: ... *ibique nota in perissa quem in alhancabut gradum attingere poterit, in eodem procul dubio solem esse recognosces*[72].

The star catalogue likewise proves to be compiled from different sources. Although it is drawn up for the epoch of the composition of the treatise (April 24, 1144 A.D.), it is obvious that Rodolfus has not adapted the coordinates of the stars for this epoch. The catalogue comprises 24 stars, 11 in the northern, and 13

correctly refers to the vertical circles, whereas the latter two (*cenit capitum*, and *zumpruz*) designate the zenith.

68 *Sed qui huius inequalitatis rationem nosse desiderat, planisperium ptolomei ab hermanno secundo translatum legat* (fol. 36[ra], ult. –[b], 2). The same place is quoted by Poulle [1], p. 84 footnote 3, from MS. Paris, B.N., lat. 16652, fol. 25[v].

69 This seems to be a mere recital of names, because no table of fixed stars by al-Ḫwārizmī has survived, neither in his treatise on the use of the astrolabe (e.g., MS. Berlin, Ahlwardt 5793, fol. 81[v]ff.; analysed, using a translation by E. Wiedemann, by J. Frank, *Die Verwendung des Astrolabs nach al Chwârizmî*, Erlangen, 1922 [Abhandlungen zur Geschichte der Naturwissenschaften und der Medizin, Heft iii]), nor in his astronomical tables (Latin version, edited by H. Suter, *Die astronomischen Tafeln des Muḥammed ibn Mūsā al-Khwārizmī* ..., Kopenhagen, 1914; English translation and commentary by O. Neugebauer, *The Astronomical Tables of al-Khwārizmī*, Kopenhagen, 1962).

70 *Arin*, from an Arabic distortion (*'ryn*, for *'zyn* = Uzayn) of the name of the Indian town of Ujjayinī. Part of the Arabic-Islamic astronomy was inspired by Indian texts translated early at the caliphs' court in Baghdad. Hence, Ujjayinī, as a place in the center of the world, on the zero meridian and on the equator, penetrated Arabic astronomical and geographical texts, and was further transmitted to the mediaeval western scholars.

71 The Paris MS., B.N. lat. 16652, which was used by E. Poulle, and of which I obtained photographs from the Bibliothèque Nationale, is less complete than the London MS. inasmuch as the drawings and the star table are omitted.

72 Haskins, p. 56, cites the Explicit as: ...*formulam tenaci memorie commendet*, from the Naples MS. (repeated by Carmody, p. 143, work no. 3). In the London MS., this is the end of the 'composition', at the bottom of fol. 37[ra]. It is followed by the 'use', Inc. *Perfecta astrolabij fabrica nunc de utilitate eius que necessaria sunt subiungamus. Ac primum de nadiar* [sic] *solis* ... (fol. 37[rb]), which ends on 37[va] as cited above.

in the southern hemisphere. The numbers in the MS. are written in Roman numerals, zero being represented by a capital *T*. Many corruptions seem to have crept into these numbers. 17 out of the 24 stars were borrowed from the oldest extant list of astrolabe stars in western texts, which was very much copied (Kunitzsch [1], type iii). Three items (nos. 3, 4, and 14) stem from a western derivation of Azarquiel's star list, which was drawn for 1066/67 A.D. (Kunitzsch [1], type xii)[73]. It is noteworthy that Rodolfus did not realize that he had entered two stars *twice* (nos. 3 and 4, from the Azarquiel tradition, which are identical with nos. 6 and 7, from the tradition of type iii). This underlines the general impression conveyed by this star table: it was of almost no practical use. The coordinates of the stars, called *longitudo* and *latitudo* as if they were ecliptical values, are partly ecliptical values (for nos. 1–4 and 12–14), and partly equatorial values (for the derivations from type iii), and all of them were selected from older tables without adapting them to the author's own epoch (not to mention the severe corruptions in many of them).

As this table represents a type of its own which was not comprised in Kunitzsch [1], I have decided to print its full text here.

(No.)	Muniri[74]	Signa	Longitudo Gradus	Longitudo Punctum	Latitudo Gradus	Latitudo Punctum	Modern designation
	In plaga septentrionali he primum collocande sunt						
1	*Munirmenelfeca*	Scorpius	10	0	49	0	α CrB
2	*Bazalhawi*	Sagittarius	5	37	36	0	α Oph
3	*Annaceralwaka*	Capricornus	1	27	62	0	α Lyr
4	*Denebadigeia*	Aquarius	23	17	63	0	α Cyg
5	*Delfin*	Capricornus	25	10	59	3	ε Del
6	*Wela*	Capricornus	1	15	72	0	α Lyr
7	*Alrifh*	Capricornus	19	3	73	7	α Cyg
8	*Alhoc*	Taurus	10	61 (sic)	71	10	β Per
9	*Alhaioc*	Gemini	3	20	74	13	α Aur
10	*Egregez*	Cancer	26	9	72	9	μ UMa
11	*Jalrucuba*	Leo	19	1	75	20	ϑ UMa
	In parte australi he similiter collocentur						
12	*Rigil argebal*	Gemini	12	57	57	31	β Ori
13	*Zoel*	Gemini	39	0 (sic)	65	0	α Car
14	*Aschare alabur*	Cancer	1	47	39	11	α CMa
15	*Aldebaran*	Taurus	27	2	59	0	α Tau
16	*Algeuze*	Gemini	18	5	60	0	α Ori
17	*Algomeiza*	Cancer	13	7	56	25	α CMi
18	*Aldiran*	Leo	14	10	52	0	(α Hya)
19	*Calbalac*	Leo	6	1	61	6	α Leo
20	*Lidineb*	Aquarius	18	20	35	7	δ Cap
21	*Deneb kairut*	Pisces	8	0	26	0	ι Cet
22	*Algurab*	Uirgo	20	0	41	0	γ Crv
23	*Alcimech*	Libra	18	11	39	50	α Vir
24	*Calbalacrab*	Scorpius	25	0	14	55	α Sco

[73] Cf. Kunitzsch, l.c. (as in footnote 15 above), p. 196f. and 201 note 18.

[74] Correctly it should be *Nomina*, be it that the scribe could not clearly read his source, or be it that he negligently wrote down a portion of the name of the first star in the list, *Munirmenelfeca*, instead.

Notes to the star table:

1. Name = Arab. *(al-)munīr min al-fakka* ("the Brilliant one of *al-fakka*"; correctly translated as *splendens* in an example given on fol. 36rb). Among the star tables edited in Kunitzsch [1], of those older than 1144 A.D., this full name only occurs in type iA no. 14 (Maslama, 978 A.D.; in its Latin translation, type i no. 17, only the abbreviated form *alfeta* is given), and the somewhat related type ii no. 16. Coordinates: *long.* is quoted in an example on fol. 36rb, as 11° Scorpius, but in the table it is 10°. If this value is meant to be the ecliptical longitude, it shows a difference against Ptolemy's value, in the *Almagest*, of +25°30 (or, 24°30, respectively). In type iii no. 2, the *mediatio coeli* of α CrB is 16° Sco, and Johannes Hispalensis (type iv no. 18, datable 1134 A.D.) has 18° Sco. *lat.* 49° comes nearer to Ptolemy's 44°30 than to the mediaeval values of declination, like 32°4 (type ii no. 16) or 30° (type iv no. 18).
2. Name = Arab. *ra's al-ḥāwī* ("the Head of the Snake-Carrier, Ophiuchus"). Coordinates: *lat.* is identical with Ptolemy's; *long.* seems to be an ecliptical longitude, which accordingly shows a difference from the *Almagest* of +10°47.
3. Name = Arab. *an-nasr al-wāqiʿ* ("the Falling Eagle", Vega); cf. the Latin derivations of Azarquiel's star table, types v no. 11 and xiii no. 10. Coordinates: *lat.* is identical with Ptolemy's; *long.* shows a difference from Ptolemy's of +14°7, exactly as in Azarquiel's table, cf. type xii no. 11.
4. Name = Arab. *ḏanab ad-daǧāǧa* ("the Tail of the Hen"); cf. type xii no. 20. Coordinates: *long.* = Ptolemy +14°7, identical with Azarquiel, type xii no. 20; *lat.* slightly corrupted from Ptolemy's (60°).
5. This star belongs to those 17 stars that are derived from the old standard star list of type iii (no. 5). The name is identical, and the coordinates are closely related to those of type iii: *long.* 25°10 Cap corresponds to the *mediatio coeli* in type iii, and *lat.* 59°3 equally corresponds to the value 59° in type iii (which, however, is not the declination; cf. the authorities quoted in Kunitzsch [1], p. 25 footnote 9, to which now have to be added O. Neugebauer, *A History of Ancient Mathematical Astronomy*, 3 vols., Berlin/Heidelberg/New York, 1975, pp. 279, 858, 892, 1082f.; and J.D. North [ed.], *Richard of Wallingford*, Oxford, 1976, vol. 3, pp. 159–161).
6. *Wela.* The same star as no. 3 above (Vega, α Lyr), but in this case derived from type iii no. 9. Its *long.*, here, is closely related to the *mediatio* there (1° Cap), and so is the *lat.* to the *altitudo* of type iii (72°).
7. The same star as no. 4 above, but here derived from type iii no. 8 (cf. the coordinates there, *mediatio* 29° Cap [here slightly corrupted], *altitudo* 73,5°). The name, also borrowed from type iii, is transcribed from Arab. *ar-ridf*, the old Arabic name for α Cyg (cf. Kunitzsch [2], p. 98 no. 248, and [3], p. 82 no. 56, and p. 143 no. 61).
8. Derived from type iii no. 24. For the name (*Alhoc*, corrupted for the usual *Algol*), cf. Kunitzsch [3], p. 69 no. 14, and p. 115 no. 24. Coordinates: *long.* closely related to type iii (*mediatio* 10° Tau), and thus the *lat.* (there, *altitudo* 71°).
9. Derived from type iii no. 25. For the name, cf. Kunitzsch [3], p. 71 no. 20, and p. 119 no. 27. Coordinates: closely related to type iii (*mediatio* 3° Gem, *altitudo* 74°).
10. Derived from type iii no. 27. For the name, cf. *ibid.* and Kunitzsch [3], p. 75 no. 28. Coordinates: closely related to type iii (*mediatio* 26° Cnc, *altitudo* 72°).
11. Derived from type iii no. 26. For the name, cf. *ibid.* and Kunitzsch [3], p. 67 no. 9, and p. 136 no. 48. Coordinates: they closely correspond to those in type iii (*mediatio* 19° Leo, *altitudo* 75°).
12. Name: correctly written *Rigilalgebar* in an explanatory example on fol. 36rb, where *long.* is given as 3° Gem (instead of 12°57, in the table); cf. type xiii no. 2 (a western derivation from Azarquiel's table, dated 577 H = 1181/82 A.D.). Coordinates: *long.*, if taken as on fol. 36rb, shows a difference of precession from Ptolemy of +13°50 [or, assuming 3°57 Gem: +14°47, which comes close to the values of type xiii]; *lat.*: perhaps 57° is a mistake of transmission, the number being repeated from the minutes of *long.*, while the minutes, 31, represented originally the degrees which would then come close to Ptolemy's value (31°30).
13. Name = Arab. *suhayl*, Canopus; cf. Kunitzsch [2], p. 106 no. 272a. Coordinates: *long.* has a difference of precession against Ptolemy of +11°50, *lat.* (65°) could be a miswriting for 75° which is Ptolemy's value. In mediaeval star tables anterior to Rodolfus' date, this star occurs

only once, in type ii no. 45, with equatorial coordinates. Thus Rodolfus' source cannot be safely ascertained in this case. (Later occurrences are: types xv no. 7; xvi no. 19; xvii no. 16).

14. For the name, cf. type xiii no. 5. The coordinates are identical with those of type xii no. 5 (with only one superfluous minute added to *lat.*, here).
15. Derived from type iii no. 19. The coordinates are almost identical in both tables (cf. type iii no. 19: *mediatio* 27° Tau, *altitudo* 41,5°, of which the value 59° in our table could easily be corrupted in Roman numerals).
16. Another derivation from type iii, no. 20. *long.* is identical with *mediatio* of type iii; *lat.* is slightly corrupted (in type III: 55°).
17. Also derived from type iii, no. 21. *long.* is slightly distorted from type iii (11° Cnc), *lat.* is almost identical (type iii: 56°).
18. Derived from type iii no. 22. The name belongs to α Gem, but the coordinates point to α Hya (see the explanation in type iii no. 22; cf. also the notes to types iA no. 8; ii no. 15; iv no. 11; and Kunitzsch [3], p. 72 no. 24). *long.* is corrupted from type iii (6° Leo); perhaps the copyist has just mixed up two lines in his source, because in the following line, with α Leo (= no. 19), he has *long.* 6°1 Leo which properly fits with no. 18 (α Hya), while 14°10 Leo of his no. 18 would better fit with no. 19 (α Leo). *lat.* is fully identical with type iii.
19. Derived from type iii no. 23. *long.* 6°1 Leo seems to be interchanged with α Hya, no. 18 above (type iii has *mediatio* 19° Leo, which in Rodolfus' table could well be represented by the value 14°10 Leo of no. 18 above, whereas 6°1 Leo of this no. 19 better fits with no. 18 above); *lat.* is almost identical with type iii (*altitudo* 61°).
20. Derived from type iii no. 18. Name = Arab. *ḏanab al-ǧady* ("the Tail of the Goat"), *Lidineb* is a corruption of its correct transliteration *denebaliedi* (cf. the notes to type iii no. 18; see also Kunitzsch [3], p. 83f., no. 59). The coordinates again show interchanges between the lines: *long.* 18° Aqr here seems to be taken from a line below in the source (type iii no. 18 has correctly 7°, but no. 19 there has 27°, for α Tau, of which 18° in our table could be an easy corruption in Roman numerals); *lat.* 35°7 seems to be taken from type iii no. 17, ι Cet (*altitudo* 35°), whereas no. 18 (δ Cap), *ibid.*, correctly has 19°.
21. Derived from type iii no. 17. Coordinates: *long.* 8° Psc, obviously interchanged with type iii no. 18 (which has 7°, of which our 8° is an easy corruption); *lat.* 26°0, is either a corruption of the correct value 35° in type iii no. 17, or of 36° in type iii no. 16 (ζ Cet), both easily explicable in Roman numerals.
22. Derived from type iii no. 13. The coordinates fit well with type iii: *long.* 18° Vir in type iii is not too far distant from 20° Vir in our table; *lat.* is identical in type iii (41°).
23. Derived from type iii no. 12. *long.* in type iii: 8° Lib, which is not too far from 18°11 in our table; *lat.* 39°50 shows a greater difference against type iii (*altitudo* 41°, the same value as in type iii no. 13).
24. Derived from type iii no. 11. The coordinates are almost identical in both tables (cf. type iii no. 11: *mediatio* 25° Sco, *altitudo* 14°).

It can be observed that the equatorial coordinates in this table mostly show slight increments on the values in its source, the star table of type iii. These increments range from 0 (identity with the values of *mediatio coeli* in type iii, in our nos. 21, 22, and 24) up to 61 minutes (no. 8), for the 'longitudes'. For the 'latitudes' there are equally such increments in the minutes, ranging from 0 (identity with the *altitudo* of type iii, in our nos. 18 and 22) to 55 minutes (no. 24). It is difficult to decide whether these increments were made purposely by Rodolfus to account for the changes in the positions of the stars in the time elapsed since the construction of the old standard star table of type iii (late 10th century, in Northeastern Spain), or whether they were simply modifications that had intruded in the process of transmission and copying of those old coordinates. The latter seems more plausible to me because the increments were not only added to the 'longitudes', but also to the 'latitudes'. And since Rodolfus did not introduce recognizable changes to the ecliptical longitudes in his catalogue, it would seem justified to assume that those slight changes in the equatorial values were equally not made purposely by him.

BIBLIOGRAPHICAL ABBREVIATIONS

Bubnov = N. Bubnov (ed.), *Gerberti postea Silvestri II papae Opera Mathematica*, Berlin, 1899.

Carmody = F. J. Carmody, *Arabic Astronomical and Astrological Sciences in Latin Translation. A Critical Bibliography*, Berkeley and Los Angeles, 1956.

Destombes [1] = M. Destombes, "Un astrolabe carolingien et l'origine de nos chiffres arabes", in: *Archives Internationales d'Histoire des Sciences*, 15, no. 58–59 (1962), 3–45.

Destombes [2] = M. Destombes, "La diffusion des instruments scientifiques du haut Moyen Age au XVe siècle", in: *Cahiers d'Histoire Mondiale* 10 (1966), 31–51.

Haskins = C. H. Haskins, *Studies in the History of Mediaeval Science*, Cambridge, Mass., 1924.

Krause = M. Krause, "Stambuler Handschriften islamischer Mathematiker", in: *Quellen und Studien zur Geschichte der Mathematik, Astronomie und Physik*, Abt. B: Studien, Bd. 3, Heft 4 (Berlin, 1936), pp. 437–532.

Kunitzsch [1] = P. Kunitzsch, *Typen von Sternverzeichnissen in astronomischen Handschriften des zehnten bis vierzehnten Jahrhunderts*, Wiesbaden, 1966.

Kunitzsch [2] = P. Kunitzsch, *Untersuchungen zur Sternnomenklatur der Araber*, Wiesbaden, 1961.

Kunitzsch [3] = P. Kunitzsch, *Arabische Sternnamen in Europa*, Wiesbaden, 1959.

Millás [1] = J. Millás Vallicrosa, *Assaig d'història de les idees físiques i matemàtiques a la Catalunya medieval*, vol. i, Barcelona, 1931.

Millás [2] = J. Millás Vallicrosa, *Las traducciones orientales en los manuscritos de la Biblioteca Catedral de Toledo*, Madrid, 1942.

Millás [3] = J. Millás Vallicrosa, *Estudios sobre historia de la ciencia española*, Barcelona, 1949.

Millás [4] = J. Millás Vallicrosa, *Nuevos estudios sobre historia de la ciencia española*, Barcelona, 1960.

Poulle [1] = E. Poulle, "L'astrolabe médiéval d'après les manuscrits de la Bibliothèque Nationale", in: *Bibliothèque de l'Ecole des Chartes* 112 (1954), 81–103.

Poulle [2] = E. Poulle, "Les instruments astronomiques de l'Occident latin aux XIe et XIIe siècles", in: *Cahiers de Civilisation Médiévale* 15 (1972), 27–40.

Sezgin = F. Sezgin, *Geschichte des arabischen Schrifttums*, esp. vol. v: *Mathematik, bis ca. 430 H.*, Leiden, 1974, and vol. vi: *Astronomie, bis ca. 430 H.*, Leiden, 1978.

Steinschneider = M. Steinschneider, *Die hebraeischen Übersetzungen des Mittelalters und die Juden als Dolmetscher*, Berlin, 1893 (repr. Graz, 1956).

Suter = H. Suter, *Die Mathematiker und Astronomen der Araber und ihre Werke*, Leipzig, 1900 (Abhandlungen zur Geschichte der mathematischen Wissenschaften mit Einschluß ihrer Anwendungen, x. Heft), with additions and corrections, *ib.*, xiv. Heft, Leipzig, 1902, pp. 155–185; further additions, by H. P. J. Renaud, in *Isis* 18, no. 52 (1932), 166–183.

Thorndike-Kibre = L. Thorndike and P. Kibre, *A Catalogue of Incipits of Mediaeval Scientific Writings in Latin*, 2nd ed., London, 1963.

Vernet = J. Vernet, *La cultura hispanoárabe en Oriente y Occidente*, Barcelona/Caracas/México, 1978.

XI

THE ASTRONOMER ABU 'L-ḤUSAYN AL-ṢŪFĪ AND HIS BOOK ON THE CONSTELLATIONS*

In the history of Arabic-Islamic astronomy an important name is Abu 'l-Ḥusayn ʿAbd al-Raḥmān ibn ʿUmar al-Ṣūfī (born in Rayy, 903; died 986). He was born in Persia and seems to have spent most of his life there, with close relations to the regional rulers of the Buwayhid dynasty. Though Persian by birth, he wrote his works in Arabic – like all the other scholars of the Islamic countries in his time[1].

Among his various works, on astronomical, astrological, and mathematical subjects, the most outstanding is his book on the constellations (*Kitāb ṣuwar al-kawākib*). In it, al-Ṣūfī treats the 48 classical constellations, which were established and transmitted in Ptolemy's *Almagest*. The description of each constellation comprises the following four sections: first, a general discussion of the constellation and its individual stars; this section also includes al-Ṣūfī's criticism of the Ptolemaic tradition and notices of his own observations. Second, a record of the indigenous Arabic star names falling under that constellation, and the exact identification of the respective stars with the corresponding Ptolemaic stars. Third, two

[1] For recent literature on al-Ṣūfī see: S. M. Stern, article ʿAbd al-Raḥmān b. ʿUmar al-Ṣūfī, in *EI*², I (Leiden, 1960; repr. 1979), 86 f.; C. A. Storey, *Persian Literature*, II, 1 (London, 1958; repr. 1972), 41 f.; F. Sezgin, *GAS* V (Leiden, 1974), 309 f.; VI (1978), 212–215; VII (1979), 168 f.; P. Kunitzsch, article Al-Ṣūfī, in *Dictionary of Scientific Biography*, XIII (New York, 1976), 149 f.; *idem*, article ʿAbd-al-Raḥmān b. ʿOmar Ṣūfī, in *Encyclopaedia Iranica*, I, fasc. 2 (London, 1982), 148 f.; G. P. Matvievskaya, "Abd ar-Rachman as-Sufi i ego rol' v istorii astronomii", *Istoriko-astronomičeskiye Issledovaniya* 16 (Moscow, 1983), 92–138.

* This article was written in commemoration of the thousandth anniversary of al-Ṣūfī's death in 986.

drawings of the constellation, one as it is seen on a celestial globe, and the other one as it is seen in the sky. Fourth, a table of the stars making up that constellation, giving a verbal description of each star's location and its longitude, latitude, and magnitude. This table closely follows the arrangement of Ptolemy's star catalogue in the *Almagest*. The verbal descriptions are mostly given in the wording of Isḥāq ibn Ḥunayn's translation of the *Almagest*[2], the longitudes are increased against Ptolemy's by 12°42′ (for al-Ṣūfī's epoch, the beginning of the year 1276, era of Alexander = October 1, 964), whereas the magnitudes consistently are those observed by al-Ṣūfī himself (*al-aqdār ʿalā mā wajadnāhā*).

The book on the constellations is of great historical interest for several reasons. It is a genuine Islamic contribution to the knowledge of the fixed stars; it gives a revision of many of Ptolemy's dates; it is a reasonable attempt to collect and identify a great number of old indigenous Arabic star names; and through its illustrations it has established a standardized typology of the images of the constellations. Being, thus, a veritable handbook for the fixed stars and their constellations, it became dominant and influential for many centuries, both in the Islamic world and in Europe.

Editions and translations. The introduction of the book was edited and translated into French by J. J. A. Caussin de Perceval, in *Notices et extraits des manuscrits*, XII (Paris, 1831), 236–276. A French translation of the entire book, including the Arabic text of the sections on the indigenous Arabic star names, was given by H. C. F. C. Schjellerup, *Description des étoiles fixes . . . par . . . Abd-al-Rahman al-Sûfi*, St.-Pétersbourg, 1874 (repr. Frankfurt, 1986). For his edition Schjellerup used a rather late manuscript kept in Copenhagen[3]. The drawings of the constellations were redrawn for the edition, not reproduced in facsimile. After completion of the translation, Schjellerup could still use a second manuscript which had only recently been acquired by the Bibliothèque Impériale de St.-Pétersbourg[4]. Since in some recent publications it has been

[2] Cf. P. Kunitzsch, *Der Almagest. Die Syntaxis Mathematica des Claudius Ptolemäus in arabisch-lateinischer Überlieferung*, Wiesbaden, 1974, 52 and footnote 132.

[3] Copenhagen, MS no. 83, dated 1010 H = 1601/2 A. D. (cf. Schjellerup, p. 27f.).

[4] See his own words, p. 28.

stated that this second manuscript, of St.-Pétersbourg (now Leningrad), was very old (1015 A. D.)[5], it should be pointed out that, on the contrary, SCHJELLERUP's Petersburg (Leningrad) manuscript was comparatively recent, like the Copenhagen manuscript. On p. 28 he describes the manuscript saying that it was in "écriture coulée (talîk)", and refers to a short description by B. DORN[6]. Neither DORN nor SCHJELLERUP mention any date given in the manuscript, which makes it clear enough that no date was to be found in it, not to speak of such an early date as 1015 A. D., which would have been worth mentioning in any case. Further, while DORN's description is rather short and does not give many details, SCHJELLERUP says that the script in the manuscript was in the *taʿlīq* style. This style of writing, of course, could not have been used in a manuscript of 1015 A. D., and therefore it is certain that SCHJELLERUP's Leningrad manuscript cannot have been the one reported as dated in 1015 A. D.[7]; in fact, if the manuscript is really written in the *taʿlīq* script, it must be dated to the 15th or 16th centuries A. D., that is, approximately the same period as the Copenhagen manuscript of 1601/2 A. D.[8]

[5] See A. KRCHŇÁK, "Die Herkunft der astronomischen Handschriften und Instrumente des Nikolaus von Kues", *Mitteilungen und Forschungsbeiträge der Cusanus-Gesellschaft* 3 (1964), 126 footnote 53, near the end; cf. also K. FISCHER, "Ein Beitrag zur Geschichte der Sternatlanten", *Bohemia – Jahrbuch des Collegium Carolinum* (München), 11 (1970), 340.

[6] The reference is to: B. DORN, *Drei . . . astronomische Instrumente mit arabischen Inschriften*, St.-Pétersbourg, 1865 (Mémoires de l'Académie Impériale des Sciences de St.-Pétersbourg, 7e série, t. IX, No. 1), p. 77 (*sub* IX).

[7] I could not check the respective manuscript (Arabskaya novaya seriya No. 191) of the Publičnaya Biblioteka at Leningrad (cited in KRCHŇÁK, see note 5, above), since the respective catalogue volumes were not available to me in Munich.

[8] Another manuscript in Leningrad is described by V. ROSEN, *Manuscrits Arabes*, St.-Pétersbourg, 1877 (Collections Scientifiques de l'Institut des Langues Orientales, I), p. 118 f., no 185. This manuscript was not yet known to SCHJELLERUP; but ROSEN, in his comments, already quotes SCHJELLERUP's work of 1874. Since both, MS 191 (said to be dated 1015 A. D.) and 185 (ROSEN), cannot be the manuscript used by SCHJELLERUP, this could perhaps be a third Leningrad manuscript, listed by B. ROSENFELD as no. 669 in the Shdanov University (LGU); cf. *Fisiko-Matematičeskie Nauki v stranach vostoka*, Moscow, 1966 (Akademiya Nauk SSR), p. 261, no. 24.

The complete Arabic text was published in 1954 by the Dairatu'l-Maʿarifi'l-ʿOsmania, Hyderabad/Dn., from five manuscripts: the basic manuscript was a manuscript once belonging to Ulugh Bēg's library (Paris, B. N. ar. 5036, datable to the middle of the 15th century); the four other manuscripts are: Istanbul, Topkapı Sarayı 3493 (dated 525/1131); Vatican, Rossi 1033 (dated 621/1224); Berlin, Ahlwardt 5658 (dated 620/1233); and Āṣafīya, Hyderabad/Dn. (no date mentioned). The drawings of the constellations, here as well as in Schjellerup's edition, were merely redrawn, and not reproduced in photographs or facsimile. The edition is deplorably rich in misprints and other errors, especially in the star tables where many of the wrongly printed coordinates have even been corrected by hand by the editors. Notwithstanding these shortcomings, from one of these faulty copies a reprint was made in Beirut, in 1981 (by Dār al-Āfāq al-Jadīda).

A really correct and reliable edition, with reproductions of the drawings in the original shape, and using some of the oldest extant manuscripts[9], remains a *desideratum.*

In the following I shall consider some of the major impacts of al-Ṣūfī's work on the constellations on later authors in the Islamic world and in mediaeval Europe.

Among the famous Islamic astronomers who have drawn upon al-Ṣūfī's book can be named al-Bīrūnī (973–1048), who in several of his works, in sections related to the fixed stars, quoted al-Ṣūfī. In one instance he even criticized him for not having corrected more of Ptolemy's stars than he did, though he was well equipped for that[10]. But a hundred years later, al-Bīrūnī was himself criticized, by Ibn al-Ṣalāḥ (d. 1154), for a similar neglect[11]. The same Ibn al-Ṣalāḥ also reproached al-Ṣūfī because, in his view, he did not go far

[9] The oldest known manuscript of al-Ṣūfī's book is the Bodleian MS Marsh 144, written by al-Ḥusayn ibn ʿAbd al-Raḥmān ibn ʿUmar ibn Muḥammad (obviously al-Ṣūfī's son), in 400/1009–10; cf. Plate no. 2 in ed. Hyderabad. A selected number of constellation drawings from this manuscript has been published by E. Wellesz, *An Islamic Book of Constellations*, Oxford, 1965 (Bodleian Picture Book, No. 13).

[10] Cf. Ibn al-Ṣalāḥ, *Zur Kritik der Koordinatenüberlieferung im Sternkatalog des Almagest*, ed. P. Kunitzsch, Göttingen, 1975 (Abhandlungen der Akademie der Wissenschaften in Göttingen, Phil.-Hist. Kl., 3. Folge, Nr. 94), 39.

[11] *Op. cit.*, 39.

enough in his criticism of Ptolemy and did not fully correct many of the dubious star positions found in his sources[12].

The sections on the constellations of the fixed stars in the famous cosmography, *'Ajā'ib al-makhlūqāt*, of Zakarīyā' ibn Muḥammad al-Qazwīnī (1203–1283)[13], are literal excerpts from al-Ṣūfī's book[14]. Some of the manuscripts of this work also contain drawings of the 48 constellations (WÜSTENFELD's edition of the Arabic text does not show any of these). The Bayerische Staatsbibliothek, Munich, houses four copies of al-Qazwīnī's *Cosmography* which I recently inspected, mostly in search of the constellation drawings. The result was: two relatively late manuscripts, cod. ar. 465 (maghrebi script, dated 1192/1778) and cod. ar. 466 (dated 1068/1658), do not contain the constellation drawings. Cod. 463 (not dated, but declared "rather recent" in AUMER's catalogue) contains the drawings, but these appear to be "degenerate". They look like paintings (mostly not highly artistic) of animals or human figures, etc., fully coloured, with golden dots, for the stars, irregularly spread over the figure, without any respect to the basic intention of these figures as images of constellations with the stars on well defined and fixed parts of their bodies. Nevertheless, the styling of several of the figures proves that even these degenerate paintings are ultimately related to the Ṣūfī-type of the constellation drawings (see especially Cassiopeia, fol. 23^v; Delphinus, 25^r; Equuleus, 25^r; Pegasus, 25^v; *al-faras al-tāmm*, 26^v; Cetus, 31^v; or Ara, 35^r).

Most convincing is the situation in cod. ar. 464, which is remarkably old (dated 678/1279–80, that is, roughly three years before the author's death) and which deserves greatest attention. Here the section on the constellations (fol. 18^v–25^r) also contains

12 *Op. cit.*, 38.

13 For the Arabic text, see *Zakarija ... el-Cazwini's Kosmographie*, ed. F. WÜSTENFELD, I (Göttingen, 1849), 29–41. The subsequent section on the lunar mansions (pp. 41–51), on the other hand, is literally extracted from Ibn Qutayba's *Kitāb al-anwā'*. Al-Qazwīnī's section on the constellations has also been edited, translated into German, and augmented by a rich commentary by L. IDELER, in his study on the history of star names: *Über den Ursprung und die Bedeutung der Sternnamen*, Berlin, 1809.

14 Cf. already a similar statement by CAUSSIN DE PERCEVAL (mentioned above, in this article), p. 239. On the author and the work, cf. C. BROCKELMANN, *GAL* I, 481; *Suppl.* I, 882; T. LEWICKI, article al-Ḳazwīnī, Zakariyyā' ..., in *EI*², IV (Leiden, 1978), 865–867.

drawings of them, in a clear Ṣūfī-style with many details of al-Ṣūfī's drawings still retained and easily to be recognized[15]. For each constellation, however, al-Qazwīnī has only reproduced one of al-Ṣūfī's two drawings. Thus it is now established, that al-Qazwīnī's *Cosmography* has borrowed the section on the constellations entirely from al-Ṣūfī's book, both in the text and the drawings. But not all of al-Qazwīnī's manuscripts have reproduced the drawings together with the text.

Next, we shall discuss the relation of Ulugh Bēg's star catalogue (epoch 1437 A. D.) to al-Ṣūfī's work. For a long time it has been known that Ulugh Bēg in the star catalogue of his *Zīj-i jadīd-i sulṭānī*[16] has largely relied upon al-Ṣūfī's book, as he states himself in his introduction[17]. As he says, he departed from al-Ṣūfī's book, but when he found many discrepancies between the star positions given by al-Ṣūfī and the actual positions of the stars at his time, he preferred to make his own observations for the catalogue. It had also been observed for a long time that the verbal descriptions of the star positions in Ulugh Bēg's catalogue are nothing but a Persian translation of al-Ṣūfī's descriptive formulas[18]. Therefore,

[15] It must be noted that, in this manuscript, fol. 21 (in the present German pencil numbering) is displaced and should be inserted before fol. 20. Further, after 22^{v} (which ends with Taurus), there follows 23^{r} (with the end of the text of Orion, and the text and drawing of Eridanus, etc.); that means that between fols. 22 and 23 several folios are lost, containing the constellations from Gemini to Orion.

[16] Cf. C. Brockelmann, *GAL* II, 212; *Suppl.* II, 298; Storey (as in note 1, above), 67–70; E. S. Kennedy, *A survey of Islamic astronomical tables*, Philadelphia, 1956 (Trans. Amer. Philos. Soc., N. S., vol. 46, part 2), 125 f. (no. 12) and Abstract 16, p. 166 f.

[17] I shall not list here all the successive editions and translations of the introductory matter and other sections of the work; for this, the readers are referred to the bibliographical sources cited in note 16, above. For my present work, I have used the old edition of Ulugh Bēg's star catalogue, accompanied by a Latin translation and commentary, by Th. Hyde, *Tabulae longitudinis ac latitudinis stellarum fixarum ex observatione Ulugh Beighi*, Oxford, 1665. A modern recension of the star catalogue, from several manuscripts existing in England, was made by E. B. Knobel, *Ulugh Beg's Catalogue of Stars*, Washington, 1917 (Carnegie Institution, Publication No. 250).

[18] In the edition of the star terminology, in my book *Der Almagest* (see note 2, above), I have continuously mentioned the Persian terms applied by Ulugh Bēg for the various Arabic terms used in the Arabic versions of the *Almagest*, by al-Ṣūfī, etc. (cf. there, p. 52 with footnote 133). The glossary

KNOBEL in his recension of Ulugh Bēg's catalogue (cf. note 17) decided to insert, in the column for the star descriptions, SCHJELLERUP's French translations of al-Ṣūfī's original Arabic formulas directly (cf. his introduction, p. 14). But did Ulugh Bēg really read, and translate, al-Ṣūfī's star tables in the original Arabic text?

Apparently not. As I shall demonstrate here, it is most probable that he read al-Ṣūfī's book in the Persian translation made about two hundred years earlier by Naṣīr al-Dīn al-Ṭūsī. Al-Ṭūsī's Persian translation of al-Ṣūfī's book was hitherto known only from the bibliographies[19], only for a few years has it been available as a printed edition[20]. In addition, the Istanbul manuscript Aya Sofya 2595 has been published in facsimile, in Tehran, 1348 Sh./1969[21]. From the colophon at the end of this manuscript it can be understood that the manuscript is the autograph of the translator, al-Ṭūsī[22].

The facsimile edition of al-Ṭūsī's Persian translation shows the complete contents of al-Ṣūfī's book, both the text and the drawings.

By comparing the formulas of the star descriptions in Ulugh Bēg's catalogue with those in al-Ṭūsī's Persian translation of al-Ṣūfī, it becomes at once evident that the two are identical. Therefore, it must be concluded that Ulugh Bēg did not read and translate the original Arabic text of al-Ṣūfī's book, but rather that he read, and copied, al-Ṭūsī's Persian translation of it. Of course, a

appended by KNOBEL to his recension (cf. note 17, above), is not reliable and contains many erroneous words or associations.

[19] See M. KRAUSE, *Stambuler Handschriften islamischer Mathematiker*, Berlin, 1936 (Quellen und Studien zur Geschichte der Mathematik, Astronomie und Physik, Abt. B, Bd. 3, Heft 4, pp. 437–532), 498 (no. 20), where it is said that the translation was completed on 25 Dhu 'l-Qaʿda 647/1250; STOREY (as in note 1, above), 41; SEZGIN (as in note 1, above) VI, 214.

[20] Edited by Sayyid Muʿizz al-Dīn Mahdawī, Tehran, 1351 Sh./1972. I could not obtain a copy of this edition in Munich.

[21] I am grateful to Mme. R. Tabatabai, of Munich, who helped me to obtain a copy of this edition from Tehran.

[22] Thus understood by KRAUSE, and accepted by STOREY. The Persian editor of the facsimile edition, PARWĪZ KHĀN NĀTILĪ, in his *Muqaddima*, expresses his doubts against this assumption, and I, after inspecting the volume, equally feel strong doubts about it. But we shall not here dwell any longer on this question, because it cannot be answered without further evidence, and it is of no major importance for our own present investigation.

few minor differences can be found in all these versions. But it has to be kept in mind that al-Ṭūsī's translation was made almost 300 years after the composition of al-Ṣūfī's book, and that from the time of al-Ṭūsī's translation until Ulugh Bēg's use of it almost another 200 years had passed. Over such long periods of time every text can suffer more or less severe corruptions and alterations. Thus, in a comparison, I found three cases where al-Ṭūsī had retained an Arabic term while Ulugh Bēg has a Persian term instead[23], and, the other way round, six cases where al-Ṭūsī offers a translated Persian term while Ulugh Bēg's text has the Arabic form[24]. (It must be mentioned that al-Ṭūsī has not converted every Arabic term, in every place, into a Persian one; very often he has retained the Arabic terms in his new, Persian context.)[25]

In sum, it can now be taken for granted that Ulugh Bēg read al-Ṣūfī's book in the Persian translation made by al-Ṭūsī, and that he

[23] In the stars UMa 3e, Peg 19, and Leo 14.

[24] In the stars CrB 1, Ser 17, Peg 13, Peg 17, Peg 18, and Leo 23.

[25] By the way, al-Ṭūsī's text offers many improvements for bad or erroneous readings in HYDE's edition of 1665. On the other hand, there are some instances where the Istanbul manuscript of al-Ṭūsī has obvious nonsense, and where the Arabic edition of al-Ṣūfī and the Persian edition by HYDE have the correct readings. For the readers' interest I shall mention the following:

Star Lyr 2 (= BAILY no. 150): al-Ṣūfī (ed. Hyderabad, p. 69): *al-ithnayn al-qarībayn*; HYDE, p. 29: *ān dō . . . ke qarīb. . . and*; but al-Ṭūsī, p. 43: *ān dō . . . ke qarīn . . . and.* (If al-Ṭūsī had met in his source copy with an undotted Arabic *al-qarī.ayn* which he converted into *dō . . . qarīn*, how could this switch over to the correct *dō . . . qarīb* in Ulugh Bēg? The graphical difference between the final *-b* and *-n* in *qarīb/qarīn* is striking.)

Star Psc 21 (= BAILY no. 694): al-Ṣūfī, p. 252: *al-thalātha al-mutawāliya*; HYDE, p. 107: *īn se kawkab-i mutawālī*; but al-Ṭūsī, p. 148: *ān se kawkab-i shamālī.*

Star Cen 10 (= BAILY no. 944): al-Ṣūfī, p. 334: *ṭaraf qaḍīb al-karm*; HYDE, p. 141: *ṭaraf-i shākh-i raz*; but al-Ṭūsī, p. 189: *ṭūl-i shākh-i raz.*

Cases like these speak strongly against MS Istanbul's being the autograph of al-Ṭūsī; it can be assumed that al-Ṭūsī was too scrupulous and precise to commit stupid errors like these.

For the incomprehensible term عصاى اهنها in the description of the star Ori 11 (= BAILY no. 744) [cf. KUNITZSCH, *Almagest*, as in note 2, above, p. 311, no. 496, and p. 372], al-Ṭūsī, p. 158, gives the same spelling عصاء اهنها which remains unexplained to me (cf. also his gloss to the star Boo 8 = BAILY no. 95: يعنى اهنها بر زده which is not in HYDE, p. 21).

adopted al-Ṭūsī's formulas for the descriptions of the stars in his own star catalogue.

Let us now switch over to the West to see which influences were exerted by al-Ṣūfī's book on the constellations in mediaeval (and Renaissance) Europe.

Here, we can speak of five successive points of contact between the Europeans and al-Ṣūfī's work. In approximate chronological order they are the following: 1. the formation of the "Ṣūfī Latinus" *corpus*; 2. the astronomical activities of Alfonso X of Castile; 3. some writings of the German astronomer PETRUS APIANUS; 4. TH. HYDE's commented edition of Ulugh Bēg's star catalogue; and 5. IDELER's commented edition of the respective sections in al-Qazwīnī's *Cosmography* which, in turn, were excerpts from al-Ṣūfī's book.

About HYDE's and IDELER's work we have already spoken above, in connection with Ulugh Bēg and al-Qazwīnī. It should yet be added that modern astronomers collected a great number of Arabic star names – both indigenous Arabic, and Arabic translated from the Greek – from these two publications and introduced them into modern astronomical use. Thus some of al-Ṣūfī's old astronomical knowledge still continues to live on in our own days[26].

As for PETRUS APIANUS (professor of mathematics in the University of Ingolstadt from 1527 until his death in 1552), it must be assumed that he used al-Ṣūfī's book in the original Arabic form. In several of his works he quoted star and constellation names from "Azophi" (= al-Ṣūfī), and he even depicted, in one of his star maps, some of the indigenous Arabic asterisms described by al-Ṣūfī. I shall not go further into this subject here, because I have treated it elsewhere[27].

Of King Alfonso X of Castile (reigned 1252–1284) it is known that he had assembled at his court a number of Christian and Jewish scholars. They completed for him some translations of

[26] Cf. on this topic, P. KUNITZSCH, *Arabische Sternnamen in Europa*, Wiesbaden, 1959.

[27] P. KUNITZSCH, "Peter Apian and 'Azophi': Arabic constellations in Renaissance astronomy", to appear in *Journal for the History of Astronomy*; *idem*, *Peter Apian und Azophi*: *Arabische Sternbilder in Ingolstadt im frühen 16. Jahrhundert*, Bayerische Akademie der Wissenschaften, Philos.-Hist. Kl., Sitzungsberichte, Jahrgang 1986, Heft 3 (München, 1986).

Arabic astronomical and astrological works, and they composed for him a collection of monographs on various astronomical instruments and other related texts which were assembled in a big handbook known as the *Libros del saber*[28]. In the final redaction of several treatises in this major work, the king took an active part. The *Libros del saber* have been edited about a hundred years ago[29], but this edition is insufficient and unreliable from a philological point of view. Further, around 1341, an Italian (Florentine) translation was made (in Sevilla) from the Old Castilian original of the *Libros del saber*[30].

The first four treatises in the *Libros del saber*[31] are dedicated to the fixed stars. For each of the 48 constellations the text gives first a general survey of the constellation and a list of the individual stars with descriptions of their location; in this section are included many of the indigenous Arabic star names mentioned by al-Ṣūfī in his book. Then follow further remarks on the character of the constellation. Each section is concluded by a "table" of the respective stars which also gives their coordinates (the longitudes show the characteristic "Alfonsine" value, i. e., Ptolemy + 17°8′, which had been calculated for the epoch year 1252, that is, the year of Alfonso's accession to the throne). These "tables" are all arranged in form of a rosette ("rueda"). A comparison shows that the first, general, section in all the constellations is derived from al-Ṣūfī's book. It is not a complete, direct translation, but all the material given proves to be derived from al-Ṣūfī. (Book IV, the fourth of these treatises, sums up once more all the constellations and their stars, and mentions again many of the Arabic names that were

[28] For the manuscripts of the *Libros del saber*, and the contents of the entire work, see G. BOSSONG, *Probleme der Übersetzung wissenschaftlicher Werke aus dem Arabischen in das Altspanische zur Zeit Alfons des Weisen*, Tübingen, 1979 (Beihefte zur Zeitschrift für romanische Philologie, Bd. 169), 57–83; A. J. CÁRDENAS, "The complete *Libro del saber de astrologia* and cod. Vat. lat. 8174", *Manuscripta* 25 (1981), 14–22.

[29] *Libros del saber de astronomía del rey D. Alfonso X de Castilla*, ed. M. RICO Y SINOBAS, I–V, Madrid, 1863–1867.

[30] The treatises on the fixed stars of the Italian version have been edited (from the unique existing manuscript, cod. Vat. lat. 8174) by P. KNECHT, *I libri astronomici di Alfonso X in una versione fiorentina del trecento*, Zaragoza, 1965.

[31] In ed. RICO Y SINOBAS, vol. I, pp. 3–145; Italian version, ed. KNECHT (239 pp.).

already listed in the first three treatises, on the northern, the zodiacal, and the southern constellations, respectively.) The star tables in the "ruedas", too, prove to be based on the respective tables in al-Ṣūfī's book, and not directly on the text of the *Almagest*. This can be readily understood, since the stars omitted by al-Ṣūfī (Aur 14, Lup 11, PsA 1e–6e) are also absent from Alfonso's "ruedas". The texts of the star descriptions, too, must be derived from al-Ṣūfī's, though they are often abbreviated (for lack of space in the "ruedas"), and sometimes slightly modified. Thus most of the astronomical material in the first four books of the *Libros del saber* has been taken from al-Ṣūfī's book. Al-Ṣūfī is once even mentioned, there, by name[32]. The star nomenclature, whose spelling was not correctly and reliably printed by Rico y Sinobas in his edition, was later edited separately, directly from three Spanish manuscripts plus the Vatican manuscript containing the Italian translation of 1341[33].

Now for the "Ṣūfī Latinus" *corpus*, which needs a more detailed discussion. In 1965 I had published a paper on the subject which was based on the knowledge of six relevant Latin manuscripts, some of which I had inspected *in loco* or on photographs and others I only knew from their descriptions in the respective manuscript catalogues[34]. In the meantime two more manuscripts have become known so that the total number of manuscripts related to the "Ṣūfī Latinus" *corpus* now amounts to eight[35]. Possessing copies from all these manuscripts I am now able to give an analysis of the "Ṣūfī Latinus" *corpus* which goes beyond my findings of 1965.

[32] The book calls him by his *kunya*, *Abolfazen* [= Abu 'l-Ḥusayn]; see ed. Rico y Sinobas, vol. I, p. 30, line 12; this became *Albofaçen* in the Italian translation, ed. Knecht, p. 44, line 10.

[33] O. J. Tallgren, "Los nombres árabes de las estrellas y la transcripción alfonsina", in *Homenaje a R. Menéndez Pidal*, II, Madrid, 1925, 633–718; "Correcciones y adiciones" to this were published in *Revista de filología española* 12 (1925), 400 f.

[34] See P. Kunitzsch, "Ṣūfī Latinus", *Zeitschr. d. Deutschen Morgenländ. Gesellschaft* 115 (1965), 65–74.

[35] I am greatly indebted to Dr. K. Fischer, of Rott, Wissembourg (France), who kindly offered me copies from his microfilms of four of these manuscripts (plus copies of another manuscript of related interest), and to Dr. N. Palmer, of Oriel College, Oxford, who supplied me with xeroxcopies from his copies of the Catania manuscript which it would have been utterly difficult to obtain otherwise.

First, I shall give a classified list of these manuscripts which will be arranged in groups, according to common characteristics of transmission[36]:

[36] I take this opportunity to mention four manuscripts of similar contents but which do not belong to the "Ṣūfī Latinus" *corpus* and, therefore, must not be confused with the Western Ṣūfī-tradition:

1. London, Brit. Lib., MS Harl. 625, fol. 1^v–6^v (mentioned by KRCHŇÁK, as in note 5, above, p. 126 footnote 53 at the end, who even wrote: "Der lateinische Sternkatalog des ... Al-Sûfi befindet sich auch (bisher nicht identifiziert) im British Museum ..."). In reality, this is a pure Ptolemaic star catalogue (the longitudes are exactly Ptolemy's, with no precession constant added) in the Latin wording of Gerard of Cremona's translation from the Arabic. No constellation drawings are added to this catalogue.

2. Formerly Königsberg, Univ.-Bibl., MS 1735 (15th cent.); registered by E. ZINNER in his *Verzeichnis der astronomischen Handschriften des deutschen Kulturgebietes*, München, 1925, under no. 8679, as a Ptolemaic star catalogue. After the second world war, the manuscript was incorporated into the University Library of Toruń (Thorn), as MS 74 (cf. K. FISCHER, as in note 5, above, 344 f.; FISCHER mentions the manuscript as a member of the Ṣūfī-family, in a larger sense). An inspection of the manuscript (Dr. Fischer was kind enough to supply me with copies from his microfilm; this film covers the portion from Aquarius, fol. 170^r, to the end, Piscis Austrinus, fol. 181^r; the beginning of the star catalogue, according to ZINNER, was on fol. 151^v) shows that here we have a star catalogue of the Alfonsine type (Latin text = *Almagest*, version Gerard of Cremona; longitudes = Ptolemy + 17°8′); the drawings of the constellations belong to a Western tradition and show no influence from the Ṣūfī type. Therefore, this manuscript must no longer be put in relation to the "Ṣūfī Latinus" *corpus*.

3. MS Vienna 5415 (around 1440), fol. 217^r–251^r (cf. ZINNER, *Verz.* no. 10238). The star catalogue was established for the epoch 1424 (applying a longitude value of Ptolemy + 18°56′), perhaps in connection with a work on the construction of the celestial globe. The text is again Gerard of Cremona's Latin version of the *Almagest*, greatly inflated with star and constellation names compiled from several sources, including texts translated from the Arabic. The constellation drawings are of an independent type. Consequently, there is no relation whatsoever between this manuscript and the "Ṣūfī Latinus" *corpus*.

4. MS Vat. Palat. lat. 1368, 51^r–56^v: star catalogue with constellation drawings. The catalogue is again in the Latin wording of Gerard of Cremona's translation of the *Almagest* from the Arabic; it has been calculated by the *magister* Johannes, Dean of Nuhusen [Johannes de Wachenheim], for the year 1420, but the longitudes in the tables show only degrees, no minutes (!). The accompanying constellation drawings were declared as being derived from Oriental models, by F. SAXL, *Verz. astrolog. u. mytholog. illustr. Handschriften des latein. Mittelalters in röm. Bibliotheken*, Heidelberg, 1915, 14 f.

I. 1. Paris, Arsenal 1036 (the manuscript will be analyzed in detail later);

II. 2. Gotha, Forschungsbibliothek, MS M II 141 (around or after 1428)[37];

3. Prag, Strahov Library, MS D. A. II, 13;

4. Berlin, Kupferstichkabinett, MS Hamilton 556[38];

5. Munich, Clm 826;

III. 6. Catania, MS Catin. 85;

7. Vienna, MS 5318;

IV. 8. Kues, Cusanus-Stift, MS 207.

(Of these, nos. 2 and 4 are those that became known to me after writing the paper of 1965. For the other six manuscripts, I may refer to the sources cited in 1965.)

The contents of these four groups of manuscripts making up the "Ṣūfī Latinus" *corpus* can be described as follows:

I. MS Arsenal 1036 is the oldest representative of the whole *corpus*. It is somehow connected with Sicily (see below). It contains a complete (Ptolemaic) star catalogue in the text of Gerard of Cremona's Latin translation from the Arabic. In the title, it gives the name of the Arabic author as *Ebennesophy*[39], and the epoch of the catalogue as 272 Hijra (= 885/6 A. D., which does not fit al-Ṣūfī's real epoch). The longitudes of the stars are = Ptolemy + 12°42′ which is indeed al-Ṣūfī's value, for his epoch 964. Each constellation is accompanied by a drawing; the drawings are visibly in al-Ṣūfī's style and doubtless derived from an Arabic Ṣūfī-manuscript[40].

But a comparison now proves that the drawings in Pal. lat. 1368 show no relationship to the Oriental (i. e., al-Ṣūfī) tradition.

[37] Cf. F. Jacobs and F. A. Ukert, *Beiträge zur älteren Litteratur oder Merkwürdigkeiten der Herzogl. öffentlichen Bibliothek in Gotha*, I (Leipzig, 1835), 196; H. J. Rockar, *Abendländische Bilderhandschriften der Forschungsbibliothek Gotha*, Gotha, 1970, p. 34 and plate 16. The complete set of constellation drawings from this manuscript has lately been published by G. Strohmaier, *Die Sterne des Abd ar-Rahman as-Sufi*, Leipzig and Weimar, 1984.

[38] Cf. H. Boese, *Die lateinischen Handschriften der Sammlung Hamilton zu Berlin*, Wiesbaden, 1966, p. 273 (no. 556).

[39] Derived from the erroneous form Ibn al-Ṣūfī, instead of al-Ṣūfī; cf. Kunitzsch (1965; as in note 34, above), 68 f.

[40] Of the 48 constellation drawings, 16 represent al-Ṣūfī's type "as seen on

II. Manuscripts 2, 3, and 4 contain the star catalogue and the drawings. But in the title they do not mention al-Ṣūfī's name, and also not the Hijra year of the catalogue's epoch. The drawings of these three manuscripts are closely related to each other and can be regarded as forming one family. They appear to be derived from no. 1, but show more signs of assimilation to a European milieu. Manuscript 5 is incomplete. It has no title. The star catalogue reaches from fol. 34^r to 41^v. From Dra onwards, the constellation titles are not entered. Most of the longitudes have not been filled in (my photographs are blank in this column, only rarely are there faint traces of numbers, e. g., in Dra, Cyg, Oph, Ser [in Ser 5–18 written in full black colour], Sge, Del [again in full black colour], and Aql [only 1e–6e, in black]). The catalogue (including the drawings) ends with Peg; the rest is missing. The drawings can be regarded as related to the family of manuscripts 2–3–4 mentioned before; there are certain differences in detail (the "Westernization" in manuscript 5 is more developed), but in comparison to the following groups III and IV, manuscript 5 must be included in group II.

III. The two manuscripts of this group again show the star catalogue as described above. They also give a title similar to that of no. 1, but the name of the Arabic author has been corrupted to *Ilbermosophim* (no. 6) and *Jeber Mosphim* (no. 7). The Hijra year of the epoch is now given as 325 (= 936/7 A. D., which could be a miswriting for an original 352 = 963/4, close to the true epoch of al-Ṣūfī's catalogue). The drawings in both these manuscripts are visibly of an identical type[41]. They are much more "Westernized" than those of group II, but have still retained the basic elements of the Ṣūfī-style. Sometimes they show elements of Western traditions that were added to the

the globe", and 32 the other type, "as seen in the sky". The Western compiler was not consistent; of al-Ṣūfī's two drawings for each constellation he chose only one, varying between the two types. (The 16 "globe type" drawings are: UMi, UMa, Boo, CrB, Her, Cyg, Per, And, Ari, Vir, Lib, Sco, CMa, Argo, Hya, and Crv.)

[41] Cf. also P. McGurk, *Catalogue of Astrological and Mythological Illuminated Manuscripts of the Latin Middle Ages*, London, 1966, 10–16 (description of the Catania manuscript).

basic figures, always identically in both manuscripts. (In no. 7, some of the figures have been reversed left to right.) Further, in this group, the original Ṣūfī complex has been augmented by two additional sections of text, a description of each constellation and an astrological note on the nature of those born under the respective constellation. These additional sections are abbreviated, sometimes with slight modifications, from the *Liber introductorius* of Michael Scot (died around 1235)[42]. The only difference between the two manuscripts lies in the arrangement of the various portions: in no. 6 the new elements (descriptions of the constellations and astrological notes, fol. 7^r–19^v) precede the star catalogue (21^r–35^r), while in no. 7 they follow (star catalogue on fol. 2^r–16^r, descriptions and notes on fol. 17^v–36^v). In both manuscripts the drawings were added to the new element, the descriptions of the constellations, instead of to the star catalogue as in manuscripts 1–5.

IV. This single manuscript (no. 8) shows elements of all the three preceding groups. It also belongs to the "augmented type", but the arrangement of the material differs again. Fol. 108^r–114^v: the descriptions of the constellations (from the *Liber introductorius*, as mentioned above)[43]; 116^v–121^v: the star catalogue, also naming the Arabic author, here as *Ebennesophus* (similar to no. 1, above!), and the year 325 (as in nos. 6/7)[44]; 122r–124r: the astrological notes (again from the *Liber introductorius*); and 124^v–135^v: the drawings of the con-

[42] Cf. the corresponding sections of the *Lib. introd.*, in MSS Munich, Clm 10268, 79rb–84va, and Vienna 2352, 7^r–26^r.

[43] This section should be described a bit closer. From 108rb to 109rb it contains elements from Scot's *Lib. introd.* (MS Vienna 2352, 2^r–7^v; Munich, Clm 10268, 77^v–79ra, etc.) that are not in the two manuscripts of group III, above. From 109rb–114vb it contains the descriptions of the constellations and the astrological notes which are also in the manuscripts of group III, but in MS Cus. they are closer to the *Lib. introd.* and show additional elements from Michael Scot that do not appear in group III.

[44] I. e., in the title (cf. Krchňák, as in note 5, above, p. 125). The *u* in *Ebennesophus* has been corrected and can be misread as *a*, or *ti*. The note at the bottom of fol. 116^v, printed by Krchňák (as in note 5, above), 126, must correctly be pieced together as follows: *Item nota ade ad motuum stellarum fixarum que verifictate sunt Anno Arabum 325 per philosophum cuius nomen Ebennesephtis* [sic], *0^S 6°16′. Et habebis Motuum stellarum fixarum Anno Domini 1427 prima die ianuarii anno perfecto.*

stellations, separately. The drawings are totally different from all the other manuscripts. While in the other manuscripts all the figures were artistically drawn, with the dresses neatly painted and many details represented, in this manuscript we have only simple linear drawings, mostly without details of decoration, etc.; all the human figures are shown naked (with the exception of Virgo), the men as youthful types, unbearded, with curled hair. But the connection with the Ṣūfī-tradition is still recognizable, cf. the two additional figures of Andromeda with the Fish, the *al-faras al-kāmil* ("the Complete Horse" – different from Eql and Peg), the strikingly long sleeve on Orion's left hand, and also the numbering of the stars in all the drawings according to al-Ṣūfī's system.

Before I proceed to a final résumé of the "Ṣūfī Latinus" *corpus*, I first have to give a detailed analysis of the basic manuscript, Paris, Arsenal 1036.

The manuscript opens on fol. 1^r with the star catalogue in which, to each constellation, a drawing derived from an Arabic Ṣūfī-manuscript is added. The title mentions *Ebennesophy* and the Hijra year 272 (= 885/6 A. D.). The longitudes are in the Ṣūfī value, i. e., Ptolemy + 12°42′. To the title is appended a notice (in the same script as that of the main text) saying that from the epoch 885 until 1210 there have to be added for each year 39″, or for every 100 years 1° 4′ 51″. The star catalogue ends on fol. 50^v (drawing of PsA; 49^r–50^r are blank). On fol. 10^v, at the bottom of the catalogue of the stars of Auriga, in a smaller but equally old hand (perhaps of the same scribe), was added the remark: *Nota quod in hac ymagine posuit pholomeus* [sic] *in almagesti .14. stellas hec nominatur* [?] *minor que est super pedem sinistrum, 1^S 30° 40′ lo., 10°20′ lat., mag. 6*. This note is especially interesting. The catalogue above indeed follows al-Ṣūfī's in so far as it only registers 13 stars for Auriga, exactly like al-Ṣūfī. But the compiler of the Latin text (who used Gerard of Cremona's Latin translation from the Arabic) had realized that there, in the *Almagest*, Auriga consisted of 14 stars. So, in his note, he added the text of the 14th star, from Gerard's version, and with Ptolemy's original coordinates (the longitude has not been transformed into al-Ṣūfī's value).[45] On fol. 50^v, below the image of PsA, there is a more recent notice saying that, for the year 1450, 6°28′ has to be added to the preceding table (which has al-Ṣūfī's longitude values).

On 51^r begins a new star catalogue (reaching to 70^v) which arranges

[45] Note that, in the longitude, the manuscript has wrongly 30°, which, of course, ought to be 20°. But most of the Latin manuscripts of the *Almagest* do really have 30° here.

the same Ptolemaic stars according to longitudes under their zodiacal signs, from Ari to Psc. According to the title, this catalogue was calculated for the year 1210 (cf. the mentioning of the same date above, on fol. 1r), and the longitudes are increased against the first catalogue (which has al-Ṣūfī's values) by 4°. At the bottom of 51r there is a notice, again in the same old hand, which points to some Sicilian connection[46].

From 71r to 72v follows the chapter on the fixed stars (I,9) from Ptolemy's *Tetrabiblos* in the Latin translation of Plato Tiburtinus made from the Arabic; cf. the edition: *Liber quadripartiti Ptholemei*, Venice, 1493, p. 14v–17r [here, also the text of the second Latin translation, by Aegidius de Tebaldis, and the commentary by ʿAlī ibn Riḍwān = "Haly" are added].

Fol. 73r–98v: astronomical tables which I have not investigated.

Fol. 99r: a table of 15 first-magnitude stars. The title, on fol. 99r, at the end, says: *In alia scriptura*, which seems to indicate that this star table (and perhaps the following also) were taken from another manuscript. The longitudes do not show a constant increment on Ptolemy's. The same table is repeated on 99v, but here each star is given its Arabic name. The Arabic transliterations used here are unique in the entire Western star name tradition, no similar forms have been noticed until now.[47]

Fol. 100r–101v: a table of 49 significant stars (said, in a notice in a different hand on top of 100r, to have been calculated for 1188), longitudes = Ptolemy + 16°30′. The wording is that of Gerard of Cremona's translation of the *Almagest*.

Fol. 102r–v: a star table of "Typ XII", as edited by P. Kunitzsch, *Typen von Sternverzeichnissen in astronomischen Handschriften des zehnten bis vierzehnten Jahrhunderts*, Wiesbaden, 1966, 72ff. Cf. also P. Kunitzsch, "Two Star Tables from Muslim Spain", *Journal for the History of Astronomy* 11 (1980), 192–201. The table in our manuscript correctly mentions the Arabic author, al-Zarqāllu: *Stelle fixe de libro azachel ebusak anacass tabularum omnium latitudinum* (the transliteration represents the fuller Arabic name, al-Zarqāllu Abū Isḥāq al-Naqqāsh;

[46] It reads: *Currente anno Christi Millesimo centesima* [sic] *Octuagesimo Tertio. Inuentum est cor leonis* [that is, the star α Leo] *per instrumenta panormi in leone Gradu .17. Minuta .46. Currente anno Christi Millesimo Centesimo Octuagesimo octauo. Indictione sexta. Inuentus fuit locus cordis leonis in ipso. Gradu .17. Minuta .47. Secunda .24. Et stella que dicitur in ermis siue stachis* [that is, α Vir] *in libra Gradu .12. Minuta .44. Secunda .33. Loca harum stellarum inuenta fuerunt ad meridiem scicilie per instrumenta certissima. Que rex Wilielmus scicilie felicis memorie In ciuitate panormi fieri fecit.* (See also Saxl [as in note 36_4, above], vol. III, 1, London, 1953, p. XXXII, note 41.)

[47] Note especially the continuous use of *æ* for Arabic *a*, as, e. g., in *ælsimæc ælæhçæl* = *al-simāk al-aʿzal*, α Vir; *mænchibu ælçæuçæ* = *mankib al-jawzāʾ*, α Ori, etc.

the reference seems to be to his work on the *Azafea* which was an instrument devised for all latitudes).

Fol. 103r: a precession table. Title: *Iste est motus octaue spere equatus in ciuitate panormi .1231. anno perfecto* . . . Here again, we have a hint to the Sicilian connections in this manuscript, but now to a later date, i. e. 1231.

Fol. 103v–104r: again the star table of "Typ XII" (as on fol. 102r–v, above), but with slightly different longitudes. According to the title[48], the table has been established for Palermo, for 1188. The value cited here, separately, for α Leo (Leo 18° 34′ 33″) differs from the one cited earlier, fol. 51r, for the same date (Leo 17° 47′ 24″). The values for α Vir are identical in both places (Lib 12° 44′ 33″).

Fol. 104v–110v: Abū Ma'shar, *De reuolutionibus annorum*, Latin version by Johannes Hispalensis[49].

Fol. 111r: a table of 15 first-magnitude stars (identical to fol. 99r, but most of the longitudes differ slightly).

Fol. 111v–112v: a table of 45 second-magnitude stars. The text is always in the wording of Gerard of Cremona's translation of the *Almagest*.

Fol. 113r: again the table of 15 first-magnitude stars (as on 99r and 111r, but here, in a notice in a different hand on top of the folio, dated to 1187, with longitudes = Ptolemy + 15°50′).

Fol. 113v–114v: again the table of 45 second-magnitude stars (as on 111v–112v), longitudes = Ptolemy + 15°50′.

Fol. 115r–121v: *Stelle Que sunt in cingulo orbis signorum*, a table of the zodiacal constellations and their stars; longitudes = Ptolemy + 16°30′.

Fol. 122r: a table of 27 stars, title: *hic est scitus*[50] *stellarum. In annis arabum .609. cum imperfecto alicbel .8. Gradus .52. Minuta.* This is a star table derived from an Arabic – obviously Maghrebine, or Spanish-Arabic – source. The longitudes are = Ptolemy + 15°30′. The Hijra year 609 is

[48] It reads: *Tabule iste directe sunt ad meridiem scicilie per instrumenta certissima que Wilielmus rex sicilie felicis memorie in ciuitate palormi fieri fecit anno domini Christi. 1188. Indictione .6. et in eodem anno inuentus fuit locus cordis leonis in ipso G. 18. M. 34. S. 33. Et stella que dicitur in ermis siue stachis in libra G. 12. M.44. et S. 33. quia uero ipse tabule sunt cum motu .8. spere ipsas latinas placuit appellare siue constantinopolitanas eo quod magister fabricator instrumentorum uenit ab ipsa ciuitate et facte fuerunt Anno domini Christi .1080. Et in omnibus .66. annis solaribus siue .70. mouentur stelle fixe G.1.* (Cf. Saxl, *loc. cit.* in note 46, above).

[49] Cf. F. J. Carmody, *Arabic astronomical and astrological sciences in Latin translation*, Berkeley and Los Angeles, 1956, 94 (*sub* 4.a).

[50] Leg. *situs*, probably originally in the plural (= *loca*), corresponding to the Arabic *mawāḍi'* which was normally used in the title of such star tables. – *alicbel*, a few words behind, is the Arabic *al-iqbāl*, sometimes used for "precession", but also "accession" (in the theory of trepidation).

= 1212/3 A. D. The table represents a type of its own that had not been noticed earlier[51]. This is the last text in the manuscript.

There are several references in the manuscript to Sicily, mentioning the years 1080 [for 1180?], 1183, 1188, King *Wilielmus* (William II, reigned 1166–1189), 1210, and 1231. It could, therefore, be possible that the construction of our star catalogue of the Ṣūfī-type (i. e., naming al-Ṣūfī in the title; using his precession value of Ptolemy + 12°42′; and adding constellation drawings that were taken from an Arabic Ṣūfī-manuscript) had been made in Sicily also. Our manuscript itself is, of course, not the first original of the entire "Ṣūfī Latinus" tradition, but it is the oldest of the extant eight manuscripts, and the closest to al-Ṣūfī's original, and to Arabic material in general. (In the catalogue of the Arsenal manuscripts it is attributed to the 14th century.)

If I now attempt to give a general survey over the eight manuscripts of the "Ṣūfī Latinus" *corpus*, I am in a better position than earlier authors each of whom only had one or two manuscripts in front of them (cf. my article of 1965, p. 72), while I now have copies from all eight manuscripts involved. This is especially important with regard to the constellation drawings. The material now allows us to subdivide the *corpus* into the four groups already mentioned above.

The oldest representative, and the closest to the Arabic original, is MS Arsenal (our no. 1 – here called "group I").

All the other groups must be traced back to this Latin manuscript (or other copies of the same character), but not directly to Arabic source material.

As an example, I shall here describe the drawing of the constellation Cepheus. Its shape in the original Arabic tradition can be seen in the booklet of E. Wellesz (cf. note 9, above), plate 2 (from the oldest

[51] The following observation is interesting: if one deduces from the present longitude value, i. e., Ptolemy + 15°30′, the precessional addition of 8°52′ mentioned in the title, one arrives at a value of Ptolemy + 6°38′ as the point of departure. There have indeed existed Arabic star tables using this very value for the longitudes, as being calculated for the year of the Hijra = 622 A. D.; see, for example, "Typ XV" (by the Spanish-Arabic astronomer Ibn Kammād, in a Latin translation by Johannes de Dumpno, made in Palermo in 1260 or 1262 – note the further possible connection to Sicily, and the date!), edited by P. Kunitzsch, *Typen* (as cited in the article, above), 99 ff.

extant Arabic manuscript, dated 400/1009–10). The Hyderabad edition of the Arabic text has, on figs. 4a–b, redrawings of Cepheus from the so-called copy of Ulugh Bēg, Paris B. N. ar. 5036. The characteristic feature of Cepheus, in the oldest Arabic manuscript, is the high cap (called τιάρα by Ptolemy, *qalansuwa* in the Arabic translations of the *Almagest* and likewise in al-Ṣūfī's book, ***kulāh*** in al-Ṭūsī's Persian translation of al-Ṣūfī's book and, subsequently, in Ulugh Bēg's star catalogue, and *pileus* in Gerard of Cremona's Latin translation of the *Almagest* from the Arabic). The left arm is raised up, the right arm bent down; also, the right leg bent backwards as if he was kneeling on that knee, the left leg stretched forward.

In MS Arsenal we find exactly the same representation, all the details of the figure, the dress, etc., are identical. The characteristic high cap has the same shape as in the Arabic manuscript, only the decoration of its outside has been modified to a form more common to the Western draughtsman.

Manuscripts 2–4 of "group II" have a figure almost identical to that in MS Arsenal. In manuscript 5, however, the figure has been given a shape more stretched in length, as if walking, or running, not kneeling. It can be recognized that no. 5 is the product of a further development out of the type shown in nos. 2–4, though still belonging to the same "group II".

Manuscripts 6 and 7, forming our "group III", show a man, upright and walking, dressed in a late mediaeval Western dress and head-dress, identical in both manuscripts. The figure has now been developed into a pure Western character and has not much in common with the original Arabic drawing, except the basic attitude of stretching the left hand and foot forward, and bending the right arm down and the right foot backwards.

The representative of "group IV", manuscript 8, has a naked male figure, with the body's attitude closer to the situation in "group II" than in "group III", and even wearing a cap resembling the sort of cap shown in "group I, or II", with its decorations reduced to the rim around its lower part and two small circles, or buttons, on its front side.

So, in judging from the drawings, our four "groups" can be classified as follows: I, closest to the Arabic original; II, closely related to group I; III, more Westernized, deriving from a model of the group I or group II-type; IV, only linear drawings, mostly naked human figures, only rarely pieces of dresses are added (the cap of Cepheus, the robe of Virgo), but still well recognizable as belonging to the Ṣūfī-*corpus*, perhaps derived from a representative of group II, or even group I.

As far as the text is concerned, the classification would be about the same: I, closest to the Arabic original (star catalogue only; naming of the author and the Hijra date – though the latter in MS

Arsenal is corrupt, which alone shows that MS Arsenal cannot have been the archetype of the "Ṣūfī Latinus" tradition, but rather that it is already one of the successive later copies); II, following I in presenting the star catalogue alone, but the old title giving the name of the Arabic author and the Hijra epoch year has been omitted; III, augmented type; it has the star catalogue, in a shape close to I (because it again names the author and the Hijra epoch year, though both in variants different from group I); additionally, it has textual elements taken from Michael Scot's *Liber introductorius*; IV, star catalogue, adding the name of the Arabic author (in the spelling of group I) and the Hijra epoch year (in a form similar to that given in group III), text elements from Michael Scot, including sections that are absent from group III.

I shall not enter here into a discussion of the dates and internal interdependence of the manuscripts in each of the four "groups". This would require closer investigation of the manuscripts, the technique of the drawings, etc., which cannot be done from microfilm pictures. Also, time and space do not suffice for such a purpose, here. My intention was to give a general survey of the material contained in the "Ṣūfī Latinus" *corpus* and to establish, for a first time, a rough classification of the eight manuscripts transmitting this *corpus*.

Regarding some special features in the constellation drawings of the "Ṣūfī Latinus" *corpus*, I should like to add a few remarks.

The figure of Canis Maior has, beneath the hind legs, a framed quadrangular form with ten stars inside, numbered from 2 to 11 (cf. STROHMAIER, as in note 37, above, p. 89): this is not a part of the figure of the Dog itself; rather these are the so-called "external" stars of that constellation which are not included in the formation of the Dog's figure but which are listed by Ptolemy under that constellation as what he calls "external" stars, ἀμόρφωτοι.

Equally, the figure of the Ship Argo shows two such framed forms underneath, near the two oars (cf. STROHMAIER, *l. c.*, p. 93). The explanation of these two forms can be found in al-Ṣūfī's original Arabic work, see, e. g., plate 20 in the booklet of E. WELLESZ (cited above, in note 9). To facilitate orientation, al-Ṣūfī had drawn to the lower side of the Ship the stars of the neighbouring constellation Canis Maior, both the stars making up the figure, 18 in number, and the "external" stars, numbers 2 to 11 (as described above) [this addition is not represented on figs. 40a–b of the Hyderabad edition]. The Western tradition has retained this additional element, but added a frame around the 18 stars of the Dog (that is, the bigger form which is pointed on its Western side) and also around the "external" stars (that is, the smaller, quadrangular form close beside the first one). So, in the drawings of the Ship Argo, these two forms are not parts whatsoever of the Ship, but they are merely the stars of Canis Maior added by al-Ṣūfī for orientation purposes.

The figure of Centaurus has been used by K. FISCHER in some of his articles[52] to derive certain historical arguments from the various shapes in which the branch (or so) in the left hand of Centaurus is styled in the respective sources. For the Arabic tradition, that is al-Ṣūfī, and equally the Western "Ṣūfī Latinus" *corpus*, I wish to stress that these drawings cannot be interpreted without taking into account the textual descriptions of the constellations. Draughtsmen who just copy the figures from one source into another, often cannot recognize certain details in a figure, and therefore more and more transform it in the course of transmission into something that appears "reasonable" or habitual to them. Thus, in our case, Centaurus carries in his one hand a θύρσος,[53] a thyrsus, which was, in classical mythology, a staff cut from a vine stock, or a staff entwined with ivy and vine leaves. Subsequently the Arabic translation of Isḥāq ibn Ḥunayn of Ptolemy's *Almagest* applied for the θύρσος the paraphrase *qaḍīb al-karm*, "the Vine Branch"; as usually, al-Ṣūfī has adopted in his star catalogue the formula used in Isḥāq's translation of the *Almagest*. Therefore, whatsoever the drawings seem to show, it must be, basically, a "vine branch", with vine leaves at its ends. The drawing on figs. 44 & 45a–b in the Hyderabad edition can be easily recognized as vine branches and leaves, while, on the contrary, the thing drawn in the oldest Arabic manuscript of 400/1009–10 does not unequivocally give the impression of vine leaves (cf. plate 21 in the booklet of E. WELLESZ, cited in note 9, above). Also, all the various shapes given to the item in the manuscripts of the "Ṣūfī Latinus" *corpus* should basically show vine leaves, but of course, the draughtsmen put in various forms just as they meant to understand the figures in their source manuscripts[54] (cf., e. g., STROHMAIER, as in note 37, above, p. 101).

As a last item in this paper I want to discuss the names under which al-Ṣūfī became known in mediaeval and Renaissance Europe. We have already seen that, in Alfonso X's *Libros del saber*, al-

[52] Article of 1970 (see above, note 5), pp. 339, 344 ff.; and "Some unpublished astrological illustrations . . .", *Journal of the Warburg and Courtauld Institutes* 27 (1964), 311–312.

[53] Cf. KUNITZSCH, *Almagest* (cited in note 2, above), p. 339 f., nos. 602–603; Ibn al-Ṣalāḥ (cited in note 10, above), 45 and 70 f. (no. 80).

[54] In this case, they could hardly be guided by the Latin text of the star catalogue, which – in Gerard of Cremona's wording – had *clipeus* ("shield") for this item, due to a wrong Arabic translation in that version from which Gerard had translated these stars' descriptions.

Ṣūfī was named by the Hispanicized form of his *kunya* (Abu 'l-Ḥusayn), as *Abolfazen* (and *Albofaçen* in the Italian translation; cf. note 32, above). While the star descriptions derived from al-Ṣūfī in the *Libros del saber* were of no known influence in the subsequent European literature, the Alfonsine text was not altogether neglected. The Jewish astronomer ABRAHAM ZACUTO (middle of 15th century) cited it in one of his astronomical works, and after him his disciple AUGUSTINUS RICIUS, in one of his works (probably written in 1517) equally mentioned al-Ṣūfī, as *Albuhassin*. From RICIUS these citations passed on to several famous Western astronomers[55].

Equally, we have already seen that in the "Ṣūfī Latinus" *corpus* al-Ṣūfī had been called *Ebennesophy*, etc., a form which was derived from an erroneous Arabic form of the name where, instead of the correct al-Ṣūfī, the wrongly styled Ibn al-Ṣūfī can be occasionally found (cf. note 39, above). But this form, as far as I know, was never cited outside the Ṣūfī-*corpus* itself.

LEO AFRICANUS, the North-African Muslim who converted to Christianity and further lived and worked in Rome (early 16th century), once called al-Ṣūfī: *Abulhusein Esophi*[56].

Much more famous, however, became a fourth Western form of al-Ṣūfī's name which ultimately seems to have been introduced by the Spanish-Jewish scholar and astronomer Abraham ibn ʿEzra (middle of 12th century): *Azophi*[57]. Ibn ʿEzra was famous as an astronomer in mediaeval Europe so that his works were much read by later generations of scholars. Therefore, also his (wrong) information on al-Ṣūfī's precession constant was several times re-

[55] Cf. the description of these texts and their interrelationship, by A. A. BJÖRNBO, "Hat Menelaos aus Alexandria einen Fixsternkatalog verfaßt?", *Bibliotheca Mathematica*, III.2 (1901), esp. pp. 197–200.

[56] See A. HAUBER, "Zur Verbreitung des Astronomen Ṣūfī", *Der Islam* 8 (1917/18), p. 48 note 1.

[57] See *El libro de los fundamentos de las Tablas astronómicas de R. Abraham ibn ʿEzra*, ed. J. M. MILLÁS VALLICROSA, Madrid/Barcelona, 1947, p. 76,10 etc. (the book was apparently written originally in Latin by Ibn ʿEzra himself). Regarding precession, Ibn ʿEzra, p. 78,9–10, writes wrongly that al-Ṣūfī had adopted a constant of 1° in 70 years. In reality, al-Ṣūfī applied the value found and used by earlier Arabic astronomers, viz. 1° in 66 years (1° in 70 years was later used by Ulugh Bēg.) Nevertheless, the wrong information about al-Ṣūfī's precession constant was later repeated by several other authors (see the next following citations in the article, above).

peated by later astronomers, such as Henri Bate in his *Magistralis compositio astrolabii* (1274): *Açophius*[58], and Jean de Murs in his commentary on the *Alfonsine Tables*: *Alzophi*[59].

Most stimulating must have been the use of the form *Azophi* by ALBRECHT DÜRER who published at Nuremberg in 1515 two maps of the stellar sky, the northern and the southern hemisphere. In the four corners of the northern map, DÜRER represented imaginative portraits of astronomers from four cultures, or nations, who made fundamental contributions to the knowledge of the fixed stars. There, in the lower right corner, we find *Azophi Arabus*[60]. Through his learned Nuremberg friends, mathematicians and astronomers (two of whom cooperated with DÜRER in the preparation of the two stellar maps), DÜRER must have got knowledge of *Azophi*, in this form of the name, which, as I have just demonstrated, seems to have originated in the book of Abraham ibn ʿEzra in the middle of the twelfth century.

After DÜRER, many more, and more recent, authors used the now popularized name *Azophi*, such as PETRUS APIANUS (cf. above, in this article), BULLIALDUS, WEIDLER, and others[61].

Also JOHANN BAPTIST RICCIOLI, who published the first detailed

[58] Cited by C. A. NALLINO, in his edition of al-Battānī's *Opus astronomicum*, I (Milan, 1903; repr. Hildesheim/New York, 1977), p. XXXV note 3.

[59] Cf. the edition by E. POULLE, "Jean de Murs et les Tables alphonsines", *Archives d'histoire doctrinale et littéraire du moyen âge* 47 (1980), p. 262, line 11 from bottom.

[60] Cf. S. G. BARTON, "Dürer and Early Star Maps", *Sky and Telescope*, vol. 6, nos. 11–12 (Sept.–Oct. 1947), also published separately as *Reprint No. 27*, by Flower Astronomical Observatory, University of Pennsylvania. For a reproduction, see also STROHMAIER (as cited in note 37, above), p. 6.

[61] As a curiosity, it may be mentioned that A. DRECHSLER, *Illustriertes Lexikon der Astronomie und der Chronologie*, Leipzig, 1881, p. 29, wrote that Azophi was a name for the constellation Draco. This is obviously the result of a misunderstanding of JOHANN BAYER's *Uranometria*, Augsburg, 1603, where, among the names for Draco, it was written: ". . ., Azophi Arabi, quinque dromedarii, etc.", meaning that for Azophi the Arab [in the Latin dative case!] this constellation was called "quinque dromedarii" (an information, by the way, that stems from PETRUS APIANUS' excerpts from al-Ṣūfī). A later, German version of the *Uranometria* (Ulm, 1720) already interpreted the Latin phrase wrongly, as: "bey den Arabern Azophi; . . ." (meaning that Draco was called Azophi with the Arabs), i. e., taking "Arabi" in the Latin nominative case.

map of the moon (drawn by F. M. GRIMALDI) in his *Almagestum Novum* of 1651, knew the name well. In his naming of the craters and other features on the moon's surface, he gave the name *Azophi* to one of the craters[62]. Together with other names introduced by RICCIOLI, *Azophi* has been adopted in the official international nomenclature of the moon. Thus al-Ṣūfī will continue to be mentioned among modern astronomers for many generations to come.

[62] Cf. MOH. A. R. KHAN, "Names of thirteen Muslim astronomers given to some natural features of the Moon", *Islamic Culture* 27 (Hyderabad, 1953), 78–85, esp. 82 f.

Notes added in proof.

Ad p. 62 n. 20: Through the kind help of Dr. J. Hamadanizadeh, Tehran, I have recently obtained a complete xerox copy of MAHDAWI's printed edition.

Ad p. 63 n. 25: The editor S. M. MAHDAWI does not contribute to the understanding of the unknown term عصاى اهنها. In the text of the star Ori 11 (p. 244) he has omitted the doubtful word and inserted the phrase *lā yuqra'* ("not readable"), instead. In the star Boo 8 (p. 50) he has omitted the gloss cited above in my note 25, without adding a notice.

Ad p. 68: A further manuscript containing the star catalogue from the "Ṣūfī Latinus" *corpus* is MS Paris, B. N. lat. 7280 (14–15th cent.), fol. 83r–100v, cf. F. S. BENJAMIN and G. J. TOOMER, *Campanus of Novara and Medieval Planetary Theory*, Madison, Milwaukee, London, 1971, p. 119 (the title cited there names *Ebennesophin* and the year 325/937). Mme. J. Sclafer of the Bibliothèque Nationale, Paris, has kindly informed me in a letter that the star catalogue in this manuscript is not accompanied by the constellation drawings. Another manuscript of the "Ṣūfī Latinus" *corpus* once existed at Forcalquier, France (the so-called *Codex Arnaldinus*); it was cited and used by I. BULLIALDUS, *Astronomia Philolaica*, I, (Paris, 1645), p. 224f. (*Ebennesophim*, 325/936–7), and by P. GASSENDI, *Opera*, vol. IV: *Astronomica* (Lyon, 1658), p. 534f. This manuscript cannot be located today; it may have been lost in the meantime, or is hidden in some private library. I am grateful to J.-F. Genest, Centre National de Recherche Scientifique, Paris, for these informations.

Ad p. 68 and 71–74, MS Arsenal 1036: In a recent article M.-TH. GOUSSET has brought forward good evidence that the manuscript was written and illuminated in Bologna, around 1270; see MARIE-THÉRÈSE GOUSSET, "Le *Liber de locis stellarum fixarum* d'Al-Sûfi . . .", *Arte Medievale* 2 (Rome, 1984), 93–108.

Further filiations of constellation drawings in al-Ṣūfī's style are discussed in a well documented article by KRISTEN LIPPINCOTT, "The astrological vault of the *Camera di Griselda* from Roccabianca", *Journal of the Warburg and Courtauld Institutes* 48 (1985), 42–70. LIPPINCOTT has classified seven manuscripts of the "Ṣūfī Latinus" *corpus* in much the same way as I have done in this paper (cf. p. 67 of her article; the Gotha MS is not mentioned in LIPPINCOTT's study). Drawings of the twelve zodiacal constellations in the style of the "Ṣūfī Latinus" *corpus* are included in an astrological treatise by Andalò di Negro, in MS London, Brit. Lib., Add. 23770 (not mentioned by LIPPINCOTT; cf. SAXL, as in note 36_4, above, vol. III,1 [by F. SAXL and H. MEIER, London, 1953], p. LXXff. with fig. 34, and 66f.). Another set of drawings of the twelve zodiacal constellations in the "Ṣūfī Latinus" style is in MS Paris, B. N. lat. 7437, fol. 1–8ʳ; cf. GOUSSET, as cited above, p. 94ff., with figs. 2, 3, 4.

The version of al-Ṣūfī's book in Alfonso X's *Libros del saber* also contains drawings of the 48 constellations in al-Ṣūfī's style; these drawings were placed in the centre of each "rueda" (freely redrawn in RICO Y SINOBAS' edition). They are repeated in the manuscript of the Italian version, Vat. lat. 8174 (for some specimens, see SAXL, as in note 36_4, above, plate VIII/fig. 17: Orion; plate IX/fig. 19: Eridanus; plate X/fig. 23: Hydra). The Alfonsine tradition was set forth, e. g., in an English manuscript, now in Los Angeles, J. Paul Getty Museum, Ludwig XII.7 (cf. LIPPINCOTT, p. 68; SAXL-MEIER, as cited above, vol III,1, p. XXX and 275f., and vol. III,2, plate LXXI, figs. 178, 179, 180). The constellation drawings in MS Oxford, Bodl. Lib., Can. Misc. 27 (astronomical tables by Jacobus Bona Dies, around 1348), cited by GOUSSET, *l. c.*, p. 94f., perhaps also belong to the Alfonsine Ṣūfī-tradition. (For a specimen, see O. PÄCHT and J. J. G. ALEXANDER, *Illuminated Manuscripts in the Bodleian Library Oxford*, I (Oxford, 1966), plate LXV/fig. 882: Sagittarius.)

XII

Zum „liber hermetis de stellis beibeniis“

Aus der großen Zahl astronomischer und astrologischer Werke, die im Mittelalter aus dem Arabischen ins Lateinische übersetzt wurden, ragt der hier zu behandelnde Text allein schon wegen des im Titel auftretenden seltsamen Wortes „beibenie“ heraus. Im Gegensatz zu vielen anderen lateinischen Übersetzungen jedoch, die längst auf bestimmte arabische Autoren und Quellentexte zurückgeführt werden konnten, sind Herkunft und Quellengeschichte dieses Traktats bisher unerkannt geblieben. Hier kann jetzt zum erstenmal die nähere Identifikation vorgelegt werden.

Für meine Untersuchungen konnte ich mich auf folgende beiden Stellen[1] stützen: erstens den Venediger Druck von 1493 des ptolemäischen „Liber quadripartitus“ (Apotelesmatika oder Tetrabiblos), welcher auf p. 118ᵃ—119ᵃ bietet: *liber hermetis capitis omnium philosophorum de iudiciis et significatione stellarum beibeniarum in natiuitatibus* (im folgenden kurz: h); und zweitens die lateinische Handschrift der Österreichischen Nationalbibliothek Wien Nr. 3124, s. XV, wo sich auf fol. 161ʳ—162ᵛ findet: *Liber hermetis de stellis fixis que dicuntur beybenie et de significationibus ipsarum in natiuitatibus Et de locis ipsarum in cello tempore ipsius hermetis; quod fuit* ⟨lacuna⟩ *annis post christum* (im folgenden kurz: H).

Wie man sieht, handelt es sich um ein Werk von relativ geringem Umfang. Es ist in die astrologischen Texte einzureihen und hat zum Gegenstand die Darstellung der astrologischen Kräfte einer Reihe von Fixsternen. Einige dieser Sterne werden mit fremdartigen Namen belegt, in denen nur zum Teil die bekannten arabischen Sternnamen erkennbar sind, während die übrigen keinerlei Assoziationen erlauben. Rätselhaft ist also vieles an dem Text, der im übrigen durch die Beziehung auf Hermes auch hinsichtlich eines ursprünglichen Autors alles im Unklaren läßt und in die schwer überschaubare pseudepigraphische Literatur einzuordnen ist.

[1] Mit dem Druck von 1493 im Wortlaut völlig identisch ist die „Quadripartitus“-Ausgabe Venedig 1519, in der der „liber hermetis“ auf p. 107ᵛ—108ᵛ steht. F. Carmody, *Arabic Astronomical And Astrological Sciences in Latin Translation*, Berkeley — Los Angeles 1956, p. 55 (Nr. 7, 4a) zählt weitere Drucke und Handschriften auf. Der Venediger Druck von 1484 enthält in dem von mir durchgesehenen Exemplar der Kölner Universitätsbibliothek übrigens den „liber hermetis“ nicht.

Daß das Buch aus dem Arabischen übersetzt wurde, ist — abgesehen von den vorhandenen arabischen Sternnamen — auch durch das Explicit in H bezeugt (fol. 162v, unten): Completus est liber hermetis de beybeniis translatus a magistro salione de arabico in latinum toleti. Die Übersetzung wurde also von SALIO in Toledo ausgeführt[2]. SALIO ist als Übersetzer aus dem Arabischen auch sonst bekannt[3].

Bevor auf die weiteren textgeschichtlichen Zusammenhänge eingegangen wird, sei zunächst die Erklärung des seltsamen Wortes „beibenie“ gegeben. Es handelt sich um ein Adjektiv, das als Attribut zu stellae auftritt und die entsprechenden Endungen annimmt[4]. Der Inhalt des Traktats bezieht sich auf die Fixsterne; diese werden auf Arabisch gewöhnlich *al-kawākib aṯ-ṯābita* genannt, wofür die gängige lateinische Wiedergabe „stellae fixae“ lautet[5]. Für die Bildung „stellae beibeniae“ (bzw. nach alter Schreibweise „stelle beibenie“) muß ein anderer Ausdruck zugrunde liegen. Es sei vorweg gesagt, daß bisher kein einziges Manuskript der arabischen Fassung dieses Textes aufgefunden wurde, so daß die Erklärung von „beibenie“ — wie auch aller übrigen Rätsel des lateinischen Textes — nicht einfach aus dem Original heraus erfolgen kann. Ältere Deutungen gingen fehl[6]. Erst 1947 führte ein Hinweis von

[2] Cf. M. STEINSCHNEIDER, *Die hebraeischen Übersetzungen des Mittelalters...*, Berlin 1893, p. 514f (§ 318); ders., *Die europäischen Übersetzungen aus dem Arabischen...*, SB. Wiener Ak. d. Wiss. 149 und 151 (Wien 1904—05), Nr. 107b; CARMODY a.a.O.

[3] STEINSCHNEIDER, *Hebr. Übers.* p. 545f (§ 339); *Europ. Übers.* a.a.O.; CARMODY a.a.O. p. 136 (Nr. 23, 1a); L. THORNDIKE, *A Third Translation by Salio*, Speculum 32 (1957), 116f.

[4] Text: stellarum beibeniarum; de beybeniis; est alia beibenia (sic h; H: est alia beiben); sunt alie beibenie. In einem spanischen Zitat des Titels aus der Mitte des 13. Jahrhunderts: Hermes... en su libro de las bebenías (*Libros del Saber de Astronomia*, ed. RICO Y SINOBAS, Madrid I—V, 1863—67; III p. 236). In der reichhaltigen Titelaufzählung zumeist aus dem Arabischen übersetzter astronomischer und astrologischer Werke bei Albertus Magnus im „*Speculum astronomicum*“ (ed. Catalogus Codicum Astrologorum Graecorum [künftig abgekürzt: CCAG] 5, 1 p. 85ff) ist dieser Titel nicht vertreten.

[5] Transkription selten; z.B. in dem Traktat „Quicumque desiderat scire certas horas...“, ed. bei J. MILLAS VALLICROSA, *Assaig d'història de les idees físiques i matemàtiques a la Catalunya medieval*, Barcelona 1931, p. 275ff., in einer Kapitelüberschrift (p. 281, Z. 158; siehe auch p. 182), in zwei von sechs benutzten Handschriften: alkevvechib et hebita; die gleiche Stelle auch in ms Bern, Burgerbibliothek, Nr. 196, s. XI, fol. 4r, col. a: alceuuechil & hebita.

[6] STEINSCHNEIDER, *Hebr. Übers.* p. 514f (§ 318), mit Anm. 99 und 100; ders., ZDMG 24 (1870), p. 374; ZDMG 50 (1896), p. 190 (§ 109, Nr. 4); ders., *Eur. Übers.* a.a.O.; C. NALLINO, *Opus astronomicum* (al-Battānī), I (Mailand 1903), p. XXX, Anm. 8.

F. CARMODY auf die richtige Spur[7]. Bei dem von ihm herangezogenen Bīrūnī steht folgende, von CARMODY nicht ganz verständlich wiedergegebene Stelle[8]: *at-tābita ... wa-summiyat bi-l-fārisīya biyābānīya li-anna l-ihtidā' fī l-falawāt yakūnu bihā* „Die Fixsterne ... und sie werden auf Persisch biyābānīya [zu *biyābān* „Wüste"] genannt, denn die Orientierung in den Wüsten geschieht nach ihnen." Zweifellos verwendete also der arabische astrologische Text, aus dem SALIO übersetzte, für die Fixsterne diese sehr seltene, von Bīrūnī erklärte Bezeichnung: *al-kawākib al-biyābānīya*.

Das gleiche Wort erscheint ebenfalls in einem anderen astrologischen Text, von dem glücklicherweise auch die arabische Version vorhanden ist[9]. Hier lautet es auf Arabisch stets: *al-kawākib al-bābānīya*, was lediglich eine leicht zu erklärende graphische Variante von *al-biyābānīya* sein dürfte[10].

Wenden wir uns nun dem Text selbst zu. Die Untersuchung kann sich hierbei nur auf die lateinische Version stützen, da bisher keine arabische Handschrift nachgewiesen ist. Der Text gibt eine astrologische Klassifizierung einiger Fixsterne, die den „Temperamenten" der Planeten zugeordnet werden. Insgesamt 31 Sterne werden auf neun Gruppen von

[7] *Leopold of Austria, Li Compilacions de la Science des Estoilles*, Modern Philology 33, No. 2 (1947), Index s. v. *biyābānī*: „Persian for desert stars", gemäß Bīrūnī, *Tafhīm*, ed. R. WRIGHT, London 1934, § 160.

[8] *Tafhīm*, ed. WRIGHT, p. 46, § 125 (!).

[9] Das sogenannte „Centiloquium" (Karpos, *Kitāb at-tamara*), das dem Ptolemäus beigelegt ist, nebst Kommentar von Aḥmad b. Yūsuf Ibn ad-Dāya (in der lateinischen Version fälschlich dem Haly = 'Alī b. Riḍwān beigelegt, der den „Liber quadripartitus" kommentiert hat). Ich zitiere hier die lateinische Version nach dem Druck Venedig 1493 (fol. 107a—116b, unmittelbar anschließend an den obengenannten „Liber quadripartitus") sowie die arabische Version nebst Kommentar nach der Handschrift Sprenger 1839 = Ahlwardt 5874 (Stiftung Preußischer Kulturbesitz, Depot der Staatsbibliothek, Tübingen):
Verbum 29, Kommentar: Sed cum fuerit aliqua ex stellis fixis que dicuntur bei bemie [Druck 1484: alebema] .i. fixe qui sunt corda fixarum.... = arab. fol. 12v—13r, *kalima* 30, *at-tafsīr*: *fa-ammā idā kānat al-kawākib al-bābānīya allatī hiya l-qulūb min at-tābita*... — Verbum 36 [hierauf bezieht sich NALLINO in der oben, Anm. 6 angeführten Stelle]: vel aliqua stella ex albemenia = arab. fol. 14v, *kalima* 36: *aw kawkab min al-bābānīya*. — Verbum 46: .... ex stellis fixis s. albehenia [arab. fol. 17v, *kalima* 47 nur: ... *min al-kawākib at-tābita*]. Die hier gebrauchten lateinischen Transkriptionen sind von der verdorbenen arabischen Schreibung *al-bābānīya* abgeleitet (latein.-span. e entspricht mit spanischer Imāla gesprochenem arab. *ā*).

[10] A. NYKL führt in Speculum 29 (1954), 85 unter verschiedenen bei 'Alī b. Abī r-Riğāl = Abenragel gebrauchten astrologischen Fachausdrücken auch auf: bebenia, bebiniya, wozu er ohne Erläuterung die arabischen Formen *baibānīya* und *bābānīya* setzt.

planetarischen „Temperamenten" aufgeteilt. Jeder Gruppe schließt sich ein kurzer Textabschnitt an, der die astrologischen Auswirkungen der betreffenden Temperamentgruppe beschreibt.

Die Angleichung von Fixsternen an die Temperamente der Planeten ist aus den klassischen Texten der „wissenschaftlichen" Astrologie, aus denen auch die Araber schöpften, wohlbekannt[11]. In diesem Zusammenhang ist die Zahl von 31 Sternen im „liber hermetis" von ausschlaggebender Bedeutung, denn sie liefert uns den Schlüssel für die text- und kulturgeschichtliche Einordnung dieses Traktats.

In der astrologischen Literatur der Antike und ihren byzantinischen Ausläufern tritt auffallend häufig eine ganz ähnliche Darstellung von ca. 30 Fixsternen in Erscheinung. Es handelt sich um die 30 sogenannten „λαμπροὶ ἀστέρες", denen die planetarischen Temperamente beigelegt und entsprechende astrologische Kräfte[12] zugesprochen werden. Eine von den vielen Versionen dieser Fixsternastrologie wird dem Ptolemäus zugeschrieben, was jedoch nach F. Boll aus mehreren Gründen ausgeschlossen ist[13]; Boll glaubt vielmehr, daß hier über Ptolemäus hinaus noch älterer babylonischer Ursprung anzunehmen ist[14]. Anders W. Gundel, der dies Verfahren und die Sternliste auf hermetische und damit letzten Endes ägyptische Herkunft zurückführt[15]. Die 30 „λαμπροὶ ἀστέρες" erscheinen im Jahre 379 n. Chr. im Werk eines in Rom schreibenden anonymen Astrologen ägyptischer Herkunft, der die ältere Lehre des Antiochus von Athen wiedergibt[16]. Mit dem Längenwert

[11] Vor allem Ptolemäus, Tetrabiblos I, 9. Cf. speziell F. Boll, *Antike Beobachtungen farbiger Sterne*, Abh. Bayer. Ak. d. Wiss., Bd. 30, 1. Abh., München 1916, p. 71ff. Den Fixsternen wird dabei meist eine Mischung aus den Kräften zweier Planeten zugelegt (Tetrab., ed. Boll-Boer, Leipzig 1940, p. 23, 3—5: . . . τὸ ποιητικὸν ὁμοίως ἔχουσι κεκραμένον τῇ τε τοῦ Ἄρεως καὶ τῇ τοῦ Κρόνου δυνάμει). Hierauf ist offenbar die Bezeichnung dieser planetären Kräfte der Fixsterne mit dem terminus technicus κρᾶσις, arab. *mizāǧ*, lat. complexio oder temperamentum, zurückzuführen.

[12] δύναμις (Ptolemäus, Tetrab.; Rhetorius Aeg.; Exc. Paris.); ἐνέργεια (Anonymus); fortitudo (Liber quadrip.; liber hermetis im Explicit 1493). Cf. zu dem Grundsatzproblem von ποιεῖν und σημαίνειν in der antiken Astrologie: E. Pfeiffer, *Studien zum antiken Sternglauben*, Stoicheia II, Leipzig-Berlin 1916, p. V.

[13] a.a.O. p. 74f. Boll bezeichnet diesen Text daher als „Pseudo-Ptolemäus".

[14] Cf. hierzu besonders den unten noch zu erwähnenden Anonymus, der ausdrücklich angibt, der διδασκαλία des Ptolemäus zu folgen, und des weiteren erläutert, daß diese Theorie und ihre Methode zuerst von den Babyloniern eingeführt worden sei (CCAG 5, 1 p. 204 mit Anm. 1).

[15] *Neue astrologische Texte des Hermes Trismegistos*, Abh. Bayer. Ak. d. Wiss., N.F. 12, München 1936, p. 124f. Cf. ebenfalls A.-J. Festugière, *La révélation d'Hermès Trismégiste*, vol. I, Paris ²1950, p. 120, 171ff.

[16] Ediert CCAG 5, 1 p. 196ff; cf. dazu ebda. p. 194 sowie 8, 3 p. 108, Anm. 3.

66

Ptolemäus + 3°40′ (was etwa dem Jahre 505 entspräche) bietet sie Rhetorius Aegyptius[17]; Theophilos Edessenus (Ṯā'ūfīl b. Ṯūmā, am Bagdader Kalifenhof, gest. 785[18]) führt sie mit dem Längenwert Ptol. +6°20′ auf[19], und ein späterer Text enthält sie mit dem Längenwert Ptol. +7°30′ für 884[20]. Dem sind noch hinzuzufügen der bereits genannte „Pseudo-Ptolemäus"[21] sowie die Exzerpte des erwähnten Rhetorius Aegyptius aus dem babylonischen Astrologen Teukros[22]. Alle genannten Stellen stimmen im wesentlichen überein. Geringfügige Abweichungen hinsichtlich der aufgenommenen Sterne sowie deren Gesamtzahl resultieren aus der weitverzweigten Überlieferung und können uns nicht hindern, alle diese Versionen als zur gleichen Familie gehörig zu erkennen. Die beiden letztgenannten Texte („Pseudo-Ptolemäus", künftig abgekürzt: p, sowie die Exzerpte des Rhetorius Aegyptius aus Teukros, künftig abgekürzt: r) enthalten die ca. 30 Sterne in Tabellenform, nach Tierkreiszeichen aufsteigend angeordnet, während die vier anderen Texte[23] eine gruppenweise Anordnung nach astrologischen Temperamenten zeigen. Dabei erweist sich die Gruppenbildung in a, R, t, e als identisch. Die hier durchgeführte Gleichsetzung der Fixsterne mit bestimmten Temperamenten ist auch in den beiden Tabellen von r und p befolgt und stimmt im übrigen weitestgehend mit den Zuweisungen in Ptolemäus' Tetrabiblos I, 9 überein.

Stellt man einmal die astrologische Gruppierung der 30 antiken

[17] Ediert CCAG 8, 4 p. 174ff; cf. dazu ebda. p. 115ff und 174, Anm. 1.

[18] H. Suter, *Die Mathematiker und Astronomen der Araber und ihre Werke*, Leipzig 1900, p. 223, Nr. 3a.

[19] Ediert CCAG 5, 1 p. 212ff; cf. dazu ebda. p. 195.

[20] „Excerpta Parisina", ediert CCAG 5, 1 p. 217ff.

[21] Ediert von Boll, a.a.O. p. 77 und 82.

[22] Ediert CCAG 7, p. 194ff. In byzantinischer Zeit ist dieses alte astrologische Fixsternverzeichnis zu einer gewöhnlichen kleinen Fixsterntafel geworden, die dem auf orientalischen Quellentexten beruhenden astronomischen Werk des Georgios Chrysokokkes eingefügt wurde (ediert in Byzantin. Zeitschr. 57, 1964, 382ff, als „Typ III"). Das ptolemäische kalendarisch-meteorologische Werk „*Phaseis*" benutzt ständig 30 Grundgestirne, je 15 erster und zweiter Größe, die im wesentlichen ebenfalls mit den „λαμπροὶ ἀστέρες" identisch sind (cf. die Tabellen in den Editionen von C. Wachsmuth: *Lydus, De ostentis* etc., Leipzig 1897, p. 278—285, sowie J. L. Heiberg: Ptolemäus, *Opera* II, Leipzig 1907, p. CLX—CLXV). Von den hermetischen Sternen fehlen hierunter Nr. 22, 24 und 27; stattdessen sind die bei Hermes und unter den „λαμπροὶ ἀστέρες" nicht vorkommenden beiden Sterne α Car [a liefert auch die Begründung für dessen Fortlassung: CCAG 5, 1 p. 204, 6—8] und α Lib vertreten.

[23] Es werden künftig folgende Abkürzungen verwendet: a = Anonymus von 379; R = Rhetorius Aegyptius, verschieden von r = dessen Exzerpte aus Teukros; t = Theophilos und e = Excerpta Parisina.

„λαμπροὶ ἀστέρες" dem mittelalterlichen Hermes-Text gegenüber, so ergibt sich mit zwingender Notwendigkeit, daß er zur gleichen Familie gehört. Die Substanz hat bei der Wanderung durch mehrfache Übersetzungsstadien (griechisch → [persisch?[24] →] arabisch → lateinisch) natürlich Entstellungen und Veränderungen erlitten. Andererseits ist es doch erstaunlich, daß sich die Zusammengehörigkeit noch mit so überzeugender Eindeutigkeit ablesen läßt.

Es seien hier die astrologischen Gruppen der antiken und der mittelalterlichen Version nebeneinander gestellt:

	R (= e, a, t)[25]		liber hermetis[26]
	α Vir[27]		α Vir
	α Lyr	Venus +	α Lyr
Venus +	α PsA*	Merkur	α Cyg
Merkur	α Cyg		
	α CrB		
		Jupiter +	α CrB*
		Merkur	———
			β Lib
	α Leo		α Leo
	α Boo		α Sco
Jupiter +	α Aql	Jupiter +	α CMa
Mars	α Sco	Mars	β Aur*
	α CMa[28]		α Aql
	β Ori		β Ori
	ε Ori*		α Gem*
Jupiter +	β Aur[29]	Jupiter +	β Per
Saturn	α Sgr	Saturn	α Sgr
	β Per[30]		α Aur
	α Aur[31]		
Mars	β Gem[32]	Mars[33]	β Gem

[24] Für eine eventuelle persische Zwischenstufe könnten, außer dem bereits behandelten Wort „beibenie", auch die weiter unten noch zu diskutierenden fremdartigen Sternnamen sprechen.

[25] Drei Sterne, die bei H/h fehlen, sind durch Stern * gekennzeichnet.

[26] Vier Sterne, die in der Temperamentzuweisung von der antiken Überlieferung abweichen, sind durch Stern * gekennzeichnet.

[27] Tetrab. p. 25, 1—2: Venus + Mars; dagegen Lib. quadrip. = R.

[28] a: zusammen mit β Gem nur Mars.

[29] Tetrab. p. 28, 9—10 und Quadrip.: Mars + Merkur.

[30] Fehlt in t. Tetrab. = R, dagegen Quadrip.: Saturn + Venus.

[31] Fehlt in a. Tetrab. p. 28, 9—10 und Quadrip.: Mars + Merkur.

[32] t: Mars + Saturn.

[33] h: „Martis & solis" (pro „Martis solius").

68

	R (= e, a, t)		liber hermetis
Jupiter +	β Lib		
Merkur	α Gem[34]		
	γ Ori		γ Ori
	α CMi[35]		α CMi
Mars +	α Ori	Mars +	δ Peg = α And
Merkur	δ Peg = α And	Merkur	α Ori
	β Peg[36]		β Vir
	β Vir[37]		———
			β Peg
Venus +	α Cen*	Venus +	ϑ Eri
Jupiter	ϑ Eri[38]	Jupiter	α Boo*
	β Leo		δ Leo
Saturn +	δ Leo	Saturn +	α Hya
Venus	α Hya	Venus	β Leo
Mars +	α Tau[39]	Mars +	α Tau
Venus		Venus	———[40]

Hieraus ist zu ersehen, daß sich in der arabischen Version eine wesentliche Abweichung nur bei der Gruppe „Jupiter + Merkur" ergeben hat. Die kurzen astrologischen Begleittexte bieten das gleiche Bild. Von den antiken Fassungen stimmen a, R, e untereinander überein, während t eine verkürzte und zuweilen leicht umgestaltete Form aufweist[41]. Die im lateinischen „liber hermetis" reflektierte arabische Version folgt hierin ebenfalls der griechischen Hauptüberlieferung, und zwar, wie sich unten aus dem Längenwert der Sterne ergeben wird, dem Rhetorius-Text (R). Die arabisch-lateinische Korruption in der Jupiter-Merkur-Gruppe läßt sich auch hier beobachten: Der Text dieser Gruppe beginnt mit Angaben, die auf die eigentlich hier anschließende Jupiter-Mars-Gruppe passen, und erst etwa in der Mitte setzt der richtige Jupiter-Merkur-Text ein. Die arabisch-lateinische Fassung endet mit der Darstellung der letzten (Mars-Venus-) Gruppe, entsprechend R in CCAG 8, 4 p. 181, 14. Ein Vergleich des Wortlauts in R und der arabisch-lateinischen Version ergibt natürlich gewisse Differenzen in der Formu-

[34] Tetrab. p. 23, 22—23 und Quadrip.: Merkur.
[35] Fehlt in r.
[36] Fehlt in a.
[37] Nur in r.
[38] Tetrab. p. 29, 7—8 und Quadrip.: Jupiter.
[39] Tetrab. p. 23, 10—12 und Quadrip.: Mars.
[40] H: Jupiter + Venus.
[41] Cf. CCAG 5, 1 p. 213.

lierung, die sich im Laufe der Wanderung durch die verschiedenen Sprachen geradezu herausbilden mußten. Andererseits ist jedoch Übereinstimmung auch im Wortlaut des Textes in überraschend eindeutiger Form häufig vorhanden.

Was ferner die 31 Sterne selbst anbetrifft, so stimmen sie nicht nur an Zahl, Auswahl und Anordnung genau zu der antiken Tradition, sondern sie haben sogar den antiken Längenwert Ptolemäus + 3°40′ bewahrt, den auch Rhetorius (R und r) benutzt. H enthält auf der ersten Seite (fol. 161r) eine Fußnote — in der gleichen Handschrift wie der Text selbst, also aus der Vorlage mit abgeschrieben —, derzufolge den alten Sternkoordinaten „aus Hermes' Zeit“ (also Ptol. + 3°40′) für das Jahr 1185 die Differenz von 11°15′ hinzuzufügen ist (woraus sich für 1185 ein Längenwert von Ptol. + 14°55′ ergibt). Da zwei andere Übersetzungen Salios auf 1218 datiert sind[42], kann das Jahr 1185 kaum etwas mit dem Zeitpunkt der Übersetzung des „liber hermetis“ zu tun haben. Welche Rolle die Jahreszahl 1185 in diesem Zusammenhang spielt, ist nicht ersichtlich. Vielleicht ist sie hinzugefügt worden in Anlehnung an einige weitverbreitete Sternverzeichnisse aus den „Tabulae Toletanae“, die durch den gleichen Längenwert, Ptol. + 14°55′, bestimmt sind[43]. Die Sterne sind im laufenden Text, zu Beginn der einzelnen astrologischen Temperamentgruppen, verzeichnet, also nicht in Form einer Tabelle aufgestellt. So konnten sich Fehler und Abweichungen besonders leicht einschleichen. Ohne das Vorbild der antiken Texte stünde man den hermetischen Sternen mit ihren auf den ersten Blick — besonders in h — konfus erscheinenden Werten hilflos gegenüber, und zwar umso mehr, als man in einem Text, dessen arabische Herkunft erkennbar ist, keinen Längenwert erwartet, der noch unter den für das Jahr 1 der Hiǧra = 622 berechneten Wert (Ptol. + 6°38′)[44] hinabreicht — vom ptolemäischen Wert im Almagest einmal abgesehen. Die Konfusion wird dadurch vermehrt, daß wahllos einige Sterne Namen tragen, andere nicht. Diese Namen sind an den Stellen eingesetzt, wo der betreffende Bearbeiter des griechischen Originals einen Stern sicher identifizieren konnte. Die meisten Sterne waren jedoch offensichtlich bereits für den alten Bearbeiter nicht identifizierbar — obwohl es sich dabei hauptsächlich um ganz bekannte Sterne der ersten Größe handelt, die sehr wohl Eigennamen besaßen! —, so daß er sie ohne Namen ließ. In den griechischen Texten sind alle diese Sterne in ptolemäischer Weise bezeichnet, d.h. mit der Positionsbeschreibung ihrer Stellung im jeweili-

[42] Cf. oben Anm. 3.

[43] P. Kunitzsch, *Typen von Sternverzeichnissen...*, Wiesbaden 1966, p. 87ff (Typ XIII, Epoche 1181/82) und p. 95ff (Typ XIV, Epoche 1180/81).

[44] Cf. Kunitzsch, a.a.O. p. 99ff (Typ XV).

gen Sternbild. Unter den im „liber hermetis“ erhaltenen Namen sind vier (s. unten die Edition Nr. 2, 3, 4, 6 und eventuell 16), die sich nicht als lateinische Transkriptionen arabischer Sternnamen identifizieren lassen. Ebensowenig scheinen sie lateinische Transkriptionen arabischer Wortgebilde zu sein, die etwa griechische Eigennamen transkribierten. In diesem Falle müßte irgendein Anklang an entsprechende griechische Namen oder Wörter vorliegen, der sich aber nicht erschließen will. Hinzukommt, daß gerade von diesen vier Sternen mindestens einer (β Lib) keinen griechischen Eigennamen besaß, der überhaupt als Vorlage

Lfd. Nr.	Länge	Größe	Name	Mod. Bez.	R	r	p	Typ III	Korrekte Länge:
									Ptol.+3°40′
1	Lib 0° 6	1	H: hasset h: alhaiselh	α Vir	1	20	18	1	Lib 0°20
2	Sgr 24°	H:1 h:2	H: brimen h: brineti .i.vultur cadens	α Lyr	2	27	25	2	Sgr 21°
3	Aqr 12°50 h: 12°15	H:2 h:1	H: siduli h: sedulli .i.cauda galine	α Cyg	4	30	29	4	H: =
4	Lib 15°20	2	cirnecir	α CrB	5	22	20	5	Lib 18°20
5	Psc H: —— h: 29°	2	humerus equi	—	—	—	—	—	———
6	Lib H: 27°50 h: 27° 5	2	H: barchat h: berthas	β Lib	18	23	21	29	Lib 25°50
7	Leo 6°10	1		α Leo	6	16	14	6	=
8	Sco H: 16°10 h: 16°20	1		α Sco	9	25	23	9	Sco 16°20
9	Gem 21°20	1		α CMa	10	11	10	10	=
10	Gem H: 6°37 h: 6°20	2		β Aur	13	10	—	13	Gem 6°30
11	Cap 9°30	H:2 h:1	 vultur volans	 α Aql	 8	 28	 27	 8	 Cap 7°30
12	Tau 23°50	1		β Ori	11	5	5	11	Tau 23°30
13	Gem 20°40	2		α Gem	19	12	11	30	Gem 27°
14	Tau 3°50	2		β Per	15	2	2	15	Tau 3°20
15	Sgr 20°40	2		α Sgr	14	26	24	14	=
16	Tau 28°40	1	H: alabra h: alchibra	α Aur	16	7	7	16	=
17	Cnc 20°	H:— h: 1		β Gem	17	13	12	17	Cnc 0°20
18	Cnc H: 27°40 h: 20°40	H:— h:2		γ Ori	20	6	6	18	Tau 27°40
19	Cnc H: 4° 4 h: 4°24	1		α CMi	21	—	9	19	Cnc 3°10 oder: 2°50
20	Psc H: 21°30 h: 12°30	2		δ Peg= α And	23	32	31	21	H: =
21	Gem 5°40	H:1 h:2	 manus geminorum	 α Ori	 22	 9	 8	 20	 =
22	Vir H: 2°20 h: 20° 7	3		β Vir	—	18	16	—	Vir 2°40
23	Sgr H: —— h: 4°	—		—	—	—	—	—	———
24	Psc 5°50	H:— h:3		β Peg	24	31	30	22	=
25	Ari H: 3°50 h: 9° 5	H:— h:3		θ Eri	26	1	1	24	H: =
26	Lib H: 0°40 h: 0° 4	H:— h:2	[H: iuxta benenas]	α Boo	7	21	19	7	H: =
27	Leo 17°18	2		δ Leo	28	—	—	26	Leo 17°50
28	Leo H: 3° h: 27°18	H:2 h:2		α Hya	29	15	13	27	Leo 3°40
29	Leo H: 28°10 h: 3° 2	H:2 h:2		 β Leo	 27	 17	 15	 25	 H: =
30	Tau H: 16°20 h: 2°20	H:1 h:2	H: aldebaran h: alolebaran	α Tau	30	4	4	28	H: =
31	Tau 23° 8	1		—	—	—	—	—	———

in Frage kommen könnte. Es drängt sich daher aufgrund dieser eigenartigen assoziationslosen Namen der Eindruck auf, daß in der Überlieferung des „liber hermetis" zwischen dem griechischen Original und der arabischen Version, die später SALIO ins Lateinische übertrug, noch eine Zwischenstufe anzunehmen ist, die — wie die Bezeichnung der Fixsterne mit dem persischen Wort *biyābānīya* = „beibenie" nahelegt — persisch gewesen sein könnte. Allerdings ergeben sich auch aus dem Persischen keine Assoziationen für die vier unerklärlichen Namen.

Ich schließe nun die Edition der Sterne aus dem „liber hermetis" an. Wie man sieht, bietet die Handschrift H meist eine bessere Überlieferung als der Druck h. Die Sterne werden in der Edition in Tabellenform angeordnet, in der Reihenfolge, wie sie der hermetische Traktat anführt. Die Aufgliederung der astrologischen Gruppen ist bereits oben gegeben. Zur weiteren Orientierung füge ich der Tabelle auch die laufenden Nummern bei, unter denen die Sterne bei R, r und p sowie in der daraus gebildeten Chrysokokkischen Tafel von Typ III[45] zu finden sind. Ferner gebe ich im Falle falscher Längenwerte auch den von Ptolemäus ausgehend korrekten Wert, weitergerechnet um 3°40′. R und r haben fast ausnahmslos den korrekten Wert! Jedem Stern ist die moderne astronomische Bezeichnung beigegeben. Anschließend folgen die notwendigen Anmerkungen.

Anmerkungen:

1: Der Name ist nach Vergleich mit anderen Parallelen[46] als Transkription von *al-aʿzal* zu erkennen, d.i. die Kurzform des arabischen Namens von α Vir, *as-simāk al-aʿzal.*

2: Name unerklärlich. In der Schreibung „Brinek" lebte das Wort über GROTIUS und BAYER als seltenere Bezeichnung für α Lyr weiter[46]. GROTIUS' Deutungsversuch aus dem alten Testament entspricht nicht der textgeschichtlichen Situation. „vultur cadens" ist die lateinische Übersetzung des arabischen Namens von α Lyr, *an-nasr al-wāqiʿ*.

3, 4, 6: Namen unerklärlich. „cauda galine" ist die lateinische Übersetzung des Ausdrucks *ḏanab ad-daǧāǧa* „Schwanz der Henne", womit auf ptolemäische Weise α Cyg bezeichnet wird.

4: Länge Rr: Lib 15°20′.

5, 23, 31: Im Verlauf der Überlieferung entstandene korrupte Nennungen, die keinen Stern bezeichnen. Die Gesamtzahl der Sterne im „liber hermetis" beträgt also in Wirklichkeit nur 28. „humerus

[45] Cf. oben Anm. 22.

[46] Cf. P. KUNITZSCH, *Arabische Sternnamen in Europa*, Wiesbaden 1959, p. 147, Anm. 1.

equi“ ist Übersetzung aus *mankib al-faras* „Schulter des Pferdes“, womit auf ptolemäische Weise β Peg bezeichnet wird, der im „liber hermetis“ als Nr. 24 vertreten ist.

9: Länge r: Gem $21^{0}40'$.

11: „vultur volans“ ist die lateinische Übersetzung des arabischen Namens von α Aql, *an-nasr aṭ-ṭā'ir*. Länge Rr: Cap $7^{0}40'$.

16: In diesem Namen wäre nur mit Mühe ein Derivat aus der lateinischen Transkription des arabischen Namens von α Aur, *al-ʿayyūq*, zu erkennen (cf. P. Kunitzsch, *Arabische Sternnamen in Europa*, Wiesbaden 1959, p. 71, Nr. 20 und p. 119f, Nr. 27).

19: Länge R: wahrscheinlich zu lesen Cnc $3^{0}10'$. Für diesen Stern ist die ptolemäische Überlieferung gespalten; der Almagest hat Gem $29^{0}30'$ bzw. $29^{0}10'$.

21: „manus geminorum“ ist die lateinische Übersetzung aus *yad al-ǧawzā'* „Hand der ǧawzā'“, dem arabischen Namen von α Ori.

22: Länge r: Vir $2^{0}26'$.

24: Cf. oben zu Nr. 5.

26: In H anstelle eines Namens eine Positionsbeschreibung: „in der Nähe von η UMa“ (benenas ist die lateinische Transkription des verkürzten arabischen Namens von η UMa, *banāt naʿš*). Länge Rr: Lib $5^{0}40'$ (das Nullzeichen falsch für 5 genommen).

27: Länge R: Leo $17^{0}18'$.

30: Der Name ist die lateinische Transkription des bekannten arabischen Namens von α Tau, *ad-dabarān*.

Zur Abrundung des Bildes trage ich hier diejenigen Sterne der antiken Überlieferung nach, die in der arabischen Version fortgefallen sind:

α PsA :	R 3 =	r 29 =	p 28 =	Chrys. 3
ε Ori :	R 12 =	r 8	——— =	Chrys. 12
α Cen :	R 25 =	r 24 =	p 22 =	Chrys. 23
ε Vir :	——	r 19 =	p 17	———
Plejaden:		r 3 =	p 3	———
ε Cnc :	——	r 14	———	———

Das Auftreten nicht identifizierbarer Sternnamen ist in der einschlägigen Literatur nicht unbekannt. Dabei kann es sich um Namen handeln, die der Übersetzer aus seiner jeweiligen Vorlage herübernahm[47], die demnach in eine feste Tradition hineingehören und aus dieser heraus erklärt werden können (und müssen!), oder um andere, die sichtbar den Charakter künstlicher Eigenschöpfung tragen[48], die also willkürlich zustande ge-

[47] Einen entsprechenden mittelgriechischen Text habe ich soeben behandelt in Byzantin. Zeitschr. 60 (1967), 309—317.

[48] Cf. Kunitzsch, *Typen* p. 51ff (Typ VIII i); in Typ XVI, ebda. p. 103ff, ist der künstliche Charakter nicht so stark ausgeprägt.

kommen und nicht ableitbar sind. Unsere hermetischen Namen sind auf jeden Fall der ersten Gruppe zuzurechnen.

Die Zuweisung dieses astrologischen Fixsterntraktats an Hermes, die SALIO zweifellos aus der arabischen Version übernommen hat, ist in den uns bekannt gewordenen antiken Fassungen nicht enthalten. Innerhalb des Arabischen kann sie als sicherer Hinweis darauf genommen werden, daß der Traktat einen aus fremden Quellen übernommenen Gegenstand behandelt, also übersetzt oder kompiliert, nicht aber rein arabisch ist.

Unter den bekannteren hermetischen Werken ist auch ein „Liber de quindecim stellis" in verschiedenen Fassungen überliefert[49], der jedoch keinerlei Beziehungen zu unserem Traktat hat[50] und scharf davon zu trennen ist. Von seinen 15 Sternen sind fünf[51] nicht unter den 30 „λαμπροὶ ἀστέρες" vertreten.

Wir kommen zum Schluß. Der unter Hermes' Namen überlieferte „liber de stellis beibeniis", der im Abendland noch bis zum Anfang des 16. Jahrhunderts sogar im Druck verbreitet wurde, gehört in eine alte astrologische Tradition. Wenn auch das arabische Zwischenglied bisher nicht aufgefunden wurde, so erweist doch der Vergleich der daraus geflossenen lateinischen Version mit den antiken Textfassungen die völlige Zusammengehörigkeit. Der Gesamtcharakter des Textes, Aufbau und Reihenfolge der astrologischen Gruppen, die Zuweisung der Temperamente an die einzelnen Sterne, der Wortlaut im astrologischen Begleittext, Auswahl und Zahl der Sterne und vor allem der mit den antiken Rhetorius-Stellen identische Längenwert der Sternkoordinaten, dazu die deutlichen Hinweise im anzunehmenden arabischen Text auf nichtarabische Herkunft — dies alles dürfte beweiskräftig genug sein, um an der Zuordnung des „liber de stellis beibeniis" zur Fixsternastrologie der 30 „λαμπροὶ ἀστέρες" jeden Zweifel auszuschließen[52].

[49] Ed. L. DELATTE, *Textes latins et Vieux Français relatifs aux Cyranides*, Liège 1942, p. 235ff; cf. auch FESTUGIERE, *La révélation*... p. 160ff.

[50] FESTUGIERE a.a.O. p. 173f.

[51] Dies sind: Alchoraya (Plejaden), Ala Corvi (γ Crv), Benenays (ηUMa), Alfeca (α CrB) und Cauda Capricorni (γ oder δ Cap).

[52] Ein ganz später, entlegener, verzerrter Reflex dieser alten Tradition findet sich bei dem arabischen Philologen Marzūqī (gest. 1030) in seinem kompilierten „*Kitāb al-azmina wa-l-amkina*", Kap. 63 (Über die bekannteren Fixsterne), Druck Hyderabad 1332 h, II p. 370, 9—12: Nach der aus der astronomischen Fachliteratur antiker Herkunft übernommenen Anführung der „Größen" der Fixsterne und Nennung der von Ptolemäus herstammenden Zahlen (in der ersten Größe 15 Sterne, in der zweiten Größe 45 — zusammen also 60) heißt es dort weiter: „Die sich mit Sterndeutung befassenden Astronomen hoben aus diesen sechzig Sternen 30 Sterne heraus, wiesen jedem davon ein 'Los' [*ḫarāǧ*] aus den Naturen der fünf Wandelsterne zu und machten sie zur Grundlage der Prognosen [*aqḍiya*], die sie zu stellen

In der Astronomie, vor allem auch in der Fixsternkunde, war bis ins siebzehnte Jahrhundert hinein im Morgen- wie Abendland Ptolemäus die höchste Autorität. Von ihm stammt auch das grundlegende astrologische Lehrwerk, aus dem anderthalb Jahrtausende ihre Kenntnisse und Theorien auf diesem Felde bezogen. Daneben scheint jedoch in der Astrologie das magische Charisma des Hermes stark genug gestrahlt zu haben, daß dieser kleine, anspruchslose, unscheinbare Text unter seinem Namen bis an die Schwelle der Neuzeit überdauern konnte.

pflegen [? — *yaḫlifūnahā*]. Gott aber tut, was er will, und setzt fest, was ihm beliebt.“ Auch bei Bīrūnī scheint auf den ersten Blick eine abgeschwächte Reminiszenz an diese Fixsternastrologie sichtbar zu werden: *Tafhīm*, ed. Wright, p. 68, § 157 (Übersetzung Wright, nach Vorwort p. XII aus der persischen Handschrift Brit. Mus. Add. 7697):... a scale of magnitudes, *qadr*, *ʿuẓm*, to the two first degrees of which astrologers give the name of glory, *šaraf*. Der gegenübergestellte arabische Text jedoch (gemäß Vorwort p. XII: ms Brit. Mus. Add. Or. 8349) besagt etwas ganz anderes: nämlich daß die Astrologen bei den 15 Sternen erster Größe für „Größe“ (statt des in der Astronomie üblichen *qadr* bzw. *ʿuẓm* = ptolem. μέγεθος) zuweilen auch den aus der astrologischen Terminologie stammenden Ausdruck *šaraf* (= ὕψωμα, exaltatio — also nicht „glory“!) benutzen (*wa-rubbamā ʿabbara aṣḥāb al-aḥkām ʿanhā* [scil. *al-aqdār* bzw. *al-aʿẓām*] *bi-š-šaraf fa-qālū fī š-šaraf al-awwal badala qawl ġayrihim al-qadr al-awwal*). Wenn sich dieser Zusammenhang bei Bīrūnī demnach also doch nicht einstellt, so bleibt immerhin interessant, daß unser besprochener „liber hermetis“ durchgehend bei den Größenangaben der Sterne statt „magnitudo“ jenes Wort „exaltatio“ hat, das durch die Bīrūnī-Stelle nun einleuchtend erklärt ist; die arabische Übersetzungsvorlage wird dort eben auch *šaraf* gehabt haben (R, r hatten das übliche μέγεθος).

Korrekturzusatz: *yḫlfūnahā* in dem obigen Zitat aus Marzūqī ist sicher aus *yaḥkumūnahā* verlesen. Man beachte hierzu die als Schlußphrase zitierten koranischen Formeln *wa-llāhu yafʿalu mā yašāʾu* [z.B. Sure 3,39] *wa- yaḥkumu mā yurīdu* [S. 5,1], deren zweite offensichtlich bewußt im Anschluß an das Verbum *yaḥkumūnahā* gewählt ist. Der gleiche Begriff erscheint bereits zu Anfang des Abschnitts in der Bezeichnung der Astrologen: *aṣḥāb al- aḥkām min al-munağğimīn*; ferner vergleiche man ganz allgemein den geläufigen Ausdruck (*al-ʿilm bi-*) *aḥkām an-nuǧūm* für “Astrologie”.

XIII

Neues zum „liber hermetis de stellis beibeniis"

Kürzlich konnte ich in dieser Zeitschrift[1] den „*liber hermetis de stellis beibeniis*", einen kleinen, aber stark verbreiteten und viel zitierten mittelalterlichen Traktat zur Fixsternastrologie, mit einem antiken astrologischen Text identifizieren. Der mittelalterliche lateinische Traktat ist aus einer arabischen Zwischenstufe übersetzt; die arabische Textfassung selbst war jedoch bisher noch nicht aufgefunden.

Soeben stieß nun Herr Dr. M. Ullmann, Tübingen, bei arabischen Handschriftenstudien in Dublin zufällig auf einen hermetischen astrologischen Text, den er mir unverzüglich mitteilte und freundlichst zur näheren Untersuchung überließ[2].

Die Untersuchung ergab, daß hier tatsächlich ein Exemplar der arabischen Fassung jener (pseudo)-hermetischen Schrift über die „*stellae beibeniae*" zu Tage getreten ist. Damit ist eine weitere Lücke in der Erforschung der mittelalterlichen Übersetzungsliteratur und ihrer Quellen geschlossen. Es verlohnt sich daher, hier ergänzend zu jenem früheren Artikel die wichtigsten Einzelheiten im Vergleich zwischen der lateinischen und der arabischen Fassung mitzuteilen.

Der arabische Text steht in der Handschrift Nr. 5399 der Chester Beatty Library in Dublin, auf fol. 206^{v}—208^{v}. Hier bricht er, fast genau in der Mitte, gegen Ende der Gruppe „Jupiter + Saturn" ab; damit sind die Sterne 1—16 der in jenem ersten Artikel auf S. 70 tabellarisch edierten Zusammenstellung erfaßt. Vermutlich handelt es sich um eine einfache Verschiebung einiger Lagen, so daß die zweite Hälfte des Traktats möglicherweise an einer anderen Stelle innerhalb der Handschrift existiert.

Die Überschrift lautet: *hāḏā kitāb waḍaʿahū Hermes ra'ʾs al-ḥukamāʾ* [vgl. lat.: *liber hermetis capitis omnium philosophorum* (h)] *wa-ntaḫalahū* [ms mit *ḥ* statt *ḫ*] *wa-ntağabahū wa-mtaḥanahū min asrār an-nuğūm* „Das ist ein Buch, das Hermes, das Oberhaupt der Weisen, verfaßt hat. Er wählte und sammelte und sichtete es aus aus den Geheimnissen der Sterne".

[1] Bd. 118, 1968, S. 62—74.

[2] Auch die Beschaffung der entsprechenden Mikrofotos besorgte Herr Dr. Ullmann, wofür ich ihm auch an dieser Stelle meinen Dank ausspreche.

Der Wortlaut der Überschrift ist nur teilweise identisch mit der lateinischen Formulierung, der Text selbst jedoch stimmt vollkommen überein. Incipit: *anā wāṣif laka min amr as-saʿāda* ... = lat. *ego dicam vobis de rebus fortune* ....

Die Bezeichnung der „*stellae beibeniae*“ wird überall ausnahmslos *(al-)bābānī(ya)* geschrieben. Die ursprüngliche persische Form *biyābānī*[3] ist hier also völlig verdrängt.

Der Planet Mars wird meist mit dem persischen Namen *bahrām* und nur zweimal mit seinem arabischen Namen *al-mirrīḫ* bezeichnet, was wohl als weiteres Überbleibsel einer ehemals persischen Vorstufe gewertet werden darf[4].

In der Beschreibung der Sterne sind die Koordinaten stets in Worten ausgeschrieben. Zu Stern 10 und 11 hat die arabische Handschrift korrektere Längenwerte (6°30 bzw. 7°30) als die beiden lateinischen Quellen. Die Breiten werden nicht in Grad angegeben, sondern es wird, genau wie auch im Lateinischen, immer nur die Himmelshälfte „südlich“ bzw. „nördlich“ vermerkt; dabei schreibt die Handschrift konsequent statt *ǧanūbī* „südlich“ schlecht und offenbar unerkannt حرى (bei Stern 3 sogar عرى), und statt *šamālī* „nördlich“ immer ماى (Sterne 9, 12, 13, 15).

[3] Cf. hierzu außer den in jenem ersten Artikel S. 63f. zitierten Stellen noch B. Dorn, *Drei ... astronomische Instrumente mit arabischen Inschriften*, Mémoires de l'Académie Impér. des Sciences de St. Pétersbourg, VIIe série, tome IX, No. 1 (1865), S. 139 und die dort erwähnten Quellen: Mullā Muẓaffar, persische Bearbeitung der *Bīst bāb* von Niẓām ad-Dīn ʿAbd al-ʿAlī al-Barǧandī, lith. Teheran 1276 = 1859 (s. dort S. 101, Anm. 1); Freytag, Lex. I 180b s.r. *byn*: *al-kawākib al-bayābānīyāt* (Dorn falsch البيبانيات); *A Dictionary of Technical Terms used in the Sciences of the Musulmans*, ed. by Mawlawies Moh. Wajih, Abd Al-Haqq and Gholam Kadir, under the superintendence of Dr. A. Sprenger and W. Nassau Lees, 1862, S. 1243 (leg. 1245).

Herr Dr. Ullmann macht mich noch darauf aufmerksam, daß schon 1942 (vor Carmody also) P. Kraus den Ausdruck *al-bayābānīya* als Bezeichnung der „Fixsterne“ bei Ǧābir festgestellt hat (*Jābir b. Ḥayyān*, vol. II, S. 285, Anm. 1). Erwähnenswert ist bei dieser Gelegenheit endlich ebenfalls der Titel einer angeblichen hermetischen Schrift (*risāla*) „*Ḥarb al-kawākib al-barbāwīya*“, zitiert in dem alchemistischen Werk *Kitāb al-ʿilm al-muktasab* von Abū l-Qāsim Muḥammad ibn Aḥmad al-ʿIrāqī, ed. u. übers. E.J. Holmyard, Paris 1923, Text S. 45, ult. — 46, 1 = Übers. S. 50: „The War of the Egyptian Stars“; aufgegriffen von J. Ruska, *Tabula Smaragdina*, Heidelberg 1926, S. 58 („Krieg oder Kampf der Tempelsterne“). Es muß dahingestellt bleiben, ob hierin die Lesung *al-barbāwīya* (Adjektiv zu arab. *barbā* „(heidnischer) Tempel“, aus kopt. *p'erpe*; cf. die Wörterbücher von Dozy und Blachère-Chouémi-Denizeau, sowie die Artikel „al-Barbā“ in der alten und der neuen Ausgabe der *Encyclopaedia of Islam*) echt ist oder ob damit ursprünglich die hermetischen *kawākib al-bābānīya* bzw. *al-biyābānīya*, „die Fixsterne“, gemeint waren.

[4] Der lateinische Text übersetzt beide Ausdrücke gleichermaßen mit *mars*.

Diese Angaben verhalten sich außer bei Stern 1, 8 und 13 stets entgegengesetzt zur astronomischen Wirklichkeit. Der Terminus für die „Größe" (*exaltatio*) der Sterne ist in der Tat *šaraf* (cf. jenen Artikel S. 73f., Fn. 52).

Die eigenartigen unkenntlichen Namen, die einige der Sterne in der lateinischen Version tragen, sind ebenso auch in der arabischen Fassung vorhanden. Sie sind hier indes genauso unkenntlich wie dort! Es lassen sich darin weder arabische, noch persische, noch griechische Gestirnnamen erkennen. Im einzelnen lauten sie wie folgt:

Stern 1: حسيد. Der lateinische Übersetzer transkribierte *hasset* [H], d.h. er las in seiner Vorlage *ḥsyd*. Die Schreibung *alhaiselh* in h ist sekundär durch Vorsetzung von *al-* in den schematisierten Typ arabischer Fremdwörter umgeformt worden. Eine Beziehung zum arabischen Namen von α Vir, (*as-simāk*) *al-aʿzal*, besteht also, wie nunmehr erwiesen ist, nicht![5] Eine sinnvolle Lesung des arabischen Schriftzuges حسيد ergibt sich nicht. Da das Wort aus einer persischen Vorstufe übernommen ist, kann der erste Buchstabe ursprünglich kein *ḥ* gewesen sein, sondern entweder *ǧ* oder *ḫ*. Eine Lesung *Ǧamšīd* (der sagenhafte erste König der Perser) kommt nicht in Frage, da dies semasiologisch sinnlos wäre und da die Handschrift die drei Zähne der Buchstaben *sīn* und *šīn* immer sorgfältig ausschreibt und niemals bloß durch eine gestreckte glatte Linie andeutet.

Stern 2: كيمان (dem *k* fehlt die Schräghaste oben rechts). Der lateinische Übersetzer transkribierte *kimen* [H; so hier besser zu lesen statt *brimen*, wie in ZDMG 118 notiert], d.h. er las den Schriftzug *kym'n*. Richtige Lesung und Bedeutung sind völlig ungewiß (*kiyān* kann „Stern" bedeuten, cf. Vullers, *Lexicon Persico-Latinum*, s.v.). Die — richtige — astronomische Gleichsetzung mit *vultur cadens* (= *an-nasr al-wāqiʿ*) = α Lyr ist in der arabischen Handschrift nicht vorhanden.

Stern 3: سيدول. Der lateinische Übersetzer transkribierte *siduli* [H], d.h. er las den Schriftzug *sydwl*. Richtige Lesung und Bedeutung des Wortes ungewiß. Die astronomische Gleichsetzung mit *cauda galine* (= *ḏanab ad-daǧāǧa*) = α Cyg ist in der arabischen Handschrift nicht vorhanden. Diesem Stern fügt der arabische Text noch eine spezielle astrologische Definition hinzu: *sahm as-saʿāda allaḏī yuqālu lahū* الهاريح = lat. [nur H] *et significat partem fortune que dicitur super omnibus esse* (der unkenntliche Eigenname ist hier als Infinitiv „*super omnibus esse*" übersetzt).

[5] Damit erübrigt sich die provisorische Gleichsetzung von *alhaiselh* mit *alaizel, alaçelh* (= *al-aʿzal*) bei P. Kunitzsch, *Arabische Sternnamen in Europa*, Wiesbaden 1959, S. 147.

Stern 4: سرماسر (weiter unten in den astrologischen Auslegungen noch einmal als: السر ماسير). Der lateinische Übersetzer transkribierte *cirnecir*, d.h. er las den ersten Bestandteil des Namens als *cir* = arab. *sirr* „Geheimnis“, den zweiten als *māsir*. Der erste Teil könnte jedoch eher pers. *sar* „Kopf, Anfang“ zu lesen sein. Die Bedeutung des Ganzen ist ungewiß.

Stern 5: In den beiden lateinischen Quellen unzulänglich überliefert und daher von mir als leeres Duplikat gedeutet. In der arabischen Handschrift ist es jedoch ein richtiger, voll zu zählender Stern; Name: نبر سم, Länge Psc 27°0 [die Stelle der Gradzahl in H leer gelassen; h hat Psc 29°0]; zweiter Größe; Breitenangabe im Arabischen und in H weggelassen, h: *septentrionalis*; Temperament Jupiter und Merkur (diese Gruppe umfaßt damit also drei Sterne). Lesung und Bedeutung des Namens sind ungewiß. Die astronomische Gleichsetzung mit *humerus equi* (= *mankib al-faras*) = β Peg in den lateinischen Quellen ist falsch; dieser Stern ist vielmehr in der Liste als Nr. 24 vertreten. Den in der arabischen Handschrift verzeichneten Namen hat die lateinische Übersetzung als einzigen nicht transkribiert. Eine eindeutige astronomische Identifikation aus den Koordinaten ergibt sich nicht (theoretisch käme in Frage δ And, dessen Länge bei Ptolemäus Psc 25°20 ist; die Zunahme um 3°40 würde auf die in h notierte Länge von Psc 29°0 führen. Doch dürfte dieser insignifikante Stern dritter Größe hier schwerlich gemeint sein).

Stern 6: برحاد. Der lateinische Übersetzer transkribierte *barchat* [H], d.h. er las den Schriftzug *barǧād*. Die richtige Lesung und Bedeutung sind ungewiß.

Stern 11: Ohne Namen, aber astronomisch gleichgesetzt mit *al-wāqiʿ* (*wa-huwa l-wāqiʿ*), d.i. der einheimische arabische Name von α Lyr. Das ist jedoch falsch, da dieser Stern bereits oben als Nr. 2 vorkommt; hier handelt es sich um *an-nasr aṭ-ṭāʾir* (= lat. *vultur volans*) = α Aql. Die lateinischen Quellen haben die richtige Bezeichnung *vultur volans*. Im Arabischen liegt offenbar eine Verwechslung der beiden *nasr* (*al-wāqiʿ* und *aṭ-ṭāʾir*) vor.

Stern 16: العبر. Der lateinische Übersetzer transkribierte *alabra* [H], d.h. er las den Schriftzug *al-ʿbr*. Im Rahmen des hier allenthalben beobachteten textgeschichtlichen Zusammenhanges kann es sich dabei nicht um ein arabisches Wort handeln; daher kann am Anfang nicht der arabische Artikel *al-* stehen (die entsprechenden Schriftzeichen wären entweder beiseite zu lassen, oder aber es verbergen sich dahinter rudimentäre Bestandteile des eigentlichen nichtarabischen Namens); das ʿ müßte ebenfalls ursprünglich ein anderer, nicht spezifisch arabischer Laut gewesen sein. Andererseits ist zu beachten, daß der Stern α Aur den griechischen Eigennamen Αἴξ, Capella trägt, den die arabischen Almagest-Versionen als *al-ʿanz* „Ziege“ (lat. Almagest: *hircus*) übertragen. Der oben zitierte Schriftzug könnte sehr gut dieses arabische Wort *al-ʿanz* darstellen, das in diesem Falle allerdings erst sekundär aus anderen arabischen Texten hierher übernommen sein müßte, da eine persische Zwischenübersetzung kaum hierauf geführt haben würde.

Der Wert dieses neuentdeckten arabischen Textstückes ist zwiefach: Auf den Einzelfall bezogen, haben wir damit die dritte Version jenes astrologischen Traktats über die dreißig λαμπροὶ ἀστέρες gewonnen: nach der antik-hellenistischen und der mittelalterlich-lateinischen Fassung nun auch die arabische, aus der jene lateinische hervorgegangen ist. Die bereits vorher getroffene Annahme eines mittelpersischen Zwischengliedes zwischen der antik-hellenistischen und der arabischen Fassung erfährt aus verschiedenen Einzelheiten der arabischen Übersetzung weitere Stützung.

Generell gesehen, kann dieser arabische Hermes-Text dazu beitragen, vorliegende isolierte Stücke der Übersetzungsliteratur mit wesentlich mehr Zutrauen zu betrachten. Wie sich in diesem Fall zu dem obskur anmutenden „*liber hermetis de stellis beibeniis*“ tatsächlich das arabische Original angefunden hat, in dem dann auch gewisse unverständliche Einzelheiten nachweisbar sind, so kann man ohne weiteres davon ausgehen, daß auch andere ähnliche ungewöhnliche und problematische Texte auf sicheren Quellen beruhen, die freilich im Einzelfall immer nur ein glücklicher Zufall zu Tage fördern wird.

Was übrigens die mittelpersische Etappe zwischen antik-hellenistischen und arabischen Texten anbetrifft, so ist unsere konkrete Kenntnis von direkten Einzelheiten daraus bisher noch mehr als dürftig. Aber es liegen immerhin auf verschiedenen Gebieten sehr konkrete Hinweise vor (z.B. die Geoponika/*filāḥa*-Literatur; astronomische Tafelliteratur — *zīǧ aš-šahriyār*; Teukrosfragmente in Abū Maʿšars *Kitāb al-madḫal al-kabīr*; etc.), und es wäre besonders wünschenswert, wenn die Forschung zukünftig gerade diesem bisher allzu stiefmütterlich behandelten Teilgebiet der Übersetzungsliteratur mehr Aufmerksamkeit widmen würde. Mit nützlichen Ergebnissen ist sicher zu rechnen.

Namenkundlich sind wir durch die Entdeckung des arabischen Textes insofern einen Schritt weiter gekommen, als wir die Transkriptionen der lateinischen Quellen (vor allem der Handschrift H) voll bestätigt finden. Doch gibt uns andererseits auch die arabische Schreibweise keinen zuverlässigen Hinweis auf die exakte Form und Bedeutung jener seltsamen, von allen bekannten Traditionen abweichenden Bezeichnungen.

XIV

Stelle beibenie – *al-kawākib al-biyābānīya*
Ein Nachtrag

In zwei Beiträgen in dieser Zeitschrift (118 [1968], 62–74, und 120 [1970], 126–130) hatte ich den *Liber hermetis de stellis beibeniis*, aus dem Arabischen ins Lateinische übertragen von Salio von Padua in Toledo, Anfang des 13. Jh.s, sowie dessen arabisches Original, von dem die erste Hälfte in der Dubliner Handschrift Chester Beatty 5399 aufgefunden wurde, besprochen. Wie ich dort nachweisen konnte, entstammt dieser pseudo-hermetische arabische Text, offensichtlich über eine persische Zwischenstufe, einer hellenistischen Tradition über die 30 λαμπροὶ ἀστέρες, d.h. also einer astrologischen Lehre, die diesen Fixsternen die 'Temperamente' der Planeten zuordnet und ihnen so einen Platz im astrologischen Lehrgebäude anweist.

In dem arabischen Text, der offensichtlich aus einer persischen Vorstufe übernommen ist, sind die Fixsterne mit dem merkwürdigen Ausdruck *al-kawākib al-bābānīya* bezeichnet (cf. 1970, S. 127). Diese auffällige Bezeichnung wurde im folgenden von mehreren anderen astrologischen und astronomischen Autoren übernommen: *Kitāb aṯ-ṯamara* (die arabische Übersetzung des pseudo-ptolemäischen Καρπός[1], zu der Aḥmad b. Yūsuf Ibn ad-Dāya [gest. 951][2] einen Kommentar gegeben hat[3]); al-Bīrūnī, *Kitāb at-tafhīm* (1029)[4]; Azarquiel

[1] Cf. 1968, S. 64, Fn. 9.

[2] Cf. M. Ullmann: *Die Natur- und Geheimwissenschaften im Islam.* Leiden 1972, S. 238f. und 327f.

[3] Nach einer Theorie von R. Lemay, der ich mich jedoch vorläufig noch nicht anschließen möchte, wäre das *Kitāb aṯ-ṯamara* von Ibn ad-Dāya selbst verfaßt und auf Ptolemäus' Namen gefälscht worden; cf. R. Lemay: *Origin and Success of the Kitāb Thamara* [*sic*] *of Abū Ja'far Aḥmad ibn Yūsuf ibn Ibrāhīm.* In: *Proceedings of the First International Symposium for the History of Arabic Science, April 5–12, 1976.* Vol. 2: *Papers in European Languages.* Aleppo 1978, S. 91–107.

[4] Cf. 1968, S. 64.

(gest. 1100)[5]; Ibn Abī r-Riǧāl (Abenragel, gest. zwischen 1041 und 1061)[6]; Ǧābir b. Ḥayyān[7]; Abraham bar Ḥiyya (gest. um 1136)[8]; der Alchimist Abū l-Qāsim Muḥammad b. Aḥmad al-ʿIrāqī (ca. Mitte 13.Jh.)[9]; az-Zanātī (ca. 12./13.Jh.)[10].

Dabei erscheint das Wort *al-bābānīya* auch in folgenden Schreibungen: *al-biyābānīya, al-bayābānīya, al-bayānīya*, und vielleicht auch *al-barbāwīya.*

al-Bīrūnī a.a.O., ausgehend von der Form *al-biyābānīya*, hatte die Bezeichnung in Anlehnung an neupers. *biyābān* „Wüste" erklärt: (*aṯ-ṯābita*) . . . *wa-summiyat bi-l-fārisīya biyābānīya li-anna l-ihtidāʾ fī l-falawāt yakūnu bihā*, „(Die Fixsterne) . . . und auf Persisch werden sie *biyābānīya* genannt, denn die Orientierung in den Wüsten geschieht nach ihnen" (cf. 1968, S. 64). Dieser von dem neupersischen Wort *biyābān* „Wüste" ausgehenden Erklärung sind auch andere orientalische und westliche Autoren gefolgt[11], und ich hatte mich ihr seinerzeit ebenfalls angeschlossen[12, 13].

[5] In seiner Schrift über die *Azafea*; cf. J. M. Millás in: Al-Andalus 1 (1933), 184 (arab. Text) und 185 mit Fn. 1 (Übersetzung). Daraus eine altspanische Übersetzung unter Alfons X. von Kastilien, *Libros del saber*, cf. 1968, S. 63, Fn. 4.

[6] Cf. 1968, S. 64, Fn. 10.

[7] Cf. 1970, S. 127, Fn. 3.

[8] *La obra Forma de la Terra de R. Abraham bar Ḥiyya ha-Bargeloni.* Trad. del. hebreo . . . por J. M. Millás Vallicrosa. Madrid-Barcelona 1956, S. 119.

[9] Cf. 1970, S. 127, Fn. 3; zum Autor vgl. Ullmann: *Nat.- u. Geheimwiss.* (l.c.), S. 225f.

[10] In einem Kairoer Druck (o.J.): *al-Aqwāl al-marḍīya fī maʿrifat al-aʿmāl ar-ramlīya*, S. 3, 14—16. Die Echtheit der Zuweisung dieser Schrift an den hauptsächlich als geomantischer Autor bekannten, im einzelnen noch nicht erforschten az-Zanātī muß vorläufig ungeklärt bleiben.

[11] Cf. Dorn, Mullā Muẓaffar, und *Dictionary of Technical Terms*, zitiert in 1970, S. 127, Fn. 3.

[12] Cf. 1968, S. 64; hiernach dann auch Ullmann: *Nat.- u. Geheimwiss.* (l.c.), S. 290. F. Sezgin: *Geschichte des arabischen Schrifttums*, IV (Leiden 1971), S. 43, Nr. 23, erwägt, ob der *Liber de stellis beibeniis* identisch sein könne mit einem *Kitāb al-kawākib as-sayyāra*, gedruckt Kairo 1297, was jedoch kaum zu erwarten ist, da unser Traktat von den Fixsternen, jenes Buch dagegen gemäß dem Titel von den Planeten handelt.

[13] Ganz anders, und offenbar ausgehend von der latinisierten Form des Wortes, deutet Abraham bar Ḥiyya die Bezeichnung: „Este movimiento se llama de acceso y receso, y es por esta razón por lo que las estrellas fijas se han llamado en las lenguas griega y latina [ב]א בינש *va, venes*, lo que quiere decir en su lengua: va y viene". Der Autor deutet das Wort hier fälschlich als (arab. bzw. hebr.) Transkription eines romanischen Ausdrucks mit der Bedeutung va y viene, der offenbar die Vorwärts- und Rückwärtsbewegung der ‘Trepidation’ umschreiben soll.

In Wirklichkeit hat das Wort *al-biyābānīya* jedoch eine ganz andere Herkunft, die durch eine Arbeit von W. B. HENNING[14] nachgewiesen ist, welche mir erst nachträglich zu Augen gekommen ist. HENNING hat dort das 2. Kapitel des *Bundahišn* übersetzt. In diesem in Pahlavi geschriebenen kosmologischen Werk vom Ende der Sassanidenzeit[15] kommt die Bezeichnung *a-wiyābān-īg* für die Fixsterne vor[16]. Der Pahlavi-Ausdruck *a-wiyābān-īg* seinerseits dient zur Wiedergabe von griech. ἀπλανής, dem geläufigen griechischen Wort für „Fixstern" (im Gegensatz zu den 'Planeten'); und aus dem Pahlavi ist dann weiter das neupers. Wort *biyābānī* im Sinne von „Fixstern" herzuleiten, das auch in unserem arabischen, aus einer persischen Zwischenstufe übernommenen pseudo-hermetischen astrologischen Traktat gebraucht ist.

Damit ist nun die richtige Etymologie von *al-kawākib al-biyābānīya* = *stelle beibenie* nachgeliefert. Sie beruht, wie man sieht, nicht auf philologischer Spekulation, sondern das Etymon ist erfreulicherweise tatsächlich in einem Pahlavi-Text nachgewiesen.

Die Kenntnis des befremdlichen Wortes blieb in der arabischen Welt nicht auf den Kreis der Astrologen und Astronomen beschränkt, auch die alten arabischen Philologen und Lexikographen haben es aufgegriffen. So finden wir es verzeichnet im *Tāǧ* (s. r. *byn*), im *Qāmūs* (s. r. *byn*) und im *Lisān al-ʿarab* (s. r. *bbn*)[17]. Ausgangspunkt dieser lexikographischen Überlieferung ist, wie auch im *Lisān* und im *Tāǧ* angegeben, der *Tahḏīb al-luġa* von al-Azharī (895/6–980/1). Hier heißt es[18]: „al-Munḏirī[19] hat mir von Abū l-Hayṯam[20] berichtet, dieser habe

[14] W. B. HENNING: *An Astronomical Chapter of the Bundahishn.* In: JRAS 1942, 229–248.

[15] Cf. HENNING a.a.O., S. 229.

[16] HENNING a.a.O., S. 232 mit Fn. 3.

[17] Aus diesen Quellen dann gekürzt bei G. W. FREYTAG: *Lexicon arabicolatinum.* 1. Halle 1830, 180b (s.r. *byn*).

[18] *Tahḏīb al-luġa,* 1–15. Kairo 1964–67; siehe Bd. 15, S. 498b,–5 ff. (s.r. *byn*): *aḫbaranī l-Munḏirī ʿan Abī l-Hayṯam annahū qāla: al-kawākib al-bābānīyāt hiya allatī lā tanzilu bihā šams wa-lā qamar, innamā yuhtadā bihā fī l-barr wa-l-baḥr wa-hiya šaʾāmiya wa-mahabb aš-šamāl minhā, awwaluhā l-quṭb, huwa kawkab lā yazūlu, wa-l-ǧady wa-l-farqadān, wa-huwa bayna l-quṭb, wa-fīhi banāt naʿš aṣ-ṣuġrā.*

[19] Zu al-Azharīs Gewährsmann al-Munḏirī (gest. Raǧab 329 = April 941) cf. C. BROCKELMANN: *Geschichte der arabischen Litteratur.* 1. Leiden 1943, 124 und Suppl. 1. Leiden 1937, 189 (Nr. 6) und die dort zitierten Quellen.

[20] Zu Abū l-Hayṯam ar-Rāzī, mit dem al-Munḏirī noch jahrelang Umgang gehabt haben soll, vgl. al-Azharī: *Tahḏīb* (l.c.), Bd. 1, S. 26 (= K. V. ZETTERSTÉEN: *Aus dem Tahḏīb al-luġa al-Azharī's.* In: MO 14 [1920], 25 f.); al-Anbārī:

gesagt: *al-kawākib al-bābānīyāt* sind diejenigen [Sterne], bei denen weder Sonne noch Mond Station machen[21]. Man orientiert sich nach ihnen zu Lande und auf See. Sie sind nördlich, und der Nordwind weht aus ihrer Richtung. Der erste von ihnen ist *al-quṭb*[22], das ist ein Stern, der nicht untergeht; ferner *al-ǧudayy*[23] und *al-farqadān*[24]; er ist zwischen (?)[25] dem Pol; zu ihm gehören (?) *banāt na'š aṣ-ṣuġrā*[26]". Dieser Wortlaut wurde auch im *Lisān* und weiter im *Tāǧ* übernommen.

Das Wort – ein Fremdwort, wie oben gezeigt mittelpersischer Herkunft – wurde von den arabischen Philologen etymologisch nicht erkannt und gewaltsam einer arabischen Wurzel (*byn*; im *Lisān* aufgrund abweichender Schreibweise *bbn*) zugeordnet. Auch der Bedeutungsgehalt ist nicht richtig erfaßt. Es wurde anscheinend soviel erkannt, daß der Ausdruck Fixsterne bezeichnen soll; er wurde dann aber nicht auf die Fixsterne schlechthin bezogen, sondern nur auf einen Teil derselben, nämlich auf die nördlichen Zirkumpolarsterne. Damit

Nuzhat al-alibbā'. Kairo 1294, 205 = ed. I. AS-SĀMARRĀ'Ī. Bagdad 1959, 102; as-Suyūṭī: *Buġyat al-wu'āt*. Kairo 1326, 410 = ed. M. A. F. IBRĀHĪM. 1–2. Kairo 1964–65, 2, 329 (Nr. 2105). Abū l-Hayṯam scheint 276 H = 889/90 gestorben zu sein (so al-Azharī: *Tahḏīb*, a.a.O. [ZETTERSTÉEN, nach seiner Hs., schlecht 206, das wäre 821/2], und as-Suyūṭī: *Buġya*, a.a.O.). al-Anbārī: *Nuzha*, a.a.O., schreibt dagegen 226 [= 840/1], was anscheinend nicht stimmen kann, da Abū l-Hayṯam laut al-Azharī a.a.O. „wenige Jahre vor dem Tod des Šimr" (welcher ebda. für 255 [= 869] angegeben wird) nach Herat gekommen sei. Bei Annahme des frühen Todesdatums 226 hätte Abū l-Hayṯam auch nie mit al-Munḏirī (gest. 941) zusammentreffen können, was dagegen als möglich erscheint, wenn man 276 = 889/90 annimmt. Cf. inzwischen noch F. SEZGIN: *Geschichte des arabischen Schrifttums*. Bd. 7. Leiden 1979, S. 346f.

21 D. h. [Fix-]Sterne außerhalb des Ekliptikbereiches, was sowohl die von der Ekliptik nach Norden wie die nach Süden zu abweichenden Sterne meint.

22 *al-quṭb* ist der Pol, also ein Punkt am Himmel und eigentlich kein Stern.

23 Zur Lesung des Namens als *al-ǧady* bzw. *al-ǧudayy* vgl. P. KUNITZSCH: *Untersuchungen zur Sternnomenklatur der Araber*. Wiesbaden 1961, S. 62f., Nr. 107a/b. *al-ǧudayy* ist der Stern α Ursae Minoris, der Polarstern.

24 *al-farqadān* ist das Sternpaar β und γ Ursae Minoris; s. KUNITZSCH a.a.O., S. 58, Nr. 96.

25 Wie der *Tāǧ* s.r. *byn* zu der Stelle erklärt, habe al-Azharī diese Stelle als Beleg für den Gebrauch von *bayna* im Sinne von *wasṭa* „inmitten von" herangezogen, was aber falsch sei, da er die Bezeichnung infolge eines Schreibfehlers *bayānīyāt* gelesen (und somit zu *byn* gestellt) habe, während die richtige Lesart *babānīyāt* sei, weshalb auch der *Lisān* den Ausdruck unter *bbn* eingeordnet habe.

26 *banāt na'š aṣ-ṣuġrā* sind die Sterne des Kleinen Bären, Ursa Minor (KUNITZSCH a.a.O., S. 48, Nr. 56), unter denen sich die oben separat genannten Sterne befinden. Statt *wa-fīhi* wäre es also sachlich richtiger, umgekehrt zu lesen *wa-hiya fī* „und diese gehören zu . . . ."

wurde die Bedeutung zwar nicht völlig verfehlt, aber andererseits auch nicht vollkommen richtig erfaßt. Abgesehen davon, ist der Wortlaut der Darstellung verworren und zeigt, daß es den Lexikographen an himmelskundlicher Sachkenntnis mangelte, so daß sie die Phänomene und die Sternnamen nicht richtig beschreiben und zueinander in Beziehung setzen konnten. Mit Abū l-Hayṯams Zeugnis bei al-Azharī haben wir nun einen Beleg für das Auftreten jener fremden Bezeichnung auch außerhalb des astrologisch-astronomischen Fachbereichs etwa in der zweiten Hälfte des 9. Jahrhunderts. Das erscheint, in chronologischer Hinsicht, durchaus als plausibel, wenn man annimmt, daß der *Liber de stellis beibeniis* bereits ebenso früh aus dem Persischen ins Arabische übersetzt wurde wie einige andere astrologische Werke (vgl. Dorotheus, ʿUmar b. al-Farruḫān, etc., vom Ende des 8./Anfang des 9. Jh.s).

XV

DIE „UNWETTERSTERNE" UND DIE „GEOMANTIE" DES ZANĀTĪ

Die Verfolgung arabisch-islamischer Spuren in der byzantinischen Astronomie führt u. a. zu einem Kalendarium, das vierundzwanzig „Unwettersterne" (περὶ ἀστέρων ποιούντων τάραξιν . . .) erwähnt, die mit eigenartigen nichtgriechischen Namen belegt sind. Die Namen dieser Sterne lauten (nach CCAG 4, 145 f. mit den Varianten von CCAG 12, 113):[1]

1.	I, 4	Γαμπίνης (var. Γαμπινού, γαμπινο͂ς [sic])
2.	I, 20	Λάμπης (var. Λάμπεις)
3.	II, 9	Σάρης (var. Σάροις)
4.	III, 9	Σαριτζε
5.	III, 25	Γαράρ
6.	IV, 5	Σιμέλ
7.	IV, 19	Βεκλεύς
8.	V, 1	Γάρφας (var. Γαράφας)
9.	V, 12	Τέμπελις (var. Παίλας)
10.	VI, 8	Ἀ[ν]ολώστας (var. Αὐολόστας)
11.	VI, 21	Γαπούς (var. Γαμπούς)
12.	VII, 5	Παπλούς
13.	VII, 19	[Name fehlt]
14.	VIII, 1	Ράφας
15.	VIII, 14	Πιστεύς (var. Πισές)
16.	VIII, 27	[Name fehlt]
17.	IX, 7	Ἀλφαράς (var. Γαρές)
18.	IX, 22	[Name fehlt in cod. Neapol.; Petropol.: Γαράφα]
19.	X, 5	Γαρεύς (var. Γιρές)
20.	X, 17	[Name fehlt]
21.	XI, 1	Ῥήκτας (var. Ρίκτας)
22.	XI, 12	[Name fehlt]
23.	XII, 9	Ἐώλας (var. Ἐάλας)
24.	XII, 17	[Name fehlt]

Das Kalendarium wird gelegentlich in der Überschrift einem „persischen Philosophen Zanates" zugeschrieben. Unter der gleichen Autorenbezeichnung ist in mehreren griechischen Handschriften[2] ein geomantischer Text überliefert.

---

[1] Ediert CCAG 4, 145 f.; dazu Varianten in CCAG 12, 113. Der gleiche Text erscheint noch an weiteren fünf Stellen: CCAG 4, 33 (ms Mutin. 11 fol. 94^{v}; in B. Z. 57 [1964], 383, Anm. 4 ist fälschlich Mutin. 12 genannt, dieser Irrtum beruht auf dem Druckfehler CCAG 4, 145, Anm. 1, wo es falsch Mutin. 12 heißt statt richtigem 11); CCAG 6, 34; CCAG 7, 23; CCAG 7, 75; HUNGER p. 282.

[2] J. BOLTE, *Wickrams Werke*, IV (Tübingen 1903), p. 289, Anm. 1, spricht von 16 Stellen. Hier seien, ohne Anspruch auf Vollständigkeit, folgende genannt: cod. Bon. 18,

Um Bedeutung und Herkunft jener eigenartigen Sternnamen aufzuklären, hätte man sich an das anzunehmende nichtgriechische Original des Textes zu halten. Zu diesem Zweck wäre zunächst die Identität des angeblichen Autors, des „persischen Philosophen Zanates“, festzustellen.

Ein gewisser Šayḫ Abū ʿAbdallāh az-Zanātī ist in der orientalischen Literatur in der Tat nachweisbar. Er stammt offensichtlich, wie der Name besagt, aus Nordafrika und wird dem Berberstamm der Zanāta zugerechnet. Die Sprache, in der er schrieb, war Arabisch. Die byzantinischen Texte bezeichnen hier also wieder, wie häufig, das Orientalische schlechthin als „persisch“. Der Name von az-Zanātī wird in arabischen Texten abweichend überliefert.[3] Nähere Einzelheiten zur Biographie sind nicht erhalten.[4] az-Zanātī dürfte im 12. oder 13. Jahrhundert gelebt haben; als terminus ante quem ist 1265/66 (6774 der Schöpfung) anzusehen; in diesem Jahr wurde die „Geomantie“ von dem Mönch Arsenios ins Griechische übertragen.[5]

Bekannt ist der Šayḫ az-Zanātī in der arabischen Literaturgeschichte als Verfasser geomantischer Abhandlungen, von denen zahlreiche Handschriften[6] und Drucke[7] bekannt sind, deren Echtheit und Verhältnis untereinander jedoch noch der Nachprüfung bedürfen.[8]

---

fol. 296v ff., nach CCAG 4, 43; dto. fol. 467f., nach ib. p. 46; cod. Neapol. 19, fol. 31 ff. [diese Stelle ist unten behandelt], nach CCAG 4, 51f.; cod. Vind. Phil. gr. 108, fol. 1, nach CCAG 6, 1; Hunger p. 213; Carra de Vaux p. 300, Anm. 3; Tannery p. 323, 372 – Liste der Namen von elf Figuren, der Anfang mit den übrigen fünf Figuren fehlt in der Handschrift; Paris B.N. gr. 2424, fol. 163v–188v, siehe Tannery p. 320ff., 354ff.; CCAG 4, 74 zu fol. 1; Paris B.N. gr. 2419, fol. 226v, siehe Tannery p. 322, 369ff.; Laur. LXXXVI, 14, fol. 1–25, nach CCAG 4, 74; cf. Tannery p. 357, Anm. 1; Laur. LXXXVI, 17, nach CCAG 4, 77. Aus den vorliegenden Angaben ist zu schließen, daß diese Texte untereinander differieren. Ferner CCAG 4, 74 Hinweis auf Laur. XXVIII, 22 (da nicht astromonischen Inhalts, die Stelle in CCAG 1, 73 nicht erwähnt); CCAG 4, 52 Hinweis auf Mediol. 11, fol. 18 ff. (dort jedoch nach CCAG 3, 7–10 nur Astrologisches); Carra de Vaux p. 300, Anm. 3 Hinweis auf Vind. Phil. gr. 179, fol. 110 (dort nach CCAG 6, 34 und Hunger p. 287 astrologische Stücke, nicht die Geomantie).

[3] Die Kairoer Drucke: aš-Šayḫ Muḥammad az-Zanātī; Brockelmann, GAL Suppl. II, 1037, Nr. 5: aš-Šayḫ Abū ʿAbdallāh az-Zanātī; ib. 1041, Nr. 40: Abū ʿAbdallāh Muḥammad b. ʿUṯmān az-Zanātī; vgl. ferner ib. 1040, Nr. 28: Abū ʿImrān az-Zanātī. Cf. auch Carra de Vaux p. 300, Anm. 3; Rosenthal p. 229, Anm. 344.

[4] Cf. die Traditionskette des Ṣāliḥī bei Carra de Vaux p. 300f.

[5] Cf. u. a. CCAG 4, 51 (aus ms Neapol. 19, fol. 31).

[6] Brockelmann, GAL a. a. O. (von der hier genannten Handschrift Nuru Osmaniye 3639 erhielt ich durch freundliche Vermittlung von Herrn Prof. W. Hartner, Frankfurt, und durch die persönliche Mühe von Dr. F. Sezgin einen Mikrofilm aus Istanbul); Steinschneider, ZDMG 25 (1871), 411; id., *Hebr. Übers.* p. 855, Anm. 59; Carra de Vaux p. 229, Anm. 1; 300, Anm. 3.

[7] Rosenthal p. 229, Anm. 344; *Enzyklopädie des Islams*, s. v. „Siḥr“ (D. B. Macdonald), IV, 446, Nr. 15 (von dem hier erwähnten Kairoer Druck fertigte mir die Case Memorial Library, Hartford/Conn., dankenswerterweise eine vollständige Xerokopie an, nach der ich zitiere; den gleichen Text, im gleichen Wortlaut, nur mit abweichender Seitenzählung, bietet der Druck Kairo 1341 = 1922/3, den die Tübinger Universitätsbibliothek besitzt).

[8] Rosenthal a. a. O. So stimmen auch die beiden von mir durchgesehenen arabischen Texte weder in Aufbau noch Anordnung noch Wortlaut überein. Die „Geomantie“ der

Die Durchsicht zweier ihm zugeschriebener arabischer Traktate über die Geomantie (der Kairoer Druck, cf. Anm. 7 und 8, genannter Autor: aš-Šayḫ Muḥammad az-Zanātī; sowie die Istanbuler Handschrift Nuru Osmaniye 3639, fol. 1ᵛ–33ᵛ, ohne Titel und ohne Nennung eines Autors, cf. Anm. 6 und 8) führte nicht zu dem Ziel, das Original des Kalendariums mit den „Unwettersternen" zu finden. Die beiden Texte sind vielmehr ausschließlich der Geomantie (*ʿilm ar-raml* „Wissenschaft vom Sande"; gräzisiert ῥαβόλιον oder ῥάμπλιον[9]) gewidmet.

*

Zanātīs „Geomantie" ist 1265/66 von dem Mönch Arsenios ins Griechische übertragen worden und in mehreren Handschriften erhalten (cf. Anm. 2 und 5). TANNERY hat sich besonders mit den beiden Pariser Handschriften B. N. gr. 2424 und 2419 beschäftigt und daraus auch die ins Griechische übernommenen arabischen Namen der sechzehn geomantischen Figuren mitgeteilt (p. 320ff.; 354ff.; Tabelle p. 410f.); elf dieser Namen zitiert er auch aus dem am Anfang defekten ms Vind. Phil. gr. 108, fol. 1ff. (p. 372).

Die sechzehn geomantischen Figuren treten in arabischen Quellen auf und sind vielleicht nordafrikanischen, berberischen Ursprungs.[10] Ihre Reihenfolge wechselt stark, dagegen werden die Namen jeder einzelnen Figur im wesentlichen einheitlich überliefert; Differenzen treten vor allem bei gewissen Alternativnamen auf.[10a] Sowohl die griechischen Transkriptionen und Übersetzungen als die übersetzten Namen in der einschlägigen lateinischen Literatur[11] beziehen sich stets eindeutig auf das gleiche arabische Original.

Istanbuler Handschrift wird von BROCKELMANN, GAL Suppl. II, 1037, Nr. 5 zu Unrecht als Werk des Zanātī bezeichnet; sie trägt keine Überschrift und nennt keinen Autor; innen werden ständig drei Autoritäten zitiert: erstens Ḥulbus bzw. aš-Šayḫ Ḥulbus (cf. den von STEINSCHNEIDER, *Hebr. Übers.* 855, Anm. 61 [ג׳ליס] und 856, Anm. 69 [Djalis, Djuleis] zitierten unverständlichen Namen, der bei BROCKELMANN, GAL Suppl. I, 893, Nr. 24a, als Ḥubayš erscheint; wahrscheinlich ist der in der Handschrift durchgehend Ḥulbus geschriebene Name ebenfalls in Ḥubayš zu emendieren. Der damit bezeichnete Autor könnte der von STEINSCHNEIDER a. a. O. 856, Anm. 69 und BROCKELMANN a. a. O. erwähnte Abū l-Faḍl Ḥubayš [b. Ibrāhīm] b. Muḥammad at-Tiflīsī, um 600 h = 1203, sein, von dem BROCKELMANN zehn, meist medizinische und therapeutische, Titel anführt [cf. übrigens auch unten Anm. 31]), zweitens aṭ-Ṭarābulusī (offenbar Abū Saʿīd aṭ-Ṭarābulusī, der von aṣ-Ṣāliḥī als Lehrer des Zanātī genannt wird; cf. STEINSCHNEIDER, *Hebr. Übers.* 567, Anm. 208 und 855, Anm. 61; BROCKELMANN, GAL Suppl. I, 908f., Nr. 2a; CARRA DE VAUX p. 301f. mit Anm. 1; TANNERY p. 330f., 403ff.) und drittens unser Zanātī. Der Kairoer Druck enthält im Anschluß an den unter dem Namen des Zanātī gedruckten Text (p. 1–42) noch verschiedene weitere kurze geomantische Kapitel unter mehreren Autorennamen (p. 42–63).

[9] Cf. TANNERY p. 322 und 357; auch u. a. CCAG 4, 46 (cod. 19, fol. 467) sowie ib. 74 (cod. 29, fol. 1).

[10] Zuweilen werden in den Tabellen auch die berberischen Namen der 16 Figuren aufgeführt (STEINSCHNEIDER, ZDMG 31; TANNERY; Ibn al-Maḥfūf im arabischen Text und in der lateinischen Übersetzung).

[10a] In dem bei A. und L. DELATTE, *Un traité byzantin de Géomancie*, in *Mélanges Franz Cumont*, Brüssel 1936, p. 616ff. edierten anonymen geomantischen Traktat aus ms Paris B.N. gr. 2419, fol. 232ᵛ–233 (cf. CCAG VIII, 1 p. 49ff.) herrscht starke Verwirrung: die Figuren, die arabischen Namen und die griechischen Bezeichnungen stimmen nicht miteinander überein; zudem sind mehrere arabische Namen verwendet, die in den geläufigen, hier in meinem Artikel herangezogenen arabischen Quellen nicht auftreten.

[11] Cf. u. a. die Tabellen bei STEINSCHNEIDER, ZDMG 31; TANNERY p. 410f.; AGRIPPA VON NETTESHEIM, *De occulta philosophia*, Lugduni (1531 ?), Buch II, Kap. 48, p. 278f.

Dies ist die Reihenfolge der sechzehn Figuren (σχήματα) in ms Neapol. 19, fol. 31 ff. nach dem Abdruck in CCAG 4, 52 (die Form der Figuren ist dort nicht mit abgedruckt. Bei waagerechter Wiedergabe erscheinen sie übrigens in arabischen Texten links-, in lateinischen rechtsläufig gerichtet). Zu jedem griechischen transkribierten Namen setze ich das arabische Original, entnommen den verschiedenen genannten arabischen Texten, die lateinische Transkription in der Ibn-al-Maḥfūf-Übersetzung sowie die lateinische Übersetzung daselbst, und schließlich die lateinischen Namen bei Agrippa von Nettesheim (1486–1535), die offenbar aus verschiedenen älteren Übersetzungen kompiliert sind (ein Stern* kennzeichnet diejenige Form, die auch bei Gerhard von Cremona gebraucht ist):

Nr.	Figur	ms Neapol. 19	Arabisch	Lat. Transkr. ms Ampl. 4° 380, 3ʳ⁻ᵛ	Latein. Übersetzg. ib.	Agrippa v. N. *De occ. phil.*
1	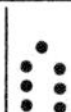 oder 	ὀφίτης ὁ καὶ λεχνεία	*al-aḥyān*, *al-ḥayān*[12]	9. elteym	caput altum	9. *Laetitia. Ridens. Sanus. Barbatus
2	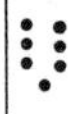 oder 	ὁ κλαγγύς	*al-inkīs*[13]	10. erueh'is (?)	caput ymum	10.* Tristitia. Damnatus. Transversus
3	 oder 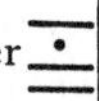	χοβροῦ	*al-ḥumra*	11. humbra	rubeus	14.* Rubeus. Ruffus
4	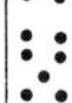 oder 	[sine titulo]	*al-bayāḍ*	12. h'ras (?)	albus	13.* Albus. Candidus
5	 oder 	νόστρα τουλχαρίτζη	[*an-*]*nuṣra* *al-ḫāriǧa*	3. [sultan, IM: *as-sulṭān*]	fortuna minor	6.* Fortuna minor. Auxilium minus. Tutela exiens

(in einem anderen Druck, ebenfalls Lugduni ohne Jahr, p. 225 ff.; in diesem Band auf p. 559–573 mit enthalten Gerardi Cremonensis *Geomantiae Astronomicae Libellus*, dessen Echtheit noch nicht nachgewiesen ist). Bolte, *Wickrams Werke* IV, p. 228, Anm. 2, verweist auf zwei Erfurter Handschriften mit geomantischem Text und Tabellen, wobei zu untersuchen bliebe, ob es sich um Zanātī handele. Wie ich inzwischen ermitteln konnte, stellt dieser lateinische Text (ms Amplon. 4⁰ 380, s. XIII–XIV, fol. 1–47; ebenfalls in 2⁰ 389, fol. 56–99) eine Übersetzung des *Kitāb fī ʿilm ar-raml* von aš-Šayḫ ʿAbdallāh b. ʿAlī Ibn al-Maḥfūf al-Munaǧǧim dar, das in der Berliner Staatsbibliothek, ms or. quart. 49 (= Ahlwardt 4200), fol. 12 ff., arabisch erhalten ist (cf. zum Autor Brockelmann, GAL Suppl. II, 367, Nr. 4, sowie auch ib. 1037, Kap. 19, Nr. 2). Der Text enthält auch die Tabelle der 16 Figuren mit arabischen und berberischen Namen (arab. fol. 16ᵛ–17ʳ, von links nach rechts verlaufend; lat. 4° 380, fol. 3ʳ⁻ᵛ, von rechts nach links verlaufend). In der *Risālat mīzān al-ʿadl fī maqāṣid aḥkām ar-raml* (cf. GAL Suppl. II, 1018, Nr. 14) von ʿAbd al-Qādir al-Ḥusaynī al-Adhamī (schrieb nach GAL Suppl. II, 393, Nr. 69b um 1207 h = 1792; hier dagegen am Ende des Textes p. 14 wörtlich 20. Šawwāl 1322 = 8. Nov. 1904, mitgedruckt im Anhang zu Teil IV von al-Būnīs *Kitāb šams al-maʿārif al-kubrā*, 4 Teile in einem Band, Kairo o. J. [1905]; die Kenntnis der Stelle verdanke ich Herrn Dr. M. Ullmann, Tübingen), werden p. 4 einige Alternativnamen erwähnt, die auffällig gut zu einigen der alten lateinischen Übersetzungen passen: *al-kawsaǧ*, cf. imberbis; *aḍ-ḍāḥik*, cf. ridens; *rāyat faraḥ*, cf. laetitia; *al-ašqar*, cf. flauus; *al-mankūs*, cf. transversus; *as-saʿāda*, cf. fortuna.

[12] Die genaue Lesung ist nicht zu ermitteln; es wird auch [*al-*]*liḥyān*[*ī*] überliefert; *liḥya* bedeutet „Bart", *al-laḥyān* „der Langbärtige", daher die lateinische Übersetzung.

[13] Lesung und Bedeutung unsicher; schon in arabischen Texten mit *mankūs* „umgekehrt" zusammengestellt (Tabelle Steinschneider, ZDMG 31; al-Adhamī als Alternativname; Carra de Vaux p. 314 nach ms Paris B. N. ar. 2731, fol. 13ᵛ, als angeblicher berberischer Name; auch bei Ibn al-Maḥfūf in die Spalte der berberischen Namen geraten), daher auch die lateinische Übersetzung.

Nr.	Figur	ms Neapol. 19	Arabisch	Lat. Transkr. ms Ampl. 4° 380,3r–v	Latein. Übersetzg. ib.	AGRIPPA V. N. *De occ. phil.*
6	 oder 	νόστρ<α> τουλταχήλ	*[an-]nuṣra ad-dāḫila*	4. embilla-siēlmi (?)[14]	fortuna maior	5.* Fortuna maior. Auxilium maius. Tutela intrans
7	 oder 	θαπίθς' καὶ ἀντιπαλχα-ρίτζη	*ṯābita,*[15] *[al-]'ataba al-ḫāriğa*[16]	16. [Name fehlt]	limes exiens	16. Cauda.[17] Limen exiens. Limen inferius
8	 oder 	χαιμὰ καὶ ἀντιπὰλ ἡ τάχηλ	*qā 'ima,*[18] *[al-]'ataba ad-dāḫila*	15. aebe-tayle	limes intrans	15. Caput.[19] Limen intrans. Limen superius
9	 oder 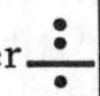	ταράχ[20] τζόταλαι	*al-ğawda-la*[21]	13. [bisara, IM: *al-faraḥ wa-l-bišāra*]	puella	12.* Puer. Flauus. Imberbis
10	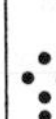 oder 	ναγκαλχάτ	*naqīy al-ḫadd*	14. negli-schad	puer	11.*Puella. Mundus facie
11	 oder 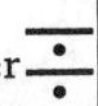	κατδιουτά-χηλ	*[al-]qabḍ ad-dāḫil*	5. cabchid-dahil	acquisicio	7.* Acquisitio. Compraehensum intus
12	 oder 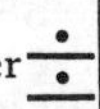	καπδθαχα-ρίτζη	*[al-]qabḍ al-ḫāriğ*	6. cabdi-harigi	amissio	8. Acquisitio.[22] Compraehensum extrà
13	 oder 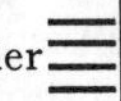	τζιάματ	*al-ğamā'a*	1. [gema .i. selen, IM: *al-ğam' wa-huwa s-salāma*]	populus	2.* Populus. Congregatio

[14] Ibn al-Maḥfūf hat abweichend von der Hauptüberlieferung: *abū l-'āqiba* [?] *wa-hiya t-tašmīr*; die lateinische Transkription läßt nicht mit Sicherheit erkennen, welches Original sie wiedergibt.

[15] Alternativname verschiedener Figuren in verschiedenen Texten; cf. STEINSCHNEIDER, ZDMG 31; CARRA DE VAUX p. 315.

[16] Ibn al-Maḥfūf hat im arabischen Text statt *'ataba* das gleichbedeutende *rakīza*, doch geht die lateinische Transkription (bei Nr. 8) auf *'ataba* zurück.

[17] GERH. V. CREM.: cauda draconis. Es handelt sich um den „Drachenschwanz", nodus descendens, der Mondbahn; cf. ms Amplon. bei Nr. 2.

[18] STEINSCHNEIDER, ZDMG 31 und CARRA DE VAUX (hier p. 315 χαισμᾶς nicht als Abkömmling von *qā'ima* erkannt).

[19] GERH. V. CREM.: caput draconis; der „Drachenkopf", nodus ascendens, der Mondbahn; cf. ms Amplon. bei Nr. 1.

[20] ταρὰχ offenbar gleich φαράχ u. ä. (TANNERY), aus *[al-]faraḥ* „Freude", in verschiedenen arabischen Texten Alternativname bei verschiedenen Figuren.

[21] Das Wort *al-ğawdala* erscheint nicht in den arabischen Lexika (cf. auch CARRA DE VAUX p. 312). Als Alternativname wird auch überliefert *al-kawsağ* „der Dünnbärtige"; hierzu offenbar die lateinische Übersetzung „imberbis" (Herr Dr. M. Ullmann, Tübingen, weist mich brieflich darauf hin, daß auch der *Vocabulista in Arabico*, ed. SCHIAPARELLI, Florenz 1871, p. 421, für „imberbis" das Äquivalent *kawsiğ* [sic] biete).

[22] Offenbar Druckfehler; die anderen lateinischen Texte, auch GERHARD VON CREMONA, haben hier stets „amissio".

Nr.	Figur	ms Neapol. 19	Arabisch	Lat. Transkr. ms Ampl. 4° 380,3r–v	Latein. Übersetzg. ib.	AGRIPPA v. N. *De occ. phil.*
14	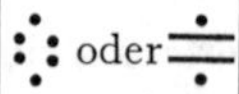	ὀκλᾶ	*al-ʿuqla*	7. vole [leg. vcle]	constrictus	4.*Carcer. Constrictus
15	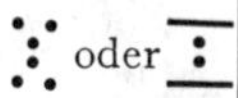	ἡτζιματί	*al-iǧtimāʿ*	8. enfama	coniunctio	3.* Coniunctio. Coadunatio [sic]
16		[. . . σχή-ματ<ος> τρίβου[23]]	*aṭ-ṭarīq*	2. tebit[24]	via	1.* Via. Iter

Die Verwendung des Artikels *al-* bei den einzelnen Namen schwankt. Im häufigen volkstümlichen Gebrauch werden die Figuren oft in der Kurzform ohne Artikel bezeichnet worden sein. Aus seinem Nichtvorhandensein darf man daher nicht schließen, daß die Vorlage möglicherweise doch persisch war statt, wie selbstverständlich anzunehmen ist, arabisch.

*

Nachdem sich erwiesen hat, daß die beiden geprüften, dem Zanātī zugeschriebenen arabischen geomantischen Texte das gesuchte Original des Kalendariums nicht enthalten, ergäbe sich als nächstes die Frage, wo nun nach diesem Original weiter zu suchen ist.

Eine Antwort hierauf läßt sich nur sehr bedingt geben. Zunächst einmal muß es als fraglich gelten, ob die griechischen Handschriften, die bei den geomantischen Textstellen mit Recht az-Zanātī als Quelle nennen, diesem Autor das Kalendarium ebenfalls rechtmäßig zuschreiben. Einige astrologische Texte, die in griechischen Handschriften ebenfalls auf einen „persischen Philosophen“[25] bzw. den „Perser Zanates“[26] zurückgeführt werden, weisen laut Inhaltsangabe im CCAG das Kalendarium jedenfalls nicht auf. Innerhalb der arabischen Literatur sind unter Zanātīs Namen bisher lediglich geomantische Texte bekannt geworden, aber keine astrologischen oder astronomischen.

Zwei Elemente im griechischen Text und Wortlaut des Kalendariums erlauben gewisse Rückschlüsse, die immerhin die Abgrenzung eines in Frage kommenden engeren Bereichs der Literatur arabischer Sprache ermöglichen, in dem jenes Original angesiedelt sein könnte.

Einmal läßt sich mit Sicherheit sagen, daß die eigenartigen abweichenden Sternnamen keinerlei Ähnlichkeit mit den auch in griechischer Transkription klar erkennbaren bekannten arabischen Sternnamen aufweisen.[27]

---

[23] τρίβος, griechische Übersetzung des arabischen Namens; cf. TANNERY p. 372, Nr. 13.

[24] Offenbar zu *ṯābita*, hier Nr. 7; Ibn al-M. hat richtig *aṭ-ṭarīq*, jedoch *ṯābita* bei Nr. 8.

[25] CCAG 3, 7 ff.: cod. Mediol. 11, fol. 18 ff.

[26] CCAG 7, 49: cod. Berlin 173, Phill. 1577, fol. 52; das gleiche incipit auch Vind. Phil. gr. 177, fol. 1v (nach HUNGER p. 282).

[27] Cf. die Tabellen von Typ I und II in B. Z. 57 (1964), 382 ff.

Aus dem gleichen Grunde kann es sich auch nicht um die Mondstationen handeln,[28] abgesehen davon, daß hier nur 24 Positionen erscheinen statt der in diesem Fall erforderlichen achtundzwanzig.

Zweitens verwendet der griechische Text die lateinischen Monatsnamen; hiernach darf man annehmen, daß auch in der arabischen Vorlage die diesen entsprechenden syrischen Monatsnamen des Sonnenjahres (*ādār* usw.) gebraucht waren, die uns auch in anderen arabischen Kalendarien und ähnlichen Texten[29] entgegentreten. Sie stammen über das Syrische aus der christlichen oder hellenistischen Geisteswelt und legen auch hier den Gedanken nicht echtarabischer Herkunft nahe.

Diese beiden Erscheinungen deuten also darauf hin, daß der dem griechischen Kalendarium zugrunde liegende Text zwar in arabischer Sprache abgefaßt war, jedoch nicht eigentlich der arabischen Literatur selbst angehörte. Vielmehr scheint es sich um einen aus einer fremden Sprache ins Arabische übersetzten Text zu handeln, dessen Sternnamen in arabischer Transkription beibehalten wurden und dann natürlich bei der Weiter- (oder Rück- ?)Übersetzung ins Griechische ihre ursprüngliche Kontur bis zur Unkenntlichkeit verloren.

Aber auch diese Abgrenzung umschließt in der Literatur arabischer Sprache noch einen Bereich von nahezu unübersehbarem Ausmaß. Durch die große, kulturgeschichtlich so bedeutende „Übersetzungsbewegung", deren Höhepunkt ins neunte Jahrhundert fällt, kamen nicht nur naturwissenschaftliche, medizinische und philosophische Werke bekannter antiker Autoren sowie indischer und persischer Herkunft in die arabische Literatur, sondern auch viele bereits in hellenistischer Zeit pseudepigraphisch überlieferte Kompilationen von vielfältigster Zusammensetzung und Herkunft. Aus diesem weiten Kreis der pseudepigraphischen Übersetzungsliteratur, die zum Teil auch über syrische und persische Zwischenstufen gewandert ist, kommen für die Suche nach der Vorlage des Kalendariums besonders die hermetischen und andere ähnliche astronomisch bzw. astrologisch orientierte Texte des Aberglaubens und der Deute-

---

[28] Cf. vier ins Griechische übersetzte, edierte Listen: CCAG V, 3 p. 90f.; ib. p. 91f.; VIII, 1 p. 217f.; und IX, 1 p. 138ff.

[29] Etwa *Liber anoe* [= *Kitāb al-anwā'*], latein. Übersetzung von Gerhard von Cremona; ed. G. Libri, *Histoire des sciences mathématiques en Italie*, I, Paris 1858, p. 393ff.; nachgedruckt von R. Dozy, Leiden 1873 (zusammen mit der arabischen Textedition, in hebräischer Schrift); Neuausgabe von Ch. Pellat, Leiden 1961 (*Le Calendrier de Cordoue*; cf. dazu J. Fück in Oriental. Literatur-Zeitg., 60 [1965] 333–337; J. Vernet, Oriens, 17, 1964, 284–286). Oder *Kitāb al-azmina* von Ibn Māsawayh, gest. 857; ed. P. Sbath in Bull. de l'Inst. de l'Égypte 15, 1932–33, 235–257. Ein gleichartiges Werk liegt auch in den ptolemäischen Φάσεις vor, und bereits aus Mesopotamien sind nach Aufbau und Inhalt ähnliche Texte bekannt (*mul*APIN, cf. u. a. W. Hartner, *The Earliest History of the Constellations in the Near East* . . ., in Journal of Near Eastern Studies, vol. 24, Nr. 1–2, Jan.-Apr. 1965). Es handelt sich also um eine Textgattung uralter Tradition. Zum Gebrauch der lateinischen Monatsnamen in arabischen Texten bzw. Inschriften cf. C. Nallino, *Opus astronomicum* (al-Battānī), II (Mailand 1907), p. 201ff.

literatur[30] sowie die weitverbreiteten *Ǧafr*- und *Malḥama*-Texte in Frage, von denen manche auch meteorologische Aussagen zum Gegenstand haben, deren Erforschung bisher jedoch über bibliographische Sammlungen[31] und unzusammenhängende Einzeluntersuchungen[32] noch nicht hinausgediehen ist.[33]

Einer der hierzu von STEINSCHNEIDER aufgezählten Texte[34] scheint nach Titel und Umfang ganz besonders für unsere Suche interessant zu sein: *ḏikr ḥawādiṯ al-ǧaww wa-l-āṯār al-ʿulwīya fī š-šuhūr ar-rūmīya ʿan Dāniyāl* ... „Darstellung der Ereignisse der Atmosphäre und der Erscheinungen der oberen Welt [= μετεωρολογικὰ φαινόμενα] in den griechischen Monaten, nach Daniel ...", in ms Bodl. Hyde 32, fol. 68ᵛ–69ʳ. Eine Fotokopie der Stelle zeigt jedoch, daß es sich dabei nur um eine ganz gewöhnliche Prognostik handelt, nach Monaten und Erscheinungen tabellarisch aufgebaut, die nichts zum Problem der „Unwettersterne" beizutragen vermag.

So wird es wohl, auch nach der oben vorgenommenen Einschränkung, ein reiner Zufallstreffer sein, wenn es einmal gelingen sollte, aus dem Meer der häufig recht unzulänglich katalogisierten arabischen Handschriften denjenigen Text herauszufischen, der uns über Herkunft und Bedeutung jener „Unwettersterne" Aufklärung geben kann.

---

[30] An einschlägigen gedruckt vorliegenden Texten wurden, neben der astrologischen und astronomischen Fachliteratur, erfolglos durchgesehen: Pseudo-Maǧrīṭī, *Ġāyat al-ḥakīm*; al-Būnī, *Šams al-maʿārif*; Ibn al-ʿAwwām, *Kitāb al-filāḥa* (cf. zu der stark antik fundierten *Filāḥa* [= Geoponika-]-Literatur, die auch zuweilen Bauernkalender und Wetterregeln enthält: *Encyclopaedia of Islam*, New Edition, s. v. „*Filāḥa*", II p. 900–902, sub I, 2 und II).

[31] Cf. vor allem M. STEINSCHNEIDER in ZDMG 28 (1874), 649–655; ib. 50 (1896), 190f., Werk Nr. 5; *Encyclopaedia of Islam*, New Edition, s. v. „*Baṭlamiyūs*", Nr. 3f., und s. v. „*al-Djafr*". Übrigens wird auch dem bereits oben in Anm. 8 erwähnten at-Tiflīsī in einer bibliographischen Notiz eine *Malḥama* zugewiesen (ZDMG 28, p. 652).

[32] Herr Dr. T. Fahd von der Universität Strasbourg macht mich brieflich darauf aufmerksam, daß 1966 bei Brill in Leiden seine umfangreiche Doktorarbeit über *La Divination arabe* erschienen sei, die mir jedoch leider vor Manuskriptschluß in Köln nicht mehr zugänglich wurde.

[33] Einer der gründlichsten Kenner der Materie, Prof. O. Neugebauer, äußerte auf meine Anfrage hin brieflich eine andere Vermutung, die davon ausgeht, daß sich in dem Kalendarium – trotz der Lücken und Verschiebungen – ein regelmäßiges vierzehntägiges Schema abzeichne, wie sie in bestimmten hellenistischen Kalendersystemen vorkommen; im Verlaufe langer Tradition und mehrfacher Übersetzung sei daraus schließlich jenes entstellte Relikt geworden, das uns heute in den byzantinischen Texten entgegentritt. Die Namen mögen dabei ursprünglich irgendwelchen Datumspatronen, Schutzheiligen, Dämonen o. ä. gehört haben; in dem uns vorliegenden Endstadium der Überlieferung seien diese, unverstanden, schließlich zu „Sternen" geworden. – Man kann hierbei an das unübersehbare antike Material der Parapegmata, Menologien usw. denken, von denen möglicherweise auch einiges in arabische Versionen weitergedrungen ist (zu den antiken Texten cf. u. a. A. REHM, Artikel „Parapegmata", RE, 36. Halbband, 1949, col. 1295 ff.; R. BÖKER, Artikel „Wetterzeichen", ib., Suppl. 9, 1962, col. 1609 ff.).

[34] ZDMG 28, p. 653, sub f.

Abgekürzt zitierte Literatur:

BROCKELMANN, GAL: C. BROCKELMANN, *Geschichte der arabischen Literatur*, Leiden I² 1943, II² 1949, Suppl. I 1937, II 1938 und III 1942.

CCAG: *Catalogus Codicum Astrologorum Graecorum*, I–XII, 1898–1953.

CARRA DE VAUX: B. CARRA DE VAUX, *La géomancie chez les arabes*, bei TANNERY (s. d.), p. 299–317.

HUNGER: H. HUNGER, *Katalog der griechischen Handschriften der Österreichischen Nationalbibliothek*, I, Wien 1961.

ROSENTHAL: F. ROSENTHAL, *Ibn Khaldûn: The Muqaddimah* (Übers.), I–III, London 1958; speziell I, 226–234, Geomancy, entsprechend ed. QUATREMÈRE I, 204–209.

STEINSCHNEIDER, ZDMG 31: M. STEINSCHNEIDER, Artikel mit Tabelle in ZDMG 31 (1877), p. 762 ff.

id., *Hebr. Übers.*: id., *Die hebraeischen Übersetzungen des Mittelalters*, Berlin 1893; Nachdruck Graz 1956.

TANNERY: P. TANNERY, *Mémoires Scientifiques*, IV (Toulouse-Paris 1920), Nr. 14 (*Le Rabolion*, 1919, posthum), p. 295 ff.; eigentlich 318–411 (cf. oben CARRA DE VAUX).

ZDMG: Zeitschrift der Deutschen Morgenländischen Gesellschaft.

XVI

Zur Tradition der „Unwettersterne“

Die astronomisch-astrologische Fachliteratur des Mittelalters umfaßt in ihren lateinischen, altspanischen und byzantinischen Texten islamisch-arabischer Provenienz vor allem eine Auswahl bekannter Standardwerke ebenso bekannter Standardautoren, die zum größten Teil bereits erkannt und dank neuerer Hilfsmittel[1] auch relativ leicht zu identifizieren sind, wenngleich die Erforschung und Edition der Einzeltexte in den meisten Fällen noch aussteht[2].

Neben solchen Standardwerken finden sich in den westlichen Handschriften gelegentlich kurze apokryphe Textstücke eingestreut, deren Identität dunkel ist und die sich nur durch glückliche Zufallsfunde einer bestimmten festen Tradition zuordnen lassen[3].

Hier möchte ich einen derartigen kleinen Text vorstellen, dessen byzantinisch-mittelgriechische Fassung seit längerem bekannt war[4] und zu dem ich inzwischen auch einen westlich-lateinischen Traditionszweig auffinden konnte. Beide diese Versionen gehen offensichtlich auf einen arabischen Text zurück, der seinerseits wiederum auf einer vermutlich antik-hellenistischen Tradition — eventuell mit persischer Zwischen-

[1] Z. B. F. C. Carmody: *Arabic Astronomical and Astrological Sciences in Latin Translation.* Berkeley — Los Angeles 1956 [cf. jedoch Fn. 2!]; L. Thorndike — P. Kibre: *A Catalogue of Incipits of Mediaeval Scientific Writings in Latin* (Revised and Augmented Edition). London 1963. Für mittelgriechisch-byzantinische Texte ist auf die 12 Bände des *Catalogus Codicum Astrologorum Graecorum.* Brüssel 1898—1953 zu verweisen.

[2] Mit dem *Introductorium maius* (= *Kitāb al-mudḫal al-kabīr*) von Abū Maʿšar z. B. hat sich jüngst R. Lemay befaßt (*Abu Maʿshar and Latin Aristotelianism in the Twelfth Century.* Beirut 1962). Seine Quellenuntersuchungen (dort S. 379ff.) ergaben gewisse Korrekturen an dem nicht immer zuverlässigen Carmody, der sich zu stark auf häufig unzuverlässige veraltete Katalogangaben stützt. Eine synoptische Edition in drei verschiedenen Versionen wird angekündigt (Lemay S. 379).

[3] Eine solche Identifikation gelang z. B. kürzlich für den *liber hermetis de stellis beibeniis,* cf. P. Kunitzsch in: ZDMG 118 (1968), S. 62—74, und 120 (1970), S. 126—130.

[4] Περὶ ἀστέρων ποιούντων τάραξιν ...; ediert *Cat. Cod. Astrol. Graec.* 4, S. 145f.; dazu Varianten aus einer Petersburger (Leningrader) Handschrift ib. 12, S. 113; weitere Vorkommen sind erwähnt ib. 4, S. 33; 6, S. 34; 7, S. 23 und S. 75; 9,1 S. 56, S. 65 und S. 74; 9,2 S. 76.

stufe — beruhen dürfte. Von der anzunehmenden arabischen Fassung war bisher noch keine Spur nachzuweisen. Doch ist zu hoffen, daß die hier erfolgende Veröffentlichung der byzantinisch-mittelgriechischen und der westlich-lateinischen Fassungen dazu beiträgt, daß dann ein aufmerksamer Leser bei anderweitigen Handschriftenforschungen zufällig auch einmal auf das entsprechende arabische Textstück stoßen wird. Eine systematische Suche wäre selbstverständlich angesichts der Situation im arabischen Handschriftenwesen ein aussichtsloses Unternehmen.

Unser je nach Seitenformat und Schriftgröße nur ca. zwei bis sechs Seiten umfassender kleiner Text besteht aus einem Kalendarium, das für bestimmte angegebene Daten das Erscheinen eines Sterns notiert, der für eine ebenfalls angegebene Anzahl von Tagen Unwetter, Stürme und Regen erzeuge. Die Zahl dieser „Unwettersterne" schwankt zwischen 24 und 44. Jeder Unwetterstern ist mit einem Namen bezeichnet, aber diese angeblichen Sternnamen zeigen keinerlei Berührung mit der üblichen und hinlänglich bekannten antiken, arabischen und lateinisch übersetzten Sternnomenklatur. Es scheint hier ein ähnlicher Fall vorzuliegen wie bei den hermetischen „*stellae beibeniae*"[5]. Besonders unter namengeschichtlichem Aspekt wäre daher die Auffindung des vermuteten arabischen Originaltextes bedeutungsvoll.

Die bereits im *Catal. Cod. Astrol. Graec.* 4, S. 145f. veröffentlichte mittelgriechische Fassung [im Folgenden bezeichnet als: G] enthält 24 Positionen, von denen bei sechs der Sternname fortgelassen ist. Einige der griechischen Handschriften, so auch die jener Edition zugrunde liegende, geben in der Überschrift ausdrücklich an, der Text stamme von einem „persischen Philosophen Zanati" (Ζανατῆ u.ä.). Dieser Name ist innerhalb der Literatur arabischer Sprache tatsächlich belegt: es ist ein Verfasser geomantischer Schriften, der Scheich Abū ʿAbdallāh az-Zanātī, von dem ein geomantisches Werk 1265/66 ins Mittelgriechische übersetzt wurde, das noch in mehreren Handschriften erhalten ist. Eine Reihe von mir durchgesehener geomantischer Schriften enthielt unser „Unwetterkalendarium" jedoch nicht[6]. Es muß als fraglich angesehen werden, ob die griechische Zuweisung des Kalendariums an Zanati echt ist. Uns kann sie zumindest als Hinweis darauf gelten, daß das Kalendarium zusammen mit den Texten, die in den Handschriften in seiner Umgebung erscheinen, zu einem Konglomerat von Übersetzungen aus dem Arabischen gehört (byzantinische Schriftsteller bezeichnen meist das Orientalische schlechthin als „persisch"). In die gleiche Richtung deuten auch die seltsamen Sternnamen, die sich nicht der überlieferten griechischen Nomenklatur zuordnen lassen.

[5] Cf. ZDMG 118 (1968), S. 69ff., und 120 (1970), S. 128ff.
[6] Cf. P. Kunitzsch in: Byzantinische Zeitschrift 60 (1967), S. 309—317.

Auf eine analoge westeuropäische Tradition stieß ich an drei Stellen. Hier wird zwar jeweils keine Quellenangabe gemacht, doch ergibt sich die völlige Identität der Texte aus dem gleichen Aufbau und Inhalt sowie der größtenteils einwandfrei erkennbaren Identität der Nomenklatur. Das vorauszusetzende arabische Original scheint also keine außergewöhnliche Rarität gewesen zu sein, wenn es sowohl im östlichen Raum wie auch im Westen verbreitet war und offensichtlich soviel Bekanntschaft oder gar Popularität genoß, daß es in beiden Bereichen sogar zu Übersetzungen, ins Mittelgriechische bzw. ins Lateinische, kam. Für Popularität scheint auch der dritte meiner westlichen Textzeugen zu sprechen, der nämlich eine deutsche Bearbeitung der lateinischen Fassung darstellt!

Die drei Fundstellen aus der westlichen Tradition sind folgende:

1. Bibliotheca Vaticana, ms. Ottob. 1826, fol. 64v [im Folgenden bezeichnet als: V][7]: *De stellis fixis que mutant tempora et commotionem faciunt in mari*. Hier haben wir es nur mit einem Fragment unserer Überlieferung zu tun. Oben auf der betreffenden Seite steht ein neunzeiliges astrologisches Stück *De Guerris*. Nach kurzem Abstand folgt darunter unser Text, der in sechs Zeilen nur die Monate März, April und Mai (mit insgesamt sechs Sternen) umfaßt und danach abbricht. Die restlichen zwei Drittel der Seite sind leer (mir lag eine Fotokopie vor). In der Überschrift, und ebenfalls noch einmal bei dem an erster Stelle erwähnten Stern, wird das erzeugte Unwetter (hier: *procella*) auf das Meer verlegt — vermutlich eine Abweichung, die sich im Verlaufe der im Westen recht freizügigen Überlieferung irgendwann eingeschlichen haben mag.
2. Paris, ms Bibl. Nat. lat. 3660 A, fol. 51r—52r [im Folgenden bezeichnet als: P][8]: *Dispositio Stellarum. Stellae sunt in coelo XXVII quae ascendunt, quarum duae faciunt in quo loco uisae, magnam tempestatem uel de uentis, uel de pluuia uel de niue uel de tonitru, uel de corruscambria*. Es folgt das vollständige Kalendarium, von März bis Februar, mit insgesamt 27 Sternen. In Daten und Namen besteht eine enge Verwandtschaft zu V, dagegen ist der Textwortlaut selbst, wie auch schon in der Überschrift, in frei abgewandelten eigenen Formulierungen gehalten. Der Schrift nach (von der Fotokopie her geurteilt) handelt es sich um ein sehr spätes Dokument.
3. Coburg, Landesbibliothek, ms Sche. 18, ca. 1485, fol. 91r—93v [im Folgenden bezeichnet als: C][9]: *Hernach ſtett geſchriben von etlichen*

[7] Cf. L. Thorndike in: Isis 49 (1958), S. 36 mit Anm. 73.

[8] Cf. Thorndike-Kibre: *Incipits* [s. Fn. 1], col. 1528.

[9] Cf. F. G. Kaltwasser: *Die Handschriften der Bibliothek des Gymnasium Casimirianum und der Scheres-Zieritz-Bibliothek*. Coburg 1960, S. 200f.

Sternen die wint vnd vngewitter machen jn jrm ſcheine vnd kumpt itzlicher auff ſeinen tag nach des manes ſchein zu Rechen. Deutsche Version des vollständigen Kalendariums, beginnend mit Januar. Die Gesamtzahl der Sterne ist auf 44 angewachsen, die Namen zeigen stärkere Deformationen und sind schwerer mit den Formen von V und P zu vergleichen. Neun Positionen scheinen, von den Namen her gesehen, Duplikate zu anderen Sternen der Urfassung darzustellen; und so dürfte die Verdoppelung des Gesamtumfanges wohl auf Kosten einer sekundären Bearbeitung gehen — ob schon im lateinischen Text oder erst in der deutschen Bearbeitung selbst, können wir nicht unterscheiden.

Der lateinischen Urfassung, d.h. also der eigentlichen Übersetzung, scheint V am nächsten zu stehen. P, und ganz besonders C, sind Beispiele aus der späteren Überlieferung, die sich — ganz offenbar im Strome einer recht breiten und wohl auch populären Tradition — ziemlich frei gestaltete. Der sachliche Grundgehalt bleibt aber weiterhin bewahrt und ist einwandfrei zu erkennen.

Im Folgenden gebe ich eine Übersicht über die Daten und Namen nach sämtlichen Textzeugen. Die Tabelle zeigt, nach dem Datum für das Erscheinen des jeweiligen Sterns, die Anzahl der Tage, für welche er Unwetter hervorruft, die Quellenangabe (GVPC, mit jeweils beigesetzter laufender Nummer innerhalb des betreffenden Textes), sowie den Namen des Sterns.

Datum Monat	Tag	Dauer Tage	Text	Name	Bemerkungen
März	8	3	C_{11}	*Serpe*	[= P_1]
	9	3	G_4	Σαριτζε	
	9	3	P_1	*scusaphe*	
	9	3	V_1	*sosafe*	
	15	1	C_{12}	*Salprotro*	[Dupl. zu C_{11}]
	15	1	P_2	*tarphe*	
	21	4	V_2	*garfe*	
	22	3	C_{13}	*Thurpe*	[= P_2]
	25	1	G_5	Γαράρ	
	25	2	C_{14}	*Trapſen*	[Dupl. zu C_{13}]

112

Datum Monat	Tag	Dauer Tage	Text	Name	Bemerkungen
April	5	5	G_6	Σιμέλ	beachtliche Einheitlichkeit von Daten u. Namen in dieser Gruppe!
	5	5	C_{15}	*Smuli*	
	5	5	P_3	*similiter* [vomKopisten verkannt]	
	5	5	V_3	*senile*	
	19	3	G_7	Βεκλεύς	
	[Petrop. 9]				
	19	3	C_{16}	*Querſin*	
	20	3	P_4	*quarse*	
	22	3	V_4	*guarse*	
Mai	1	7	G_8	Γάρφας	
	1	4	C_{17}	*Turpen*	
	1	3	P_5	*Arpha*	
	1	3	V_5	*Arsa*	
	10	3	P_6	*crehil*	
	12	4	G_9	Τέμπελις [var. Παίλας]	
	12	3	C_{18}	*Equiell*	
	21	3	V_6	*cireil*	
	24	4	C_{19}	*Edell*	[Dupl. zu C_{18}]
Juni	8	3	G_{10}	Ἀ[ν]ολώστας [var. Αὐολόστας]	
	8	3	P_7	*Suelle*	
	9	3	C_{20}	*Saulle*	
	17	7	C_{21}	*Saulem*	[Dupl. zu C_{20}]
	21	1	G_{11}	Γαπούς [var. Γαμπούς]	
	[Petrop. 13]				
	23	1	P_8	*Neapra*	
	24	1	C_{22}	*Achaim*	
	24	2	C_{23}	*Solam*	[Dupl. zu C_{22}]
Juli	5	1	G_{12}	Παπλούς	
	5	1	P_9	*Holde*	
	5	1	C_{24}	*Beldem*	

Datum Monat	Tag	Dauer Tage	Text	Name	Bemerkungen
	8	1	P_{10}	*beca*	
	10	1	C_{25}	*Debeett*	[Dupl. zu C_{24}]
	19	1	G_{13}	[Name fehlt]	
	19	1	C_{26}	*Ebochi*	[Datum: *XX.*, korr. in *19.*]
August	1	1	G_{14}	Ῥάφας	
	1	1	P_{11}	*tarse*	
	2	2	C_{27}	*Tarpe*	
	5	3	C_{28}	*Innemine*	
	13	3	C_{29}	*Sadicett*	
	14	2	G_{15}	Πιστεύς [var. Πισές]	
	14	2	P_{12}	*copisseuite*	
	17	2	C_{30}	*Tibien*	
	19	4	C_{31}	*Ebochill*	[Dupl. zu C_{26}]
	26	*diebus multis*	P_{13}	*locabim*	
	27	4	G_{16}	[Name fehlt]	
	27	3	C_{32}	*Lachabio*	
Sept.	7	7	G_{17}	Ἀλφαράς [var. Γαρές]	
	9	1	C_{33}	*Iperchacietali*	
	9	10	P_{14}	*Alfureali*	
	12	9	C_{34}	*Ephatoel*	
	22	3	G_{18}	[Name fehlt, var. Γαράφα]	
	22	3	P_{15}	*ebfaran et micade*	
	29	3	C_{35}	*Elforacrima*	
Oktober	5 [Petrop. 3]	1	G_{19}	Γαρεύς [var. Γιρές]	
	5	1	C_{36}	*Rachaboliach*	
	8	1	P_{16}	*Boctone*	
	17	3	G_{20}	[Name fehlt]	
	17	3	C_{37}	*Fatam*	
	19	3	P_{17}	*Sareti*	
	21	3	C_{38}	*Gaicho*	

114

Datum Monat	Tag	Dauer Tage	Text	Name	Bemerkungen
Nov.	1	1	G_{21}	῾Ρήκτας [var. Ρίκτας]	
	1	1	C_{39}	*Elbotham*	
	2	1	P_{18}	*Elbuta*	
	12	5	G_{22}	[Name fehlt]	
	12	5	C_{40}	*Querchem*	
	12	5	P_{19}	*Octurie*	
	16	1	C_{41}	*Ehedi*	[Dupl. zu C_{42}]
	21	1	C_{42}	*Elieudi*	
	22	1	P_{20}	*Eltedi*	
Dez.	9	4	G_{23}	᾿Εώλας [var. ᾿Εάλας]	
	9	6	C_{43}	*Elecha*	
	9	6	P_{21}	*electa*	
	17	3	G_{24}	[Name fehlt]	
	23	3	C_{44}	*Elethera*	
	26	3	P_{22}	*Elcera*	
Januar	2	3	C_1	*Abnoxe*	
	3	5	P_{23}	*Eldra*	
	4	5	C_2	*Glira*	
	4	5	G_1	Γαμπίνης [var. Γαμπινού]	
	17	7	P_{24}	*Ecaphe*	
	20 [Petrop. 25]	3	G_2	Λάμπης	
	22	6	C_3	*Teſatobell*	[Dupl. ?]
	23	5	C_4	*Adra*	[= P_{23}?]
	30	8	C_5	*Enanem*	
	30	3	P_{25}	*Sudele*	
Februar	9	4	G_3	Σάρης	
	12	8	C_6	*Gelbem*	
	12	7	P_{26}	*Sebim*	
	17	3	C_7	*Giebi*	[Dupl. zu C_6]
	18	2	C_8	*Agnen*	
	20	3	C_9	*Leembie*	
	22	3	P_{27}	*Elzierben*	
	24	1	C_{10}	*Carabi*	[Dupl. zu C_9]

Über den Charakter dieses eigenartigen Unwetterkalendariums läßt sich nichts Positives aussagen. Es kann sich dabei jedenfalls um keinen herkömmlichen astronomischen bzw. astrologischen Text handeln, da die benutzten Gestirnnamen keiner bekannten Tradition angehören. Auch ein Mondstationensystem kommt nicht in Betracht, einmal da die Zahl der Nennungen in keinem Falle mit der Zahl der 28 Mondstationen übereinstimmen würde, und zweitens da auch hier die Namen keinerlei Anklang an die hinreichend bekannten Mondstationennamen[10] aufweisen[11]. Ebenso wenig scheint eine Verbindung zu den diversen Systemen von „verworfenen Tagen" und *dies Aegyptiaci*[12] zu bestehen; diese fallen auf andere kalendarische Daten, und sie stehen vor allem auch nicht in Abhängigkeit von gewissen — geschweige denn gar namentlich bezeichneten — Sternen. Die Intervalle zwischen den einzelnen Daten sind ungleichmäßig und ergeben keine erkennbare Periodizität.

*

Exkurs

In THORNDIKE-KIBRE's *Incipits* (cf. Fn. 1), col. 1437, wird noch ein weiterer einschlägiger lateinischer Text aufgeführt, dessen Incipit lautet: *Sequuntur stelle fixe aerem turbantes in singulis mensibus*[13]. Es handelt sich um eine Handschrift „Miscellanea astrologica" in Katowice, Biblio-

[10] Für die arabischen Namen der Mondstationen cf. die Aufstellung bei P. KUNITZSCH: *Arabische Sternnamen in Europa.* Wiesbaden 1959, S. 55f.

[11] Einige Ausnahmen wären eventuell in C aufzuspüren: z. B. C_7 *Giebi* (evtl. =*al-ǧabha,* 10. Station); C_{20} *Saulle* und C_{21} *Saulem* klingen an *aš-šawla* (Station 19) an, jedoch gehört mindestens C_{20} zur ursprünglichen Überlieferung (cf. P_7 *Suelle*) und kann infolgedessen vom Prinzip her kein Mondstationenname sein; C_{24} *Beldem* scheint an *al-balda* (Station 21) anzuklingen, gehört jedoch ebenfalls zur ursprünglichen Überlieferung (cf. P_9 *Holde*); endlich C_{39} *Elbotham* klingt sehr stark an die Latinisierungen des Namens der 2. Station (*al-buṭayn*) an, gehört aber auch zum ursprünglichen Text (cf. P_{18} *Elbuta*). Diese — auch nur vereinzelten — Ähnlichkeiten lassen jedenfalls auf kein System schließen und scheinen rein zufällig zu sein.

[12] Cf. G. KEIL in: Sudhoffs Archiv f. d. Geschichte d. Medizin u. d. Naturwissenschaften 41 (1957), S. 27—58.

[13] Als Fundort ist dort ein Wiener Antiquariatskatalog von 1936 angegeben. Nach ausführlichen Nachforschungen gelang es mir zu ermitteln, daß die bewußte Handschrift seinerzeit nach Kattowitz verkauft wurde, und nach weiteren Bemühungen konnte ich aus der noch jetzt dort vorhandenen Handschrift Fotokopien der entsprechenden Blätter erhalten. In diesem Zusammenhang möchte ich der Direktion der Biblioteka Śląska sowie auch Herrn Prof. Dr. B. OLSZEWICZ von der Polnischen Akademie der Wissenschaften für ihre freundliche Hilfe besonders danken.

teka Śląska, geschrieben ca. 1493 an der Universität Krakau; das in Rede stehende Unwetterkalendarium nimmt darin die folia 146v—147r ein. Hier begegnen wir einem Kalendarium, das in seiner Anlage grundsätzlich verschieden ist von dem oben beschriebenen Typus. In insgesamt 42 Positionen[14] werden von 17 verschiedenen Gestirnen Abend- und Morgenauf- bzw. -untergänge verzeichnet, die angeblich Wetterstörungen hervorrufen. Die Sterne sind ebenfalls mit Namen bezeichnet, in denen man auf Anhieb bekannte Benennungen aus der antik-abendländischen Tradition erkennt: *delphin, lira, pleyades, orion* usw. Daneben erscheinen so befremdliche Formen wie *macalis, hiriphi* usw., die orientalischen Einfluß widerzuspiegeln scheinen. Ein Vergleich mit echten, rein antiken Kalendarien[15] belehrt uns, daß der Kattowitzer lateinische Kalender einer alten, auf griechische Quellen zurückgehenden Tradition angehört: Aufbau, Inhalt, Daten und Namen sind weitgehend identisch. Es wirkt freilich höchst seltsam, daß der Text einerseits die alten griechischen Namen vollkommen originalgetreu schreibt und anwendet[16], andererseits dagegen die abenteuerlichsten Wortgebilde aufweist[17]. Diese ließen sich teilweise (*yces* = Αἴξ, *hiriphi* = Ἔριφοι) aus der mittelalterlichen Aussprache des Griechischen erklären, während die Entstehung anderer (*adictiron* = Ἀρκτοῦρος, *macalis* = Στάχυς) rätselhaft bleibt. Weder die griechische noch die lateinische Paläographie bietet hier Ansätze. Eine arabische Zwischenstufe mag man aber auch nicht annehmen, da dann wiederum die anderen griechischen Namen nicht in so vorzüglichem

[14] Die Sternangaben fallen auf folgende Daten: I 4. 5. 28 (zwei Angaben). 25; II 1. 25; III 18; IV 1. 18. 24. 26; V 6. 7. 19. 24; VI 2. 6. 28; VII 3.—.28. 25; VIII 28 (drei Angaben); IX 7. 4. 18; X 6. 7. 16. 23; XI 6. 13. 21. 28 (zwei Angaben); XII 1. 10. 21.

[15] Vgl. die edierten Texte in *Cat. Cod. Astr. Graec.* 9,1 S. 128—137; 11,2 S. 168—173 [Excerpta ex calendario Clodii]; 12, S. 109—112 [περὶ τῶν λη′ ἀστέρων ἀπλανῶν τῆς ὀγδόης σφαίρας . . .].

[16] Auch hier gibt es gelegentlich Divergenzen: *arctos* steht teils für Ἀετός, α Aql (I, 5; VI, 2; VII, 25), teils für Ἀρκτοῦρος, α Boo (VI, 6; IX, 4; dieser Stern heißt daneben einmal auch *adictiron*: II, 25); *prothrios* = Προκύων, α CMi, ist lateinisches Schreiberprodukt (sonst üblicherweise *prochion* u. ä.); *lampades* (I, 25) entspricht in den griechischen Kalendarien ὁ λαμπρὸς ἐν τῷ Λέοντι u. ä. und bezeichnet demnach α Leo; *ipos* (III, 18) entspricht ὁ λεγόμενος Ἵππος, Identifikation unklar.

[17] *yces* = Αἴξ, α Aur (V, 6. 24; IX, 7; XII, 21); *hyſtre* = Ὀϊστός, Sagitta (VIII, 28); *macalis* = Στάχυς, α Vir (IX, 18); *zopherios* = Στέφανος, [α] CrB (X, 6; derselbe bei XI, 28: *ſachrinos*); *hiriphi* = Ἔριφοι, ζη Aur (X, 7; derselbe Name bei XII, 10: *herſa*); *andravis* (XI, 28; wahrscheinlich = Orion, der freilich sonst richtiger *yrion* [VI, 28] bzw. *orion* [VII, 3] heißt). Diese Identifikationen ergeben sich aus dem Textvergleich mit den alten griechischen Kalendarien. Ein Stern (*iares*; II, 1 — Abenduntergang) ist dort nicht vertreten und kann daher nicht erklärt werden.

Erhaltungszustand vorlägen. Die Geschichte des Kattowitzer Kalendariums bedarf also für sich weiterer Nachforschungen, wozu die Auffindung neuer Textzeugen eine der wichtigsten Voraussetzungen wäre. Mit unserem „Unwetterkalendarium“ des oben beschriebenen Typs hat es jedenfalls nichts zu tun.

*

Mit dieser Veröffentlichung soll nicht nur der neuentdeckte westlich-lateinische Zweig der Tradition der Unwettersterne bekannt gemacht werden. Es verbindet sich damit vielmehr, wie bereits oben ausgesprochen, auch der Zweck, die Aufmerksamkeit von Orientalisten, falls sie sich mit einschlägigen Handschriften befassen, auf diesen Text hinzulenken, in der Hoffnung, daß so einmal durch glücklichen Zufall das unscheinbare kleine versprengte Textstück in einer arabischen Quelle aufgefunden wird. Damit wäre der Traditionskette dann ein weiteres, und sehr wichtiges, Glied hinzugefügt, das freilich, wie zu vermuten ist, für sich auch noch nicht den letzten Ausgangspunkt der gesamten Überlieferung darstellt, die letztlich in hellenistisch-antike Zusammenhänge hinabreichen dürfte.

XVII

Abū Maʿšar, Johannes Hispalensis und Alkameluz

Zu den wenigen Sternnamen arabischen Ursprungs, die sich einer überlieferungsgeschichtlich abgesicherten Deutung bisher entzogen haben, gehört die durch H. Grotius und J. Bayer weitergetragene Bezeichnung *Alkameluz* für den Stern α Bootis (Arkturus)[1]. Weder das ursprüngliche Auftreten des Wortes im westlich-lateinischen Bereich noch gar sein fremdes, vermutlich arabisches Original ließen sich fixieren.

Nun will es scheinen, als ob der Ursprung jenes Namens (wie auch einiger anderer) im engeren Umkreis gewisser astrologischer Texte zu suchen ist. Wir betreten damit wieder einmal das Gebiet der mittelalterlichen arabisch-lateinischen Übersetzungsliteratur, das aus Mangel an erschlossenem Textmaterial immer noch voller Rätsel und Überraschungen steckt und nur selten einen geschlossenen, schlüssigen Überblick über einen größeren Problemkreis freigibt. Auch in unserem Fall gelingt nur die Annäherung und nähere Einkreisung, und leider nicht die lückenlose Aufklärung.

Ausgangspunkt und wesentliches Funktionselement der Astrologie bilden die Planeten mit ihren ständig wechselnden Positionen und den sich daraus unablässig neu ergebenden Konstellationen. Das hinderte jedoch nicht, daß zur Vervollkommnung des Systems auch die Fixsterne mit in die astrologischen Theorien einbezogen wurden; über das Vergleichsmoment der Farbähnlichkeit oder -gleichheit wurde ihnen die Wirkungskraft(„Temperament", κρᾶσις, *mizāǧ*, *complexio*; auch „Charakter", *ṭabīʿa*, *natura*) des bzw. der betreffenden Planeten zugeschrieben. Diese Lehren sind bereits aus hellenistischer Zeit bekannt, und Listen von Fixsternen mit ihren astrologischen Temperamenten gehören zum Grundbestand astrologischer Handbücher jener Zeit[2]. Mit dem gesamten System übernahm die islamische Welt auch diesen Zweig der Astrologie, und so wird hier häufig in Sternverzeichnissen jedem Stern neben den astronomischen Koordinaten auch sein astrologisches „Temperament" (*mizāǧ* oder *ṭabīʿa*) beigeschrieben. Ebenso gliedern astrolo-

[1] Cf. P. Kunitzsch, *Arabische Sternnamen in Europa*, Wiesbaden 1959, S. 133ff. (Nr. 46).

[2] Für Beispiele cf. zuletzt P. Kunitzsch in *Byzantin. Zeitschrift* 57, 1964, S. 382ff., speziell S. 406—411, und *ZDMG* 118, 1968, S. 62—74, sowie die dort zitierte ältere Literatur.

gische Autoren ihren Schriften gern ein Kapitel über die astrologischen Wirkungen der Fixsterne ein, das zuweilen dann auch eine entsprechende Aufzählung von Sternen enthält.

Im Werk des berühmtesten und fruchtbarsten islamischen Astrologen, Abū Maʿšar al-Balḫī (geb. vor 786, gest. 886)[3], sind solche regelrechten Sternlisten nicht leicht aufzufinden. Vor allem das Handbuch *Kitāb al-mudḫal al-kabīr* (lateinische Übersetzung des Hermannus Dalmata von 1140: *Introductorium in astronomiam Albumasaris abalachi*, gedruckt Augsburg 1489, Venedig 1506 u.ö.) behandelt zwar zweimal die Fixsterne — in Buch II, Kap. 1 in der Reihenfolge und Terminologie der 48 ptolemäischen Sternbilder, sowie in Buch VI, Kap. 1 nach dem spätantiken System der Paranatellonta —, gibt jedoch nirgends eine Aufzählung bestimmter Einzelsterne. Das gleiche gilt von dem großen, thematisch freilich rein den Planeten gewidmeten Werk „*De magnis coniunctionibus*" (Augsburg 1489 u.ö.)[4]. Fündig werden wir dagegen in der kleinen, ebenfalls lateinisch übersetzten, offenbar einst stark verbreiteten Schrift „*Flores astrologiae*", deren arabisches Original kürzlich J. VERNET in zwei Handschriften nachweisen konnte[5]. Eine zweite, hiervon unabhängige Liste von „Unglückssternen" ist nur aus Übersetzungen bzw. Exzerpten bekannt und scheint im Original verschollen zu sein.

Von Abū Maʿšar aus führt dann ein Weg zu Johannes Hispalensis, einem der bekanntesten arabisch-lateinischen Übersetzer Spaniens in der Mitte des zwölften Jahrhunderts, der selbst mehrere Werke Abū Maʿšars übersetzt hat und bei dem wir Abū Maʿšars Material dann auch „selbständig" — d.h. in eigenen Kompilationen — verarbeitet finden.

Ursprünglicher Ausgangspunkt dieser drei verschiedenen astrologischen Sternverzeichnisse, von denen hier zu handeln ist, scheint das Verzeichnis der „*Flores astrologiae*" gewesen zu sein, das wir im folgenden als I bezeichnen wollen (22 bzw. 24 Sterne). Hiervon erscheint später eine erweiterte Fassung mit 29 Sternen, genannt I'. Als II bezeichnen wir die Liste von 14 „Unglückssternen" bei Abū Maʿšar, und mit III schließlich kennzeichnen wir die Kompilation des Johannes Hispalensis. Jedes dieser Verzeichnisse birgt für sich eine eigene Problematik, die

[3] Cf. BROCKELMANN, *GAL* I, 221; Suppl. I, 394; *Encyclopaedia of Islam*, new edition, Artikel *Abū Maʿshar* (J. MILLAS VALLICROSA).

[4] Die Bibliographie von Abū Maʿšars Werken und deren lateinischen Übersetzungen ist noch nicht geschrieben und muß als dringendes Desiderat angesehen werden. Hier kann aus begreiflichen Gründen auf übergreifende bibliographische Probleme nicht eingegangen werden. Cf. auch Fn. 5.

[5] ms Escorial 918 und Paris B.N. ar. 2588; cf. seinen Artikel *Problemas bibliográficos en torno a Albumasar*, in der in Barcelona erscheinenden Zeitschrift *Biblioteconomia*, 1952, S. 12—17.

jeweils hier jedoch nicht geklärt werden kann, da das zugängliche Material weiterreichende Schlüsse noch nicht zuläßt. Uns kommt es vorderhand darauf an, die Zusammengehörigkeit dieses Komplexes zu zeigen, zu dessen sichtbarer kulturgeschichtlicher Nachwirkung u.a. die Einführung und Weitergabe einiger Sternnamen gehört, deren bekanntester wohl *Bellatrix* (γ Orionis) ist.

I.
Flores astrologiae (*Kitāb an-nukat*)

Arabischer Text *Kitāb an-nukat*, nachgewiesen durch J. Vernet (cf. Fn. 5) in ms Escorial 918 [= E], ohne Kapitelzählung, fol. 5ᵛ—6ʳ: *wa-ammā maʿrifat al-kawākib aṯ-ṯābita*; und in ms Paris, B.N. ar. 2588 [= P], mit Kapitelüberschriften, aber ohne Zählung, fol. 15ᵛ—16ʳ: *bāb fī maʿrifat al-kawākib aṯ-ṯābita*. Die gleiche Liste zitiert, laut D. Pingree[6], der etwas jüngere as-Siǧzī (um 1000)[7] unter verschiedenen Auszügen aus Abū Maʿšar in seinem *Kitāb al-qirānāt wa-taḥāwīl sinī l-ʿālam*; Pingree druckt die Liste ab auf S. 77 nach ms Istanbul, Nuruosmaniye 2795 [hier = I, bei Pingree = N], fol. 60ᵛ—61, und ms Paris, B.N. ar. 2581 [hier = S, bei Pingree = P], fol. 60ʳ. Die lateinische Übersetzung, von Johannes Hispalensis (Mitte des 12. Jhd.), existiert in zahlreichen Drukken[8]; die Kapitel sind auch hier nicht gezählt. Der Druck Augsburg 1488 (nachgedruckt Zwickau 1928) [= F] enthält auf S. b 4 das entsprechende Kapitel *De stellis fixis quid operantur in reuolutionibus ac natiuitatibus*; im wörtlich gleichlautenden Druck Venedig 1506 (?)[9] steht es auf S. c 4. Für die Untersuchung lagen vor EPS in Fotokopien[10], F im Original (1488 und 1506); I kann nur nach Pingrees Wiedergabe zitiert werden.

Die Liste umfaßt 22 bzw. 24 Sterne, deren Ekliptiklängen in Grad und Minuten angegeben werden; für die Breiten wird lediglich die Himmelshälfte notiert, und auch das nur bei einigen Sternen, nicht durchgehend. Hinzu tritt als wichtigstes die planetäre Temperamentangabe. Zwei Dinge fallen an der Liste besonders auf: Bis auf Nr. 3 (Plejaden) und Nr. 4 (Aldebaran, α Tauri) sind alle Positionen o h n e N a m e n gelassen; und die verzeichneten Sterne sind fast ausnahmslos Träger der Temperamente der Unglücksplaneten Mars und Saturn. Es handelt sich

[6] *The Thousands of Abū Maʿshar*, London 1968, S. 70ff., speziell 76f.

[7] Cf. Brockelmann, *GAL* I, 219; Suppl. I, 388.

[8] Cf. F. J. Carmody, *Arabic Astronomical and Astrological Sciences in Latin Translation*, Berkeley—Los Angeles 1956, S. 92ff. (Nr. 3).

[9] Die Jahreszahl 1506 erscheint nur bei dem gemeinsam hiermit gebundenen, bei demselben M. Sessa gedruckten *Introductorium maius*.

[10] P ist sehr nachlässig und verständnislos geschrieben und weist über längere Strecken Verschiebungen innerhalb von Zeilen und Kolumnen auf.

also um eine Liste unüblicher Art, in der nicht wie sonst die bekanntesten astronomisch relevanten Sterne zusammengefaßt sind — diese sind ja sämtlich durch eigene Namen ausgezeichnet, und so wären sie dann auch hier, analog zu Nr. 3 und 4, individuell benannt worden; zweitens sind diese sehr verschiedenen Temperamenten zugeordnet[11] und würden nicht in das vorliegende Mars-Saturn-Schema passen. Demnach muß es sich hier um eine Zusammenstellung speziell im Hinblick auf die Unglück bringenden Mars- und Saturn-Temperamente handeln, eine Zusammenstellung, die also genötigt war, vorwiegend auch unbekannte, in der Tradition namenlos gelassene Sterne zusammenzusuchen und aufzunehmen. Bei sehr subtiler kritischer Auswertung der leider nur allzu nachlässig überlieferten Koordinatenangaben läßt sich viermal ein Längenwert von Ptolemäus + $6^0 45$ ermitteln, der bei Zugrundelegung der antiken Präzessionskonstante von 1^0 in 100 Jahren ganz grob auf eine Epoche um 820 führt, in eine Zeit also, die rund ein Jahrzehnt vor dem Zeitpunkt liegt, zu dem sich Abū Maʿšar der Astrologie zugewendet haben soll[12]. Das läßt darauf schließen, daß Abū Maʿšar mit dieser Liste einer älteren, vermutlich hellenistischen Tradition folgt. Es ist übrigens bezeichnend, daß sich Abū Maʿšar im Rahmen der antik fundierten astrologischen Tradition hält und nicht etwa die seit den „*Tabulae probatae*" (*az-zīǧ al-mumtaḥan*, 214 h = 829/30) vorhandenen, von den Arabern wesentlich verbesserten astronomischen Präzessionswerte von 1^0 in 66 bzw. 70 Jahren aufgreift! Bei den übrigen Positionen der Liste schwanken die überlieferten Werte so stark, daß man die Sterne nicht eindeutig unter den 1025 Sternen des ptolemäischen Katalogs nachweisen kann, zumal bei gleichzeitiger Berücksichtigung der Temperamentangaben. Auf mittelalterliche Leser und Autoren muß dieses Verzeichnis, das unter 22 bzw. 24 Positionen nur zwei bekannte, leicht erkennbare Gestirne aufweist, sehr befremdlich gewirkt haben. Hieraus ist wohl die Erweiterung zu erklären, die die Liste nachträglich erfuhr (I′).

I′.

An drei zufällig und willkürlich aufgefundenen Stellen stoßen wir auf die astrologische Sternliste I′, die gegenüber I um neun neue Positionen erweitert ist (unten in der Edition mit a-i bezeichnet). Die neu hinzugefügten Positionen umfassen zumeist jene bekannten hellen Sterne, die in der Liste I — gemäß ihrem Zweck, Unglück bringende Sterne aufzuzählen — absichtlich fortgelassen waren, die man also später vermißte

[11] Cf. die Übersicht in *ZDMG* 118, 1968, S. 67f.

[12] Das soll im Alter von 47 Jahren geschehen sein; bei seinem Tode 886 soll er über hundert Jahre alt gewesen sein. Für seine Bekehrung zur Astrologie ergäbe sich daraus 832 oder früher (cf. Brockelmann a.a.O.).

und zur eigenen Orientierungshilfe nachtrug. Die erweiterte Liste I′ erscheint an folgenden drei Stellen: ms Florenz, Conv. Soppr. J III 28 (= S. Marci 180), s. XIV, fol. 79ʳ [= M]. Die Handschrift ist eine astrologische Sammelhandschrift und enthält zahlreiche Traktate und Fragmente aus dem Kreis übersetzter arabischer Astrologie; nach einer kurzen Einleitung beginnt unser Kapitel aus Abū Maʿšar unvermittelt, also als Exzerpt; es wird gefolgt von der entsprechenden Gebrauchsanweisung, wie in F vorhanden; danach setzen weitere Exzerpte aus anderen Autoren ein. Zweitens: ms Wien 3124, s. XV, fol. 163ʳ⁻ᵛ [= W]; Einleitung und Text stimmen überein mit M. Der kleine Text wird hier als Exzerpt des Guido Bonatti[13] aus Abū Maʿšar bezeichnet. Als Präzessionswert wird ausdrücklich angegeben 1⁰ in 100 Jahren; das ist der antike Wert, den später arabische Astronomen auf 1⁰ in 66 bzw. 70 Jahren verbesserten. Drittens: ms Bern, Burgerbibliothek 483, s. XV, fol. 130ʳ [= B][14]; hier findet sich nur das bloße Verzeichnis ohne erläuternden Beitext; laut Überschrift sind die Koordinaten berechnet für 1497; der Vergleich zeigt, daß die Längen gleichmäßig um 13⁰0 gegenüber MW vermehrt sind. Die drei Textzeugen MWB lagen in Fotokopien vor. Die neu hinzu gekommenen Gestirne tragen jene spezifischen Benennungen, die sonst nur noch bei Johannes Hispalensis in II und III auftreten. Die Originalitätsfrage dürfte so zu beantworten sein, daß diese in der Tat höchst seltsamen Bezeichnungen zuerst bei Johannes Hispalensis erschienen und später von Kompilatoren (vielleicht Guido Bonatti?) in die erweiterte Liste I′ eingebracht wurden.

Sternliste I/I′

(Quellen: arabisch EPSI, lateinisch FMWB)[y]

Lfd. Nr.	Zeichen	Länge	Himmelshälfte	Temperament	Name	Mod. Bez.
1	Aries	13⁰25 S 13⁰24 M 13⁰44 W 13⁰45 → B 26⁰45	N	♂ ♄	EP *fī ra's al-ḥamal* FMW *in capite arietis*	γ Ari[a]

13 Kompilierender Astrologe, gest. 1297; cf. Carmody a.a.O. S. 172.

14 Cf. E. Zinner, *Verzeichnis der astronomischen Handschriften des deutschen Kulturgebietes*, München 1925, Nr. 311, mit Anm. am Ende (die Natur der Tabelle ist hier nicht erkannt).

Lfd. Nr.	Zeichen	Länge	Himmelshälfte	Temperament	Name	Mod. Bez.
2		14°25 S 27°25 I 24°25 M 14°15 W 14°45 → B 27°45	N	♂ ♄	(wie oben Nr. 1)	β Ari[b]
3	Taurus	9°55 bis 13°0 EP 9°15 S unpunktiert[d] I 9°26 → B 22°55 P: bis 10°40	N	♂ ☽ SIMWB: ♂ [MW: ☿ *et* ☽ *participant*]	EP *aṯ-ṯurayyā* FM *pliades*, W *pleiades*, B *plyades*	Plejaden[c]
a		MW 13°0 → B 26°0	N	♂		. . .[e]
b		M 14°45 → B 27°45 W 14°55		♂	M *diabolus*, WB *dyabolus* [cf. II, 5]	(p¹ Per)
c		MW 15°0 → B 28°0		♂	*caput diaboli* [WB *dyaboli*] *uel gorgonis* [cf. II, L² und III, 2]	(β Per)
4		19°15[f] P 19°55, punktiert SI 19°54 ES unpunktiert[d] → B Gem 2°15	—— SI: N	♂	EP *ad-dabarān* FMW *aldebaran*, B *aldeberam*	α Tau
d	Gemini	M[g]W 8°0[h] → B 21°0		♂ ♄	*humerus canis* [cf. II, 9 und III, 5]	(α Ori)
e		M[g]W 10°15[i] → B 23°15		♂	*bellatrix* [cf. II, 5 und III, 4]	(α Aur)

Lfd. Nr.	Zeichen	Länge	Himmelshälfte	Temperament	Name	Mod. Bez.
5		$17^{0}55$ F $18^{0}55$, *in alio libro* $16^{0}0$ EPS un- punktiert[d] → B Cnc $0^{0}55$	S	⊙ M: ♄ WB: nicht ange- geben		?
f		MW $18^{0}52$ → B Cnc $1^{0}52$		⊙ ♂ B: ⊙ ♂ *et* ♄	*malefica*	?
g	Cancer	MW $2^{0}3$ → B $15^{0}3$		♄ ☽	WB *camelus*, M *cammellus*	?
6[k]		$0^{0}55$ PS unpunk- tiert[d] E $17^{0}55$, unpunktiert[l] F $12^{0}55$ MW $17^{0}55$ → B Leo $0^{0}55$[m]	N S:S	♂		?
7[k]		E $7^{0}55$[n] P $4^{0}55$, un- punktiert[d] SI $3^{0}55$, un- punktiert[d] F: nicht vorhanden MW $7^{0}55$ → B $20^{0}55$[m]	N	♂ MWB: nicht ange- geben		?
h		MW $13^{0}0$[o] → B $26^{0}0$		⊙ ☽	*occidens* *camelum* [M *cammelus*] [cf. III, 6]	(α Car)
i		MW $13^{0}0$[o] → B $26^{0}0$	MB: N	♄	WB *pes canis* [cf. III, 7], M *pes cammeli*	(α CMa)
8[k]		E $15^{0}15$, punktiert P $15^{0}55$, punktiert F $2^{0}55$		♄		?

Lfd. Nr.	Zeichen	Länge	Himmelshälfte	Temperament	Name	Mod. Bez.
9	Leo	EP 1°15 SI 1°25[p] FMW 15°55 → B 28°55	N MWB: nicht angegeben	♄		?
10[q]		E 7°25 F 7°22	N	♄		?
11	Virgo	E unleserlich P 20°55, unpunktiert[d] SI 13°55, unpunktiert[d] F 7°1, *in alio* 4. MW 7°11 → B 20°11		♂		?
12		EF 25°0 P 25°1 SI 24°1 MW 15°0 → B 28°0		♄		?
13	Libra	EPI 26°1 S[r] FMW 26°0 → B Sco 9°0		♄		?
14	Scorpius	1°2 FMW 1°3 → B 14°3		♂		?
15		PSI 8°9 EFMW 9°0 → B 22°0[s]		♂		?
16		SI 28°7 P 28°4 EFMW 8°7 → B 21°7[s]		♂		?
17	Sagittarius	19°2 → B Cap 2°2 W 19°3		♄		?
18[t]		21°1 → B Cap 4°1 P 21°2 SI 21°3		♄		?

Lfd. Nr.	Zeichen	Länge	Himmelshälfte	Temperament	Name	Mod. Bez.
19		ESIF 25°8[u] P 24°8		♄		?
20	Capricornus	27°2 →B Aqr 10°2 P 25°40		♄	(*mala*)[v]	?
21		29°5 →B Aqr 12°5 E[w] P 29°2		♄		?
22	Aquarius	9°4 → B 22°4 P 9°2 SI 9°7		♄ E: nicht angegeben MB: ♂ ♄		?
23[x]		29°0		♄		?
24	Pisces	E 4°4 P 17°4 F 4°7, *in alio* 4. M 4°7 → B 17°7 SI 3°4		EPF: nicht angegeben M: ♂ B: ♂ ♄ S: ♄		?

Anmerkungen:

[a] Länge bei Ptolemäus Ari 6°40; Zunahme um 6°45 führt genau auf das vorhandene 13°25.

[b] Länge bei Ptolemäus Ari 7°40; Zunahme um 6°45 führt genau auf das vorhandene 14°25.

[c] Der 30. Stern des Taurus, einer aus dem Sternhaufen der Plejaden, hat bei Ptolemäus die Länge Tau 2°30. Zunahme um 6°45 führt auf 9°15, wie offenbar die richtige Lesung des Wertes in der Fassung I ist.

[d] Bei fehlender Punktierung ist nicht zu unterscheiden zwischen $y = 10$ und $n = 50$; so kann die unpunktierte Notierung des Minutenwertes 15′ oder 55′ gelesen werden.

[e] Abū Maʿšar gibt den Plejaden bei Nr. 3 eine Ausdehnung über mehrere Grad. Oberflächliche Auswertung durch Kompilatoren machte die zweite Begrenzungszahl (oben: 13°0) zur Länge eines eigenen Sterns.

[f] Durch den beigeschriebenen Namen ist der Stern als α Tau gesichert. Seine Länge bei Ptolemäus beträgt Tau 12°40; Zunahme um 6°45 führt auf 19°25. Die anderen Werte (19°15, daraus weiter 19°55) scheinen hieraus korrumpiert zu sein.

g Hier sind d und e als éin Stern behandelt (Natur: ♂).

h Länge von α Ori bei Ptolemäus: Gem 2°0; Zunahme um 6°45 würde also auf 8°45 führen. Der in MW vorhandene Wert könnte damit identisch sein, unter Wegfall des Minutenbetrages.

i Länge von α Aur bei Ptolemäus: Tau 25°0; Zunahme um 6°45 würde auf Gem 1°45 führen. Der in MW vorhandene Wert ließe sich als im lateinischen Bereich daraus entstandene Korruption erklären (Gradwert 1 zu 10 abgewandelt, Minutenwert 45 in 15 verschrieben). Anderenfalls könnte es sich um eine nachlässige Übernahme des Wertes 10°55 aus III, 4 handeln.

k Hier herrscht Unstimmigkeit in den Quellen. EP führen insgesamt drei, SIF nur zwei Sterne im Zeichen Cnc auf.

l Offenbar ist der Wert des vorangehenden Sterns Nr. 5 irrtümlich wiederholt.

m In MWB ist die Reihenfolge: g, 7, h, i, 6.

n Der unpunktierte Wert ist vom Schreiber falsch zur Breite des vorangehenden Sterns Nr. 6 abgewandelt worden.

o 13° vermutlich aus 3°5 von III, 6 (bzw. 3°35 von III, 7) degeneriert.

p In S in der folgenden Zeile wiederholt mit Länge Lib 1°25.

q Nur in EF vorhanden.

r S hat hier den Längenwert des folgenden Sterns; diese Verschiebung setzt sich fort bis zum Ende der Tabelle.

s In MWB steht Nr. 16 vor Nr. 15.

t Mit diesem Stern bricht die Fotokopie von W ab. Die folgende Seite lag nicht vor.

u Fehlt in MWB.

v E gibt diesem Stern ausnahmsweise eine kurze Charakterisierung: (*kawkabān*) *aḥaduhumā azraq wa-huwa naḥs* „(zwei Sterne), einer davon ist blau, und es ist ein Unglücksstern". Wortgetreu heißt es in F: (*due*) *quar'm una mala*. In der erweiterten Fassung I′ ist daraus ein Name geworden: M *que nominatur mala*, B *que dicitur mala*.

w Die Handschrift verwendet hier eine unübliche Ligatur; die Gradzahl ist am ehesten als *kd* = 24° oder *kh* = 25°, aber nicht als *kṭ* = 29° zu lesen. Die Minutenzahl scheint eher eine Null als *h* = 5′ zu bedeuten.

x Erscheint nur in P.

y Während der Drucklegung dieses Artikels erhielt ich von der Bodleian Library aus Oxford Fotokopien der von L. Thorndike — P. Kibre, *A Catalogue of Incipits of Mediaeval Scientific Writings in Latin* (Revised and Augmented Edition, London 1963), mit dem Incipit *De stellis fixis et de potestatibus earum* verzeichneten Passage aus ms Digby 114, 14. Jhd., fol. 105r—107v. Die Untersuchung ergab, daß es sich dabei um das hier besprochene Fixsternkapitel aus den *Flores Albumasaris* handelt. Da eine Einarbeitung in die Tabelle zu diesem Zeitpunkt nicht mehr möglich war, sei hier nur das Wichtigste nachgetragen: ms Digby folgt der Textordnung von F, die Sternliste — demgemäß also in der Fassung I — steht auf fol. 105r. Die Koordinaten entsprechen bei Stern 1—4 F (Schreibung des Namens von 4: *aldabaram*), 5 = Hauptüberlieferung, 6—7 = EMW, 8 = P (also drei Sterne im Zeichen Cancer), 9—10 (d.h. das Zeichen Leo) vom Schreiber versehentlich ausgelassen, 11 = F, 12 = EF, 13 = FMW, 14 = Hauptüberlieferung, 15—16 = EFMW, 17—22 = Hauptüberlieferung, 23 wie überall, so auch hier nicht vorhanden, 24 = FM.

II.

14 „Unglückssterne", nach Abū Ma'šar.

Diese Zusammenstellung umfaßt neben den rötlichen marsähnlichen Sternen α Tau und α Sco vor allem auch sieben „neblige" Sterne[15], die schon in antiken astrologischen Texten als schädlich oder unglückbringend (σινωτικοί, σινοποιοί) aufgezählt wurden[16]. Auch diese Liste scheint demnach, wie schon I, letzten Endes auf antike Quellen zurückzugehen. Sie ist bisher nur in Übersetzungen und Exzerpten, aber noch nicht in Abū Ma'šars Original zutage getreten. Die arabische Fassung finden wir wiederum in Auszügen des bereits genannten as-Siǧzī, und zwar in Oxford, Bodleiana, ms Huntington 437 [= A], fol. 49v[17]. Die Epoche ist hier anscheinend auf as-Siǧzīs Zeit umgerechnet, nämlich 330 Yazdaǧird = 962; der Längenwert beträgt einheitlich Ptolemäus + 11°12 (cf. jedoch weiter unten). In der Nomenklatur herrscht eine gewisse Verwirrung, die teilweise innerhalb der arabischen Überlieferung entstanden sein mag, hauptsächlich aber darauf zurückgehen dürfte, daß einige Bezeichnungen bereits im Griechischen leicht von der Terminologie des Almagest abwichen und sich im Verlauf der Übersetzungen dann immer mehr von dem festen Almagest-Standard entfernten. Das zweite Zeugnis findet sich in einer mittelgriechischen Übersetzung ἀπὸ τῆς τῶν ἐναλλαγῶν τῶν γενεθλίων βίβλου τοῦ 'Απομάσαρ[18]. Die Übersetzung scheint aber nicht direkt aus einem Originaltext von Abū Ma'šar selbst gemacht zu sein, sondern aus Exzerpten, die einem σοφώτατος Μούγνης beigelegt werden. Nach bisheriger Deutung handelt es sich hierbei um den Werktitel *al-Muġnī* (*fī n-nuǧūm*) des arabischen Astrologen Ibn Hibintā, der etwa gleichzeitig mit as-Siǧzī geschrieben zu

[15] Bis auf $\mu^{1,2}$ Sgr sind sie auch im Almagest von Ptolemäus als νεφελοειδής bezeichnet.

[16] Cf. u.a. *Catalogus Codicum Astrologorum Graecorum* V, 1 S. 169f.; ib. S. 208f.; V, 3 S. 129f.; VII S. 111; VIII, 4 S. 190f.

[17] Hierauf verweist M. Destombes, *Note sur le catalogue d'étoiles du Calife Al-Mamoun, Actes du VIIIe Congrès International d'Histoire des Sciences* (*Florence 3—9 Sept. 1956*), S. 309—312, speziell S. 311 (hier irrtümlich ms Marsh 437, was richtig Huntington 437 heißen muß).

[18] ἐναλλαγαί entspricht *taḥāwīl*, τὰ γενέθλια entspricht *al-mawālīd*. Die bibliographische Identifikation steht noch aus. In der lateinischen Fassung von *De magnis coniunctionibus* sowie in der lateinischen und der arabischen Fassung des *Introductorium* = *Kitāb al-mudḫal* habe ich diese Stelle jedenfalls nicht gefunden. Die bei Brockelmann, *GAL* Suppl. I, 395, Werk 8, genannte Berliner Handschrift or. oct. 3090 enthält eine persische Fassung von Abū Ma'šars *Kitāb taḥwīl sinī l-mawālīd*, in dem aber das Verzeichnis der 14 Unglückssterne ebenfalls nicht enthalten ist.

haben scheint[19]. Das Verzeichnis steht, zusammen mit vier anderen Sternlisten[20], in ms Vat. graec. 1056, s. XIV (Kopie aus einer Handschrift des 12. Jhd.) [= G], fol. 31v. Die gleichen Sterne sind auch dem dort auf fol. 31r stehenden, aus mehreren Quellen kompilierten Sternverzeichnis eingegliedert, fallen dort aber mit ihren Koordinaten aus dem System heraus: während die anderen Sterne dort auf dem Verzeichnis der arabischen „*Tabulae probatae*" (*az-zīǧ al-mumtaḥan*) von 214 h = 829/30 beruhen und gegenüber diesen um 5°5 vermehrte Längenwerte aufweisen (mit der Epoche 1161), sind für die 14 eingeschobenen Unglückssterne Werte benutzt, die von A auszugehen scheinen und gegenüber A um 4°25 vermehrt sind. Das wäre dann also = Ptolemäus + 15°37 und würde auf eine Epoche um 1160, d.h. genau 1161, das Jahr der Übersetzung oder der Redaktion der Quellenhandschrift, hinweisen[21]. Diese Zahlen stimmen gut zueinander. In A dagegen ist der Präzessionswert Ptolemäus + 11°12 für die Epoche 962 zu niedrig angesetzt (cf. aṣ-Ṣūfī: Ptol. + 12°42 für 964; Maslama al-Maǧrīṭī: Ptol. + 12°40 für 978; Kūšyār ibn Labbān: Ptol. + 12°0 für 933); vielleicht war dort die Jahreszahl ursprünglich zu lesen 230 Yazdaǧird = 862 (statt 330 Y. = 962), was dann zugleich dem Präzessionswert Ptol. + 11°12 wie auch der astrologischen Schaffensperiode Abū Maʿšars entspräche und mit der Längendifferenz von 4°25 weiter auf die Epoche um 1160 in G führen würde. Der dritte Fundort ist die astrologische Kompilation *Epitome totius astrologiae* des vor allem als Übersetzer hervorgetretenen Johannes Hispalensis[22], die innen mehrfach auf 1142 datiert ist; sie lag mir vor im Druck Nürnberg 1548 [= L], herausgegeben von Ioachim Heller.

[19] Zur Datierung: Ibn Hibintā schrieb nach 329 h = 940/41; cf. C. Nallino, *Opus astronomicum* (al-Battānī), vol. I, Mailand 1903, S. LXVI; ferner D. Pingree in *Dumbarton Oaks Papers*, No. 18, 1964, S. 139.

[20] Cf. *Cat. Cod. Astrol. Graec.* V, 3 S. 11. Die Stelle lag mir zunächst in einer Abschrift von Prof. D. Pingree vor, die er mir im Zusammenhang mit einer anderen Arbeit freundlichst zur Verfügung gestellt hatte; für diese jetzige Untersuchung erhielt ich aus Rom eine Fotokopie. Zwei längere Listen aus diesem Konglomerat von fünf Sternverzeichnissen werden von mir im nächsten Heft dieser Zeitschrift besprochen.

[21] Gemäß den Auszügen von Prof. D. Pingree erscheint die geschlossene Liste der 14 Unglückssterne in dieser Sammlung übersetzter Sternverzeichnisse noch ein zweites Mal, auf fol. 33r; hier ist der Längenwert um 18′ kleiner als auf fol. 31v, also Ptol. + 15°19′; die Nomenklatur weicht von derjenigen auf fol. 31v etwas ab. Insgesamt sind die 14 Unglückssterne damit hier dreimal vertreten: selbständig auf fol. 31v (Länge Ptol. + 15°37′), eingeschoben in das längere Verzeichnis auf fol. 31r (Länge ebenfalls Ptol. + 15°37′), und noch einmal selbständig auf fol. 33r (Länge Ptol. + 15°19′).

[22] Cf. Carmody a.a.O. S. 168f., Werk 1; ferner M. Alonso in *Al-Andalus* 18, 1953, S. 17—49 und L. Thorndike in *Speculum* 34, 1959, S. 20—38.

Aufbau, Inhalt und Quellengeschichte dieser Kompilation sind bisher noch nicht untersucht; es zeigt sich nun, daß auf jeden Fall Schriften von Abū Maʿšar darin mitverwertet sein müssen[23]. Hier finden wir in Buch II (*De nativitatibus*) das Kapitel VII: *De locis stellarum fixarum,* in welchem 13 Sterne aufgezählt werden, von denen 11 mit den 14 Unglückssternen von Abū Maʿšar identisch sind. Der Längenwert ist Ptolemäus + 15°55 (bzw. A + 4°43) für die Epoche des Buches, 1142. Die Nomenklatur weicht hier leicht von A ab und zeigt vor allem die gleichen charakteristischen Eigenarten, die am stärksten in III ausgeprägt sind. In L sind die Sterne abweichend von AG nach Längen aufsteigend angeordnet.

Sternliste II

(Quellen: A, arabisch; G, mittelgriechisch; L, lateinisch)

Lfd. Nr.		Name	Länge	Länge bei Ptolemäus	Breite	Mod. Bez.
1	A	*ʿayn at-tawr* „Auge des Stiers"	Tau 23°52	Tau 12°40	— 5°10	α Tau
	G	ὁ ὀφθαλμὸς τοῦ ταύρου	22°17[a]		„	
	L[4]	*Oculus Tauri*	28°35		„	
2	A	*qalb al-asad* „Herz des Löwen"	Leo 13°42	Leo 2°30	0°15	α Leo
	G	ὁ ἐπὶ τῆς καρδίας τοῦ λέοντος	12°7[a]		0°10	
	L[7]	*Cor Leonis*	18°25		0°10	
3	A	*qalb al-aqrab* „Herz des Skorpions"	Sco 23°52	Sco 12°40	— 4°0	α Sco
	G	ὁ ἀντάρης	22°17[a]		— 5°10	
	L[9]	*Cor Scorpij*	28°..[b]		— 4°0	
4	A	*mankib al-faras* „Schulter des Pferdes"	Psc 13°22	Psc 2°10	31°0	β Peg
	G	ὁ ἐν τῷ ὤμῳ τοῦ ἵππου	17°7[c]		31°0	
	L[13]	*Humerus equi maioris dexter*	18°5		3°[d]	

[23] Auch unter den Übersetzungen des Johannes Hispalensis ist Abū Maʿšar mehrfach vertreten, cf. Carmody a.a.O. S. 38ff., Werke Nr. 1a (*Introductorium*), 2 (*De magnis coniunctionibus*), 3a (*Flores*), 4a (*De reuolutionibus annorum,* auch genannt *Liber experimentorum*).

Lfd. Nr.	Name	Länge	Länge bei Ptolemäus	Breite	Mod. Bez.
5	A *ra's al-ġūl* „Kopf der *ġūl*"[e]	Tau 5°52	Ari 24°40	20°40	16, p^1 Per (bei Ptolemäus: der 3. externe des Perseus)
	G ὁ προηγούμενος τῶν ἀστέρων τῶν ὄντων ἐν τῇ κεφαλῇ τῆς γοργόνης	10°16[f]		„	
	L^3 *oculus bellatoris siue diaboli*[g]	10°35		„	
6	A *kawkab al-ǧabbār* „der Stern des Gewaltigen"[h]	Leo[i] 5°32	Cnc 24°20	52°2[k]	μ Leo
	G ὁ ἐν τῇ κορυφῇ τοῦ λέοντος	9°57		12°0	
	L^8 *Nubes*[l] *quae est in capite Leonis*	10°15		12°0	
7	A *kawkab āḫar* „ein anderer Stern"[m]	Leo[n] 5°22	Cnc 24°10	9°30	ε Leo
	G ὁ ἐν τῇ κορυφῇ τοῦ λέοντος ἕτερος ἀστήρ	11°47[o]		„	
	L ———————[p]				
8	A *al-šǧ'ʿ allaḏī fī ṭaraf baršāwuš*[q]	Tau 7°52	Ari 26°40	47°30[r]	χh Per
	G ὁ ἐπὶ τὸ ἄκρον τῆς δεξιᾶς χειρὸς τοῦ περσέως	12°17		40°30	
	L ———————[p]				
9	A *allaḏī ʿalā ra's al-ǧabbār* „derjenige, der auf dem Kopf des Gewaltigen [= Orion] steht"	Gem 8°12[s]	Tau 27°0	—43°0[t]	λ Ori
	G ὁ ἐπὶ τῆς κεφαλῆς ὠρίωνος	12°35[u]		—13°50	
	L^6 *Caput Canis ualidi*[v]	12°55		—18°50[w]	
10	A *allaḏī fī baṭn as-saraṭān* „derjenige, der auf dem Bauch des Krebses steht"[x]	Cnc 21°32	Cnc 10°20	40°15[y]	ε Cnc
	G ὁ ἐπὶ τοῦ στήθους τοῦ καρκίνου	25°57		0°40	
	L^5 *Nubes in pectore cancri*	26°15		— 0°40	

Lfd. Nr.		Name	Länge	Länge bei Ptolemäus	Breite	Mod. Bez.
11	A	*alladī yatbaʿu š-šawla* „derjenige, der dem Skorpionstachel folgt"	Sgr 12°22	Sgr 1°10	—13°45	G Sco
	G	ὁ ἐπακολουθῶν τῷ κέντρῳ τοῦ σκορπίου	16°47		—13°15	
	L	————[p]				
12	A	*alladī ʿalā ʿayn ar-rāmī* „derjenige, der auf dem Auge des Schützen steht"	Sgr 26°22	Sgr 15°10	4°40[z]	$\nu^{1,2}$ Sgr
	G	ὁ ἐπὶ τὸν ὀφθαλμὸν τοῦ τοξότου	Cap 0°47		0°40	
	L[10]	*Nubes in Sagittario*	Cap 1°5		0°45	
13	A	*alladī ʿalā zuǧǧ nuššābat ar-rāmī* „derjenige, der auf der Spitze des Pfeils des Schützen steht"	Sgr 16°52[aa]	Sgr 6°40	2°30	$\mu^{1,2}$ Sgr
	G	ὁ ἐπὶ τῷ ξίφει τοῦ ἰστοῦ τοῦ τοξότου	21°17		2°50	
	L[11]	*Sagitta Sagittarij*	22°32[bb]		2°50	
14	A	*alladī ʿalā rukbat ad-daǧāǧa al-yumnā* „derjenige, der auf dem rechten Knie der Henne steht"[cc]	Aqr 23°22[dd]	Aqr 12°10	63°45[ee]	ω^{1-3}Cyg
	G	ὁ ἐπὶ τῷ δεξιῷ γόνατι τῆς ὄρνιθος	27°47		„	
	L[12]	*Poples gallinae*	28°5		„	
—	L[1]	*Caput caudae Arietis*[ff]	Tau 7°15	Ari 21°20	4°50	ε Ari
—	L[2]	*caput diaboli*[ff,gg]	Tau 15°25[hh]	Ari 29°40	23°0	β Per

Anmerkungen:

[a] Bei diesen drei Sternen hat G einen Längenwert, der genau um 6° zu klein ist gegenüber dem sonst in G üblichen Wert. Der Katalog auf fol. 31ʳ hat die richtigen Werte: 28°17, 18°7, 28°17.

[b] Die Zahl der Minutenangabe ist im Druck ausgefallen.

[c] Hier besteht eine Unstimmigkeit. Zu erwarten wäre 17°47, bzw. mit der bei Stern 1—3 vorhandenen Minderung um 6°: 11°47. Eventuell ist der

Wert der Handschrift 12°7 zu lesen, da die Handschrift für β ein Zeichen setzt, das dem ζ sehr ähnlich sieht; in diesem Fall wäre 12°7 der um 6° verminderte Wert 18°7, welcher seinerseits durch irgendeine Fehllesung an der Stelle des zu erwartenden 17°47 eingetreten war. Im Katalog auf fol. 31r erscheint 17°7 (statt 17°47).

d Druckfehler für 31°.

e Die arabische Bezeichnung ist hier stark gekürzt und lautet gleich wie der bekannte Name von β Per (*Algol*).

f Sic ms statt 10°17; ebenfalls so im Katalog auf fol. 31r.

g L reflektiert hier eine Verwirrung in den Benennungen der Sterne 5—8, die auch in A sichtbar wird. A ist jedoch im Wortlaut nicht identisch mit der unmittelbaren Vorlage von L. Zu *bellator* cf. unten sub q.

h *al-ǧabbār* „der Gewaltige" ist die übliche arabische Paraphrase für den griechischen Sternbildnamen *Orion*. An dieser Stelle ist die Bezeichnung jedoch falsch gebraucht, da es sich bei Stern 6 und 7 um Sterne im Kopf des Löwen handelt (cf. auch unten sub v).

i Die Handschrift notiert als Tierkreiszeichen *ǧ* = 3 statt *d* = 4; es handelt sich um einen leicht möglichen Irrtum rein graphischer Natur.

k Ausdrücklich als 52° punktiert, statt richtigem 12°.

l Der Begriff des „nebligen" ist von den Sternen 5—6 und 10—12 (in L) falsch hierher übertragen.

m Verkürzung der ursprünglichen Bezeichnung, cf. G.

n Die Handschrift notiert als Tierkreiszeichen *w* = 6 statt *d* = 4; es handelt sich wie schon bei Stern 5 um einen reinen Schreibfehler.

o Die Längen der Sterne 6 und 7 sind nur um 10′ unterschieden; der Gradwert 11° ist hier fehlerhaft. Auch der Katalog auf fol. 31r hat 11°47.

p Diese drei Sterne sind in L nicht vertreten (cf. den Anhang am Ende der Tabelle).

q Das erste Wort könnte gelesen werden *aš-šuǧā'*, mit welcher Bezeichnung die arabischen Almagest-Versionen das ptolemäische Sternbild *Hydra* belegen, oder *aš-šuǧā'* bzw. *aš-šaǧā'* „der Tapfere", was in dieser Bedeutung in der astronomischen Nomenklatur nicht belegt ist. Hierauf scheint die Übersetzung *bellator* (L oben bei Stern 5) zurückzugehen (cf. ferner III, 4). Die richtige Lesung des Wortes an dieser Stelle wäre *as-saḥābī* „der neblige". Ferner ist hinter *ṭaraf* zu ergänzen *yad* „Hand". Somit hat die richtige Bezeichnung des Sterns auf Arabisch zu lauten: *as-saḥābī allaḏī fī ṭaraf yad baršāwuš* „der neblige [Stern], der auf dem Ende der Hand des Perseus steht", in Übereinstimmung mit den arabischen Almagest-Versionen.

r Ptolemäus: 40°30.

s ms: 8°2.

t Ptolemäus: — 13°30.

u Sic statt 12°37 auch im Katalog auf fol. 31r.

v Die gleiche Bezeichnung gebraucht Johannes Hispalensis auch bei Aufzählung der achtundzwanzig Mondstationen für die 5. Station (*Epitome*, Lib. I, cap. XI und Lib. IV, cap. XVIII f), welche die Araber *al-haq'a* und auch *ra's al-ǧawzā'* nennen (cf. P. Kunitzsch, *Untersuchungen zur Sternnomenklatur der Araber*, Wiesbaden 1961, Nr. 241). *ualidus* dient zur Wiedergabe des arabischen Begriffs *al-ǧabbār* „der Gewaltige" (cf. oben sub h). Wie Johannes Hispalensis auf *canis* kam, ist dagegen nicht zu erkennen. In einem anderen Text ist *canis* einmal lediglich eine lateinische Schreiberkorruption aus *orionis* (cf. P. Kunitzsch, *Typen von Sternverzeichnissen* ..., Wiesbaden 1966, Typ V, Nr. 2 und 4), doch läßt sich diese Erklärung auf unseren Fall nicht anwenden. Den auffälligen Ausdruck *caput canis validi* hat Petrus de Abano von hier in seine lateinische Übersetzung von Abraham ibn ʿEzras *Principium Sapientiae*, Venedig 1507, übernommen: Kapitel II (S. VII[rb]), Sterne im Zeichen Gemini.

w Las in der arabischen Vorlage *yḥ* = 18° statt *yǧ* = 13°.

x Nach der üblichen ptolemäischen Definition steht er vielmehr auf der „Brust", arab. *ṣadr*.

y Ptolemäus: 0°20.

z Ptolemäus: 0°45.

aa Arabischer Schreibfehler *yw* = 16° statt *yz* = 17°; lag auch G vor.

bb Sic statt 22°35.

cc Die Handschrift hat falsch *al-yusrā* „(auf dem) linken (Knie)".

dd Die Fotokopie zeigt vom Schriftzug der Zahl 23° (*kǧ*) nur das auslaufende Ende des *ǧ*.

ee Der Almagest hat 64°45; laut Heibergs Apparat hat jedoch die Handschrift B anstelle von 64° in der Tat 63°.

ff Die drei Sterne 7, 8 und 11 erscheinen nicht in L. Stattdessen hat L die beiden hier nachgetragenen Sterne, welche in A und G nicht vorkommen. Die Gesamtzahl der Sterne beträgt bei L nur 13.

gg Cf. auch III, 2. Diese Benennung ist ebenfalls in das *Principium Sapientiae* eingegangen (S. VI[ra], Sterne im Zeichen Taurus; cf. oben sub v).

hh Druckfehler statt 15°35.

III.

Johannes Hispalensis

Die bereits sub II erwähnte astrologische Kompilation *Epitome totius astrologiae* von Johannes Hispalensis wird eingeleitet von einer *Isagoge*, die in 12 Kapiteln die zwölf Zeichen des Tierkreises behandelt. Am Ende jedes Kapitels werden ein oder einige wichtige in das betreffende Zeichen fallende Fixsterne aufgeführt. Insgesamt ergeben sich so 18 Sterne, die jeweils mit Länge, Breite und Größe notiert sind. Der Längenwert ist, für das Jahr der Abfassung 1142, = Ptolemäus + 15°55, wie auch schon oben bei II L. Auffällig sind auch hier wieder gewisse eigenartige, von der traditionellen Nomenklatur völlig abweichende Benennungen, über deren Entstehung man nur Vermutungen anstellen kann. Johannes Hispalensis hat die *Epitome* gewiß kompiliert. Es gibt kaum eine andere Erklärung als die, daß er seine Exzerpte nicht sorgfältig genug gemacht hat oder daß irgendeine Störung seine Aufzeichnungen durcheinander brachte, so daß er schließlich diese höchst seltsamen, zum Teil regelrecht sinnwidrigen Bezeichnungen anwandte[24]. Selbstverständlich sind auch Fehler und Nachlässigkeiten in den arabischen Vorlagen mit anzunehmen, ähnlich jenen, die uns in II A begegnen. Der Terminus für die „Größe" der Sterne ist *honor*, was auf die in der Astrologie übliche arabische Bezeichnung *šaraf* (statt *ʿuẓm* oder *qadr*) zurückgeht[25]. Im Druck Nürnberg 1548 [= N] sind die Sterne am Rande vom Herausgeber Iochim Heller [= H] identifiziert und mit den allgemein üblichen Namen gleichgesetzt worden[26]. Diese Tierkreiszeichenbeschreibung mit den Sternangaben erscheint auch in ms Oxford, Bodleiana, Digby 38, s. XIV [= O], fol. 84ʳ—85ᵛ[27]; der Text ist nachlässig behandelt; die Schrift ist verblichen und zumal auf der Negativ-Kopie schwer lesbar; die Jahreszahl lautet hier 1102. Drittens sind diese 18 Sterne herausgehoben in einem separaten Kapitel *De his stellis hoc modo continetur in tractatu magistri johannis yspalensis* in der astrologischen Sammelhandschrift Paris B.N. lat. 7316 A, s. XIV [= P], fol. 52ᵛ—53ʳ[28]; hier heißt die Jahreszahl 1146. Im Druck N scheinen die Koordinaten vom Herausgeber I. Heller purgiert worden zu sein; die beiden handschriftlichen Fassungen zeigen demgegenüber teilweise starke Abweichungen, vor allem O. P hat Werte, die niedriger sind als in N, am häufigsten um ca. 50′.

[24] Eigenwillige Übersetzungen für arabische Sternnamen kennen wir auch aus seinem selbständigen Astrolabtraktat (Inc. *Dixit Johannes: Cum uolueris facere astrolabium accipe auricalcum optimum* ...); cf. P. Kunitzsch, *Typen von Sternverzeichnissen* ..., Wiesbaden 1966, Typ IV.

[25] Cf. *ZDMG* 118, 1968, S. 73f., Fn. 52.

[26] Eine ähnliche Behandlung der Tierkreiszeichen findet man in Abraham ibn ʿEzras fast gleichzeitig mit der *Epitome*, nämlich 1148 verfaßtem *Principium Sapientiae* (lateinische Übersetzung des Petrus de Abano, 1293, nach der altfranzösischen Version von Hagin le Juif, 1273), Venedig 1507, Kapitel II (S. IVᵛᵇ—XVᵛᵇ). Hier sind eindeutig zwei der charakteristischen Benennungen von Johannes Hispalensis weiter verwendet (cf. II sub v und gg).

[27] Cf. L. Thorndike in *Speculum* 34, 1959, S. 34 (hier falsch fol. 83ᵛ). Weiter vorn, auf fol. 72ᵛ, enthält die gleiche Handschrift eine Liste der 28 Mondstationen, ebenfalls nach Johannes Hispalensis' eigenwilliger Terminologie, wie in der *Epitome* vorhanden. (Die vorliegende Fotokopie bricht auf fol. 84ᵛ unten nach Stern 12 ab). [Dieselben Mondstationennamen erscheinen auch in einer 1367 zu Tournay zusammengestellten Übersicht über die Mondstationen, und zwar in einer dritten Kolumne, überschrieben „*Nomina earundem latina*", in ms Paris B.N. lat. 7277, fol. 1ᵛ; cf. dazu auch L. Thorndike in *Journal of the Warburg and Courtauld Institutes* 20, 1957, S. 115 (Nr. 5). Auch Leopold von Österreich (um 1270; cf. Carmody a.a.O. S. 170f.) hat die Mondstationenliste in Johannes' Version in seine *Compilatio de astrorum scientia* aufgenommen: Druck Augsburg 1489, Tractatus I].

[28] Cf. L. Thorndike in *Journal of the Warburg and Courtauld Institutes* 20, 1957, S. 133 (Nr. 30). Thorndike gibt hier versehentlich an, es werde für jedes der zwölf Zeichen ein Stern notiert. In Wirklichkeit handelt es sich um die 18 Sterne aus der *Isagoge* der *Epitome totius astrologiae*.

Sternliste III
(Quellen: N (H), O und P)

Lfd. Nr.	Zeichen	Länge	Länge bei Ptolemäus	Breite	Name	Mod. Bez.
1	Aries	N 16°5 O 3°32 P 19°36	Ari 0°10	—53°0 —19°60 —19°20	*littus fluminis*[a] *extremitas ann' . . . alio uocatur sat' fluminis* *extremitas annis* H: *lucida in extremo fluminis*	ϑ Eri
2	Taurus	N 15°35 O 14°45 P ,,	Ari 29°40	23°0 4°0 20°0	*caput Diaboli* *capud diaboli* *caput dyaboli* H: *Caput Algol*	β Per
3		N 28°35 O 27°45 P ,,	Tau 12°40	— 5°10 ,, ,,	*oculus Tauri sinister* *sinister tauri* *oculus sinister tauri* H: *Aldebaram*	α Tau
4	Gemini	N 10°55 O 10°5 P 10°15	Tau 25°0	22°30 ——— 22°30	*Bellator* *sidus bellicosum* . .[b] *sidus bellicosum* H: *Hircus, Alhajoth*	α Aur
5		N 17°55 O ———— P 17°5	Gem 2°0	—17°0 — 7°0 ,,	*humerus canis magni*[c] *(sidus . . . dictum) humerus canis magni*[b] *humerus canis magni* H: *in dextro Orionis humero, Bedelgeuze*	α Ori
6	Cancer	N 3°5 O 2°20 P 2°30	Gem 17°10	—69°0 — 3°0 —75°0	*Camelus occidens* *occidens camelus*[d] *occidens camelus* H: *Canopum dicit*	α Car
7		N 3°35 O ——— P 2°45	Gem 17°4	—39°10 ,, ,,	*Pes canis*[e] ——————[d] *pes canis* H: *alias Alhabor seu Canicula*	α CMa

Lfd. Nr.	Zeichen	Länge	Länge bei Ptolemäus	Breite	Name	Mod. Bez.
8		N 15°25	Gem 29°10, var. 29°30	—16°10	*oculus Leonis*	α CMi
		O ——		——	——[f]	
		P 14°46		—16°10	*oculus leonis* H: *alias Algomeyza et Procyon*	
9	Leo	N 18°25	Leo 2°30	0°10	*Cor leonis*	α Leo
		O 17°43		,,	,, ,,	
		P ,,		,,	*os leonis*[g]	
10	Virgo	N 10°25	Leo 24°30	11°50	*cauda leonis*	β Leo
		O 9°34		—°5	,, ,,	
		P ,,		11°5	,, ,,	
11	Libra	N 12°35	Vir 26°40	2°0	*Inermis*	α Vir
		O 11°23		—11°0	*sidus sine lancea*	
		P 11°43		— 2°0	,, ,, ,,	
12		N 12°55	Vir 27°0	31°30	*Lancea*	α Boo
		O 12°4		31°0	,,	
		P ,,		,,	,,	
13	Scorpius	N 28°35	Sco 12°40	— 4°0	*cor Scorpij*	α Sco
		P 27°44		— 3°0	*cor scorpionis*	
14	Sagittarius	N ..°45[h]	Sco 24°50	36°0	*caput Victi*	α Oph
		P 9°33		,,	,, ,, H: *super caput Ophiuchi*	
15	Capricornus	N 3°45[i]	Sgr 17°20	62°0	*Aquila Cadens ... quam Amigdalam Ptolomęus dicit*[k]	α Lyr
		P 2°23		40°0	*aquila cadens, ... amidola*	
16		N 19°45	Cap 3°50	30°10	*aquila uolans*	α Aql
		P 17°44		—29°10	,, ,,	
17	Aquarius	N 25°5	Aqr 9°10	60°0	*Cauda gallinae*	α Cyg
		P 24°14		,,	*galine cauda*	
18	Pisces	N 18°5	Psc 2°10	31°0	*Humerus equi*	β Peg
		P 17°14		,,	,, ,,	

Anmerkungen:

[a] Ursprüngliche Lesung wohl *finis fluminis,* gemäß Ptolemäus; die andere Lesart dürfte *extremitas amnis* sein.
[b] Stern 4 und 5 sind in O zu éinem zusammengezogen.
[c] Orion heißt hier *canis magnus,* in II, 9 sowie als Mondstation in der *Epitome* (und entlehnt im *Principium Sapientiae*) dagegen *canis ualidus.*
[d] Stern 6 und 7 sind in O zu éinem zusammengezogen.
[e] Dieser abweichende Name dürfte leicht als Korruption innerhalb der lateinischen Überlieferung zu erklären sein: Ursprünglich muß er *os canis* gelautet haben, in Übereinstimmung mit der Definition des Sterns im Almagest. Durch Schreiberirrtum wurde dann *os* zu *pes* verschrieben.
[f] In O übergangen.
[g] *os*: lateinischer Schreibfehler für *cor.*
[h] Die Zahl der Gradangabe ist im Druck weggefallen.
[i] Zu erwarten wäre 3°15.
[k] *Amigdala* „Mandel" entstand aus der Verlesung von arab. *al-lwr'* bzw. *al-lwrh* (*al-lūrā* bzw. *al-lūra,* als Transkription des griechischen Sternbildnamens λύρα) in das arabische Wort *al-lawza* „Mandel".

***

Die Aufhellung aller Probleme dieses Komplexes stellt der künftigen Forschung eine ganze Reihe von Aufgaben: Das Werk Abū Ma'šars ist eingehend zu erforschen; seine Quellen sind nachzuweisen, vor allem im Hinblick auf antikes Gut; die lateinischen Übersetzungen sind ihren arabischen Originalen zuzuordnen. Ähnlich ist auch das Werk des Johannes Hispalensis zu untersuchen, seine selbständigen Schriften sind auszusondern und speziell im Hinblick auf ihre Quellen zu durchforschen. Dabei wird, als Nebenprodukt, auch die Erklärung der eigenartigen, von der Tradition abweichenden Nomenklatur in I'/II/III abfallen.

Wir stellen nun noch einmal diejenigen Namen zusammen, die uns hier als fremd und ungewöhnlich entgegentreten:

diabolus I b; in Verbindung zu bringen mit dem kleinen Perseus-Stern II, 5. Dort ist in L, also bei Johannes Hispalensis in der *Epitome,* *diabolus* richtig die Übersetzung von arab. *al-ġūl,* womit die Araber üblicherweise das antike γοργόνιον paraphrasieren. Aus dem ursprünglichen *ra's* „Kopf", auch in G κεφαλή, ist bei L abweichend *oculus* geworden.

caput diaboli I c; II (L^2); III, 2. Die traditionelle westliche Bezeichnung behielt den arabischen Terminus in Transkription bei: *caput algol* (= *ra's al-ġūl*). *diabolus* scheint Johannes' eigene Übersetzung von *al-ġūl* zu sein. Der Wanderweg des Wortes beginnt also bei Johannes, in II oder III, und führt von da aus zu I'. In seinem Astrolabtraktat verwendet Johannes, wiederum abweichend, die Übersetzung *draco* (also dort: *caput draconis* = *ra's al-ġūl*, β Per; cf. P. Kunitzsch, *Typen* [s. Fn. 24], Typ IV, Nr. 2).

canis I d; *canis ualidus* II, 9; *canis magnus* III, 5. Hiermit ist überall Orion gemeint. Das Beiwort *ualidus* bzw. *magnus* mag aus der arabischen Paraphrase *al-ǧabbār* „der Gewaltige" herzuleiten sein. Für *canis* selbst gibt es keine greifbare Erklärung (cf. die Stellen oben).

bellatrix I e; *bellator* II, 5; III, 4. In II, 5 setzt Johannes *bellator* gleich mit *diabolus* = *al-ġūl* (s.o.). Wir wissen nicht, ob er damit eine Variante seiner arabischen Quelle wiedergibt oder ob der Ausdruck auf andere Weise in jene sowieso nicht korrekt weitergegebene Benennung eingedrungen ist. In III, 4 erscheint *bellator* völlig unbegründet für den Stern α Aur, der sonst als *Kapella*, arab. *al-ʿayyūq* = lat. *alaioc* u.ä. allgemein bekannt war. Mit dieser Identifikation tritt der Name schließlich in der Femininform *bellatrix* in I e auf. Spätere Astronomen wurden sich offenbar der Unangemessenheit der Übertragung des Namens auf den traditionell anders benannten Stern α Aur bewußt und gaben die Benennung *Bellatrix* dem in der Tradition nicht benannten Stern γ Ori (zu beobachten seit der ersten Hälfte des 15. Jahrhunderts in der Wiener Astronomenschule, Johann von Gmunden; cf. P. Kunitzsch, *Typen* [s. Fn. 24], S. 11 und Typ XVII, Nr. 11; id., *Arabische Sternnamen* [s. Fn. 1], S. 148f., Nr. 70). Damit ist der auch heute noch bekannte Sternname *Bellatrix* nunmehr bis zu Johannes Hispalensis und in den Umkreis astrologischer Abū Maʿšar-Texte zurückverfolgt.

camelus I g. Es gibt keine greifbare Erklärung; es ist nicht einmal sicher, ob ein Zusammenhang mit dem nächstfolgenden Namen besteht oder ob es sich hier um ein vollkommen unabhängig und aus ganz anderer Wurzel entstandenes Gebilde handelt.

occidens camelum I h; *camelus occidens* III, 6 (N); *occidens camelus* III, 6 (OP). Auch hierfür ist keine schlüssige Erklärung zur Hand. Es ist nicht zu erkennen, ob *occidens* transitiv sein soll („tötend") oder intransitiv („untergehend", evtl. „westlich", von Gestirnen). Der einheimische arabische Name von α Car war *suhayl*; es ist aber kaum anzunehmen, daß Johannes versucht haben sollte, diesen an sich unübersetzbaren alten Gestirnnamen als *camelus* zu übertragen (in mittelalterlichen europäischen Sternverzeichnissen und einschlägigen Texten wurde der Name *suhayl* üblicherweise in Transkription beibehalten). Auch als lateinische Korruption aus „*Canopus*" ist *camelus* nicht zu erklären (Handschriften der Alphonsinischen Tafeln haben dafür gelegentlich *campus* oder *tempus*). Vermutlich liefert das Wort *camelus* — sei es von I g, sei es von I h — den Ursprung der späteren Bezeichnung *Alkameluz* für α Boo (cf. oben die Eingangsbemerkung): das Wort mag durch Zusatz des Artikels *Al-* äußerlich arabisiert worden sein; die Übertragung auf α Boo geschah rein willkürlich, da man aus der traditionellen Nomenklatur keinen Stern kannte, dem die Bezeichnung *camelus* (*occidens*) zukam.

pes canis I i; III, 7. Sehr wahrscheinlich lateinische Korruption aus ursprünglichem *os canis*, gemäß Ptolemäus.

oculus leonis III, 8. Diese Bezeichnung ist nicht aus der astronomischen Tradition heraus zu erklären; bei Ptolemäus im Almagest ist kein Stern des Löwen auf dessen „Auge" lokalisiert. Dagegen ist innerhalb der einheimischen arabischen Astrothesie *aṭ-ṭarf* bzw. *ṭarf al-asad* „Blick des Löwen" der Name der 9. Mondstation (bestehend aus κ Cnc + λ Leo), die auch von Johannes Hispalensis in der *Epitome* an den entsprechenden Stellen *oculus leonis* genannt wird. Damit ist die Bezeichnung selbst belegt; fragwürdig bleibt die Übertragung auf den in der Tradition seit ältester Zeit mit festem Namen benannten Stern α CMi (*Procyon*; *Algomeiza* u.ä.).

caput Victi III, 14. Die aus der ptolemäischen Definition abgeleitete arabische Bezeichnung von α Oph war *ra's al-ḥawwā'* bzw. *ra's al-ḥāwī* „Kopf des Schlangensammlers, -beschwörers". Es ist nicht ersichtlich, auf welchem Assoziationswege oder durch welche Fehllesung Johannes hier aus dem arabischen Wort einen „Besiegten" herauslas.

XVIII

AL-ḲUṬB (A.), pole, as an astronomical term

In Arabic, *ḳuṭb* covers nearly the same field of semantic aspects as Greek ὁ πόλος: a pivot around which something revolves (in Arabic, especially the pivot for mill stones), which was extended to the revolution of the sky, designating the axis of the celestial east-west movement and, more specifically, its two "poles" (*LA*, Beirut 1955, i, 681b f.; Lane, *Lexicon*, s.v. *ḳuṭb*; Liddell and Scott, *Greek-English Lexicon*, s.v. πόλος). Hence later, in Arabic translations of Greek works, *ḳuṭb* was often used as a rendering of πόλος, (e.g. Ptolemy, *Almagest*, i, 3 (both translations, al-Ḥadjdjādj b. Yūsuf b. Maṭar, and Isḥāḳ b. Ḥunayn); but Aristotle, *De caelo*, has *al-falak* in all instances, cf. Badawī's edn., Cairo 1961, 232 f., 280, 294).

The celestial pole was a known feature among the Arabs already in their "pre-scientific" period, i.e. before their contact with the scientific literature of the Greeks. Certain traditions concerning the location and nature of the pole have been transmitted in some philological and astronomical books (Ibn Ḳutayba, *K. al-Anwāʾ*, Hyderabad 1956, 122 f.; Abū Ḥanīfa al-Dīnawarī, *apud* al-Marzūḳī, *K. al-Azmina wa 'l-amkina*, Hyderabad 1332, ii, 7 f., 371 ff.; Ibn al-Adjdābī, *al-Azmina wa 'l-anwāʾ*, Damascus 1964, 65 f.; Ibn Manẓūr, *K. Nithār al-azhār*, Istanbul 1298, 146, 173; Abu 'l-Ḥusayn al-Ṣūfī, *K. Ṣuwar al-kawākib*, Hyderabad 1954, 27 f. (repeated by al-Ḳazwīnī, *Kosmographie*, ed. Wüstenfeld, i, Göttingen 1849, 29; tr. H. Ethé, *El-Kazwînî's Kosmographie*, Leipzig 1868, 62); al-Bīrūnī, *K. al-Tafhīm*, ed. and tr. R. R. Wright, London 1934, 77 [§ 163]; idem, *al-Ḳānūn al-Masʿūdī*, Hyderabad, iii (1956), 995 f.). Here it is stated that the [north] pole was not a star, but merely a spot (*nuḳṭa, buḳʿa*) in the sky. Its place was assumed to be in the middle of the figure of a fish (*samaka*; al-Bīrūnī has instead: *shakl halīladjī*, "an elliptical figure") formed by two curved lines of stars, one consisting of the five stars β, ζ, ε, δ, α Ursae Minoris, and the opposite one of very small and faint stars (most of them not registered by Ptolemy in his *Almagest*), among which Fl. 5 and 4 Ursae Minoris and 32 H Camelopardalis. In some places, this statement is corrected, with the addition that the true place of the pole is not in the very middle of that figure, but more in the direction of α Ursae Minoris, near one of the small stars in the curved line of faint stars, (al-Marzūḳī, Ibn al-Adjdābī, al-Ṣūfī, *locc. cit.*). This difference in location obviously reflects the effect of precession, due to which the place of the north pole was near the star 32 H Camelopardalis in A.D. 1000, and still more southwards in A.D. 500. It was also observed that the opposite point of the sky, the south pole, was invisible in the lands of the Arabs (Ibn Ḳutayba, *loc. cit.*, 122; Ibn Manẓūr, *op. cit.*, 173)

After the introduction of scientific astronomy, it was known that this pole belonged to the equatorial system, hence it was occasionally called ***ḳuṭb muʿaddil al-nahār***, "pole of the equator", besides its other designations, such as simply ***al-ḳuṭb***, or ***al-ḳuṭb al-shamālī*** (and ***al-djanūbī*** respectively), ***ḳuṭb al-djudayy***, ***ḳuṭb banāt naʿsh***, ***ḳuṭb al-falak***, ***ḳuṭb al-kull***, etc. Al-Bīrūnī seems to be the first to declare α Ursae Minoris (= "Polaris") to be the nearest bright star to the [north] pole, at his time, and hence to serve as a substitute for the pole *wa-yanūbu ʿani 'l-ḳuṭbi li-annahū fī zamāninā aḳrabu 'l-nayyirati ilayhi* (*Tafhīm*, 77 [§ 163]). At the same time, the ecliptical system was assigned two poles, as well, which were called *ḳuṭb*[*ā*] *falak al-burūdj* and *ḳuṭb*[*ā*] *dā'irat al-burūdj*, "pole[s] of the ecliptic" (al-Ṣūfī, *loc. cit.*, 25 f.; al-Bīrūnī, *Tafhīm*, 55 f. [§ 139]; idem, *Ḳānūn*, iii, 993, 995 ff.; al-Ḳazwīnī, *op. cit.*, 27, tr. 59, and 52 f., tr. 108).

North and south pole became of actual value for the Islamic navigators of the Indian Ocean (*ca.* A.D. 1500), who used both of them for altitude measure-

3

ments in order to fix their position and determine their routes. In their terminology the north pole was generally called *al-djāh* (a word of Persian origin, used both for the pole itself and for the Pole Star). Travelling south of the equator, they also acquired a complete knowledge of the south pole, *ḳuṭb suhayl* ("the pole [in the region] of the star α Carinae"), or simply *al-ḳuṭb* (as opposed to *al-djāh*), which they observed—in contradistinction to the north pole—not to be marked by a bright star near it, but having the two Magellanic Clouds at some distance (Aḥmad b. Mādjid and Sulaymān al-Mahrī; in English tr. see G. R. Tibbetts, *Arab navigation in the Indian Ocean*, London 1971, the glossaries and indexes, 518 and 547 [s.v. *al-djāh*], 538 [s.v. *ḳuṭb*], 609 [s.v. Polaris; Pole, North-South]).

The poles also seem to have excited the fantasy of astrologers, for two lists of magic virtues (*khawāṣṣ*) appertaining to both of them are repeated by Ibn Manẓūr (*op. cit.*, 146-53; partially also in al-Ḳazwīnī, *op. cit.*, i, 30 f., tr. 64 f. and 40, tr. 83 f.).

In modern terminology, the terrestrial poles are also called *ḳuṭb* (with adjective *ḳuṭbī*, "polar").

Apart from this, in the construction of the astrolabe *al-ḳuṭb* signifies the central pivot, or axis (elsewhere also called *al-miḥwar*), which keeps together its different discs, the spider, and the rule (*Alhidade*), latinised (since the 10th century) as *Alchotob*, *Alchitob*, *Alchitoth*, etc. See ASṬURLĀB; al-Khwārazmī, *Mafātīḥ al-ʿulūm*, 235; al-Battānī, *Opus astronomicum*, ed. Nallino, i, Milan 1903, 141 and 319; al-Bīrūnī, *Tafhīm*, 195 [§ 325], etc. Latin: J. M. Millás Vallicrosa, *Assaig d'història de les idees físiques i matemàtiques a la Catalunya medieval*, i, Barcelona 1931, 278, 66; 288, 4; 289, 29. 32. 36; N. Bubnov (ed.), *Gerberti postea Silvestri II papae opera mathematica*, Berlin 1899, 123, 10; Herimannus, *De mensura astrolabii*, in Migne, *Patrologia latina*, cxliii, Paris 1882, 387 A; (Ps.-) Messahalla (Māshāʾallāh), *De compositione astrolabii*, ed. R. T. Gunther, *Chaucer and Messahalla on the astrolabe*, Oxford 1929, 201, 202, etc.

XIX

AL-**MADJARRA**, the Galaxy or Milky Way. This remarkable celestial phenomenon was well-known to the peoples of the Islamic world. Its popular assimilation to the traces of spilt milk seems to be of Greek origin (cf. τὸ γάλα [Aristotle], ὁ τοῦ γάλακτος κύκλος [Euclid, Geminus], ὁ γαλακτίας [κύκλος] [Ptolemy], ὁ γαλαξίας [κύκλος] [other authors]; see Liddell and Scott, *s.vv.*), whereas in the Near East the image of traces of lost straw, or chaff, prevails (cf. Pers. *rāh-i kāhkas͟hān*, Turk. *samanyolu*, etc., and already Syriac *s͟hbhīlā d-t͟hebhnā* (A.D. 660), also colloquial Arabic *darb al-tabbāna*, etc.; see Wiedemann [1], 1 f., and [2]; and Eilers [1], 108-11; [2], 15 f.; [3], 4; for the Syriac, Severus Sēbhōk͟ht, ii, 2 [p. 349]; xi, 1.3 [pp. 392-3]).

In Arabic, the indigenous designation is *al-mad͟jarra*, which is derived both by classical Arabic lexicographers and by modern scholars from the root *d͟j-r-r* "to pull, or draw" and would designate a place, or path where something is drawn (Ibn Ḳutayba, *Anwā'* 123; al-D͟jawharī, in *LA*, iv, 129a; Ibn Sīda, *Muk͟haṣṣaṣ*, 8; al-Marzūḳī, ii, 9; Ibn Manẓūr [2], 173; Eilers [1], 111). The mediaeval Latin translator of the Arabic *Almagest*, Gerard of Cremona, generally retained the Arabic designation *al-mad͟jarra* as *almaiarati* etc.; but in one instance he added a gloss: *id est arca que mouetur*, in which he understood *al-mad͟jarra* as a drawer that can be pulled (cf. Kunitzsch, *Der Almagest*, 139). Occasionally there occurs in Arabic texts a confusion between the name of the Milky Way, *al-mad͟jarra*, and the graphically closely-similar word *al-mid͟jmara*, which is the Arabic version for the Greek constellation name τὸ Θυμιατήριον, the Altar, *Ara* (see Kunitzsch, 203); cf., e.g., *al-mad͟jarra* for *al-mid͟jmara* on a celestial globe by Ibrāhīm b. Saʿīd al-Sahlī, dated 478/1085, described by J. M. Millás Vallicrosa, *Assaig d'història de les idees físiques i matemàtiques a la Catalunya medieval*, i, Barcelona 1931, 55 ff.; the same confusion of *al-madjarra* for *al-midjmara*, in *Les mansions lunaires des Arabes. Texte arabe en vers de Moh'ammed El-Moqri*, trad. et ann. par A. de C. Motylinski, Algiers 1899, 99 (no. 11) and n. 2, where the translator did not realise and correct the mistake; cf. also ms. d of the Arabic *Almagest* in Kunitzsch, 346, no. 640. A literal Arabic translation from the Greek, in Ptolemy's *Almagest*, was *al-falak al-labanī* (al-Ḥadjdjādj b. Yūsuf b. Maṭar), and *al-dā'ira al-mushabbah lawnuhā bi-lawn al-laban*, or shortly, *al-dā'ira al-labaniyya* (Isḥāḳ b. Ḥunayn) (see Kunitzsch, *Der Almagest*, 139).

The Milky Way was known to the Arabs in their classical period; cf. a collection of classical and post-classical Arabic verses naming *al-madjarra*, in Wiedemann [1], 673-5 (see also Ibn Manẓūr [2], 118 f.). In lexicographical texts, the positions of several fixed stars are described in relation to the Milky Way. Abū Ḥanīfa al-Dīnawarī (d. 282/895) gave a description of the course of the Milky Way in the sky as seen by the Arabs in their "pre-scientific" period (Abū Ḥanīfa, in al-Marzūḳī, ii, 9 f.; tr. Wiedemann [1], 664-6).

Further, it was observed and mentioned by the Arabs that the position of the Milky Way in the sky changed in different seasons according to the annual revolution of the sky (Abū Ḥanīfa, in al-Marzūḳī, ii, 10 f. = tr. Wiedemann [1], 666-8; Ibn Ḳutayba, 123; al-Ḳazwīnī, 21 = tr. Ethé, 44). A late description based on personal experience was given around A.D. 1490 by Aḥmad b. Mādjid, the famous navigator of the Indian Ocean (see Aḥmad b. Mādjid, 149 f.; tr. Tibbetts, 140). It should be noted that some of these non-scientific authors understand the Milky Way as a dense accumulation of faint stars that appears to the observer's eye as a nebulous mass, a theory which seems to have originated from independent observations not influenced by the teachings of Greek philosophy (cf. Ibn Ḳutayba, 124; al-Ḳazwīnī, 21 = tr. Ethé, 44; Aḥmad b. Mādjid, 149 = tr. Tibbetts, 140).

A scientific astronomical description of the Milky Way was given by Ptolemy in his *Almagest* (book viii, ch. 2), which became known in Arabic through a series of translations from the end of the 2nd/8th to the end of the 3rd/9th centuries. Abū Ḥanīfa's description, however, apparently was not influenced by the *Almagest*, since he uses descriptive elements of pure Arabic origin only. Also, his description of the galactic circle begins and ends in the constellation of Scorpius, near the ecliptic, whereas Ptolemy's description begins and ends in the constellation of Centaurus, in the southernmost region of the sky visible to him in Alexandria.

Cosmological theories of the Milky Way were conveyed to the Arabs by translations and paraphrases of Greek philosophical works, as, e.g., Aristotle's *Meteorology* (for the chapter on the Milky Way, see *al-Āthār al-ʿulwiyya*, ed. Badawi, 12-15; ed. Petraitis, 23-6 [Arabic]; ed. Schoonheim, 66/67-70/71; ed. Daiber, 62 f.), and their commentaries. While Aristotle taught that the Milky Way is a vaporous phenomenon belonging to the sublunar sphere, other philosophers such as Democritus, and younger commentators as Olympiodorus, etc., believed the Milky Way to be far out in space, similar to the fixed stars, and belonging to their sphere (cf. Petraitis, 23 [Arabic], n. 2; Daiber, 93-5). Muslim scholars apparently favoured these latter theories against Aristotle's one (see the texts cited by Wiedemann [1], 669-71, and Daiber, 93-5, and the following quotations from Ibn al-Haytham and al-Bīrūnī). Especially interesting are the scientific deductions of Ibn al-Haytham (d. *ca.* 432/1041), who arrived at the conclusion that the Milky Way does not form part of the air (the sublunar sphere) but must be far out in the space (see Wiedemann [1], 672, and the literature quoted there). Also, al-Bīrūnī in his *al-Ḳānūn al-Masʿūdī* (written around A.D. 1030) says that, contrary to the teachings of Aristotle and his followers, *al-madjarra* is of the same height (above the earth) as the fixed stars because the moon and the planets are not influenced by it when crossing in front of it, as should happen if the Milky Way was below them, and because it shares all the peculiarities of the fixed stars as described by Ptolemy, *viz.* that it is found in their sphere and unvariably keeps the same relations in distance and shape just as the fixed stars do (al-Bīrūnī [2], iii, 992, 16-993, 1; nearly identical with this is the quotation from Barhebraeus (d. A.D. 1286) in Wiedemann [1], 671).

Bibliography: Aḥmad b. Mādjid, *Kitāb al-Fawā'id fī uṣūl ʿilm al-baḥr wa 'l-ḳawāʿid*, ed. I. Khoury, Damascus 1971; Eng. tr. G. R. Tibbetts, *Arab navigation in the Indian Ocean before the coming of the Portuguese*, London 1971; Aristotle, *Meteorology*, Arabic paraphrase *Kitāb al-Āthār al-ʿulwiyya*, ed. A. R. Badawi, Cairo 1961 (see also Daiber, Petraitis, Schoonheim); Bīrūnī [1], *Kitāb al-Tafhīm li-awā'il ṣināʿat al-tandjīm*, ed. and tr. R. R. Wright, London 1934, § 167 (p. 87); Bīrūnī [2], *al-Ḳānūn al-Masʿūdī*, i-iii, Hyderabad-Dn., 1954-6; H. Daiber, *Ein Kompendium der aristotelischen Meteorologie in der Fassung des Ḥunain ibn Isḥâq*, Amsterdam-Oxford 1975; W. Eilers [1], *Stern—Planet—Regenbogen. Zur Nomenklatur der orientalischen Himmelskunde*, in *Der Orient in der Forschung. Festschrift Otto Spies*, Wiesbaden 1967, 92-146; Eilers [2], *Die vergleichend-semasiologische Methode in der Orientalistik*, Wiesbaden 1974 (Akad. d. Wiss. u. d. Lit. Mainz, Abh. Geistes- u. Sozialwiss. Kl., Jahrg. 1973, Nr. 10); Eilers [3], *Sinn und Herkunft der Planetennamen*, Munich 1976 (*SB Bayer. Ak.*, Philos.-Hist. Kl., Jahrg. 1975, Heft 5); Ibn Ḳutayba, *Kitāb al-Anwā'*, Hyderabad-Dn. 1956; Ibn Manẓūr [1], *Lisān al-ʿarab*, iv, Beirut 1955 (p. 129a, s.r. *dj-r-r*); Ibn Manẓūr [2], *Kitāb Nithār al-azhār fi 'l-layl wa 'l-nahār*, Istanbul 1298; Ibn Sīda, *Kitāb al-Mukhaṣṣaṣ*, ix, Cairo 1319; Ḳazwīnī, *Kosmographie*, ed. Wüstenfeld, i, Göttingen 1849; German tr. H. Ethé, *Zakarija ... el-Kazwînî's Kosmographie. Die Wunder der Schöpfung*, Leipzig 1868; P. Kunitzsch, *Der Almagest, Die Syntaxis Mathematica des Claudius Ptolemäus in arabisch-lateinischer Überlieferung*,

Wiesbaden 1974; al-Marzūḳī, *Kitāb al-Azmina wa 'l-amkina*, i-ii, Hyderabad-Dn. 1332; C. Petraitis, *The Arabic version of Aristotle's Meteorology*, Beirut 1967; P. L. Schoonheim, *Aristoteles' Meteorologie in arabischer und lateinischer Uebersetzung*, Leiden 1978; Severus Sēbhōkht, *Le traité sur les constellations*, ed. and tr. F. Nau, in *Revue de l'Orient Chrétien*, xxvii (1929-30), 327-410, and xxviii (1931-2), 85-100; E. Wiedemann [1], *Über die Milchstrasse bei den Arabern*, originally printed 1926-7, repr. in *Aufsätze zur arabischen Wissenschaftsgeschichte*, ii, Hildesheim 1970 (the pagination of the reprint is quoted); Wiedemann [2], art. *al-Madjarra* in *EI*[1] (mostly verbatim identical with Wiedemann [1], except for the translation from Abū Ḥanīfa and the collection of verses, which are not in this article).

XX

AL-MANĀZIL (A.) or more fully *manāzil al-ḳamar*, the lunar mansions, or stations of the moon (sing. *manzil* or *manzila*), a system of 28 stars, groups of stars, or spots in the sky near which the moon is found in each of the 28 nights of her monthly revolution.

The system seems to be of Indian origin (see Scherer; Pingree [1] and [2]; Billard). Babylonian origin has sometimes been suggested (cf. Hommel), but could never be established from the documents. The "stars in the moon's path", in the ^{mul}APIN text (cf. van der Waerden [1], 77; recently re-dated to 2300 B.C., cf. van der Waerden [2]) are 17 or 18 in number and rather represent an early stage in the development of the zodiac. The system of the lunar mansions was adopted by the Arabs, through channels as yet unknown, some time in the pre-Islamic period, since the term *manāzil* is already mentioned in the Ḳurʾān (X, 5; XXXVI,39). To the single mansions, the Arabs applied names already found with them previously, and originally used to designate their *anwāʾ* [see ANWĀʾ]. A complete list of the 28 mansions is reported by ʿAbd al-Malik b. Ḥabīb (d. 238/852) on the authority of Mālik b. Anas (d. 179/795); nearly contemporary to this is the list drawn up by the astronomer al-Farg͟hānī. Items of information concerning the lunar mansions were collected by the Arabic philologists in their *kutub al-anwāʾ* (see the printed works of Ibn Ḳutayba, al-Marzūḳī, Ibn Sīda, Muḥammad al-Muḳriʾ, al-Ḳazwīnī, Ibn al-Ad͟jdābī, Ibn Manẓūr/al-Tīfās͟hī, and Aḥmad b. Mād͟jid), and by astronomers, who took pains in identifying these mansions astronomically (see al-Farg͟hānī, al-Battānī, al-Ṣūfī, al-Bīrūnī [1], [2], and [3]). Whereas the scientific astronomers of the Arabic-Islamic period did not actually use the lunar mansions, these apparently were of some importance for the distribution of the ecliptic (besides of the zodiac) in earlier times. Later, they were often used by astrologers and others for different systems of divination (see e.g. *Picatrix*, and ʿAlī b. Abi ʾl-Ridjāl; cf. also Savage-Smith, for the lunar mansions in relation to geomancy). Hence they were engraved, together with other calendric and astrological items, on the back of many Islamic and some western astrolabes (cf. Hartner, 2549 f.; Michel, 42; Mayer, pls. XV, XVIB, XX, XXII-XXV). Through Latin translations of Arabic works, the list of the 28 lunar mansions and their Arabic names became known to mediaeval Europe from the late 10th century (see Millás, 251 ff.), and, later, were much used for divinatory purposes (see e.g. Steinschneider, Vian, Svenberg, Weidemann, Lutz and Müller). Lists of the 28 lunar mansions, with their Arabic names, have also penetrated Byzantine astrological texts (see CCAG).

The names of the 28 mansions, and their astronomical identifications, are as follows (for the names, see the individual entries in Kunitzsch [1], where sources and further details are given):

1. *al-sharaṭān* (also: *al-naṭḥ*), βγ, or βα Arietis.
2. *al-buṭayn*, εδρ Arietis.
3. *al-thurayyā*, the Pleiades.
4. *al-dabarān*, α Tauri.
5. *al-haḳʿa*, $\lambda\varphi^{1.2}$ Orionis (according to the *Almagest*, one nebulous object, the first star of Orion; but registered as three individual stars by al-Bīrūnī [3]). Alternatively also, *al-maysān*, which properly would be one of the two stars of no. 6.
6. *al-hanʿa*, γξ Geminorum; also *al-taḥāyī*, ημν Geminorum (either separately, or together with γξ Geminorum). Al-Bīrūnī [3] has it νγξ Geminorum.
7. *al-dhirāʿ*, αβ Geminorum. There is confusion in the sources as to whether this *dhirāʿ* is *al-dhirāʿ al-maḳbūḍa* or al-*dhirāʿ al-mabsūṭa*.
8. *al-nathra*, ε Cancri, or εγδ Cancri (Ibn Ḳutayba, and al-Bīrūnī [3]).
9. *al-ṭarf*, κ Cancri + λ Leonis.
10. *al-djabha*, ζγηα Leonis (included with this station is

the star α Leonis, "Regulus", which had no individual name in the classical Arabic star lore).

11. *al-zubra* (also: *al-kharātān*), δθ Leonis.
12. *al-ṣarfa*, β Leonis.
13. *al-ʿawwāʾ*, βηγε Virginis, sometimes δ Virginis is also added to these.
14. *al-simāk* (i.e. *al-simāk al-aʿzal*), α Virginis.
15. *al-ghafr*, ικλ Virginis (al-Bīrūnī [3] has ικ only).
16. *al-zubānā*, αβ Librae.
17. *al-iklīl*, βδπ Scorpii.
18. *al-ḳalb*, α Scorpii.
19. *al-shawla*, λυ Scorpii. Sometimes, *al-ibra*, or *ibrat al-ʿaḳrab*, is given as an alternative designation, but some authors refer this name to a different object, viz. the nebulous cluster following behind *al-shawla*, i.e. M 7 Scorpii.
20. *al-naʿāʾim*, the two groups of four stars each, γδεη + σφτζ Sagittarii; alternatively, *al-waṣl*, the space between these two groups.
21. *al-balda*, a region void of stars, between stations nos. 20 and 22.
22. *saʿd al-dhābiḥ*, $\alpha^{1,2}$ νβ Capricorni.
23. *saʿd bulaʿ*, με Aquarii, to which some authors add Fl. 7, or ν Aquarii, as a third star.
24. *saʿd al-suʿūd*, βξ Aquarii + c^1 Capricorni.
25. *saʿd al-akhbiya*, γπζη Aquarii.
26. *al-fargh al-muḳaddam* (also *al-fargh al-awwal*), αβ Pegasi.
27. *al-fargh al-muʾakhkhar* (also *al-fargh al-thānī*), γ Pegasi + α Andromedae.
28. *baṭn al-ḥūt* (also *al-rishāʾ*). β Andromedae.

Some authors additionally register the names of some stars, or spots in the sky, near which the moon is seen when failing to reach her proper mansion, whereas the interstices between two mansions, generally, are called *furdja* (see Ibn Ḳutayba, 86; al-Marzūḳī, 196 f.; Ibn Sīda, 12; Ibn Manẓūr, 180 = al-Tīfāshī, 205; al-Bīrūnī [1], 351 f., tr. 353 f.).

The knowledge of the 28 lunar mansions has lived on into modern times, and agricultural calendars formed according to them are still today found in various regions of the Arabic-speaking world and its neighbourhood (cf. Landberg, Cerulli, Monteil, Serjeant, Hiskett, Galaal, and the literature cited there; less so consistent are the observations reported by C. Bailey, *q.v.*). Such calendars are already known from mediaeval times (see *Liber anoe* [= *Kitāb al-anwāʾ*, Spain, 961 A.D.; translated into Latin by Gerard of Cremona], and Ibn al-Bannāʾ); they appear to continue older astro-agricultural traditions as paralleled in Ptolemy's *Phaseis* and the Babylonian mulAPIN texts.

Bibliography: 1. Arabic sources: ʿAbd al-Malik b. Ḥabīb, *Risāla fī maʿrifat al-nudjūm*, ms. Ait Ayach, Ḥamzawiyya 80/4, p. 188 (cf. F. Sezgin, *GAS*, vii, 346, 373); Aḥmad b. Mādjid, *Kitāb al-Fawāʾid fī uṣūl ʿilm al-baḥr wa 'l-ḳawāʿid*, ed. I. Khoury, Damascus 1971, 31 ff., Eng. tr. G.R. Tibbetts, *Arab navigation in the Indian Ocean before the coming of the Portuguese*, London 1971, 79 ff.; ʿAlī b. Abi 'l-Ridjāl, *Kitāb al-Bāriʿ fī aḥkām al-nudjūm*, Latin tr. *Praeclarissimus liber completus in iudiciis astrorum*, Venice 1485, see pars vii, cap. 101; al-Battānī, *Opus astronomicum*, ed. and tr. C.A. Nallino, i-iii, Milan 1899-1907, see ch. 51 (text, iii, 187 ff.; tr., i, 124 ff.; comm., i, 295 ff.); Latin tr., printed Nuremberg 1537, and Bologna 1645; in the Old Spanish tr. this chapter is missing, cf. G. Bossong (ed.), *Los Canones de Albateni*, Tübingen 1978, 84; al-Bīrūnī [1]: *al-Āthār al-bāḳiya*, ed. E. Sachau, Leipzig 1878, 336 ff., Eng. tr. E. Sachau, *The chronology of ancient nations*, London 1879, 335 ff.; idem [2]: *Kitāb al-Tafhīm li-awāʾil ṣināʿat al-tandjīm*, ed. and tr. R.R. Wright, London 1934, §§ 164-6; idem [3]: *al-Ḳānūn al-Masʿūdī*, i-iii, Hyderabad 1954-6, see ix, 8 (pp. 1139 ff.); al-Farghānī, *Elementa astronomica*, ed. J. Golius, Amsterdam 1669, see ch. 20; two Latin translations, by Johannes Hispalensis (A.D. 1135), ed. F.J. Carmody, *Al-Farghānī, Differentie scientie astrorum*, Berkeley 1943, and by Gerard of Cremona, ed. R. Campani, *Alfragano: Il "libro dell'aggregazione delle stelle"*, Città di Castello 1910; Ibn al-Adjdābī, *al-Azmina wa 'l-anwāʾ*, Damascus 1964, 60 ff. (lacuna in the ms., supplied from other sources); Ibn al-Bannāʾ, *Le Calendrier d'Ibn al-Bannâʾ de Marrakech*, ed. and tr. H.P.J. Renaud, Paris 1948; Ibn Ḳutayba, *Kitāb al-Anwāʾ*, Hyderabad 1956, 16 ff.; Ibn Manẓūr, *Kitāb Nithār al-azhār fi 'l-layl wa 'l-nahār*, Constantinople 1298, 174 ff., re-edited, under the name of the original author Aḥmad b. Yūsuf al-Tīfāshī, *Surūr al-nafs bi-madārik al-ḥawāss al-khams*, Beirut 1980, see 199 ff.; Ibn Sīda, *Kitāb al-Mukhaṣṣaṣ*, ix, Cairo 1319, 9 ff.; al-Ḳazwīnī, *Kosmographie*, ed. F. Wüstenfeld, i, Göttingen 1849, 41 ff.; German tr. H. Ethé, *Zakarija ... el-Kazwîni's Kosmographie*, Leipzig 1868, 87 ff.; *Liber anoe: Le Calendrier de Cordoue*, ed. R. Dozy, Leiden 1873, new ed. Ch. Pellat, Leiden 1961; al-Marzūḳī, *Kitāb al-Azmina wa 'l-amkina*, i-ii, Hyderabad 1332, see i, 184 ff. and 310 ff.; Muḥammad al-Muḳriʾ (cf. Brockelmann, S II, 364: Abū Miḳraʿ al-Baṭṭūwī), *Les mansions lunaires des arabes*, ed. and tr. A. de C. Motylinski, Algiers 1899; *Picatrix* (Ps.-Madjrīṭī, *Kitāb Ghāyat al-ḥakīm*), ed. H. Ritter, Berlin-Leipzig 1933, see i,4; German tr. H. Ritter and M. Plessner, *"Picatrix". Das Ziel des Weisen*, London 1962; al-Ṣūfī, *Kitāb Ṣuwar al-kawākib...*, Hyderabad 1954.

2. Modern studies: C. Bailey, *Bedouin star-lore in Sinai and the Negev*, in *BSOAS*, xxxvii (1974), 580-96; R. Billard, *L'astronomie indienne*, Paris 1971, 15, 18, 40; *CCAG* (= *Catalogus codicum astrologorum graecorum*, i-xii, Brussels 1898-1953), see v, 3, pp. 90 f. and 91 f.; viii, 1 p. 217 f.; ix, 1 p. 138 ff.; E. Cerulli, *Le stazioni lunari nelle nozioni astronomiche dei Somali e dei Danākil*, in *RSO*, xii (1929-30), 71-8, xiii (1931-2), 76-84; M.H.I. Galaal, *The terminology and practice of Somali weather lore, astronomy, and astrology*, Mogadishu 1968; W. Hartner, *The principle and use of the astrolabe*, in A.U. Pope (ed.), *A survey of Persian art*, Oxford 1938-9, iii, 2530 ff. (repr. separately as *Astrolabica*, no. 1, Paris 1978); M. Hiskett, *The Arab star-calendar and planetary system in Hausa verse*, *BSOAS*, xxx (1967), 158-76; F. Hommel, *Ueber den Ursprung und das Alter der arabischen Sternnamen und*

insbesondere der Mondstationen, in *ZDMG*, xlv (1891), 592-619; P. Kunitzsch [1], *Untersuchungen zur Sternnomenklatur der Araber*, Wiesbaden 1961; idem [2], *Arabische Sternnamen in Europa*, Wiesbaden 1959, 53-7; C. von Landberg, *Glossaire Daṯînois*, ii, Leiden 1923, 1092 ff.; B.F. Lutz, *Das Buch "Alfadol"*, diss. Heidelberg 1967, 131, 321 ff. (appendix by P. Kunitzsch); L.A. Mayer, *Islamic astrolabists and their works*, Geneva 1956; H. Michel, *Traité de l'astrolabe*, Paris 1947, repr. Paris 1976; J.M. Millás Vallicrosa, *Assaig d'història de les idees fisiques i matemàtiques a la Catalunya medieval*, i, Barcelona 1931; V. Monteil, *La toponymie, l'astronomie et l'orientation chez les Maures*, in *Hespéris*, xxxvi (1949), 189-219; U. Müller. *Deutsche Mondwahrsagetexte aus dem Spätmittelalter*, diss. Berlin 1971; C.A. Nallino, *Raccolta di scritti*, v, Rome 1944, 175 ff.; Ch. Pellat, *Dictons rimés*, anwāʾ *et mansions lunaires chez les arabes*, in *Arabica*, ii (1955), 17-41; D. Pingree [1], in *Isis*, liv. no. 176 (1963), 229 f.; idem [2], in *Viator*, vii (1976), 144, 146, 174 ff.; L. de Saussure, in G. Ferrand, *Instructions nautiques et routiers arabes et portugais des XVe et XVIe siècles*, iii, Paris 1928, 143 ff.; E. Savage-Smith and M.B. Smith, *Islamic geomancy and a thirteenth-century divinatory device*, Malibu, Calif., 1980; A. Scherer, *Gestirnnamen bei den indogermanischen Völkern*, Heidelberg 1953, 151 ff.; R.B. Serjeant, *Star-calendars and an almanac from South-West Arabia*, in *Anthropos*, xlix (1954), 433-59; M. Steinschneider, *Über die Mondstationen (Naxatra) und das Buch Arcandam*, in *ZDMG*, xviii (1864), 118-201, with additions in xxv (1871), 378-428; E. Svenberg, *Lunaria et zodiologia latina*, Göteborg 1963 (esp. 45 ff.); B.L. van der Waerden [1], *Erwachende Wissenschaft*, ii, *Die Anfänge der Astronomie*, Basel-Stuttgart 1968; idem [2], *On pre-Babylonian mathematics. II*, in *Archive for History of Exact Sciences*, xxiii (1980), 36 (confirming the theory of W. Papke, in his unpublished doctoral diss. of 1978, Tübingen); R. Vian, *Ein Mondwahrsagebuch*, Halle 1910; B. Weidemann, *"Kunst der Gedächtnüss" und "De Mansionibus", zwei*

7

frühe Traktate des Johann Hartlieb, diss. Berlin 1964; Ph. Yampolsky, *The origin of the twenty-eight lunar mansions*, in *Osiris*, ix (1950), 62-83.

XXI

JOHN OF LONDON AND HIS UNKNOWN ARABIC SOURCE

In my 1959 book on Arabic star names,[1] I made out a group of names which could be traced back only to an "unbekannte Quelle" ("unknown source"), but not beyond.[2] Some years later, it had become clear that the "unknown source" for this group of Arabic star names in Western texts was a star table established by John of London in Paris, in 1246.[3] But John of London's star table offered, in its turn, several names for which no evidence in Arabic original texts or on instruments could be found. Having continued, in the years that followed, the study of related Arabic and mediaeval Western sources, I think it appropriate now to sum up and report on John of London's star table and his "unknown Arabic source".

For the fixed stars, the basic and most comprehensive table was that incorporated in Ptolemy's *Almagest*, Books VII-VIII (containing 1025 stars), available to the mediaeval West since 1175 in Gerard of Cremona's Latin translation made from the Arabic.[4] Apart from that, for practical use on instruments, especially the astrolabe, shorter tables of basic stars (mostly containing from c. 20 to 50 stars) were drawn up in the Latin West from the late tenth century onwards, partly obtained, through translation, from Arabic sources, and partly compiled by Western authors after such translated models.[5] One such table was drawn up in 1246 by John of London in Paris. It contains 40 stars with ecliptic coordinates (longitude and latitude) and their magnitudes. John of London's table is of great historical interest for several reasons: in itself, it is an outstanding document of astronomical observation because John has not, as many others, merely computed the coordinates by adding a constant, for precession, to the values of the longitudes in the *Almagest*, but he has observed all his stars *per instrumentum armillarum* ("with an armillary sphere") and thus found their coordinates *by observation*, which is very rare in mediaeval times.[6] Further, his table influenced the compilation of another star table which was appended to the astrolabe treatise of Pseudo-Messahalla and, in this context, was widely diffused.[7] Eventually, partly directly and partly through the Pseudo-Messahalla table, John's star nomenclature became influential in modern astronomy down to our own day.

John's name as the author, and the date and place of the observational work, are transmitted in the respective manuscripts, in the title of the star table.[8] The correctness of these indications is corroborated by a letter of John of London to a certain *magister* R. de Guedingue in which he answers six astronomical and related questions asked by the latter.[9] In one manuscript[10] there is a re-examination of the star table made four years later (i.e., in 1250) with an instrument devised by an otherwise unknown Rogerus Linconus who calls himself a *discipulus* of the "famous astronomer" John of London. Fontès suggests that our John of London is identical to one *Jo. Lond.* sometimes mentioned by Roger Bacon.[11] The letter to R. de Guedingue shows John as a learned and well-read man; among his authorities he cites: *Albategni* (al-Battānī), *Thesbith* (Thābit ibn Qurra), *Ptholemeus* and *liber Almagesti*, *Abrachis* (Hipparchus, in Gerard of

Cremona's spelling taken from the Arabic spelling *'.b.r.kh.s* in the Arabic *Almagest*), *Pythagoreus* (cited for astrological evidence, obviously a pseudepigraph), *Arzachel* (al-Zarqāllu), and *Aomar* (the astrologer ʿUmar ibn al-Farrukhān).

While John of London's astronomical contribution – in observing a number of fixed stars and transmitting their coordinates[12] – is completely evident, the textual descriptions of his stars raise a number of questions. The observation of the stars and the finding of their coordinates were elements for which John was personally responsible, whereas the naming and verbal description of the stars could not be invented by John but had to be gleaned by him from the sources current in his time.

A close examination of John's stellar nomenclature in his table of 1246 shows that, on the one hand, he of course had knowledge of and access to the most prominent source on stars and constellations in his time, Ptolemy's *Almagest* in Gerard of Cremona's Latin translation from the Arabic; but that, on the other hand, he introduced several non-traditional descriptions derived from his own personal view of the sky and that, most interestingly, he added Arabic material not found elsewhere in the preceding and contemporary Western literature and, to some extent, not even in all the Arabic original sources known today.

Relations to, and verbal borrowings from, the *Almagest* can be recognized in the descriptions of the following stars:[13] no. 2 (α And, 1st star of Pegasus in *Almagest*; John has substituted *umbilicus* and *andromeda* for the *Almagest's sumen* ("navel") and *mulier catenata* ("the Chained Woman"), i.e. the Arabic paraphrase of Ptolemy's Andromeda); no. 4 (β And, 12th star of And in *Almagest*; while Gerard's text has the correct transliteration *mizar* (also spelled *miçar* etc. in the manuscripts) of the original Arabic term *al-mi'zar* ("the girdle", or "loin-cloth"), John has the spelling *mirach*, with an erroneous metathesis of the *z*- and *r*-sounds, which lives on into modern times as the star name Mirach for β And); no. 6 (α Ari, 1st external star of Aries in *Almagest*; here, Hipparchus's name is retained as *abrachis*, in Gerard's transliteration taken from the Arabic spelling *'.b.r.kh.s.*); no. 11/13 (β/α Ori; *pes/humerus audacis*, where *audax* is Gerard's translation of the Arabic *al-jabbār* ("the Giant"), the common Arabic paraphrase for Ptolemy's Orion); no. 14 (α CMa, 1st star of CMa in *Almagest*: *in ultimitate luminis* is directly from *Almagest* where it was the Latin and Arabic translation of Ptolemy's ὁ λαμπρότατος ("the brightest"), referring to α CMa as being the brightest of the fixed stars; one MS also gives *transitus*, as a translation,[14] by John himself or by a later reader, of the indigenous Arabic name of this star, *al-ʿabūr*, normally transliterated into Latin as *alhabor*, etc.); no. 18 (α UMa, 16th star of UMa in *Almagest*: *super dorsum*, directly from *Almagest*); no. 23 (η UMa, 27th star of UMa in *Almagest*); no. 25 (α Vir, 14th star of Virgo in *Almagest*: *super palmam sinistram*, directly from *Almagest*); no. 26 (α Boo, the external star of Boo in *Almagest*: *non est in forma*, from the extra heading for the external star in *Almagest*); no. 34 (γ Del, 7th star of Del in *Almagest*: *in rumbo*, derived from *Almagest*); no. 37 (ε Peg, 17th star of Peg in *Almagest*: *musida* ("the muzzle"), following *Almagest*; the same also in no. 6, α Ari); no. 38 (δ Aqr, 18th star of Aqr in *Almagest*: *crus*, from *Almagest*); no. 39 (β Peg, 3rd star of Peg in *Almagest*: *humerus*, from *Almagest*); and no. 40 (β Cet, 22nd star of Cetus in *Almagest*: *in caude ceti ramo meridiano*, from *Almagest*).

Non-traditional individual descriptions by John himself become visible in no. 3 (ζ Cet, not consistent: the description points to τ or π Cet while the coordinates refer to ζ Cet which was the star traditionally mentioned in such star lists); no. 15 (α Gem, stating that α Gem is less bright, *obscurius*, than β); no. 17 (ρ Pup, said to be the star following behind Sirius and assumed to belong to the Ship); no. 18 (α UMa, said to be the brightest one in the quadrangle of UMa); no. 24 (γ Crv, not consistent: the description means β Crv while the coordinates point to γ Crv which is the star traditionally mentioned in such lists); no. 28 (δ Oph, expanding on its position in the constellation figure and stating that they were two stars close together, *et sunt due coniuncte*, i.e. δ and ε Oph; for "hand", John uses the common *manus*, whereas *Almagest* had *palma* instead); no. 31 (γ Dra, the Dragon dividing the two Bears, from a classical, non-Ptolemaic tradition); no. 34 (γ Del); no. 35 (δ Cap, the more easterly of the two stars close together, viz. γδ Cap); and no. 39 (β Peg, the brightest star in the Pegasus square).

If we now come to the Arabic material, first we must make the general remark that to nearly all of the 40 stars John adds an Arabic name, or an Arabic descriptive term that later came to be used as a proper name for the respective star. All of these names seem to have been read in Arabic and transliterated into Latin characters by John himself; that is, even for those stars which had had "Arabic names" in the Western tradition since long before John's time, he gave a new spelling of his own, different from the older spellings handed on in the written tradition, in texts and on instruments.

Apart from the revised spellings of several pre-existing names, he also introduced a number of new Arabic names (and descriptive terms) not found in earlier Western sources, many of which continued to live on and are used as star names even at the present time; as, for example, *sceder* (α Cas, now spelt Schedir or Schedar), *menkar* (α Cet), *bedalgeuze* (α Ori, after a Renaissance "explanation" spelt Betelgeuze), *yed* (δ Oph), *sceath* (δ Aqr, now mostly spelt Skat; also transferred, in the 15th century, to β Peg, where it is now spelt Scheat), and *enif* (ε Peg), to name but some.[15]

In the course of his descriptions, he sometimes also mentioned Arabic names of constellations (not stars), some of which later stuck to the respective stars as a proper name. Among these are *alpheraz*/*alferaz* = Pegasus (in his nos. 2, α And, and 37, ε Peg; of today's Alpheratz for α And it cannot be said with complete certainty whether it goes back to John's form, or to a similar form used in the West since the late tenth century for β Peg where it was an abbreviation of the full Arabic name *mankib al-faras* ("the Horse's Shoulder")); *alkeb* (from *al-kalb* ("the Dog"), in MS g only, in no. 16, for Canis Minor); *markeb* (see later); *edubh* (also spelt *dubhe*, in no. 18, α UMa, for Ursa Maior; the modern name Dubhe for α UMa has its origin in John's star list); *alichil* (only MS f, in no. 27, α CrB, from *al-iklīl* [*al-shamālī*] ("the [Northern] Crown"), Arabic for CrB[16]); and *aldigege* (in no. 36, α Cyg, from Arabic *al-dajāja* ("the Hen"), for Cyg).

Clearly John must have had access to original Arabic material, either written texts or inscribed instruments, from which he could gather all those new names and words that appear in Europe for the first time in his star table. The only major question in connection with this would be whether John of London knew Arabic himself well enough to collect these new materials from the Arabic sources, or whether a helper assisted him in this who had a good command of Arabic. In any case, the names are there, and correct Latin translations are added to most of them.

One feature, however, casts a shadow over the tradition of John's star table: the manuscripts are not always uniform. In certain instances some of the manuscripts show either more complete, or even quite different names for some stars, such as no. 6 (where d and g add *enif* = Arabic *anf* ("nose")), no. 8 (d has a fuller *algeb ymin* for the usual *algemb/algenib*), no. 15 (d and f have full form for the short and incomplete *raz*, *caput*, of the other MSS), no. 28 (g has full form as against the short and incomplete *yed* of the other MSS), no. 37 (d has full form as against the short *enif* of the other MSS), no. 39 (here two entirely different Arabic star names were used; the majority of the MSS used in the 1966 edition having *bedalferaz*, from Arabic *yad al-faras* ("the Horse's Fore-foot"), while d and g have *menkeb alferaçh/mekebalferaz* instead, from Arabic *mankib al-faras* ("the Horse's Shoulder") – both Arabic names being well documented in Arabic original sources).[17] In short, there are a number of corruptions of different kinds, and additions from other sources, in the manuscripts of John's table, which therefore must be described as being in a relatively poor state of transmission. We cannot exclude the possibility that some of the differences in the Arabic material are due to modifications made by later readers who had Arabic material at their disposal and were in sufficient command of Arabic to have wished to modify, or even supply, some of the details in John's star list as they received it.

The handling of the Arabic material was usually correct. As far as I have noticed there remain only two doubtful cases (no. 7, α Cet, and no. 1, α Cep), and of these more shortly.

Now we come to the most outstanding and exciting feature in John's star table, his "unknown Arabic source". While all the Arabic names and words discussed above are well documented in Arabic sources, we encounter a small number of genuine Arabic forms that are not found in any of the Arabic original sources, texts and instruments, so far discovered and studied. These are the following: no. 6, α Ari, *enif* (Arabic *al-anf* ("the nose")[18]); no. 9, γ Eri, *algetanar* (from Arabic *ʿarjat al-nahr* ("the bend of the river")[19]); no. 17, ρ Pup, *markeb* (John here mentions *markeb* as the name of the Ship, Ptolemy's constellation of Argo; in all Arabic sources, however, this Ship is unanimously called *al-safīna*,[20] and the word *markab* is never used, though this indeed means "ship" and doubtless underlies John's form *markeb*[21]); no. 37, ε Peg, *enif* (Arabic *anf* ("nose"), whereas the Arabic tradition has other terms here,[22] and *anf* in relation to Pegasus remains undocumented in the Arabic). Perhaps the following two names also belong to this group: no. 7, α Cet, *menkar* (from Arabic *minkhar* ("nose")[23]), and no. 1, α Cep. For this star the manuscripts give an Arabic name variously spelt *aldramin*, *alderaimin*, *aldheraymin*, and explained in Latin as *dextrum adiutorium cephei*. Here again, the details do not all fit together when compared to the entire previous tradition. In the *Almagest*, α Cep (4th star with Ptolemy) is located "above the right shoulder", which latter term is given in the Arabic texts as *katif* or *mankib*, the second of which again was translated *humerus* by Gerard of Cremona.[24] John's *adiutorium* is well known from Gerard's translation of the *Almagest* where it invariably stands for "upper arm"; in this sense, the word occurs in the *Almagest*, in Cepheus, with the 8th star (ι Cep, on the *left* arm) where the Arabic has the usual *ʿaḍud* (this was translated by Gerard), and in another version *sāʿid* ("forearm"; "forearm" is equally given as *al-dhirāʿ*, for ι Cep, in an astrological source, viz. Abū Maʿshar). Therefore, John's Latin form diverges from the traditional description of α Cep. His Arabic forms similarly cannot be harmonized either with his own Latin description ("right upper arm"), or with Ptolemy's location ("right shoulder").[25]

Some of the manuscripts conclude the title of the star table with the phrase "And here are written their [i.e., the fixed stars'] names by the table's editor himself, in the way in which they are correctly put on the astrolabe" (*sicut vere in astrolabio ponuntur*). Three MSS, however, instead conclude the phrase as follows: "... as they are correctly named in Arabic" (*sicut vere in arabico nominantur*).[26] Both these versions are in fact correct, because the stars in the table were certainly intended for the construction of astrolabes and the Arabic names here are given in a new, more expanded and more correct form than in most of the older tables current until John's time. But it must be admitted that the styling of the lengthy title prefixed to the table in the manuscripts was not the work of John of London himself, but rather that of the various readers and copyists of the table. Therefore, it is possible that the two versions of the last phrase cited are *both* "authentic", in that they were originally conceived by different writers. Otherwise, it could be that one of the two is a corrupted, or modified, variant of the other. Anyhow, it is evident that John's contribution to the introduction and transmission of Arabic star names in the West was already recognized in late mediaeval times.

For us, there remains the problem of John's "unknown Arabic source" for those names (*enif*, *algetanar*, *markeb*, and perhaps *aldramin* and *menkar* also) that are undocumented in the entire Arabic tradition hitherto known. The stellar nomenclature used in works of astronomy and astrology, and also in poetry and literature, in the Arabic-Islamic world, in and since mediaeval times, is based on two components: the indigenous Arabic star lore and traditions from 'pre-scientific' times, and the nomenclature formed, through translation, after classical Greek models.[27] Both these branches of tradition are well documented, the first group in the lexicographic collections of mediaeval Arabic philologists, largely controlled and recognized by astronomers like al-Ṣūfī and al-Bīrūnī, and the second group in the numerous Arabic translations of scientific Greek works on astronomy and astrology and later Arabic works in these fields. The *corpus* of Arabic star names was transmitted in written texts and on inscribed instruments. All this material has been edited and studied over a period of hundreds of years in great quantity, so that it can be assumed that our present knowledge of it is quite comprehensive and complete. In our explanation of those three (or perhaps five) undocumented names, therefore, we cannot reach beyond this point. Either John of London had at his disposal obsolete Arabic material of which no traces have so far been found (but in view of the uniform tradition in all our known sources, this is hardly probable), or – and this cannot be excluded – he, or his helper, has freely substituted well-known common Arabic nouns (*anf*, *markab*, ʿ*arja*) in place of the fixed terms of the traditional sources.

REFERENCES

1. P. Kunitzsch, *Arabische Sternnamen in Europa* (Wiesbaden, 1959).

2. See *ibid.*, 45, 92-94, and the individual entries; *cf.* also pp. 39f of P. Kunitzsch, *Typen von Sternverzeichnissen in astronomischen Handschriften des zehnten bis vierzehnten Jahrhunderts* (Wiesbaden, 1966).
3. Edited, as "Typ VI", in Kunitzsch, *Typen*, 39-46. While John of London's original list of "Typ VI" contains 40 stars, there exists also an abbreviated recension of the list containing 31 stars only, edited in the same work as "Typ VII", pp. 47-50.
4. An edition of the Arabic versions and of Gerard of Cremona's Latin translation is being prepared by the present writer. For a study of the various texts and translations *cf.* below, ref. 14.
5. Seventeen "types" of such tables have been edited by Kunitzsch, *Typen*.
6. This also becomes evident by comparing John's coordinates (both the longitudes and latitudes) to Ptolemy's: in both cases they are always different. One of the late mediaeval users has also realized this, but his conclusion was quite contrary to ours; to him, John's table was *corrupta tabula*, because it was not in harmony with the *Almagest* (see Kunitzsch, *Typen*, 39).
7. Kunitzsch, *Typen*, "Typ VIII" (compiled from the two tables of "Typ III" and "Typ VI"), continued in "Typ IX-XI". Since John's table was an isolated item not forming part of a relevant context, it could easily be included by later compilers and writers in various other texts, such as Pseudo-Messahalla's astrolabe treatise or the *Toledan tables*. Also, some mediaeval Western astrolabes have used John's star nomenclature; see Kunitzsch, *Typen*, 40, n. 4.
8. My edition (in Kunitzsch, *Typen*) was made from seven manuscripts, now in Vienna, Paris, Erfurt, and Madrid. Two British manuscripts are mentioned by L. Thorndike, "Notes on some less familiar British astronomical and astrological manuscripts", in *Journal of the Warburg and Courtauld Institutes*, xxii (1959), 157-71, p. 161. E. Zinner, in his *Verzeichnis der astronomischen Handschriften des deutschen Kulturgebietes* (Munich, 1925), erroneously calls the author of the star table Johannes Sacrobosco instead of Johannes de Londoniis; *cf.* his nos. 4655-9.
9. The letter was edited and discussed by M[onsieur] Fontès, "Le manuscrit de Jean de Londres", *Bulletin de l'Académie des Sciences, Inscriptions et Belles-Lettres de Toulouse*, i (Toulouse, 1897-98), 146-60; for reference to the star observations see p. 149, lines 18-19, and p. 153, lines 5-2 from bottom. *Cf.* also L. Thorndike, *op. cit.* (ref. 8), and his *History of magic and experimental science*, ii (New York and London, [1]1923; [2]1964), 95. The same manuscript, Paris B.N. lat. 7413 (2), which contains the letter on fol. 19v-21r, later on contains also John's star table, on fol. 36r (followed, on fol. 36v, by a different star table, dated 1233, belonging to another tradition; edited in Kunitzsch, *Typen*, 67-71, "Typ XI").
10. Ms Erfurt, Amplon. 4°369 (= g in the edition cited in ref. 3).
11. Fontès, *op. cit.* (ref. 9), 148, n. 1, and 159.
12. Additional evidence for John's personal derivation of the coordinates, independently of the classical traditions, can be deduced from the fact that nearly all of the longitudes and about half of the latitudes consist of full degrees, with no minutes added, which seems to result from the limitations of his observations. (The same again applies to the *mediationes coeli* and declinations added to the table in 1250 by John's pupil Rogerus Linconus.)
13. Reference is always made to the edition of John's star table cited above in ref. 3 ("Typ VI").
14. *Cf.* P. Kunitzsch, *Der Almagest: Die Syntaxis Mathematica des Claudius Ptolemäus in arabisch-lateinischer Überlieferung* (Wiesbaden, 1974), 320, no. 531.
15. For full details *cf.* the works cited above in refs 1 and 2.
16. A connection to the 17th lunar mansion, called *al-iklīl* and located in βδπ Sco, as suggested in Kunitzsch, *Typen*, 46, note to no. 27, is not indicated as can now be deduced from the evidence in the Arabic *Almagest*.
17. For the full wording the reader is again referred to the edition cited in ref. 3.
18. For the traditional Arabic term used for this star see Kunitzsch, *op. cit.* (ref. 14), 264, no. 265.
19. For the several "bends" in the River, see Kunitzsch, *op. cit.* (ref. 14), 314ff., nos. 512, 515, and 518. Of these, Isḥāq's translation of the *Almagest* uses *al-munʿaraj*, "the bend, turn" (root *ʿ-r-j*) for 512 (2nd star of Eri in *Almag.* = β Eri), and for 518 (29th star of Eri in *Almagest* = 43, d Eri), but *rajʿat al-nahr*, "the return of the river" (root *r-j-ʿ*) for 515 (18th star of Eri in *Almagest* = τ^1 Eri). Neither of these forms is identical to John's assumed *ʿarjat al-nahr*, "the bend of the river" (root *ʿ-r-j*). Further, γ Eri, the star described by John (10th star of Eri in *Almagest*), is not located in any "bend" or "turn", but rather in a διάστασις ("interval"), Arabic *al-buʿd* or *al-masāfa*, Gerard *spacium* (*cf.* Kunitzsch, *loc. cit.*, 315, no. 514). Therefore, both John's name of the star and its astronomical identity are different from the classical traditions and undocumented in the sources. John's name has lived on in some modern star atlases and name lists as Angetenar, for τ^2 Eri (19th star of Eri in *Almagest*).

20. *Cf.* Kunitzsch, *op. cit.* (ref. 14), 198.
21. Markeb has lived on as a name for the star called κ Vel in modern astronomy.
22. See Kunitzsch, *op. cit.* (ref. 14), 259, no. 237. In the Arabic tradition, the term *anf* ("nose") only occurs with ρ Ser (Kunitzsch, *loc. cit.*, 251, no. 186), γ Tau (*ibid.*, 266, no. 276), and λ Cet (*ibid.*, 307, no. 472), and additionally in the combination *anf al-sāq*, lit. "nose of the shin" (for Ptolemy's ἀντικνήμιον, the fore part of the shin). John's *enif* is still used today as a name for ε Peg.
23. The situation in John's table is confused. According to the coordinates, the star indicated is α Cet, and consequently Menkar has lived on as this star's name until the present time. According to Ptolemy, however, α Cet is situated on the snout, at the end of the jaw (Latin of Gerard: *muscida/extremitas mandibule; cf.* Kunitzsch, *op. cit.* (ref. 14), 307, nos. 473/74), whereas the "nose", Arabic *al-minkhar* (from which John's word *menkar* is derived), is represented by λ Cet (*ibid.*, 307, no. 472). John puts his α Cet on the nose, *menkar*, and adds: *et secundum quosdam est in guttere eius* ("and according to certain [people] it is in its [*scil.* the animal's] throat [or gullet, or neck]"). On the one hand, therefore, it would seem that in his alternative designation he imitates Ptolemy's twofold description of α Cet, whereas, on the other hand, his terms are not in harmony with Ptolemy's. We cannot, therefore, safely understand whether John here follows a hitherto "unknown Arabic source", or simply has confused the material in the well-known available sources.
24. See Kunitzsch, *op. cit.* (ref. 14), 225, no. 46.
25. The most probable seems to be that the Arabic forms are variants of the name of α Gem on astrolabes and in astrolabe star lists (Arabic *muqaddam al-dhirāʿ ayn* ("the preceding one of the two forearms")), and equally of the 7th lunar mansion, *al-dhirāʿ*, consisting of αβ Gem, of both of which Latin transliterations circulated in the West since late tenth century in great variety. The first use of these forms, applied to α Cep, is in John's table. (Today, α Cep is still called Alderamin.) An often assumed derivation of John's Arabic forms from Arabic *al-dhirāʿ al-yamīn* ("the right [fore]arm") appears highly doubtful because in Ptolemy's descriptions no "right arm" is mentioned for Cepheus (there is only a "left arm", which however is ι Cep, not α); further, in the Arabic, *al-dhirāʿ* is mostly constructed as a feminine noun, so the correct composition would be *al-dhirāʿ al-yumnā* (which no longer allows the relation of the last part in the Latinized forms, -[*y*]*min*, to an Arabic *yamīn*; and which also, with the repeated article *al* in front of the adjective *al-yumnā*, would not fit to the bare Latinized [*y*]*min*). If my hypothesis, to understand *aldramin* etc. in α Cep as a displaced derivate of the true name of α Gem or the 7th lunar mansion, αβ Gem, is rejected, then this would mean that John, in forming the name of α Cep, must have followed the "unknown Arabic source" because all the existing known Arabic sources do not have the underlying terms in connexion with α Cep. It may be inferred that in no. 8, α Per, MS d (which is a Spanish MS, Madrid, Bibl. Nac. 10053) cites an Arabic form *algeb ymin* (from Arabic *al-janb al-ayman* ("the right side, or flank")) where the element *ymin* represents the Arabic masculine form *al-ayman* correctly used in combination with the masculine noun *al-janb*; whereas in α Cep, if *al-dhirāʿ* was really involved, a feminine form is required for the adjective. It cannot be excluded that the spelling *-ymin* for α Cep in MS d was a deliberate harmonization of the name of α Cep to that of α Per by a writer who had only a slight knowledge of Arabic and who took it upon himself to "correct" the name of α Cep whose historical development (as being derived ultimately from the names of α Gem, or the 7th lunar mansion) he was not able to determine from the sources. (It should be noticed that John obviously knew the lunar mansions well because in no. 10, α Tau, *aldebaran*, he explicitly adds the remark that this star is one of the stars making up the 4th lunar mansion: *et est de quarta mansione lune*.)
26. I.e., MSS c and f of the edition cited above in Kunitzsch, *Typen*, and MS Bodl., Ashmole 191.I, fol. 39v, cited by Thorndike, *op. cit.* (ref. 8), 161.
27. *Cf.* P. Kunitzsch, "Arabische Sternnamen – Sternnamen der Araber: Zur Begriffsbestimmung", *Sudhoffs Archiv*, lxi (1977), 105-17.

XXII

THE STAR CATALOGUE COMMONLY APPENDED TO THE ALFONSINE TABLES

In earlier works on the history of star names[1] the Alfonsine Tables, or "the authors of the Alfonsine Tables", or summarily "the Alfonsinians", were often made responsible for the introduction of new Arabic material into the *corpus* of star names traditionally transmitted, and continuously used until today, in mediaeval Western, and modern, astronomy.

Owing to the knowledge we now have of a considerably larger number of original sources, both Oriental and Occidental, and our better understanding of their historical interdependence, we are now in a position to examine and revise the judgement on the star catalogue included in the Alfonsine Tables. It may be remembered that the Alfonsine Tables were a set of astronomical tables for chronology, for the Sun, Moon, and planets, etc., accompanied by an explanatory introduction (quite in the manner of the Arabic-Islamic *zījes*), which were compiled and composed by order of King Alfonso X of Castile (reigned 1252-84). Of the Old Castilian original version of the tables made in Castile around 1263 to 1276, only the introductory text has survived.[2] The Old Spanish text was of no major influence outside Spain. In the first decades of the fourteenth century, however, a Latin version of the Alfonsine Tables was constructed in Paris which later became the most influential handbook of practical astronomy in Europe until the sixteenth century or even later.[3] In what follows, when we speak of the Alfonsine Tables, we always refer to their Latin version.

Like every complete *zīj*, or work of astronomical tables, the Alfonsine Tables also contain a star catalogue. When we examine it we find at once that the star catalogue accompanying the Alfonsine Tables is nothing else but Ptolemy's star catalogue in the *Almagest*, in the wording of Gerard of Cremona's Latin translation made in Toledo about 1175 from the Arabic, save for the values of the longitudes (in the coordinates). These were increased against Ptolemy's, for precession, by 17°8′ for an assumed epoch 1252, i.e., adopting the value used in several star tables and in the complete Ptolemaic star catalogue included in Alfonso's *Libros del Saber*.[4] The Alfonsine Tables proved very popular and were circulated in innumerable manuscripts. The star catalogue is included in many of these copies, but it is also transmitted separately in many manuscripts (where naturally it is easily recognized from the 'Alfonsine' longitude value, Ptolemy + 17°8′).[5] The popularity and dissemination of the Alfonsine Tables were certainly increased by several printed editions which, later on, became highly influential in the formation of the nomenclature of the stars. Of these editions I have examined seven: *Alfontij regis castellę illustrissimi cęlestium motuum tabulę*, Venice 1483 [henceforth called *at1*]; *Tabule astronomice Alfonsi Regis*, Venice 1492 [henceforth called *at2*]; *Tabule astronomice Diui Alfonsi Regis*, Venice 1518 (at the end of the book: 1521); *Alfonsi Hispaniarum Regis Tabule*, Venice 1524 [henceforth called *at4*]; *Divi Alphonsi... Regis Astronomicae tabulae*, Paris 1545; the same, Paris 1553 (identical in the text to the preceding edition of 1545); and

Tabulae Alphonsinae perpetuum motuum coelestium, Madrid 1641. In addition, I have looked through a number of manuscripts kept in Munich and Vienna.

Coming back once more to the earlier assumption of "Alfonsine authors", we now know, for the star catalogue, that no collaborators of Alfonso's were involved in the establishment of this catalogue. Moreover, some of the later Latin compilers have simply added to the table *corpus* the *Almagest* star catalogue in Gerard of Cremona's translation. This catalogue is itself known from an old edition[6] and a number of manuscripts.[7] That means that, for the history of the stellar nomenclature, the basic text was the *Almagest*, and the Alfonsine Tables did not introduce any contribution of their own.

In comparing, further, the editions of the Alfonsine Tables we find that, while in *at1* the star catalogue is totally identical to Gerard's version of the *Almagest* (and, thus, rather poor in Arabic star names), in *at2* it contains a number of additional, new star names always connected to the standard text by the formula ...*et dicitur*... ("and it [*scil.* the respective star] is called...").[8] We shall, however, see soon that no "Alfonsine authors" were involved in the introduction of these new names. As for the succeeding editions, especially *at4*, etc., they retained only some of these additional names and omitted the rest.

In the following I give a list of the 29 'new' names in *at2* and shall afterwards discuss their history and ultimate sources.

1. α UMi (1st star in UMi): *Alrucaba*
2. α UMa (16th star in UMa): *Dubhe*
3. η UMa (27th star in UMa): *Elkeid & Bennenazc*
4. γ Dra (5th star in Dra): *Rasaben*
5. α Cep (4th star in Cep): *Alderaimim*
6. α Her (1st star in Her): *Rasalheti*
7. α Lyr (1st star in Lyr): *wega*
8. α Cyg (5th star in Cyg): *Denebadigege*
9. α Cas (2nd star in Cas): *Scheder*
10. α Per (7th star in Per): *Alchemb*
11. α Oph (1st star in Oph): *Rasalaugue*
12. α Tau (14th star in Tau): *Aldebaran id est oculus vel cor Tauri*
13. β Gem (2nd star in Gem): *Rasalgense*
14. β Leo (27th star in Leo): *Denebalezeth*
15. α Sco (8th star in Sco): *Cabalatrab*
16. γ Cap (23rd star in Cap): *Denebalchedi*
17. α PsA (42nd star in Aqr): *Fomahant*
18. α Cet (2nd star in Cet): *menkar*
19. ζ Cet (14th star in Cet): *Batenkaiton*
20. β Cet (22nd star in Cet): *Denebcaiton*
21. α Ori (2nd star in Ori): *Beldelgenze*
22. γ Ori (3rd star in Ori): *Bellatrix*
23. β Ori (35th star in Ori): *& dicitur Algebar. Nominatur etiam Rigel*
24. τ^2 Eri (19th star in Eri): *Angetenar*
25. θ Eri (34th star in Eri): *Acarnar*
26. k Pup (6th star in Argo): *markeb*
27. α Hya (12th star in Hya): *Alphart*

28. α Crt (1st star in Crt): *Alhes*
29. γ Crv (4th star in Crv): *Algorab*

This is a considerable number of 'new' star names that are not found earlier in the catalogue tradition, i.e. the *Almagest* and its continuation in the Alfonsine Tables (*at1*). After continuous study of relevant source texts I am now able to give here the historical analysis of these 29 'new' names, going in this beyond my earlier findings of 1959.[9]

Whenever we find Arabic material in mediaeval Western texts we can be sure of two things: such material can only be brought in through translation from the Arabic; and a text using such material must, therefore, be itself either a translation from the Arabic, or a work based on a translation. In no other way could Arabic material be introduced into Latin writings. This, of course, also applies to our 'new' star names, all of which (with the single exception of no. 22) are visibily Arabic.

Now, the translation period from the Arabic in mediaeval Europe had ended in the twelfth century, or at the latest in the time of Alfonso (who had translations normally made into Old Spanish, whence some texts were then translated into Latin). On the other hand, the compilation and composition of the Latin versions of the Alfonsine Tables in Paris falls in the first decades of the fourteenth century, and we do not know exactly when the star catalogue was added to the *corpus*.[10] Certainly, it would have been well after the end of the translation period. Therefore, all Arabic material found in the Alfonsine star catalogue beyond that which was already contained in Gerard's translation of the *Almagest* catalogue, must be expected to derive from earlier translated material that lies outside the *Almagest* and Alfonsine Tables tradition. In my 1959 study I could only trace these 'new' names back to an "unbekannte Quelle" ("unknown source"), and I offered the suggestion that this unknown source must be sought among the mediaeval literature on the astrolabe.[11] This "unknown source" I afterwards located in an astrolabe star list composed, in conjunction with his own star observations, by John of London in Paris in 1246.[12] But John's table does not contain all of our 29 'new' names, and so further sources must lie behind them.

Looking around, I found that several Alfonsine and related star catalogues in old manuscripts show a similar inflation of star names to that of our *at2*. Especially rich in such nomenclatory additions is a star catalogue in MS Vienna 5415 (datable to 1444), fol. 216v-251r, computed for the epoch 1424, longitudes = Ptolemy + 18°56′, i.e. Alfonsine longitudes + 1°48′. It is a complete catalogue of the 1025 Ptolemaic stars in the well-known Gerard version, apparently copied in relation to the Alfonsine Tables since the longitude difference to *Alfoncius* is explicitly mentioned in the title. The outstanding feature of this catalogue is that its author has compiled numerous Arabic names, both for the constellations and for individual stars, from a great many pre-existing translated or derived texts. The texts, or star tables, involved are "Typ III", "Typ VI", "Typ VIII", "Typ XIII", and "Typ XV";[13] the chapter on the constellations of Abū Ma'shar's *Introductorium maius* in the Latin translation of Johannes Hispalensis;[14] an astrological text which is still not fully explored; and perhaps a very old astrolabe treatise. In this inflated catalogue, with a few exceptions all our 'new' names of *at2* are represented. But Vienna 5415 is not the only specimen of such inflated

Alfonsine star catalogues; others certainly exist.[15]

This leads us to the conclusion that the editor of *at2* must have used for his version of the star catalogue a manuscript of the inflated type described above which contained all the 29 'new' names registered here.

It is perhaps not by mere chance that our specimen manuscript of the inflated type is at Vienna (i.e., Vienna 5415), and made for the epoch 1424. Another Vienna MS, no. 5311 (14/15th century), contains four of the five 'types' of star tables involved here,[16] and the fifth 'type' exists in two other Vienna manuscripts.[17] All, or most, of these manuscripts fall into the first four decades of the fifteenth century, which was the life and working period of John of Gmunden (d. 1442), creator and well-known representative of the first Vienna school of astronomers. In the list of lectures delivered by him in the University of Vienna[18] we see that in 1424 he taught "Sphaera materialis", that is the celestial globe, for which star catalogues of the sort mentioned above (Vienna 5415 and its Munich relatives, for epoch 1424) were needed, and perhaps directly composed. In 1434 he gave lectures on the astrolabe. And similarly we see in the various copies of his star table devised for the astrolabe that here too he uses most of those 'new' names found in the Vienna 5415 catalogue and later in *at2*.[19]

Taking all this together, therefore, it would seem that the compilation of those 'new' names and their addition to the great star catalogue of the *Almagest*-Alfonsine type, and also their further use in shorter astrolabe star tables, was the work of John of Gmunden or some of his pupils or collaborators. In this way, he — or they — did something similar to what has been done later over and over again by other astronomers who collected and compiled star names from older texts or philological studies and introduced them into astronomical usage, astronomers such as Johannes Bayer (1603) and Giuseppe Piazzi (1814).

Next we shall analyse the historical background of the 29 'new' names, of which — as should be noted — many have lived on into our own time.

A major group of these names was first applied in the star table of "Typ VI" and afterwards also incorporated into "Typ VIII",[20] so that the fifteenth-century compiler may have taken them from either of the two, or from the two together; to this group belong our nos. 2, 4,[21] 9, 10, 11,[22] 12,[23] 13,[24] 14, 16, 18, 19, 20, 21,[25] 24,[26] 26,[27] and 27.

One name, no. 15, occurs in identical spelling in "Typ III", "Typ VI", and "Typ VIII", so that it cannot safely be determined which of these sources was in fact used by the compiler. Some names occur in similar spelling in "Typ III" and "Typ VIII" so that one or other of these, or both, could be the source for the compiler: these names are nos. 1,[28] 3,[29] 7, and 29.

"Typ XIII" is at the base of the names of our nos. 8,[30] 17,[31] and 23.[32] No. 25 is derived from "Typ XV" (existing, e.g., in MS Vienna 5311).[33]

Four names require special treatment. No. 6 is also found in the Vienna tradition (John of Gmunden; Vienna 5415, etc.; the 1468 astrolabe of Regiomontanus kept in the Germanisches Nationalmuseum at Nuremberg, reg. no. WI 5), but in no earlier source; it must, therefore, be assumed that the Vienna authors themselves 'constructed' this name, compounding the well-known element *Ras* ("head", mostly written by them, owing to a confusion in the name of α Dra, as *Razd*) with the Arabic constellation name of Her as transmitted in the *Almagest*-Alfonsine tradition, *algiethi*, their spellings being such as *Razd algethi/algeti/algieti/alchieti*, *razalgeti*, etc. No. 22 is a Latin name occurring for the first time in astrological texts related to Abū Ma'shar and Johannes Hispalensis, and was originally used for α Aur; the transfer to γ Ori seems to have been made in the Vienna school of astronomers.[34] No. 28 was again first applied in the Vienna school where, originally, *alhes* was the name of the constellation Crater;[35] it was collected from Johannes Hispalensis's Latin translation of Abū Ma'shar's astrological work *Introductorium maius*, in the chapter on the description of the celestial bodies and the constellations of the fixed stars.[36] The transfer of *Alhes* to the star α Crt is first visible in the star lists of John of Gmunden (Vienna 5268, 2nd list; Melk 1099 = St Florian XI. 619, etc.); Regiomontanus's Nuremberg astrolabe of 1468 has the star as *Alies*. (The star list in Vienna 2352, fol. 101r (*cf*. ref. 10, above) also mentions the name, but still as the constellation name: *Alhes id est vas vel Crater*, intended to accompany γ Crt (2nd star of Crt in *Almagest*) whose descriptive text, however, was wrongly placed in the line above.)

The last remaining name, no. 5, is the most difficult to explain. The Typ III-tradition has a star called *aldiraan*, etc. (III, no. 22), whose name basically designates α Gem, but the star is mostly confused with α Hya or with a star in the "forehead" of Leo.[37] Typ VI (no. 1) later applied *aldramin*, etc., for α Cep, but it is not evident whether here the older spelling of Typ III was living on, or whether the author of Typ VI had formed a new and independent name from another Arabic original.[38] Typ VIII, again later, has retained both names, the form from Typ III (= VIII, no. 18), and the form from Typ VI (= VIII, no. 42). The use of the name for α Cep, and the spelling, in the tradition of the Vienna school and in *at2*, follow the use of Typ VI, but we cannot ultimately confirm the etymology of the name in Typ VI.[39]

Thus, the history of all the 29 'new' names in *at2* is established. But it has become evident that no "Alfonsine authors" were involved at all. The tradition in the texts seems to indicate that most of these 'new' names were added to the traditional *corpus* in the first half of the fifteenth century by the Vienna school of astronomers, who compiled them from a number of older sources. The role of the Alfonsine Tables in the history of star nomenclature was in the propagation and dissemination of certain names, but they did not play an active and productive part in this history. The erroneous assumption of earlier authors that the "Alfonsinians" were the authors, or creators, of those 'new' names may have been formed under the impression that a greater number of astronomers from all the contemporary cultures co-existing in Spain, Christian, Jewish, and Muslim, were assembled at Alfonso's court where they constructed the star catalogue and thereby brought in elements of their foreign nomenclature. That the number of Alfonso's collaborators was limited is now known,[40] and that these were not in any way involved in the formation of the specifically Alfonsine star nomenclature (found in *at2*, etc.) has been demonstrated here.

For the sake of completeness I should add one more detail. From *at4* on, there was added to the editions of the Alfonsine Tables an astrological survey of the fixed stars.[41] This survey contains also a number of Arabic names, of certain constellations and individual stars, which were much discussed in Renaissance times but most of which are now obsolete.[42] This survey was derived from Ptolemy's astrological work *Tetrabiblos* (Book I, ch. 9), in the Latin translation of

Plato Tiburtinus made from the Arabic.[43,44]

The reader should note that no etymologies of the Arabic names have been given in this article. For these, the reader is referred to the works cited in references 9 and 13.

Addendum. Since we are here discussing the contribution of the Vienna school of astronomers to the formation of the *corpus* of modern star names, I should like to take this opportunity to comment on two more puzzling star names, in the formation of which the Vienna school was heavily involved (but which, however, were not transmitted within the Alfonsine Tables).

In modern astronomy α Peg is commonly called Markab, and β Peg Scheat. This arrangement is already found with John of Gmunden and in the other relevant Vienna texts mentioned above. More exactly, α Peg is there called *Markab Alferaz id est humerus equi*, and β Peg *Scheat Alferaz id est crus equi.*

β Peg is a star mentioned in almost all mediaeval astrolabe star lists and on most of the existing astrolabes. In Typ III, no. 6 we have it as *alferat*, etc., abbreviated from its full Arabic name *mankib al-faras*, "the Horse's Shoulder". Typ VI, no. 39 gives the full name as *menkeb alferaçh* and *mekebalferaz*,[45] but most of the manuscripts have a parallel name, *bedalferaz*,[46] instead. Typ VIII, later, has adopted the star from both III (no. 46 in VIII, equally called *alferaz*) and VI (no. 47 in VIII, corrupted in its spelling to *mentichel*, etc.[47]). Here lies the reason for greater confusion to come because later readers of course assumed that nos. 46 and 47 in Typ VIII represent two different stars whereas, in reality, the compiler of Typ VIII has inserted the same star, β Peg, twice, not recognizing its identity because of the different names in his two source star tables, viz. Typ III and Typ VI. In Typ XIII, no. 21 we have β Peg correctly as *mankab alferaz.*

Further, Typ VI, no. 38 — that is, immediately preceding β Peg, no. 39 — had listed *sceath id est crus...*, δ Aqr.[48] This star was also taken over by the compiler of Typ VIII and inserted immediately before his *alferaz*, i.e. VIII, no. 45; the text reads *sceach* (in the column for the Arabic names), and *crus* (in the column for the Latin descriptions).[49]

Afterwards, readers and users of these tables were evidently confused as to the correct identifications of these stars, especially in Typ VIII where the twofold mention of the same star, β Peg, under different names had happened, this being overlooked by later readers who were, of course, not in command of Arabic and so could not interpret the Arabic names correctly. Therefore, the Typ VIII-stars no. 45 (*sceach*, δ Aqr) and 46 (*alferaz*, originally β Peg) were conflated into one new star, now called with a seemingly-complete Arabic name *scheat alferaz*, and assumed to designate β Peg (for which also the Latin equivalent of the former *scheat* in Aqr, *crus*, "the shin", was retained).[50] Subsequently, the second *alferaz*-name of Typ VIII, no. 47, and of the other sources cited above, in its full spelling as *markab*[51] *alferaz*, and with its traditional correct Latin equivalent *humerus equi* (originally designating β Peg in the texts),[52] was applied to α Peg. Thus, in the Vienna school of astronomers and down to modern times, both these stars were called by their wrong names, α Peg as *Markab* (which originally was the Arabic *mankib*, "shoulder", of β Peg), and α Peg as *Scheat* (which originally was the Arabic *sāq*, "shin", of δ Aqr).[53]

REFERENCES

1. There may be mentioned, as the most influential, L. Ideler, *Über den Ursprung und die Bedeutung der Sternnamen* (Berlin, 1809), and R. H. Allen, *Star-names and their meanings* (New York, 1899; reprinted as *Star names, their lore and meaning* (New York, 1963)). Allen largely depends on Ideler as far as Arabic matters are concerned.
2. Printed in *Libros del Saber de Astronomia del Rey D. Alfonso X de Castilla*, ed. by M. Rico y Sinobas, iv (Madrid, 1866), 111-83 ("Libro que a por nombre el de las taulas Alfonsies"). This text originally was not part of the *Libros del Saber*, but was included by Rico in his monumental edition; *cf.* G. Bossong, *Probleme der Übersetzung wissenschaftlicher Werke aus dem Arabischen in das Altspanische zur Zeit Alfons des Weisen* (Tübingen, 1979; Beihefte zur Zeitschrift für romanische Philologie, 169), 62 (no.4) and 69.
3. See J. D. North, "The Alfonsine Tables in England", in *ΠΡΙΣΜΑΤΑ. Naturwissenschaftsgeschichtliche Studien: Festschrift für Willy Hartner* (Wiesbaden, 1977), 269-301; *cf.* also O. Gingerich, "The astronomy of Alfonso the Wise", *Sky & telescope*, lxix (1985), 206-8. In a useful edition (*Les tables alphonsines avec les canons de Jean de Saxe* (Paris, 1984)) E. Poulle has recently presented what can be taken as the nucleus of the Alfonsine Tables and which consists of chronological and planetary tables, always in harmony with the accompanying canons of John of Saxonia; the star catalogue, as being part of the "membra adjuncta", was not included in this edition; *cf.* there, pp. 6f., 26 (at *c*), 224. For a Hebrew translation, made in 1460 from the Latin, see B. R. Goldstein, "The survival of Arabic astronomy in Hebrew", *Journal for the history of Arabic science*, iii (1979), 31-39 (esp. pp. 36f.). Most striking is the lack of reference to the Alfonsine Tables in the work of the Hebrew astronomer Levi ben Gerson (in southern France, 1288-1344), *cf.* B. R. Goldstein, *The astronomical tables of Levi Ben Gerson* (Transactions of the Connecticut Academy of Arts and Sciences, xlv; New Haven, Conn., 1974), 21.
4. 1 June 1252 was the date of Alfonso's accession to the throne and was afterwards applied as the epoch for his Tables; *cf.* North, *op. cit.* (ref. 3), 270. While the Alfonsine precession value was adopted by the compilers of our 'Alfonsine' star catalogue, the Spanish text of the Ptolemaic catalogue in the *Libros del Saber* remained without any visible influence on the Latin text of our catalogue which is pure Gerard of Cremona. Gerard's Latin version of the *Almagest* star catalogue was similarly included in what could be called the "mediaeval Latin tradition of al-Ṣūfī", here using the longitude value of al-Ṣūfī, i.e., Ptolemy + 12°42′ (*cf.* P. Kunitzsch, "Ṣūfī Latinus", *Zeitschrift der Deutschen Morgenländischen Gesellschaft*, cxv (1965), 65-74). The star catalogue in Copernicus's *De revolutionibus orbium coelestium* (Nuremberg, 1543) (at the end of Book II, 46v-62v) is an exact replica of Ptolemy's catalogue in the *Almagest* (with the only difference that the longitudes of the stars are smaller by 6°40′ than those in the *Almagest*, because Copernicus started the counting at Ptolemy's 1st star of Aries, γ Ari — thus avoiding the problem of precession, while Ptolemy started with the vernal equinox as Ari 0°0′, and therefore in his system the longitudes are subject to precession); but the wording of the star descriptions in Copernicus's catalogue is related neither to Gerard's translation from the Arabic nor to Georgius Trapezuntius's translation from the Greek (printed in Venice, 1528; Basel, 1541 and again in 1551), nor to the Latin translation made from the Greek in Sicily around 1160 (I am grateful to Prof. M. Folkerts of Munich who allowed me to examine his microfilm of MS Vat. lat. 2056; R. Lemay of New York is about to present new evidence for the origin of this translation). L. A. Birkenmajer earlier had found a series of parallels between *De revolutionibus* and George Valla's *De expetendis et fugiendis rebus...* (Venice, 1501), which includes a Latin version by Valla of Ptolemy's star catalogue made from the Greek, *cf.* L. A. Birkenmajer, *Stromata Copernicana* (Cracow, 1924), 154-68, esp. pp. 161-2 (in Polish; I have used a German translation: "Dienstliche Uebersetzung der Publikations-Stelle in Berlin-Dahlem", ausgeführt von Gymnasial-Prof. Bassmann (no date [c. 1942], reproduction from the original typescript); I am indebted to Dr H. M. Nobis who put the copy kept in the Copernicus-Forschungsstelle, Deutsches Museum, Munich, at my disposal); see also J. Dobrzycki, "Katalog gwiazd w *De revolutionibus*", *Studia i materialy z dziejów Nauki Polskiej*, Seria C, Z. 7 (Warsaw, 1963), 109-52 (English summary on p. 153). In order to verify these contentions, I recently made my own comparison between Valla's star catalogue (from a xerox-copy kept in the Copernicus Research Centre mentioned above; the catalogue is found in vol. i, Book XVII, pp. ddii^r-eevi^r, of Valla's *De expetendis...* (Venice, 1501)) and Copernicus's catalogue (from a new critical edition: Nicolaus Copernicus, *Gesamtausgabe*, ii: *De revolutionibus libri sex*, ed. by H. M. Nobis and B. Sticker (Hildesheim, 1984); the catalogue is here to be found on pp. 114-79). This comparison proved that Copernicus's verbal descriptions of many stars are identical to those of Valla, and many others are obviously inspired by Valla's formulae. In several other stars, however, Copernicus's text deviates from Valla's to a greater or less extent (or is correct where Valla's is defective), so that it must be assumed that Copernicus had at his disposal other source material in addition to

Valla's work. The source background of Copernicus's star catalogue, therefore, still needs more detailed investigation.

5. An easy means in star catalogues to find the longitude applied in the catalogue, is to examine the star α Hya (12th star of Hydra in *Almagest*), which in the *Almagest* has longitude Leo 0°0′ so that the addition can directly be read off (e.g., in the Alfonsine Tables: Leo 17°8′).
6. *Almagestum Cl. Ptolemei Pheludiensis Alexandrini Astronomorum principis...* (Venice, 1515)—to be distinguished from other editions containing the Renaissance translation of Georgius Trapezuntius made from the Greek.
7. There are c. 35 manuscripts of Gerard's version that are complete with regard to the star catalogue. A critical edition is presently prepared by P. Kunitzsch, to appear later as vol. ii of a major edition project.
8. Even in cases where Arabic names were already used by Gerard, the addition of *at2* can be clearly discerned; *cf.* α Cyg (5th star of Cygnus in *Almagest*): *Lucida que est in cauda: et est aridef*[*at 2: Arided*]*: et dicitur Denebadigege*, where the element opened by the words *et dicitur* is 'new'.
9. P. Kunitzsch, *Arabische Sternnamen in Europa* (Wiesbaden, 1959).
10. John of Saxonia in his canons, datable to around 1327, speaks of a recent verification of the coordinates of the fixed stars by *Alfontius* (Canon 26, ed. Poulle (ref. 3), 100-1) which seems to be a reference to what we here call the Alfonsine star catalogue (*cf.* also Poulle, *loc. cit.*, 6f. and 224). Further there exist some longer excerpts from the Alfonsine Tables' star catalogue dating from the fourteenth century: 276 stars, by Jean de Lignères, one of the first commentators of the Alfonsine Tables (in the 1320s—the table is found, e.g., in MS Paris, BN lat. 10264, fol. 36v-38v; the longitudes are Alfonsine, i.e. Ptolemy + 17°8′. Of the 'new' names, it has already two: *benenarum* = no. 3 above, and *Razcaben* = no. 4); 60 stars for epoch 1340, long. = Ptol. + 18°0′, by Heinrich Selder, another commentator of the Alfonsine Tables (he wrote in 1364), found, e.g., in MS Munich, Clm 27, fol. 178r-179r (none of our 29 'new' names appears, but instead two others both of which are taken from one of the sources from which the 'new' names were also borrowed); 226 stars, anonymous, for epoch 1338, long. = Ptol. + 17°52′, found, e.g., in MS Erfurt, Ampl. 2° 395, fol. 104v-105v, and Munich, Clm 26667, fol. 46v-47v (here, many of the stars are without textual description; the names are often utterly corrupted; *cf.* ref. 44, below); 87 stars of 1st, 2nd and 3rd magnitudes, anonymous, for epoch 1357, long. = Ptol. + 18°15′ in most of the stars (in a few stars the minutes end in 1′, 2′, or 4′), MS Vienna, 2352 (datable to 1393), fol. 100r-v (of the 'new' names there appear *Edub* = no. 2, *Benenan* = no. 3, *Raschaben* = no. 4, *vega* = no. 7, and *rigel Algebar* = no. 23; η Peg is curiously called *Adam*, or "in the right knee of *Adam*", of unknown connection. On fol. 101r-v there is a further list of 51 1st, 2nd and 3rd magnitude stars, now with long. = Ptol. + 18°18′, partly containing the same stars as the first list, and mentioning some more of the 'new' names, but the additional nomenclature cited here in the last two columns is often in wrong arrangement).
11. See Kunitzsch, *Arabische Sternnamen in Europa*, 45 and 92-94.
12. See P. Kunitzsch, "John of London and his unknown Arabic source", *Journal for the history of astronomy*, xvii (1986), 51-57.
13. Reference is here made to certain types of mediaeval star tables as edited by P. Kunitzsch, *Typen von Sternverzeichnissen in astronomischen Handschriften des zehnten bis vierzehnten Jahrhunderts* (Wiesbaden, 1966).
14. Not edited; I compared MSS Munich, Clm 374 and Clm 122.
15. From Munich, I mention the star catalogues of MSS Cgm 595 (15th cent.), fol. 44v-59v (long. = Ptol. + 17°8′, i.e. the Alfonsine value); Clm 10662 (15th cent.), fol. 129v-146r (long. = Ptol. + 18°56′, as in Vienna 5415); and Clm 24103 (15th cent.), fol. 141r-155r (epoch 1424, long. = Ptol. + 18°56′, also as in Vienna 5415). But even the older MS Munich, University Library 4° 740 (13/14th cent.), fol. 1r-17r, with its 'pure' Ptolemaic-Alfonsine catalogue (long. = Ptol. + 17°8′) has at least added two of our 'new' names: *Edib* (= our no. 2), and *benenaz* (= no. 3). *Cf.* also ref. 10, above.
16. I.e., "Typ XV" (fol. 129v; on the top of the page, another [later?] hand has added a longitude computation for the year 1428), "Typ III" (fol. 130rb), "Typ VI" (fol. 130v), and "Typ XIII" (fol. 131r; again computation for the year 1428 added on top). Evidence for the use of 5311 by the compiler of the inflated star catalogue in 5415 can be gleaned from the mention of *alhadib* among the names of the constellation Lupus in 5415 (and its Munich relatives Cgm 595 (here added to the 1st star of Lupus), Clm 14583 and 24103 (in these two MSS spelled *alhadibh*)). *alhadib* is the traditional Arabic name of β Cas and was transmitted to the West in the star table of "Typ III" (in Kunitzsch's edition cited in ref. 13, above). In some branches of the "Typ III"-tradition, *Alhadib* was explained in a Latin gloss as *in telo* (which was incorrect because the star has no relation to *telum* which could hardly be anything other than the northern constellation of Sagitta). In 5311, fol. 130rb, instead of the traditional gloss *in telo*, to the right

of *alhadip* is written *lupus*. This was the source for the compiler of the catalogue in 5415 who, subsequently, entered *alhadib* among the names of the constellation Lupus.

17. I.e., "Typ VIII", in MSS 2323 (14th cent.), fol 80v-81r, and 2367 (15th cent.), fol. 194r.
18. See R. Klug, *Johannes von Gmunden, der Begründer der Himmelskunde auf deutschem Boden* (Vienna and Leipzig, 1943; Akad. d. Wiss. Wien, Phil.-hist. Kl., Sitzungsberichte, 222. Band, 4. Abhandlung), 18.
19. As a specimen we may use the two neighbouring star tables in MS Vienna 5268, fol. 30v-31r, written in 1437 by John of Gmunden himself; the first: 44 stars computed for the epoch 1436, long. = Ptol. + 19°2′ (identically found in MS Munich, Univ. Libr. 4° 738, fol. 10r-v, but here the longitudes are 1′ greater, for the year 1438), and the second: 41 stars computed for 1432, with *mediatio coeli* and declination (identically found in Munich, Univ. Libr. 4° 738, fol. 11r; Munich, Clm 14504, fol. 226r-227r; and Nuremberg, Cent. VI 18, fol. 73r (autograph by Regiomontanus)).
20. The reference is to the edition cited above, ref. 13.
21. Already found with Jean de Lignères; *cf.* above, ref. 10.
22. It should be mentioned that Typ III (followed later by Typ VIII) and Typ VI had different types of spelling. Later, with John of Gmunden, in Vienna 5415 etc., we find that these two traditions were merged and confused. The Vienna authors applied *Razd alhawe* (Typ III-spelling in the second word) for α Oph, and *Razd alangue* (thus spelt, from the Typ VI-tradition) for α Ser (perhaps mislead by the assonance of *alangue* to Latin *anguis*, "serpent"; but in reality *alangue*, originally spelled *alaugue*, was but a Latinized rendering of Arabic *al-ḥawwā'*, "the Snake Collector", the traditional Arabic term in the *Almagest* for Ophiuchus). In modern times, Rasalhague is used for α Oph, in a 'corrected' Renaissance spelling.
23. The Latin gloss, *id est oculus vel cor Tauri*, is found in VI, no. 10 (MS d), and in VIII, no. 9.
24. Originally, the Arabic name was added to α Gem (VI, no. 15, MSS d and f; VIII, no. 14; John of Gmunden (ref. 19), first table, no. 16; second table, no. 16; MS Munich, Cgm 595 (ref. 15)), while in some texts it was transferred to the next star, β Gem (thus in Vienna 5415). α and β are both located on "heads", each on the head of one Twin.
25. Here, only Typ VI has been involved; Typ VIII has another name that stems from Typ III.
26. While Typ VI (and subsequently Typ VIII) and most of the fifteenth-century texts gave this name to γ Eri (10th star in *Almagest*), *at2* has applied it more fittingly to the 19th star = τ^2 Eri.
27. Originally given to ρ Pup (2nd star of Argo in *Almagest*), but transferred to the 17th star (ζ Pup) in Vienna 5415, Cgm 595, Clm 10662 (here added both to the 2nd and to the 17th star on the outer margin), and Clm 24103. Not in John of Gmunden's two star tables (ref. 19), but in some other tables by him, e.g., MSS Melk 1099 (dated 1445), p. 228, and St Florian, XI. 619, fol. 192v (no. 14) — here for the 1st star of Argo (= 11 Pup); equally so in Clm 10662, fol. 99v (computed for the year 1430). In *at2*, eventually, given to the 6th star of Argo (k Pup). In modern times, Markeb lives on as a name for the star κ Vel.
28. In III and VIII given to θ UMa. The transfer to α UMi is effected in Vienna 5415, Cgm 595, Clm 24103, and John of Gmunden (ref. 19).
29. The element *Bennenazc* occurs in all three types, III, VI, and VIII. But the element *Elkeid* only occurred in some of the oldest Western astrolabe treatises (MS Paris, BN lat. 7412, fol. 19v: *Alkais*, *cf.* Kunitzsch, *Arabische Sternnamen in Europa*, 90; and MS Ripoll 225: *Alkaid*, *cf. ibid.*, 78, no. 40) and must have been borrowed from that tradition. Typ III as such is also related to those oldest texts.
30. The other sources use another Arabic name or, in VI, only the constellation name *adigege*, without the *deneb* element. But XIII, no. 20, is complete.
31. Some manuscripts still have retained the fuller form of XIII, no. 12, as, e.g., Vienna 5415: *fomahant algeista*, similarly also Clm 10662 and Clm 24103. The modern form is Fomalhaut, derived, with a slight change, from a 'corrected' Renaissance spelling.
32. Here, the basic standard form *rigil algebar* (XIII, no. 2) has been broken up and rearranged contrary to the original Arabic formula. But John of Gmunden's second table (*cf.* ref. 19), no. 12, has the correct *Rigil Algebar* (whereas his first table, no. 9, applies *Rigil Algeuze*, after the Typ VI-tradition). *Rigil algebar* also appears in further star tables of John of Gmunden, e.g. in MS Melk 1099, and St Florian XI. 619, Munich Cgm 739, and Clm 10662.
33. Typ VIII, no. 5 contains the same star, calling it by the Latin name *finis fluxus* alone (without an Arabic parallel name), but it is not known from which source it was derived. (The declination (called "latitudo"), 4°30′, must be a mistake for 40°30′; MS f alone has 32°30′.)
34. For details *cf.* P. Kunitzsch, "Abū Ma'šar, Johannes Hispalensis und Alkameluz", *Zeitschrift der Deutschen Morgenländischen Gesellschaft*, cxx (1970), 103-25. A Vienna MS showing this strange application (to α Aur) is Vienna 3124 (15th cent.), fol. 163r. The original Arabic form lying behind *Bellatrix* cannot be guessed; corresponding Arabic texts still must be found.

35. E.g., in Vienna 5415, Munich Clm 10662, Clm 24103; *cf.* also Vienna 2352 cited above in the present article.
36. Book II, ch. 1. *Cf.* MSS Munich, Clm 374 (13th cent.), fol. 15rb-va (spelling abbreviated: *al'es id est uas*), and Clm 122 (spelling: *albes id est uas*). The translation of Hermannus de Carinthia has *crater libri patris* (!) instead, *cf.* the edition of Venice, 1506, p. b3[r].
37. I.e., the 10th lunar mansion of the Arabs consisting of ζγηα Leo. The confusion about this star is already observed in the oldest Spanish-Arabic star table, by Maslama, *cf.* Kunitzsch, *op. cit.* (ref. 13), 18, note to no. 8. The many various spellings of the name sometimes show a relationship to our no. 5 (*cf. Aldirahemin*, MS Paris, BN lat. 7412, fol. 19v, cited in Kunitzsch, *Arabische Sternnamen in Europa*, 91 [*sub* q]; *Aldiranemin*, *ibid.*, 72, no. 24).
38. See Kunitzsch, *op. cit.* (ref. 12), 55 and ref. 25.
39. Apart from the possible implications of Typ III, there is also the 7th lunar mansion of the Arabs consisting of αβ Gem, called in Arabic *al-dhirāʿ*, whose name was spread in the West in countless translations and varieties among which some are quite close in spelling to the Typ VI-spelling, as, e.g., *Alderaaim* (MS Berlin, Phill. 1830 (12th cent.), fol. 3r); *Alderaam* (see E. Svenberg, *Lunaria et Zodiologia Latina* (Göteborg, 1963), 49ff., from MS London, Brit. Libr. Egerton 821 (12th cent.), fol. 17vff.).
40. *Cf.* North, *op. cit.* (ref. 3), 269.
41. In *at4*, pp. 106r-107v; in the edition of Paris, 1545, pp. 208-11.
42. With the exception of *açubene* (1545: *acubenae*), now used as Acubens for α Cnc; *almucedie alaraph*, now sometimes mentioned separately as Almuredin for ε Vir, and Alaraph for β Vir; and *alascha* which gave rise to an erroneous conjecture in Renaissance times as a result of which the 'corrected' spelling Lesath is now used for υ (and sometimes λ) Sco.
43. The text of Plato's translation can be consulted in some early printed editions, the fullest being that of Venice, 1493, which gives the two translations of Plato Tiburtinus and of Aegidius de Tebaldis as well as the commentary of 'Alī ibn Riḍwān (Haly Heben Rodan), the three texts for each section being assembled together. In Kunitzsch, *Arabische Sternnamen in Europa*, Plato's version is abbreviated as *Pq*, and the derived survey in *at4*, etc. as *Pqg* (*cf. ibid.*, 39 and 43).
44. The anonymous list of 226 stars (mentioned in ref. 10, above) is also related to the *Tetrabiblos* chapter, but some of the Arabic names used here are derived from another, anonymous, Latin translation from the Arabic dated 1206, found, e.g., in MS Wolfenbüttel, Gud. lat. 147 (14th cent.), fol. 162ff. (the section on the fixed stars is on fol. 166v-167v).
45. Spelled *menkeb alpharaz*, and twisted into *alferaz mentel*, etc., in Typ VII, no. 30, a direct derivate from Typ VI.
46. Also authentic, from Arabic *yad al-faras*, "The Horse's Fore-foot".
47. But MSS o and p have the 'better' spelling *mekebalfera.*
48. Again continued in Typ VII, no. 29, in corrupted spellings such as *scenath*, etc.
49. Continued in Typ IX, no. 24, as *straach*, *scheat*, etc. The same star is spelled *Cenok* in a related star table in MS Cambridge, Univ. Libr. Add. 6860 (14th cent.), fol. 71r.
50. In reality, the term *crus*, "shin" is not used for any star of Pegasus in the *Almagest*, but it was used there for δ Aqr (the 18th star in Aqr).
51. Thus corrupted from the correct forms *mankab*, *menkeb*, etc.
52. Also in the *Almagest* tradition, *humerus* occurs in the description of β Peg.
53. This erroneous Vienna arrangement of the two names was also followed by Johannes Stöffler in the star tables of his *Elucidatio fabricae ususque astrolabii* (Oppenheym, 1512; at the end of the book: 1513), p. XX[v] (second table, giving ecliptical coordinates). His first star table (giving *mediatio coeli* and declination) is incomplete in the edition, for it breaks off after α Aql (p. XVIII[r]); but the rest of the table exists in the manuscripts, as, e.g., MS Cologne, Historisches Archiv, GB fol. 64 (17th cent.), fol. 265r, where the same Vienna arrangement is found.

XXIII

PETER APIAN AND 'AZOPHI': ARABIC CONSTELLATIONS IN RENAISSANCE ASTRONOMY

Pictorial representations of the stellar sky are known, in various forms, from Antiquity. Most frequent among these was the celestial globe on which the constellations were depicted. Celestial globes were known in Ptolemy's time (around A.D. 150) and most probably even earlier, in the time of Hipparchus (around 130 B.C.) or before. In his *Almagest* (Book viii, chap. 3), Ptolemy gives exact rules for the construction of the celestial globe, and it is quite likely that he himself used a globe when composing his star catalogue that is included in the *Almagest*. An Arab historian (around A.D. 1230) even reports, on the authority of an eye witness, that in the year 435 Hijra (= A.D. 1043-44) there existed, in the Library of Cairo, a celestial globe made of copper (or rather, brass), made by Ptolemy himself — a report which seems highly improbable and has possibly been distorted in the course of transmission.

Maps of the sky drawn on a plain ground, although less frequent, are also known. Manuscripts of the Latin Germanicus-tradition (i.e., commentaries on the Latin translations of Aratus's Greek poem on the constellations, the *Phaenomena*) sometimes contain a sky map where all the 48 classical constellations are drawn, as mere figures and without individual stars or the coordinates, in a single chart. This tradition has continued into mediaeval and Renaissance times, but there were introduced, in the course of time, scientific improvements, such as drawing the fundamental circles for the coordinates, setting the constellations according to their true positions, adding individual stars to all the figures, and sometimes dividing the device, for greater clarity, into two separate charts, one for the northern and the other for the southern celestial hemisphere. For both types of such sky maps, the one-map-type and the two-maps-type, outstanding examples have survived, in manuscripts as well as in printed form.

As an especially famous specimen of the two-maps-type, I may refer to Albrecht Dürer's two woodcuts of the Northern and Southern Skies, printed in Nuremberg in 1515.[1]

A fine specimen for the one-map-type is the star chart of Peter Apian. Peter Apian (to be distinguished from his son Philipp, famous cartographer and map maker) was born in or around 1495. From 1527 until his death in 1552 he was a professor of mathematical sciences in the University of Ingolstadt (which later was transferred to Landshut in 1800, and then to Munich in 1826). Apian's map contains the 48 classical constellations, and it is apparent that his drawings of the individual figures were strongly influenced by Dürer's. This map was printed identically twice, a first time in 1536, at Ingolstadt, separately on a single sheet (see our Figure 1, *Imagines syderum coelestium*). Here the titles of the constellations and some individual star names are printed in Gothic letters. The second

FIG. 1. Peter Apian, *Imagines syderum coelestium*, chart of the traditional 48 constellations as devised and catalogued by Ptolemy in his *Almagest* (c. A.D. 150), printed (on a single sheet) at Ingolstadt, 1536 (reprinted, Munich, 1927). The constellations appear as seen on a celestial globe. Reduced size. Photo: Bayerische Staatsbibliothek, Munich.

time, the map was printed in 1540, again at Ingolstadt, and was included in Apian's *Astronomicum Caesareum*. Here, the drawings of the figures are exactly the same as in the earlier, separate edition of 1536, but the titles of the constellations are now printed in Roman type. Also, here two more Arabic star names were added, viz. Angentenar in Eridanus (besides Acarnar which was already there in 1536), and Yed in Ophiuchus; equally, in 1540 two Latin star names were added to Leo, Ceruix and Dorsum, which did not exist in the 1536 edition. Most of the figures in the work of 1540 have been hand-coloured (therefore I chose for Figure 1 the pure black-and-white printing of 1536). For a

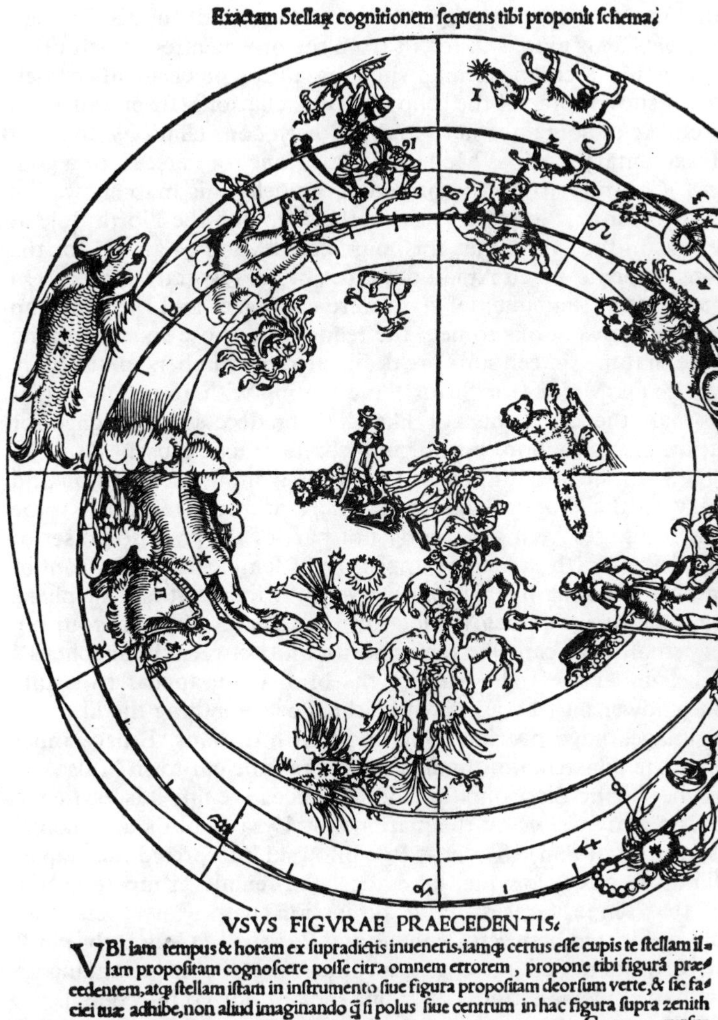

FIG. 2. Star map from Peter Apian, *Horoscopion generale* (Ingolstadt, 1533). The map was designed for the identification of 16 numbered fundamental stars, along the ecliptic. The constellations appear as seen in the sky. Near the centre, a few indigenous Arabic constellations are represented. Photo: Bayerische Staatsbibliothek, Munich.

modern observer Apian's star map may appear strange at a first glance — it must be kept in mind that the map shows the sky as seen on a globe; that is, all the figures and their mutual relations are reversed!

But there exists a second star map by Peter Apian which was also printed identically twice, in 1533 at Ingolstadt, included in his work *Horoscopion generale* (see our Figure 2), and equally in 1533 at Ingolstadt, in his German work *Instrument Buch*. This map is different from the one mentioned before, in several respects. It is not a complete map showing all the 48 classical constellations, but rather it shows some of the ecliptical constellations (from Taurus to Scorpius) and a choice of some northern and southern constellations, some of which are only fragmentary (such as Medusa's head alone for Perseus, or a goat alone for Auriga). Contrary to all the preceding models, this map shows the constellations in their natural order, as seen in the sky, with the North Pole at the centre). The reason for this is that this map was devised as an aid for the users of certain instruments which Apian describes in the respective books. On these instruments, sixteen fundamental stars were engraved, and the star map was added to the respective books to help the reader find those stars. Looking closely, one can see that the sixteen stars are designated by numbers on the map, starting with 1 in Canis Maior (for Sirius), and ending with 16 in Orion (for Betelgeuse). This map, therefore, was not, like all its predecessors, designed for mere contemplation or instruction, but for immediate practical use!

One thing in the map, however, appears to deviate from the convention and turns out contrary to the practical intention: there are depicted on it some figures, or groups of figures, that are clearly not part of the traditional set of constellations. Near the North Pole there appear three female figures in front of a woman sitting on something like a chair, or sofa; to their left is a shepherd with a dog and a flock of sheep; below the group of females there are four big camels with a very small baby camel in their middle; and between the shepherd's group and Cygnus, under the right wing of the bird, there appear two stars surrounded in their lower part by a ring of feathers or something the like.

These strange images have puzzled some modern historians. Ernst Zinner wrote: "... some of the constellation figures are of an unusual form" (also, he erroneously took the shepherd's group for Bootes, while in reality it is located in Cepheus; Bootes is plainly visible on the map, below Ursa Maior's tail, having Corona Borealis in front of him);[2] Tadeusz Przypkowski interpreted the map as "une sorte de fantaisie cosmologique" and took the female group replacing Ursa Minor as the Three Graces, in a discussion with Cassiopeia (but Cassiopeia should be located between Cepheus and Perseus!).[3] Only Deborah J. Warner has correctly identified these extraordinary figures as belonging to "Bedouin rather than Ptolemaic tradition", but she did not go into details.[4]

To understand these unorthodox constellation figures correctly we must not only look at Apian's maps, but also read some of his books. Here we find it mentioned definitely that he knew the *Book on the constellations* of 'Azophi', i.e. the Islamic astronomer Abu 'l-Ḥusain al-Ṣūfī (A.D. 903-986). Al-Ṣūfī was Persian by birth, but wrote his works in Arabic. He became known in Europe, from medieval times, in several traditions one of which (in the twelfth century) produced the spelling 'Azophi' for his name (al-Ṣūfī). Dürer added to the northern of his two star maps of 1515, in the four corners of the page, imaginative portraits of four famous astronomers representing the most important nations that had contributed to the science of the stars: Aratus Cilix (as a classical Greek author), M. Manilius Romanus (as a Roman author), Ptolemeus Aegyptius (as an Hellenistic author), and Azophi Arabus (as an Arabic-Islamic author).

Apian followed Dürer in using the form 'Azophi' for al-Ṣūfī. But it must be stated at once that in his knowledge of al-Ṣūfī's work Apian was totally independent of any of the preceding Western traditions on al-Ṣūfī.

Al-Ṣūfī, in his *Book on the constellations* (written around A.D. 964), gave a detailed account of the 48 classical constellations (which the Arabs knew through translations of Greek astronomical works and through pictorial representations on globes and from other sources), complemented by records of star names of indigenous Arabic origin which he took pains to identify astronomically with the respective Ptolemaic stars.

It was from this section on the indigenous Arabic star names in al-Ṣūfī's book that Apian extracted a number of names and mentioned them in some of his own writings, most of them in the chapter on the constellations in his *Astronomicum Caesareum* of 1540.

Comparing the Arabic constellations, or asterisms, on the map of 1533 to those mentioned by name in the *Astronomicum Caesareum*, we find that there is no complete agreement between the two. Thus, the first of the Arabic asterisms on the map of 1533, the group of three females standing in front of a sitting woman, is not found among the Arabic star names in Apian's text. But it can be clearly identified. The Arabs called the seven stars of the Wain (i.e., in Ursa Maior) *banāt na'sh* (hence our Benetnasch), and then extended the same name to the similarly shaped seven stars of Ursa Minor, distinguishing the two as the Greater and the Lesser *banāt na'sh*, respectively. This could be found by Apian directly in al-Ṣūfī's book. Al-Ṣūfī also goes into more detail adding that, in each of the two figures, the four stars are the *na'sh*, and the three stars the *banāt* (i.e., "Daughters"). Here, also, Apian could find the popular Arabic interpretation of the name as "the Daughters of the Bier" (while, in reality, the true meaning of *na'sh*, in this connection, is unknown). So, in place of Ursa Minor, he drew the three *banāt* as three female figures or maidens, while he interpreted the *na'sh* as something to sit on, and added another woman sitting on it.

To the left of the female group, in the place of Cepheus, there is the group of the Shepherd, the Dog, and the Sheep. These are mentioned in the *Astronomicum Caesareum* as *pastor*, *canis*, and *oves*, respectively, and they correspond completely to similar names, assigned to the same region of the sky, by al-Ṣūfī.

Under the female group, there is the formation of four big camels and a baby camel. These too are mentioned in Apian's book, as *quinque dromedarii*, whereas al-Ṣūfī had them separately as four mother camels, and a young camel in their middle seeking protection. As a name, Apian had simply added them up to give him "Five Camels", but he knew al-Ṣūfī's text well and therefore drew them, according to the Arabic conception, as four big (mother) camels and a baby camel in their midst. (The four mother camels are located in $\beta\nu\gamma\xi$ Draconis.)

Thus, out of thirteen Arabic star names mentioned in the text of the *Astronomicum Caesareum* (and a fourteenth in another book), Apian has selected for pictorial representation on the map of 1533 only four (*pastor*, *canes*, *oves*; *quinque dromedarii*), plus the female group (*banāt na'sh*) which is not contained in the text of his book.

There remains the enigmatic figure under the right wing of Cygnus which I cannot identify or explain; it equally escaped interpretation by D. J. Warner. Al-Ṣūfī mentions only two indigenous Arabic asterisms in Cygnus neither of which leads to a similar figure. Under Cepheus, Apian mentions (apart from the Shepherd group) another Arabic asterism, *duae alae* ("The Two Wings"), formed by two stars, and it is not at all improbable that by that enigmatic figure (which is visibly marked by two stars on his map!) he wanted to represent these "Two Wings". On the other hand, Apian's "Two Wings" cannot be identified in al-Ṣūfī's book. There are indeed some additional Arabic star names registered by al-Ṣūfī under Cepheus, but none of these could be translated as "Two Wings". It could, however, be that Apian here (as in several other cases) has misinterpreted the Arab original, or that in his book there is a misprint (perhaps *alae*, "Wings", originally was another word, designating something to which the enigmatic figure under the right wing of Cygnus would better conform).

Attention should also be paid to the figure of Lyra (directly under the "Five Camels" on the map), which is drawn as a big bird with a musical instrument, like a viol, in front of its belly. Here also we have Arabic influence: the star α Lyrae alone, and equally α Lyrae together with some neighbouring stars, were called by the old Arabs *al-nasr al-wāqiʿ* (hence our Vega!), commonly rendered into Latin as *Vultur cadens*, "The Falling Eagle". Consequently, here, the figure of an eagle was drawn (which, basically, was Arabic), to which was added a musical instrument in order to retain the classical idea of Lyra. But this was not Apian's invention. Exactly the same representation of Lyra had been given earlier by Dürer (1515), and it was merely repeated by Apian on his both star maps, the "classical" one (*cf.* Figure 1), and the "extraordinary" map of 1533. Dürer, in his turn, did not understand Lyra in the classical sense of the word, but rather he took the instrument as the contemporary Italian stringed instrument of Renaissance times called *lira*, or more completely *lira da braccio*, which was an instrument of the violin family and a forerunner of the various modern types of violins.[5] Thus he depicted the constellation of Lyra as this extraordinary combination of the old Arabic conception of *Vultur cadens* and the Renaissance Italian *lira da braccio*. Johannes Bayer repeated a similar figure, for Lyra, in his *Uranometria* of 1603, but he drew the musical instrument as a real lyre, not a Renaissance *lira*.

Several of the Arabic star names quoted by Apian (in Latin) from Azophi have later been repeated by others, such as Bayer (1603),[6] Schickard (1623),[7] Caesius (1662),[8] Lach (1796),[9] and Allen (1899),[10] but they never came into general use in modern astronomy.

Among the things quoted by Apian from al-Ṣūfī, there is also an Arabic proverb referring to the little star g (80) Ursae Maioris (mentioned by al-Ṣūfī under Ursa Maior), "I show him [the star] *al-suhā*, and he shows me the Moon". It is quoted in several of Apian's works, in different wording, but he invariably calls the little star by the name *Alcor* or *Alkor* (which is still used today). This is another doubtful case among Apian's Arabic borrowings, for the name of the star in Arabic is *al-suhā* from which nothing like *Alcor* could be derived. Even more, in one instance Apian translates *Alcor* as *parvus eques*, "The Little Rider", which was plainly the popular German name of the little star ("das

Reiterlein") and has no relation whatsoever to *al-suhā* (which could be translated something like "The Forgotten, or Neglected One") or to *Alcor* (except that it designates the same star). In reality, *Alcor* is not a name of its own given to g UMa; moreover, it is simply a corrupt spelling of the Arabic name of the star ε UMa, *al-jaun*, used in the Latin translation of the *Almagest* from the Arabic and spelt, subsequent to a misreading of the Arabic letters, as *alioze* which afterwards degenerated into *aliore* and other odd spellings. *Alcor* has developed out of *aliore*.

Altogether, Apian's knowledge and use of al-Ṣūfī's book appears problematic and raises a number of questions. We see that Apian has quoted Arabic star names from Azophi only under five constellations, at the beginning of the complete series of Ptolemy's (and al-Ṣūfī's) 48 classical constellations. Even in these five constellations, he has not quoted all the names mentioned by al-Ṣūfī in those places, but only a small portion of them. Of the thirteen names mentioned by Apian, only eight can be regarded as correct translations from al-Ṣūfī's Arabic forms, while the others cannot even be recognized in the Arabic text. Among the pictorial representations of Arabic constellations on the map of 1533, there is one (the *banāt na'sh* group) which is not mentioned textually. The use that Apian had made of al-Ṣūfī's work, therefore, was rather limited, and not always correct.

Did Apian himself know Arabic so as to be able to understand al-Ṣūfī's book in the original text? No evidence can be found for such an assumption. On the other hand, at his time there existed no Western translations of al-Ṣūfī's book that could have supplied him with all the details quoted in his *Astronomicum Caesareum* and elsewhere. The only explanation remaining for us in our present state of knowledge, therefore, is to assume that Apian had a helper who was in command of some Arabic and who supplied him with some selected details; but the cooperation, as it seems, did not continue.

Was there available to Apian, in his days and at Ingolstadt, a copy of al-Ṣūfī's book in the Arabic original text? This too is hard to believe. At any rate, at present the libraries of Leipzig and Vienna (where Apian spent some years as a student) and of Munich (the heir to old Ingolstadt) do not possess Arabic copies of al-Ṣūfī's book (a very recent copy at Munich is not relevant in this connection).

Weak though the background may seem, there remains the fact that a number of Arabic names from al-Ṣūfī's book were quoted in some of Apian's works and that some Arabic asterisms were actually depicted on one of Apian's star maps. But still more: in a *Practica* for the year 1532, published in 1531, Apian gives a list of titles and books that he intended to edit and print in the future (he had established a printing-office of his own at Ingolstadt); equally, on 3 July 1532, Emperor Charles v issued him a printing privilege (which afterwards was printed at the beginning of several of his books, such as the *Horoscopion generale* of 1533, the *Instrumentum primi mobilis* of 1534, and the *Astronomicum Caesareum* of 1540) for a long list of titles "reserved" for Apian and his press for a number of years: both in the list of 1531 and in the imperial privilege of 1532, among the numerous titles reserved for Apian, there is also "*liber Azophi Astrologi vetustissimi*" ("*the book of the ancient astronomer Azophi*"). It is hardly possible to imagine that he ever had applied for, and was granted, the imperial privilege of editing a book that was not within reach of his hands.

So we have to assume that Apian really had a copy of al-Ṣūfī's book somewhere in the background, and that he had a person nearby who was able to assist him in reading the Arabic text. But this connection seems to have ended before the project could be carried out. The few quotations from al-Ṣūfī are not always reliable, and they break off very soon. Neither in the remaining twenty years of Apian's lifetime, nor after his death (from the papers he left), was Azophi's book ever published.

Could it be that Apian had realized that his helper's knowledge of Arabic was too weak to complete the entire book of al-Ṣūfī, so that after a while he preferred to dismiss the whole project?

One more detail, but one unrelated to the Arabic implications, should be added at this point. On both his star maps, the "classical" one and the "extraordinary" one, Apian has added some hounds to the figure of Bootes. On the "classical" map, there are three hounds, connected to Bootes's raised left hand by long leashes, whereas on the "extraordinary" map there are two hounds, connected by shorter leashes to Bootes's drooped right hand. This seems to be an early predecessor of the Canes Venatici introduced as a regular constellation about one-and-a-half centuries later by Hevelius, in his (posthumous) *Prodromus astronomiae* of 1690. Apian's hounds are approximately in the right place for those Canes Venatici. But I cannot say which myth, or which source, Apian followed in adding the hounds to the figure of Bootes. Two hounds are also drawn, between the feet of Bootes and connected by leashes to his right hand, in a Florence manuscript apparently of the same age (16th century).[11]

REFERENCES

A detailed study, *Peter Apian and Azophi: Arabische. Sternbilder in Ingolstadt im frühen 16. Jahrhundert*, by the present author was published in 1986 in the *Sitzungsberichte* of the Bayerische Akademie der Wissenschaften at Munich.

1. See, e.g., the plates illustrating the article of S. G. Barton, "Dürer and early star maps", *Sky and telescope*, vi (1947), no. 11, 6-8 and no. 12, 12-13; reprinted separately, as Reprint no. 72, by Flower Astronomical Observatory of the University of Pennsylvania.
2. E. Zinner, *Deutsche und niederländische astronomische Instrumente des 11.-18. Jahrhunderts* (Munich, 1956; 2nd edn, 1967), 171f.
3. T. Przypkowski, "Premières cartes modernes du ciel", *Archives internationales d'histoire des sciences*, xiv (1961), 305-15, esp. p. 310.
4. D. J. Warner, *The sky explored: Celestial cartography 1500-1800* (New York and Amsterdam, 1979), 9.
5. The conclusive hint to the Italian *lira* of Renaissance times came from my esteemed colleague, Prof. Th. Göllner of Munich.
6. J. Bayer, *Uranometria* (Augsburg, 1603).
7. W. Schickard, *Astroscopium* (Tübingen, 1623).
8. Ph. Caesius, *Coelum astronomico-poeticum* (Amsterdam, 1662).
9. F. W. V. Lach, *Beytrag zur Orientalischen Sternkunde* (in: Eichhorn's Allgemeine Bibliothek der biblischen Litteratur, vii.3, 4; Leipzig, 1796).
10. R. H. Allen, *Star-names and their meanings* (New York, 1899); reprinted as *Star names: Their lore and meaning* (New York, 1963).
11. See the reproduction on Plate IV (d) in P. McGurk, *Catalogue of astrological and mythological illuminated manuscripts of the Latin Middle Ages*, iv: *Astrological manuscripts in Italian libraries (other than Rome)* (London, 1966).

XXIV

A NOTE ON STAR NAMES, ESPECIALLY ARABIC, AND THEIR LITERATURE

Compared to the countries of German language, it seems that in the Anglo-American world star names are a much more popular topic. This can be gathered not only from the relatively great number of publications, and published lists, of such names, but even more from the several successively published handlists with pronunciations of star names, which obviously indicate that these names are still much used in the English-speaking world.

One of the major sources for the knowledge, and explanation, of star names was since long the voluminous book by Richard Hinckley Allen: *Star-Names and Their Meanings*, New York, etc., 1899, and which recently was reprinted by the Dover Publications, Inc., New York, 1963, under the slightly modified title: *Star Names. Their Lore and Meaning*.

So some words should occasionally be given on the value and reliability of Allen's book. It seems Allen did not carry out investigations of his own into the field of star names. Moreover he compiled the book from all sorts of (secondary) sources available up to his time, and informations on the celestial nomenclatures of different peoples and civilizations. Regarding Arabic star names, the last author before Allen, who had given detailed and source-bound informations on the names, their history, derivations and meanings, was Ludwig Ideler in his book *Untersuchungen über den Ursprung und die Bedeutung der Sternnamen*, Berlin, 1809. In the following 90 years, as it seems, no further progress in this field was made. So Allen could not but repeat the informations given by Ideler on ancient and Arabic stellar nomenclature and by others on the celestial nomenclatures of other civilizations. His book may be taken as a handbook summing up the state of knowledge arrived at by his time.

As far as the Arabic matter is concerned, his book suffers from three major drawbacks, linguistic and historical. First, Allen had no own knowledge of Arabic. He could rely, for parts of his book, on the help of Professor R.J.H.Gottheil in this matter[1] but this did not save the book from innumerable mistakes in the Arabic names. These mistakes belong to simple grammar[2] as well as to the transliteration of the Arabic words[3]. Second, he introduced, indiscriminately, all the Arabic terms he could gather in his sources, without differentiation as to whether such terms were really used in Arabic texts, or merely conjectured by western authors. Third, in identifying the 'Arabic star names' in western, Latin, use he blindly followed his predecessors and thus passed on a great number of erroneous identifications which those had not been able correctly to establish due to the lack of original source material in their times[4]. So, with regard to the Arabic matter, Allen's book was, and still is, of rather restricted value.

A careful comparison of a selected series of pages containing characteristical mistakes and misprints, in both editions of the book[5], shows that the wording is completely identical in both editions[6], which means that the 1963 edition does not show any progress in the given informations as against the first print of 1899, of which it is a straight reprint.

As Arabic stellar nomenclature, and its western derivations, have been the subject of a series of critical studies and analyses in recent years, so, for the benefit of all those interested in this domain, it might be worth while naming once more the most important of the respective references (which are all written in the German language): Western adaptations of Arabic star names have been studied by P.Kunitzsch: '*Arabische Sternnamen in Europa*', Wiesbaden, 1959. The indigenous Arabic stellar nomenclature was displayed by the same author: '*Untersuchungen zur Sternnomenklatur der Araber*', Wiesbaden, 1961. For the classification of 'Arabic star names' and the stellar nomenclature of the Arabs see also a paper by P.Kunitzsch: 'Arabische Sternnamen – Sternnamen der Araber. Zur Begriffsbestimmung', in *Sudhoffs Archiv* 61 (1977), pp. 105–117. The Ptolemaic stellar nomenclature and terminology, in the Greek-Arabic-Latin tradition, was lately analysed and edited from all available Arabic versions and Gerard of Cremona's Latin translation, again by the same author: *Der Almagest. Die Syntaxis Mathematica des Claudius Ptolemäus in arabisch-lateinischer Überlieferung*, Wiesbaden, 1974 (especially pp. 172–212 for the 48 Ptolemaic constellations, and pp. 217–370 for individual stars).

NOTES

1. Cf. Allen's own words in the introduction, p. xii of the Dover edition.

2. Up to p. 265 (constellation of Lepus) he usually gives wrong genitive constructions, of the type *Al Kursiyy al Jabbār*, or *Al 'Arsh al Jauzah*, instead of the correct *Kursiyy al Jabbār*, or '*Arsh al Jauzā*' (p. 265). In the later part of the book, and in the 'Arabic Index' (pp. 537–547) this sort of grammatical mistake has disappeared.

3. This means that in many cases the diacritical marks in the Arabic words are not correctly put in, or are missed out. (So *Jauzah*, above in note 2, should correctly be spelt *Jauzā*'.) This sort of mistake was mostly retained in the Arabic Index at the end of the book.

4. For an example comprising all three sorts of mistakes, see p. 197 where it is stated that the western name *Azelfafage* for π^1 Cygni is probably ultimately derived from Arabic *Al Ṭhīlf al Faras* ('the Horse's Foot or Track'). In this, the first *Al* is grammatically wrong and should be omitted; second, *Ṭhīlf* should correctly be spelt *Ṭhilf* (in Allen's system, = *ẓilf* in modern transliteration). Third, the whole expression never occurred in any Arabic source; the true origin of *Azelfafage*, moreover, is the Arabic term *al-sulaḥfāh* ('the tortoise'), one of several alternative Arabic designations for the Ptolemaic constellation of Lyra. Hermannus Dalmata, the mediaeval Latin translator of the ultimate source, Abū Ma'shar's *Introductorium maius*, misinterpreted Lyra, in this place, as '*cignus*', and this caused the error of Johann Bayer who picked up the name *Azelfage* from Abū Ma'shar and, in his *Uranometria* of 1603, attributed it to some stars in the constellation of Cygnus. (In Allen's Arabic Index, p. 546, the spelling appears in a corrected form as *Ṭhilf al Faras*.)

5. I am glad to express my thanks to Dr D.W.Dewhirst, Chairman of the Library Committee of the Royal Astronomical Society, for supplying me with Xerox-copies of the specimen pages which I had selected for examination. Copies of the 1899 edition seem to have become very rare in European public libraries, as, for several years, I was not able to obtain such a copy, not even through the international book loan service.

6. Some minor changes refer to the pagination of the 'Introduction' (1963: v–xiv, against 1899: xi–xx. In the list of authors, pp. 555 ff., however, the old numbering has been left unchanged!), and on the verso of p. 563 the author's 'Envoi' has been omitted.

XXV

Zur Namengebung Kairos (al-Qāhir = Mars?)

Anton Spitaler zum 65. Geburtstag

Die Astrologie war bekanntlich ein ständig präsentes und wirksames Element im Leben des Altertums und des Mittelalters, wie im hellenistisch-römisch-westlichen so dann natürlich auch im islamisch-arabischen Kulturkreis. Das Privatleben des Einzelnen konnte davon ebenso berührt oder gar bestimmt werden wie große politische Entscheidungen und historische Entwicklungen. Notwendigerweise finden denn auch Erzählungen astrologischen Einschlags immer wieder Aufnahme in die Berichte zeitgenössischer Chronisten und später folgender Historiker. Es kann nicht wundernehmen, daß hierbei Historisches, Anekdotisches und Abergläubisch-Hintersinniges oft eine nur noch schwer oder gar nicht mehr auflösbare Verbindung eingehen.

Auf eine solche astrologisch gefärbte Anekdote stoßen wir in der Fatimidengeschichte, im Zusammenhang mit der Gründung der neuen Fatimidenresidenz vor den Toren von Fusṭāṭ und deren Benennung als *al-Qāhira*, Kairo.

Die Historiker scheinen sich seit längerem mit einer historisch halbwegs plausiblen Erklärung der Angelegenheit zufrieden gegeben zu haben. Wer dagegen etwas näher mit der astronomisch-astrologischen Fachliteratur vertraut ist, wird sich schwer daran stoßen, daß hier angeblich der Planet Mars den Namen *al-qāhir* „der Siegreiche" trägt.

Das soll uns Anlaß geben, dieser Frage erneut nachzugehen und die Berichte der Quellen speziell im Hinblick auf das *qāhir*-Problem einer kritischen Überprüfung zu unterziehen.

Es fällt von vornherein auf, daß die bekanntesten unter den älteren zuständigen Historikern anläßlich der Behandlung der fatimidischen Eroberung Ägyptens und der Gründung der neuen Residenz die umstrittene *qāhir*-Anekdote nicht aufweisen, vgl. Ibn al-Aṯīr (gest.

1233)[1]) und ʿAlī b. Mūsā Ibn Saʿīd (1214—1274 bzw. 1286)[2]). Ebenso führen auch die ältesten europäischen Fatimidenhistoriker dieses Detail an den entsprechenden Stellen nicht an: É. QUATREMÈRE (1836)[3]) und F. WÜSTENFELD (1881)[4]).

Diejenigen arabischen Autoren, bei denen sich die *qāhir*-Anekdote findet, sind demgegenüber relativ jung: al-Maqrīzī (1364—1442)[5]), Ibn Taġrībirdī (1411—1469)[6]) und as-Suyūṭī (1445—1505)[7]). Noch spätere können hier wohl übergangen werden. Unter diesen drei gibt allein as-Suyūṭī eine Quelle an: das Buch „*as-Sukkardān*" (verf. 757 = 1356) des Ibn Abī Ḥaǧala (1325—1375)[8]), das somit die bisher älteste greifbare Stelle wäre, im ganzen aber freilich auch nicht wesentlich älter als alle übrigen Quellen und nahezu vierhundert Jahre von dem eigentlichen angeblichen Ereignis entfernt.

Die *qāhir*-Anekdote erzählt, der Fatimidenfeldherr Ǧawhar habe nach dem Fall von Fusṭāṭ am 17. Šaʿbān 358 = 6. 7. 969 weisungsgemäß mit der Grundsteinlegung für die Mauer der neu anzulegenden Residenz begonnen. Um der Stadt ein günstiges Schicksal zu sichern, berief er Astrologen, die den günstigsten Augenblick für den Baubeginn ermitteln sollten. Eine unmittelbare und gleichzeitige Benachrichtigung aller Bauleute sollte dadurch gewährleistet werden, daß er außen rings um die vorgesehene Mauer auf Pfosten ein System von Schnüren aufstellen ließ, an denen Glocken befestigt waren. Vermittels der Glocken sollten die Astrologen den Bauleuten den Arbeitsbe-

---

[1]) *al-Kāmil*, Druck Bulak (mit Masʿūdī, *Murūǧ*), VIII 194,—4/3.

[2]) *an-Nuǧūm az-zāhira fī ḥulā ḥaḍrat al-Qāhira* (*al-qism al-ḫāṣṣ bi-l-Qāhira min Kitāb al-muġrib fī ḥulā l-maġrib*), ed. Ḥ. NAṢṢĀR, Kairo: Maṭbaʿat Dār al-Kutub 1970, vor allem 22; 38—45; 101—106.

[3]) *Vie du khalife Moëzz-li-din-Allah*, in: *Journal Asiatique* IIIe série (1836), t. III, 49; 74; 165f.

[4]) *Geschichte der Fatimiden Chalifen* (Abh. d. Gesellsch. d. Wiss. Göttingen 27, 1881), 2. Abtheilung, S. 12; 23f. Auch bei C. H. BECKER, *Beiträge zur Geschichte Ägyptens unter dem Islam*, Straßburg 1902, wird auf das *qāhir*-Problem nicht eingegangen (cf. später seinen *EI*-Artikel „Cairo"!).

[5]) *Ḫiṭaṭ*, Kairo 1324, II 204 = Bulak I 377 [hiernach übersetzt von P. CASANOVA, *Description historique et topographique de l'Egypte*, 4e partie (Mém. ... Inst. Français d'Archéol. Orientale du Caire, t. IV), Kairo 1920, S. 82]; *Ittiʿāẓ al-ḥunafā'*, ed. Ǧ. AŠ-ŠAYYĀL, Kairo 1387 = 1967, 112.

[6]) *an-Nuǧūm az-zāhira fī mulūk Miṣr wa-l-Qāhira*, ed. JUYNBOLL, II (Leiden 1855/56), 415,14—416,13 = ed. Kairo: Dār al-Kutub, IV (1933), 41f.

[7]) *Ḥusn al-muḥāḍara fī ta'rīḫ Miṣr wa-l-Qāhira*, Kairo [1299], I 12,8—19 = ed. M. A. F. IBRĀHĪM, Kairo: Dār Iḥyā' al-Kutub al-ʿArabīya 1387 = 1967, I 25f.

[8]) Cf. BROCKELMANN, *GAL* II 12 (Nr. 17; Werk 2); S II 5.

ginn signalisieren, sobald ein günstiger Sternstand eingetreten war. Unvorhergesehenerweise ließ sich indes plötzlich ein Rabe auf einer der Schnüre nieder und löste dadurch zur Unzeit das entscheidende Glockensignal aus, woraufhin die Arbeiter unverzüglich mit den Bauarbeiten einsetzten. „Da schrien die Astrologen: *al-qāhir* steht im Aszendenten! Aber die Sache nahm ihren Fortgang, und das, was sie angestrebt hatten, ging ihnen verloren. Wie es heißt, stand Mars [*al-mirrīḫ*] bei Beginn der Grundsteinlegung im Aszendenten, indem er *qāhir al-falak* 'Beherrscher des Himmels' war [oder: „d.h. er war *q. al-f.*"], und so nannten sie sie *al-Qāhira*"[9]). Bei Ibn Taġrībirdī und as-Suyūṭī zeigt der Wortlaut gewisse Abweichungen: „Da schrien die Astrologen: Nein, nein! *al-qāhir* steht im Aszendenten! . . . Aber es geschah, daß Mars [*al-mirrīḫ*] im Aszendenten stand, und der wird bei den Astrologen *al-qāhir* genannt". Da habe man erkannt, daß diese Stadt der Türkenherrschaft unterworfen sei[n werde] und daß diese sie notwendigerweise in Besitz nehmen würden. Als al-Mu'izz eintraf, der (vollkommene) Kenntnis der Sterndeutung besessen habe, und hierüber unterrichtet wurde, habe er ihnen darin beigepflichtet, daß die Türken die Oberherrschaft über diese Stadt gewinnen würden, und so habe er deren ursprünglichen Namen abgeändert und sie *al-Qāhira* („die Obsiegende, Siegreiche") genannt[10]). Hier wäre noch ein Zitat aus Ibn Duqmāq[11]) (ca. 1349—1407)[12]) anzufügen, das Ḥ. I. Ḥasan beibringt: „Sie wurde so benannt, da ihre Grundfeste angelegt wurde beim Aufgang eines Sterns, den einer der sieben Weisen beobachtet hatte, die in Ägypten lebten, und zwar war es ein Stern, den man *al-qāhir* nennt"[13]).

Diese Stellen — vor allem die bereits im vorigen Jahrhundert in Drucken zugänglichen *Ḫiṭaṭ* von al-Maqrīzī und Ibn Taġrībirdīs *an-*

---

[9]) al-Maqrīzī, *Ḫiṭaṭ* und *Itti'āẓ* (wie Fn. 5): *fa-ṣāḥa l-munağğimūn: al-qāhir fī ṭ-ṭāli' fa-maḍā ḏālika wa-fātahum mā qaṣadūhu wa-yuqālu inna l-mirrīḫ kāna fī ṭ-ṭāli' 'inda btidā' waḍ' al-asās wa-huwa qāhir al-falak fa-sammawhā l-Qāhira.*

[10]) (Wie Fn. 6 und 7): *fa-ṣāḥa l-munağğimūn lā lā! al-qāhir fī ṭ-ṭāli' . . . fa-waqa'a anna l-mirrīḫ kāna fī ṭ-ṭāli' wa-huwa yusammā 'inda l-munağğimīn al-qāhir . . .*

[11]) Im Text bei Ḥasan *Ibn Dq'q*, Druckfehler für Ibn Duqmāq.

[12]) Cf. Brockelmann, *GAL* II 50; S II 49.

[13]) Ḥasan (wie Fn. 14), S. 528: *fa-qāla Ibn Duqmāq innahā summiyat ka-ḏālika li-anna asāsahā šuqqa 'alā ṭulū' kawkab raṣadahū aḥad al-ḥukamā' as-sab'a allaḏīna kānū bi-diyār Miṣr wa-huwa kawkab yuqālu lahū l-qāhir.* Das Zitat stammt, was Ḥasan nicht vollständig angibt, aus Ibn Duqmāq, *Kitāb al-intiṣār li-wāsiṭat 'aqd al-amṣār*, ed. Vollers, Kairo 1314 = 1893, II 35,—3/2.

Nuǧūm az-zāhira — lieferten die Meinungsgrundlage für eine lange Reihe von orientalistischen Autoritäten, die sich zu diesem Problem äußerten[14]).

Astrologische Maßnahmen gerade im Zusammenhang mit Stadtgründungen sind keine Seltenheit und werden häufig berichtet[15]). Im Prinzip besteht also keine Veranlassung, die *qāhir*-Anekdote von vornherein in Zweifel zu ziehen. Die Kritik der in Anm. 14 genannten Autoren bezieht sich daher zumeist auch nur auf die historischen oder historiographischen Begleitumstände, ohne die technisch-astrologische Seite des Problems überhaupt zu berühren. Das gängige Meinungsspektrum läßt sich am übersichtlichsten so zusammenfassen, wie es Šayyāl (cf. Fn. 5) S. 111 formuliert: Nach den Berichten der Historiker (al-Maqrīzī, *Ḫiṭaṭ* und *Itti'āẓ;* Ibn Taġrībirdī, *Nuǧūm*) erhielt die Stadt entweder bei der Gründung zunächst die Benennung *al-Manṣūrīya*, in Anlehnung an das früher von al-Mu'izz' Vater gegründete Manṣūrīya bei Kairouan, und wurde erst vier Jahre später, bei al-Mu'izz' Einzug (7. Ramaḍān 362 = 11. 6. 973), in *al-Qāhira*

---

[14]) In chronologischer Reihenfolge: P. Ravaisse, *Essai sur l'histoire et sur la topographie du Caire d'après Maḳrîzî* (Mém. Miss. Archéol. Franç. au Caire, t. I, fasc. 3, Paris 1887), S. 420; St. Lane-Pool, *A History of Egypt in the Middle Ages*, London [1901 =] ⁴1925, S. 103; ders., *The Story of Cairo*, London [1902 =] 1924, S. 118 [Zitat aus *History* S. 103, mit vereinfachter Transkription]; C. H. Becker, Artikel „Cairo", *EI*¹ (1913), S. 856; A. R. Zakī, *al-Qāhira*, Kairo: Ḥiǧāzī, I (1934), S. 57; K. A. C. Creswell, *Ta'sīs al-Qāhira*, in: *al-Muqtaṭaf* 85 (Nov. 1934), 306f.; G. Wiet, *L'égypte arabe* (= t. IV von G. Hanotaux, *Histoire de la nation égyptienne*), Paris 1937, S. 185f.; F. Faraǧ, *al-Qāhira*, Kairo: Maṭba'at al-Ma'ārif, III (1946), S. 411; Ḥ. I. Ḥasan, *Ta'rīḫ ad-dawla al-fāṭimīya*, Kairo: Maktabat an-Nahḍa al-Miṣrīya ²1958, S. 528f.; Ǧ. Šayyāl, ed. *Itti'āẓ al-ḥunafā'*, 1967 (cf. Fn. 5), S. 111; Ḥ. Naṣṣār, ed. Ibn Sa'īd, 1970 (cf. Fn. 2), S. 22; O. V. Volkoff, *Le Caire 969—1969. Histoire de la ville des »Mille et une Nuits«* [Public. Inst. Franç. d'Archéol. Orient. du Caire, Bibliothèque Générale, Tome II], Kairo 1971, S. 48; Janet L. Abu-Lughod, *Cairo. 1001 years of the city victorious*, Princeton N.J. 1971, S. 18.

[15]) Vgl. z.B. bezüglich Bagdad, bei H. Suter, *Die Mathematiker und Astronomen der Araber und ihre Werke*, Leipzig 1900, Nr. 2 und 8; J. Vernet, in: *Rev. Inst. Estud. Islam.* Madrid 15 (1970), 92; ders., in: Accad. Nazion. dei Lincei, Atti dei Convegni 13 (Convegno Internazionale 9—15 Aprile 1969: *Oriente e occidente nel medioevo: Filosofia e scienze*), Rom 1971, 752f. (hier auch Stellenangaben für Byzanz und Kairo); M. Ullmann, *Die Natur- und Geheimwissenschaften im Islam*, Leiden-Köln 1972 (Hdb. d. Or., 1. Abt., Erg.-Bd. VI, 2. Abschn.), S. 273; D. Pingree, in: *Journ. Amer. Orient. Soc.* 93 (1973), 37b; al-Mahdīya in Tunesien, bei Ullmann *a.a.O.*; Aleppo, bei F. Rosenthal, *A History of Muslim Historiography*, Leiden ¹1952, S. 98ff., speziell 99¹ = ²1968, S. 110ff., speziell 111³. S. ferner weiter unten al-Mas'ūdī zu Alexandria.

umbenannt[16]). Oder aber die Neugründung erhielt, wie in der *qāhir*-Anekdote bei al-Maqrīzī beschrieben, von Anfang an den Namen *al-Qāhira*.

Daneben gibt es eine einfachere, nichtastrologische, immerhin aber ebenfalls wunschbestimmte, ätiologische Erklärung des Namens al-Qāhira, bei Ibn Saʿīd S. 22 aus al-Bayhaqī[17]), mitangeführt bei Ḥ. I. Ḥasan S. 529, derzufolge die Stadt so benannt sei, damit sie gemäß ihrem Namen künftig ihre Widersacher und Abtrünnigen stets niederzwingen und eine umfassende Herrschaft über die Völker ausüben möge.

Völlig aus dem Rahmen fällt demgegenüber ein Ableitungsversuch bei Farağ S. 411: al-Muʿizz, gelehrt und hochgebildet in Geschichte und Sprachen, habe lange nachgedacht und sich schließlich daran erinnert, was er über die lange Geschichte von Ägypten und Memphis gelesen habe, Memphis — „*qāhirat al-wağhayn*" (die Bezwingerin der beiden Landesteile [d.h. Ober- und Unterägyptens]), und hiernach habe er befohlen, der bisher al-Manṣūrīya genannten Stadt den neuen Namen al-Qāhira zu geben[18]).

Verlassen wir hiermit den Bereich der Deutungen und Meinungen und versuchen wir, den Vorgang noch einmal im Lichte des Tatsachen- und Quellenmaterials zu analysieren.

Dann ergibt sich etwa folgender Ablauf: Die am 17. Šaʿbān 358 = 6. 7. 969 von Ğawhar neugegründete Stadtanlage bei Fusṭāṭ scheint zunächst, in bewußter pietätvoller Angleichung an eine andere fatimidische Stadtgründung in Nordafrika, al-Manṣūrīya genannt worden zu sein[19]). Das Aufkommen des Namens al-Qāhira ist nicht genau fixierbar; nach gewissen Quellenangaben soll diese Benennung vier Jahre später durch al-Muʿizz selbst bei seinem Einzug in die Stadt am 7. Ramaḍān 362 = 11. 6. 973 geprägt worden sein. Faktisch nachgewiesen ist der Name al-Qāhira, soweit bisher bekannt, auf Münzen

---

[16]) Cf. Ravaisse 420²; Lane-Pool, *History* 103¹; Becker, Art. „Cairo" in *EI*¹; Creswell 306f.; Farağ 411; Ḥasan 528.

[17]) Es ist nicht sicher, welcher al-Bayhaqī hier gemeint ist; cf. in diesem Sinne auch Brockelmann, *GAL* I 324; S I 557 (dieser Bayhaqī lebte von 1105—1169).

[18]) In dem *Wörterbuch der ägyptischen Sprache* von A. Erman–H. Grapow finde ich keine Bezeichnung für Memphis, die der von Farağ angeführten Bedeutung entspräche; vgl. lediglich II 130: *mḫꜣ.t-tꜣwj* „die Waage der beiden Länder" (zu *ḫꜣj* „messen").

[19]) Cf. die eindringlichen Parallelen, die Creswell S. 306 zusammengestellt hat!

einmal erst 394 = 1003/4, danach wieder 508 — 524 = 1114—1130, und dann ab 1168 unter den Aijubiden ständig[20]).

Der Name ist nun also vorhanden, aber die Begründung scheint noch zu fehlen. Es wird nicht umsonst sein, daß, ähnlich wie die schon erwähnten älteren Historiker, auch der Geograph Yāqūt (1179—1229) in seinem *Muʿǧam al-buldān* wohl den Namen anführt und die Geschichte der Stadt al-Qāhira kurz umreißt, über die Herkunft des Namens jedoch kein einziges Wort verliert[21]). Ebensowenig enthalten die arabischen Nationallexika hierzu irgendeine einschlägige Angabe[22]).

Als nächste Etappe der Entwicklung des *qāhir*-Mythus stoßen wir auf die ätiologische Erklärung, die Ibn Saʿīd aus al-Bayhaqī kennen will (s.o.). Hierin erhält der Name al-Qāhira eine wunschbestimmte Ausdeutung, die im eigentlichen Grunde rein von den etymologischen Gegebenheiten ausgeht und die Namengebung in Anlehnung an die Bedeutung der Wurzel *qhr* („mächtig sein, bezwingen, obsiegen, usw.“) geschehen läßt. Ein astrologischer Einschlag ist hier noch nicht vorhanden.

Dieser begegnet vielmehr in einer weiteren Etappe, die Ḥ. I. Ḥasan bei Ibn Duqmāq aufgespürt hat (cf. das obige Zitat). Die bewußte

---

20) Lane-Pool, *History* 103¹. Der Name al-Qāhira tritt ferner in Erscheinung um 985 in der Länderbeschreibung von al-Muqaddasī sowie 1046 bzw. danach in der Reisebeschreibung von Nāṣir-i Ḫusraw (gemäß J. L. Abu-Lughod [wie Fn. 14], S. 19); die Stellen sind: *Bibl. Geogr. Arab.*, ed. M. J. de Goeje, Pars III: *Descriptio Imperii Moslemici auctore Al-Moqaddasī*, Leiden ²1906, S. 200,11—15; Ch. Schefer, *Sefer Nameh. Relation du voyage de Nassiri Khosrau*, Paris 1881, Übers. S. 124 und 126 = Text S. 42,9—10 und 43,7—8 (mit einer eigenen, offensichtlich spekulativen Deutung der Namengebung: „... parce qu'il [scil. Mouïzz] s'était rendu maître du pays à main armée“).

21) Ed. Wüstenfeld IV 22.

22) *Lisān al-ʿarab*; *Tāǧ*; *Qāmūs* (s. vv. *qhr* und *mrḫ*). In dem *Arabic-English Lexicon* von Lane heißt es s. v. *qhr* zwar: "[*al-qāhiru* The planet Mars.]" (sic), aber ohne Quellenangabe und in eckige Klammern gestellt, was gemäß Lanes Preface S. XXVI bedeutet, daß es sich dabei um „interwoven additions of my own“ handelt. Da Lane hierfür keine schriftliche Quelle namhaft macht (wie sonst mehrfach bei Sternnamen, z.B. Qazwīnī/Ideler u. dgl.), scheint er die Wortgleichung auf anderem Wege, am wahrscheinlichsten mündlich von einheimischen ägyptischen Gewährsmännern, in Erfahrung gebracht zu haben. Diese ihrerseits dürften *al-qāhir* in der Pseudo-Bedeutung als „Mars“ wiederum nur aus den von mir oben mehrfach zitierten Historikern (al-Maqrīzī, Ibn Taġrībirdī, as-Suyūṭī, eventuell noch jüngere daran sich anschließende) gekannt haben; andere Quellen sind hierfür ja nicht nachweisbar. Die einschlägige Notiz von Lane ist daher in ihrer Aussagekraft eingeschränkt und darf keine allgemeine Gültigkeit beanspruchen.

Stelle ist freilich geeignet, auf Anhieb das größte Mißtrauen zu erwecken. Wohl wäre man geneigt, den Topos „Stadtbenennung nach einem Stern, der bei Baubeginn gerade aufging" als begründet und, im Sinne der Zeit, „vernünftig" hinzunehmen. Stark verunsichert wird der gesamte Bericht aber dadurch, daß der Autor nicht genau anzugeben weiß, um was für einen Stern (Fixstern oder Planeten?) es sich dabei handelte; daß er in sehr unbestimmten, laienhaften Worten davon redet, es habe sich um einen Stern gehandelt, der *al-qāhir* genannt werde; und daß dessen Beobachtung bei Baubeginn durch einen der „sieben Weisen" (*al-ḥukamā' as-sab'a*), die im Lande Ägypten lebten, erfolgt sei. Daß die „sieben Weisen" ein Reflex antik-hellenistischer Tradition sind und selbstverständlich nicht das Geringste mit dem fatimidischen Ägypten zu tun haben, bedarf keines weiteren Wortes[23]). Desgleichen ist aber auch die Notiz über den Stern namens *al-qāhir* Unsinn, denn aus der gesamten bisher erschlossenen astronomisch-astrologischen Fachliteratur arabischer Sprache ist kein Stern dieses Namens bekannt!

In einer weiteren, dritten Etappe endlich finden wir diejenige Ausgestaltung des Mythus, die den Planeten Mars ins Spiel gebracht hat (cf. oben die Zitate aus al-Maqrīzī, Ibn Taġrībirdī und as-Suyūṭī).

Wir haben es also gewissermaßen mit einer Klimax zu tun: Benennung *al-Manṣūrīya* durch Ǧawhar, 969 — Einführung von *al-Qāhira* durch al-Mu'izz selbst, 973 — Aufkommen eines ätiologischen Mythus, und zwar auf etymologischer Grundlage (al-Bayhaqī, 12. Jh.?/ Ibn Sa'īd) — Aufkommen eines astrologischen Mythus, noch stark unbestimmt (Ibn Duqmāq) — Spezialisierung des astrologischen Mythus auf Mars (al-Maqrīzī, Ibn Taġrībirdī, as-Suyūṭī).

Mit diesem Faktenmaterial, wie es uns Texte und archäologische Befunde (Münzen) liefern, haben wir uns also auseinanderzusetzen.

Unterziehen wir daher vor allem erst einmal die speziell auf Mars bezüglichen Stellen einer genaueren Kontrolle.

Bei al-Maqrīzī wird zuerst die aufgeregte Reaktion der Astrologen auf den vorzeitigen Baubeginn vermerkt; sie rufen aus: „*al-qāhir* steht im Aszendenten". Das klingt nach einer korrekten astrologischen Angabe: Der Aszendent, d.h. der in dem kritischen Augenblick am Osthorizont aufgehende Punkt der Ekliptik bzw. das dazugehörige 'Haus' (*aṭ-ṭāli'*), ist als Fachterminus bekannt[24]); was dagegen mit

---

[23]) Cf. z.B. *Fihrist* ed. Flügel 245, 13; übersetzt bei F. Rosenthal, *Das Fortleben der Antike im Islam*, Zürich-Stuttgart 1965, S. 64.

[24]) Ravaisse 420 und Lane-Pool, *History* 103 erfassen den Ausdruck richtig; Creswell 306 unten schreibt dagegen befremdlicherweise: *wa-kāna* ...

al-qāhir hier gemeint ist, geht aus dem Satz nicht hervor. In seltsam kontrastierendem, unpersönlich referierendem Ton folgt ein Stück weiter, anscheinend ganz unabhängig, die chronistische Notiz: „Wie es heißt, stand Mars [*al-mirrīḫ*, korrekt mit seinem bekannten klassischen Namen genannt!] bei Beginn der Grundsteinlegung im Aszendenten"; hieran schließt sich ein mit *wa-huwa* eingeleitetes Satzglied, das als *ḥāl*-Satz auf „Mars" bezogen („indem er *qāhir al-falak* 'Beherrscher des Himmels' war") oder dem vorangehenden Satz „Mars stand im Aszendenten" explikativ angehängt sein kann („d.h. er war [dadurch, infolgedessen, in diesem Augenblick, in dieser Stellung o.ä.] *qāhir al-falak* 'Beherrscher des Himmels'"). Dieser chronistische Satz scheint der eigentliche Träger der Mars-Information zu sein. Er ist weitgehend fachlich korrekt: es wird klar ausgesprochen, daß bei Baubeginn der Planet *al-mirrīḫ* = Mars im Aszendenten stand, d.h. also im Osten aufging[25]). Das Wort *qāhir* wird

---

al-mirrīḫ fī l-awǧ „Mars ... kulminierte, stand ganz hoch am Himmel" (wobei der Ausdruck *awǧ* selbst in diesem Sinne ebenfalls unfachmännisch schlecht gebraucht ist; richtig eigentlich = „Apogäum, Aphel").

[25]) Nach den neuesten Computerberechnungen in den Tafeln von B. TUCKERMAN, *Planetary, Lunar, and Solar Positions A.D. 2 to A.D. 1649* [Mem. Amer. Philos. Soc., Vol. 59], Philadelphia 1964, S. 502, stand Mars am 4. Juli 969 bei 65°.47 Länge (und am 14. 7. 969 bei 72°.18), so daß für den 6. 7. 969 der Wert 66°.81 zu interpolieren ist. An denselben Daten stand die Sonne bei 106°.94 bzw. 116°.50, d.h. am 6. 7. 969 interpoliert bei 108°.85. Mars lief am 6. 7. 969 also um rd. 42° vor der Sonne her; in Zeit ausgedrückt, ging er grob $2^h\ 50^{min}$ vor der Sonne auf. Nehmen wir für die Sonne am 6. 7. 969 eine mögliche Aufgangszeit von ca. 4.10 Uhr an, so wäre Mars demnach in der Nacht vom 6. zum 7. Juli 969 etwa gegen 1.20 Uhr rechnerisch aufgegangen und je nach Klarheit des Osthimmels wenig später sichtbar geworden. Die Behauptungen der *qāhir*-Anekdote stehen also nicht in Widerspruch zu den astronomischen Fakten. Fügen wir hinzu, daß nach TUCKERMAN *a.a.O.* auch die auffälligen Planeten Saturn (interpoliert ca. 22°.79 vor Mars) und Jupiter (ca. 11°.13 hinter Mars) in jener Nacht in derselben Himmelsgegend standen und ca. 1½ Std. vor bzw. 45 Min. nach Mars aufgingen. Innerhalb eines Zeitraums von ca. 2¼ Std. waren also drei bekannte und gut sichtbare helle Planeten aufgegangen, von denen jener „Unglücksrabe" bei seinem verirrten Niedersitzen auf den Drähten angeblich ausgerechnet den mittleren, den Mars, erwischte. Ein hyperkritischer Interpret könnte aus dem Planetenstand herauslesen, daß die Astrologen vermutlich die Aufgänge der beiden „Unglücksplaneten" Saturn und Mars vorübergehen lassen wollten, um sodann beim Aufgang des „Glücksplaneten" Jupiter — der wenig später, rd. 2 Std. vor Sonnenaufgang erfolgt wäre — das Signal für den Baubeginn zu geben. Diese Absicht hätte es gewesen sein können, die der Rabe vereitelt hat. — Über die realistischen Aspekte des nächtlichen Umherfliegens eines Raben mögen übrigens Ornithologen ein Urteil fällen!

dabei nicht als Namensform für den Planeten benutzt! Vielmehr dient der mit *wa-huwa* angehängte Satz dazu, die astrologische Funktion des in einem gegebenen Augenblick im Aszendenten aufgehenden Planeten zu umschreiben: in der bewußten Position ist er nämlich *qāhir al-falak* „Beherrscher des Himmels". Aus der astrologischen Fachliteratur ist mir dieser Ausdruck selbst zwar nicht bekannt[26]), doch drängt sich die Parallele auf zu so klassischen astrologischen Begriffen wie *haylāǧ* (latinisiert *hyleg* etc.) „Herr des Lebens", *kadḫudā* (οἰκοδεσπότης; latinisiert *alcochoden* etc.) „Herr des Hauses", und ganz besonders gewissen mit -κράτωρ zusammengesetzten Bildungen (cf. χρονοκράτωρ, ἐνιαυτοκράτωρ, ὁριοκράτωρ), die bestimmte astrologische Mächtigkeiten von Planeten in besonderen Positionen bezeichnen. Bei der Vielfalt und Verworrenheit astrologischer Deute- und Bezugssysteme vermag ich nicht auf Anhieb anzugeben, welcher klassischen Funktion der *qāhir al-falak* hier genau entspricht. Nicht auszuschließen wäre überdies, daß dabei ein eventuell zugrunde liegender echter astrologischer Terminus womöglich laienhaft-literarisch abgewandelt wurde, um den erstrebten Bezug auf den existierenden Stadtnamen al-Qāhira zu erzielen. Klar ist auf jeden Fall, daß an unserer Stelle mit *qāhir al-falak* nicht Mars, als ein ganz bestimmter Planet, gemeint ist, sondern vielmehr die Funktion eines Planeten im Aszendenten zu einem gegebenen Zeitpunkt, wobei es im Prinzip gleichgültig ist, welcher der sieben klassischen „Planeten" gerade dort steht.

Das dürfte der sachliche Kern der Mars-Anekdote sein, wobei, wie wir soeben gesehen haben, der funktionale Begriff *qāhir al-falak* grundsätzlich von dem namentlich erwähnten Objekt, dem Planeten *al-mirrīḫ* = Mars, abgehoben ist.

Sternkundlichen Nichtfachleuten mag das alles etwas undurchsichtig erschienen sein, und so kommt es nahezu zwangsläufig sekundär zu der sachlich falschen Gleichsetzung *al-qāhir* = Mars. Das scheint sich bereits bei al-Maqrīzī anzudeuten, wenn er in der Anekdote die Astrologen ausrufen läßt: *al-qāhir fī ṭ-ṭāli'* „*al-qāhir* steht im Aszendenten!", während in der referierenden chronistischen Notiz bei ihm der Planet *al-mirrīḫ* = Mars und die astrologische Funktion als *qāhir al-falak* „Beherrscher des Himmels" noch getreulich voneinander getrennt sind.

---

[26]) Cf. die Aufzählung einschlägiger Wörterverzeichnisse und Begriffssammlungen bei P. Kunitzsch in: *ZDMG* 124 (1974), 164f.

Noch weiter schreitet der falsche Gleichsetzungsprozeß bei den auf al-Maqrīzī folgenden späteren Schriftstellern Ibn Taġrībirdī und as-Suyūṭī voran, wo es im Endstadium regelrecht heißt: *wa-huwa yusammā ʿinda l-munaǧǧimīn al-qāhir* „und der [d.h. *al-mirrīḫ* = Mars] wird bei den Astrologen *al-qāhir* genannt".

Davon kann indes überhaupt keine Rede sein. Die astronomisch-astrologische Fachliteratur arabischer Sprache kennt weder einen Planeten noch einen Fixstern des Namens *al-qāhir*. Auch die Lexika (cf. Fn. 22) verzeichnen unter *qhr* bzw. *mrḫ* nichts davon, daß etwa Mars (auch) *al-qāhir* hieße. Wohl sind aus der arabischen Fachliteratur für die fünf (eigentlichen) Planeten neben ihren klassisch-arabischen Namen noch gewisse jüngere, eventuell hellenistisch beeinflußte Alternativnamen bekannt: Mars hat neben seinem klassischen Namen *al-mirrīḫ* so noch die Bezeichnung *al-aḥmar* „der Rote"[27]) bzw. *al-kawkab al-aḥmar* „der rote Stern"[28]). Auch von dieser Seite her ist also kein *al-qāhir* für „Mars" nachweisbar[29]).

In diesem Zusammenhang müssen wir noch auf zwei abgelegene Stellen eingehen, die die Existenz von *al-qāhir* zu belegen scheinen — aber nur vermeintlich, wie man sogleich sehen wird.

---

[27]) Cf. P. Kunitzsch in: *Zeitschr. f. deutsche Sprache* 25 (1969), 172f.; ders. in: *Wolfram-Studien*, hg. v. W. Schröder, Bd. 2, Berlin 1974, S. 29 mit Anm. 74.

[28]) *WKAS* 443a 12f.

[29]) Dasselbe gilt für gewisse Nebenbereiche, in denen die Planeten ebenfalls funktional verwertet werden: In der Alchimie dienen ihre Namen als Decknamen für bestimmte Metalle; cf. bei A. Siggel, *Decknamen in der arabischen alchemistischen Literatur*, Berlin: Akademie-Verlag 1951, S. 10—12 und das alphabetische Verzeichnis S. 33ff., wo zwar alle klassischen wie auch jene Alternativnamen und sogar die persischen und originalgriechischen Planetennamen vorkommen, wo jedoch *al-qāhir* nicht verzeichnet ist. Auch in den Schriften des *Corpus Ǧābirianum*, die ismāʿīlitischen Geistes sind und der schiitischen Fatimidendynastie innerlich näherstehen könnten, tritt, soweit ersichtlich, *al-qāhir* als Bezeichnung des Mars nicht auf (cf. P. Kraus, *Jābir b. Ḥayyān* I 74 [Bibliogr. Nr. 294] und die Indices zu beiden Bänden seines Werkes). Dazu kommt endlich noch ein Fall aus der Fixsternsphäre: Der rötliche Stern α Scorpii trug in der Antike den Namen Antares (᾿Αντάρης), d.h. etwa „dem Mars [an Farbe] gleich". Hierzu gibt es zwei etymologisierende Übersetzungen in der arabischen Fachliteratur: *ḫilqat al-mirrīḫ* „Beschaffenheit des Mars" (Naṣīr ad-Dīn aṭ-Ṭūsī) und *ṭabīʿat al-mirrīḫ* „Natur, Charakter des Mars" (al-Bīrūnī), zuerst nachgewiesen von C. A. Nallino, cf. zuletzt P. Kunitzsch, *Der Almagest. Die Syntaxis Mathematica des Claudius Ptolemäus in arabisch-lateinischer Überlieferung*, Wiesbaden 1974, S. 292. Alle diese Sonderfälle erbringen also ebenfalls keinerlei Hinweis auf die Bezeichnung des Mars als *al-qāhir*.

In seinem Aufsatz „The Pseudoplanetary Nodes of the Moon's Orbit in Hindu and Islamic Iconographies" (1938)[30]) erwähnt W. HARTNER u.a. auch eine Tigris-Brücke in Ǧazīrat ibn ʿUmar, die 559 h = 1164 von einem gewissen Ǧamāl ad- Dīn erbaut wurde und welche in Wandnischen die Bilder der Tierkreiszeichen und ihrer Hausplaneten zeigt. Die Brücke ist von E. HERZFELD ausführlich beschrieben worden, der dabei im Steinbock gelesen haben will: *al-qāhir šaraf al-ǧady* (was richtig lauten müßte: ... *šarafuhū al-ǧady* „Mars, seine 'Erhöhung' ist der Steinbock")[31]). HARTNER selbst bezweifelt die Richtigkeit von HERZFELDS Lesung, wie ich meine, völlig zu recht. Auf der Abb. 2 bei HARTNER, in Feld 3, ist die Schrift des Feldes leider nicht mehr klar zu erkennen. Es darf aber mit größter Sicherheit angenommen werden, daß in HERZFELDS ohnehin inkorrekter Lesung dieses Schriftfeldes das erste Wort auf keinen Fall *al-qāhir* war.

Noch eindeutiger ist eine zweite Stelle aufzuklären: H. PIGEONNEAU bespricht in seinem Buch *Le cycle de la croisade et la famille de Bouillon*, St. Cloud 1877, die Namen gewisser „heidnischer" (d.h. pseudo-islamischer) Gottheiten, die in den altfranzösischen Kreuzzugsepen zuweilen den „Heiden" (= Muslimen) zugeordnet wurden. Eine derselben heißt *Cahu* (übrigens auch bei Wolfram von Eschenbach, *Willehalm* 358, 13: *Kâhûn*). Diesen pseudo-islamischen Götternamen *Cahu* will PIGEONNEAU S. 107 und 112 irrig herleiten aus *el-kahir* [d.h. also *al-qāhir*], le dieu Mars, le démon des batailles. Dem ist die sehr viel glücklichere, zutreffende Ableitung entgegenzustellen, die H. GRÉGOIRE geliefert hat: nämlich von der Gottheit *Chaos* bei Lactanz und Dracontius[32]).

Somit wäre immerhin geklärt, daß *al-Qāhira* „Kairo" jedenfalls nicht die „Stadt des Mars" ist.

Natürlich kann der Name der Stadt nicht aus dem Nichts heraus entstanden sein. Es müßte möglich sein, Assoziationen ausfindig zu machen, die an dieser Namengebung (bzw. später an der Entstehung der ätiologischen *qāhir*-Anekdote) mitgewirkt haben. Ich will im folgenden einige Beobachtungen mitteilen, die dazu vielleicht einiges beizutragen vermögen.

---

[30]) In: *Ars Islamica* vol. V, part 2, Ann Arbor 1938; nachgedruckt in: W. HARTNER, *Oriens-Occidens, Festschrift zum 60. Geburtstag*, Hildesheim 1968, S. 349ff.

[31]) *A.a.O.* S. 114 = *Oriens-Occidens* S. 351, mit Anm. 4.

[32]) H. GRÉGOIRE, „Des Dieux Cahu, Baraton, Tervagant ...", in: *Annuaire de l'Institut de Philologie et d'Histoire Orientales et Slaves*, t. VII, New York 1939—44, S. 451—472.

Der Wortstamm *qhr* erscheint im Umkreis von al-Muʿizz mehrmals in unübersehbarer Akzentuierung. In — wie mir scheint sehr glücklicher — Assoziation weist Ḥ. I. Ḥasan S. 529 auf den missionarischen Satz hin, den al-Muʿizz seinem Feldherrn Ǧawhar mit auf den Weg gegeben haben soll, als er ihn zur Eroberung Ägyptens ausschickte: *wa-la-tadḫulanna fī ḫarābāt Ibn Ṭūlūn wa-tabnī madīnatan taqharu d-dunyā* ,,Du sollst einziehen in Ibn Ṭūlūns Trümmerfeld und eine Stadt erbauen, die die Welt beherrscht!"

Ebenso prononciert erscheint eine Form des Stammes *qhr* in dem wegen seiner blasphemischen Übertriebenheit häufig zitierten Anfangsvers eines Lobgedichtes von Ibn Hāniʾ al-Andalusī (gest. 973) auf al-Muʿizz:

mā šiʾta lā mā šāʾati l-aqdāru *
fa-ḥkum fa-anta l-wāḥidu l-qahhāru

,,Solange Du magst — nein, solange die Geschicke es wollen *
herrsche, denn Du bist der Einzig[artig]e, der
mächtige Bezwinger"[33]).

Der Gedanke erscheint daher verlockend, daß tatsächlich, wie bei einigen Historikern vermeldet, die bis dahin al-Manṣūrīya genannte Stadt bei al-Muʿizz' Einzug 973 den Namen *al-Qāhira* erhielt, mochte er von al-Muʿizz selbst vorgebracht worden sein oder von jemandem aus seiner engeren Umgebung. Das Wort jedenfalls scheint, wie die beiden obigen Zitate beweisen, um ihn herum ,,in der Luft gelegen" zu haben. In diesem Fall stünde es dann auch außerhalb allen Zweifels, daß *al-Qāhira* von Anbeginn an nichts anderes war als ein echter politisch-dynastischer Segenswunsch für die neue Residenz: sie sollte ,,die Mächtige, Siegreiche" sein für al-Muʿizz und alle Nachfahren seines Hauses[34]).

---

[33]) *Dīwān* 62 [gemäß Notiz des Herausgebers von *Ḥusn al-muḥāḍara*, also wohl nach einer Kairiner Ausgabe, vielleicht Kairo 1276, wie in *GAL* I 91 notiert] = ed. K. al-Bustānī, Beirut: Maktabat Ṣādir 1952, 76, v. 1 [= ders., ib. 1384 = 1964, S. 146. Freundliche Auskunft von Herrn Prof. Dr. A. Spitaler, München]. Zitate z.B. bei Ibn al-Aṯīr (wie Fn. 1), VIII 205; as-Suyūṭī, *Ḥusn al-muḥāḍara*, Kairo [1299] II 12,13 = ed. Ibrāhīm I 599,—2. Die Epitheta *al-wāḥid* und *al-qahhār* gehören nicht nur jedes für sich zu den *asmāʾ al-ḥusnā*, sie kommen im Koran sogar sechsmal gemeinsam in dieser Kombination vor! Cf. M. F. ʿAbdalbāqī, *al-Muʿǧam al-mufahras li-alfāẓ al-qurʾān al-karīm*, Kairo: Maṭābiʿ aš-Šaʿb 1378, S. 554a.

[34]) In diesem Sinne faßt auch Wilhelm von Tyrus, der berühmte Chronist der Kreuzzüge und hervorragende westliche Historiker des Vorderen Orients, um 1182 die Namengebung Kairos auf: ... *constructoque Cahere,*

Daß sich dieser ursprüngliche Gedanke später verloren haben sollte, so daß Gelegenheit für die Bildung der ätiologischen astrologischen Anekdote gegeben war, braucht kein bloßer Zufall zu sein: Die überwiegend sunnitische Geschichtsschreibung könnte wenig Interesse daran gehabt haben, am Ruhme des aus der Reihe fallenden ägyptischen Schiitenstaates weiterzuspinnen und den *qahr* seiner Hauptstadt fortzurühmen.

Die faktische Entstehung und Bedeutung des Namens *al-Qāhira*, Kairo könnte insoweit als aufgeklärt gelten.

Man hätte sich dann noch einige Gedanken zu machen, wie oder warum es später zur Entstehung der ätiologischen astrologischen Anekdote gekommen sein mag.

Auch hierzu bieten sich mehrere Ansätze zu fruchtbarer Assoziation.

Angenommen, die tatsächliche Entstehung des Namens war im Schatten der Vergangenheit versunken, so boten sich einschlägigen Autoren verschiedene Möglichkeiten der Anknüpfung.

Eine sehr naheliegende wäre die etymologische Spekulation[35]). Eine Probe davon sahen wir oben in dem Bericht des Ibn Saʿīd nach al-Bayhaqī. Freilich läßt sich nicht eindeutig entscheiden, ob al-Bayhaqī noch die tatsächliche Ursprungsgeschichte des Namens im Sinn hatte (cf. deren vorangegangene Darstellung) oder ob er lediglich von dem nackten Namen ausging und ihn logisch-spekulativ ausdeutete. Im letzteren Falle wäre seine Erklärung als freundlich und positiv gegenüber der Stadt Kairo bzw. dem Lande Ägypten bzw. der Fatimidendynastie oder gar dem Schiitentum allgemein zu bewerten.

Nicht minder naheliegend ist die astrologische Deutungsmöglichkeit, zumal, wie im vorliegenden Falle, im Hinblick auf Stadt-

---

quod interpretatur vincens; *eo quod domini eorum praecipui et summi principis omnia vincentis futurum erat domicilium* (*Recueil des Historiens des Croisades, Historiens Occidentaux*, Bd. 1, Teil 2 [Paris 1844], S. 917,1—2 = Migne, *P. L.* 201, S. 767 A) = altfranzös. Übersetzung: ... *et fist apeler celui* [ed. Paris: *le leu*] *Cahere qui vaut autant comme* veinqanz, *porce que c'estoit sieges de celui qui tot veincroit* (*Rec. Hist. Crois.*, a.a.O., frz. Z. 12—13 = *Guillaume de Tyr et ses continuateurs*, ed. P. Paris, Bd. 2 [Paris 1880], S. 283 unten).

[35]) Zum „etymologischen Denken" vgl. auch E. Wagner, *Abū Nuwās. Eine Studie zur arabischen Literatur der frühen ʿAbbāsidenzeit*, Wiesbaden 1965, S. 380 mit Anm. 6; demselben Phänomen in der mittelalterlichen Literatur des Westens widmet E. R. Curtius den Exkurs XIV: *Etymologie als Denkform*, in seinem Buch *Europäische Literatur und lateinisches Mittelalter*, Bern/München [5]1965, S. 486—490.

gründungen (cf. oben und Fn. 15). Hier bieten sich mehrere auslösende Faktoren an, die bei der Entstehung der astrologischen Anekdote mitgewirkt haben können: Erstens ganz allgemein das Prinzip astrologischer Konsultation und Planung bei Stadtgründungen (s.o.). Sodann der Ruf al-Mu'izz', selbst stark astrologiegläubig gewesen zu sein[36], und zu einem gewissen Teil wahrscheinlich auch die astronomisch-astrologischen Interessen und Unternehmungen und die geheimnisumwitterten Lebensumstände seines Enkels al-Ḥākim (reg. 996—1021).

Einen weiteren ganz besonders wichtigen Einzelzug hat, soweit ich sehe, zuerst Creswell[37]) ins Blickfeld der Forschung gehoben: al-Mas'ūdī (gest. 956, also 13 bzw. 17 Jahre vor dem Ereignis von Kairo!) berichtet eine nahezu ähnliche Anekdote von der Gründung Alexandrias[38]). Alexander wollte ebenfalls einen günstigen Moment für den Baubeginn der neu zu errichtenden Stadt wählen. Zur Benachrichtigung der Arbeiter ließ er genau so ein System von glöckchenbehängten Schnüren aufbauen, wie es die *qāhir*-Anekdote für Kairo erzählt. Bevor der entscheidende Augenblick herannaht, fällt Alexander jedoch in einen Schlaf. Währenddessen setzt sich ein Rabe (!) auf die Schnüre, das Signal erfolgt, der Bau beginnt zur Unzeit. Die genannten Züge finden sich später in fast derselben Form in der *qāhir*-Anekdote wieder. Ein Element freilich fehlt in al-Mas'ūdīs Bericht: Bei ihm hat die Erzählung zwar ebenfalls astrologischen Charakter, doch wird bei ihm kein Stern namhaft gemacht, der dabei im Aszendenten (oder in einer anderen bedeutungsvollen Position) gestanden hätte.

Betrachten wir die verschiedenen Erscheinungsformen unserer *qāhir*-Anekdote im Lichte der genannten Assoziationsansätze, so ergeben sich folgende Beobachtungen.

Die allgemeinste Form hat die astrologische Anekdote bei Ibn Duqmāq (s.o.); sie ist hier fast nur auf den klassischen Ausgangstypus „Aufgang eines Sterns bei Beginn der Bauarbeiten" reduziert. Ange-

---

36) Cf. Zakī I 57; Ḥasan 529; Ibn Taġrībirdī und as-Suyūṭī (wie Fn. 6—7).

37) Wie Fn. 14, S. 307; hiernach auch angeführt von Ḥ. I. Ḥasan S. 529 [falsch auf al-Maqrīzī bezogen statt auf al-Mas'ūdī!] und Ǧ. Šayyāl S. 111[1]. Ebenfalls bei J. L. Abu-Lughod (wie Fn. 14), S. 18, Anm. 13, hier jedoch nach einem späteren Werk von Creswell: *The Muslim Architecture of Egypt: Ikhshīds and Fāṭimids, A.D. 939—1171*, Oxford 1952, S. 23.

38) *Les Prairies d'Or*, Trad. franç. de Barbier de Meynard et Pavet de Courteille, revue et corrigée par Charles Pellat, t. II (Paris 1965), S. 314 (chap. XXXII, § 829) = alte Ausgabe, Paris 1861ff., II 423—425.

reichert ist sie durch Einführung der „Sieben Weisen" als Vollzieher der astrologischen Himmelsbeobachtung (ein Element, das bei den übrigen Autoren fehlt. Dort ist immer nur generell von *al-munaǧǧimūn* „den Astrologen" die Rede) und durch Beigabe des angeblichen Namens des bewußten aufgehenden Sterns, *al-qāhir*. Bei letzterem muß offen bleiben, ob wir darin einen Reflex der *qāhir*-Anekdote bei al-Maqrīzī oder gar im *Sukkardān* des Ibn Abī Ḥaǧala (cf. o.) zu erblicken haben oder aber eine von jener unabhängige etymologische Spekulation, die einfach von dem existierenden Namen al-Qāhira ausging.

Die größte astrologisch-sachliche Korrektheit zeigt die Anekdote in der bei al-Maqrīzī überlieferten Form, wo zumindest in einem chronistisch-referierenden Satz noch zuverlässig zwischen dem aufgehenden Planeten (*al-mirrīḫ* = Mars) und dessen astrologischer Funktion in dieser Stellung (als *qāhir al-falak* „Beherrscher des Himmels") unterschieden wird.

Freilich wird im rein anekdotischen Teil auch bei al-Maqrīzī schon die dann später bei Ibn Taġrībirdī und as-Suyūṭī explicite vollzogene falsche Umwandlung des Terminus *al-qāhir* in einen regelrechten Eigennamen des Planeten Mars sichtbar. Sie ist natürlich zu erklären aus der astrologisch-fachlichen Unkenntnis der Autoren. Ursprünglich — wie noch in jenem chronistischen Maqrīzī-Satz erkennbar — hatte es also wohl bedeuten sollen, daß die Stadt ihren Namen von der beherrschenden Stellung und Funktion des gerade aufgehenden Planeten erhalten habe. In der verderbten Endfassung dagegen ist ihre Benennung abgeleitet von einem in Wirklichkeit nicht existierenden vermeintlichen Namen des Planeten Mars.

Es bleibt zu überlegen, wie weit das nicht nur — oder nicht so sehr — astrologischer Unverstand als vielmehr bewußte, ja vielleicht böse Absicht der Autoren war. Im Gegensatz zu dem neutralen Ton der Anekdote bei Ibn Duqmāq oder gar dem positiv-freundlichen Tenor der etymologischen Deduktion bei al-Bayhaqī/Ibn Saʿīd wird der Fassung bei den drei anderen Autoren nämlich eine offensichtlich stark anti-fatimidische (-schiitische?, -ägyptische?, -kairiner?, eventuell auch anti-türkische?) Tendenz gegeben.

Mars ist, mit Saturn, in der Astrologie einer der beiden *naḥsān* „Unglücksplaneten". Ihn bei der Gründung im Aszendenten zu haben und danach benannt zu werden, wäre für die Stadt also ein äußerst schlechtes Omen. Daß ein Rabe den vorzeitigen Baubeginn auslöst, ist sicherlich ebenfalls kein Zufall: Nicht nur wir reden im Volksbrauch vom „Unglücksraben", auch in der islamischen Welt genoß dieser schwarze Vogel keinen guten Ruf[39]). Darüber hinaus sprechen die Autoren ihre schwarzen Gedanken im Zusammenhang mit dieser Namengebungsanekdote auch ganz unverhüllt aus: *wa-qtaḍā naẓaruhum annahā lā tazālu taḥta l-qahr* ,,Ihre [d.h. der Astrologen] Ausdeutung [des Gestirnstandes] führte zu dem zwingenden Schluß, daß sie [d.h. die Stadt Kairo] immerfort unter (Fremd-?, Gewalt-?) Herrschaft stehen wird" (al-Maqrīzī, *Ḫiṭaṭ*); *Itti'āẓ* dafür am Ende *taḥta ḥukm al-atrāk* ,,unter der Türkenherrschaft". Im Hinblick auf Türkenherrschaft interpretieren auch Ibn Taġrībirdī und as-Suyūṭī die astrologische Anekdote[40]).

Hier offenbart sich vielleicht ein weiteres Motiv, das diese Autoren bewegt haben mochte: im Gewande eines gängigen astrologischen Topos eine bittere pessimistische Klage über die seit Beginn der Mamelukenzeit — wenn man will, auch sogar schon seit den Aijubiden — über Kairo und Ägypten lastende türkische Fremdherrschaft anzubringen[41]). Die Armen konnten nicht wissen, daß sie damit

---

[39]) Cf. A. von Kremer, *Culturgeschichte des Orients unter den Chalifen*, II (Wien 1877), 252 (Beispiele aus *Aġānī* und *Ḥamāsa*).

[40]) Der Planet Mars weist nicht in jedem Falle eindeutig auf die Türken hin; die Angaben der einschlägigen Autoren differieren hier erheblich! So werden bei al-Bīrūnī im astrologisch-geographischen Bezugssystem dem Mars vielmehr zugeordnet ,,das 3. Klima" bzw. an benannten Ländern aš-Ša'm, ar-Rūm, aṣ-Ṣaqlab und ,,the North Western countries" (*man kāna fī-mā bayna l-maġrib wa-š-šamāl*). Die Türken (*at-turk*; Wright dafür befremdlicherweise "the country of the Teviks"!) dagegen sind bei ihm dem Jupiter unterstellt; cf. al-Bīrūnī, *Kitāb at-tafhīm li-awā'il ṣinā'at at-tanǧīm*, ed. u. übers. R. R. Wright, London 1934, S. 241 (§ 402ff.) und 242 (§ 407f.). Andererseits lesen wir bei 'Alī ibn Abī r-Riǧāl, *Kitāb al-bāri' fī aḥkām an-nuǧūm*, latein. Übers. von Aegidius de Tebaldis: *Liber completus in iudiciis astrorum*, Druck Venedig 1485, S. 150ra (lib. VIII, cap. 36): *Mars secundum dicta romanorum habet septimum clima terram turchie terram promissionis et romaniam. Et secundum dicta indorum habet terram alexandrie et terram promissionis.* Nach Abū Ma'šar, *De magnis coniunctionibus*, Augsburg 1489, lib. IV a.E., wiederum zählt *terra thurcorum*, im Zeichen Leo, zum 4. Klima und untersteht somit der Sonne. Bei Durchsicht weiterer Quellen kämen noch mancherlei andere derartige Angaben hinzu, cf. z.B. E. Honigmann, *Die sieben Klimata* ..., Heidelberg 1929, S. 141f., wo bei Abū Ma'šar, *De magn. coni.* II,4 (nach ,,persischer Lehre", im Unterschied zu derjenigen der Griechen [*romanorum*]) und bei al-Kindī als Erkennungsname für das 3. Klima, das dem Mars unterstellt ist, *thurthia* und *at-Turk*, also die Türken erscheinen.

[41]) In diesem Zusammenhang ist auf ein Kapitel *Fī sabab kawn ahl Miṣr aḏillā' yaḥmilūna ḍ-ḍaym* (,,Warum die Bewohner Ägyptens [so] unterwürfig, verächtlich sind und Unrecht hinnehmen") hinzuweisen, das as-Suyūṭī eigens seinem *Ḥusn al-muḥāḍara* eingefügt hat (wie Fn. 7, Druck 1299, II 181—183 = ed. Ibrahim II 336—339).

gleichzeitig auch eine Prophezeiung über Ägyptens Schicksal während der nächsten fünf Jahrhunderte aussprachen!

Soweit glaube ich, in der Aufklärung und Interpretation der *qāhir*-Anekdote vordringen zu können. Ich habe mich dabei auf die bislang gedruckt vorliegenden Quellentexte gestützt, die freilich alle erst aus dem 14.—15. Jahrhundert stammen, aus sehr später Zeit also. In Zukunft sind auch die bis jetzt noch unerschlossen in Handschriften ruhenden viel älteren Quellentexte zur Fatimidengeschichte[42]) daraufhin zu untersuchen, ob sich darin Vorstufen oder Zwischenglieder finden, welche die Bildung der astrologischen *qāhir*-Anekdote begünstigt oder veranlaßt haben könnten.

Im übrigen ist die *qāhir*-Anekdote ein weiteres kleines Beispiel für die Verwendung klischeehafter Topoi in der islamisch-arabischen Geschichtsschreibung, ein Phänomen, auf das in jüngster Zeit A. Noth mehrfach in dankenswerter Weise aufmerksam gemacht hat[43]).

---

[42]) Vgl. die Quellenaufzählungen bei Becker (wie Fn. 4); E. Blochet in: *Revue de l'Orient Latin* VI (Paris 1908); A. R. Zaki, *A Bibliography of the Literature of the City of Cairo* (*Marāǧiʿ ta'rīḫ al-Qāhira munḏu inšā'ihā ilā l-yawm*) [engl.-arab.], Le Caire 1964. Gewisse aufgrund ihrer Titel von mir durchgesehene Werke enthalten kein einschlägiges Material zu unserem Problem, so z.B. A. R. Guest, *The Foundation of Fustat and the Khittahs of that Town*, in: *Journ. of the Royal Asiatic Society* 1907, 49ff.; und Šaraf ad-Dīn Yaḥyā Ibn al-Ǧīʿān, *at-Tuḥfa as-sanīya bi-asmā' al-bilād al-miṣrīya*, Kairo: al-Ahlīya 1898.

[43]) Cf. seine aufschlußreichen Aufsätze in ZDMG 118 (1968), 274—296, und *Der Islam* 47 (1971), 168—199, sowie letzthin sein Buch *Quellenkritische Studien zu Themen, Formen und Tendenzen frühislamischer Geschichtsüberlieferung*. Teil I: *Themen und Formen*, Bonn 1973 [Bonner Orientalist. Studien, N.S. Bd. 25].

XXVI

Der Sternhimmel in den „Dichterischen Vergleichen der Andalus-Araber"

Die vor einigen Jahren bekannt gewordene Anthologie dichterischer Vergleiche aus der spanisch-arabischen Welt von Muḥammad b. al-Ḥasan Ibn al-Kattānī (gest. um 420 h = 1029) hat in kurzer Zeit großes Interesse gefunden, wurde bereits zweimal ediert[1] und liegt auch in einer vollständigen deutschen Übersetzung, von W. HOENERBACH, vor[2]. Der Reichtum der darin vorgeführten Bilder ist überwältigend. Trotz mancher Unklarheiten im Text und einzelner Fehlgriffe in der Übersetzung ergeben sich darin doch soviele Fakten, daß HOENERBACHS These erhärtet erscheint, daß auch in der spanisch-arabischen Poesie weitgehend Motive und dichterisch-sprachliches Material aus der klassischen, ostarabischen Dichtung weiter verwendet wurden. (Wie weit vielleicht die bewußte Auswahl durch Ibn al-Kattānī an diesem Gesamtbild beteiligt ist, wäre eine andere Frage.) Aber dies ist letzten Endes keine isolierte Erscheinung. Auch von der wissenschaftlichen Fachliteratur her ist ja bekannt, daß die westlichen Araber mit den Erzeugnissen ihrer östlichen Landsleute und Glaubensbrüder wohl vertraut waren: teils verbrachten sie selbst Studienaufenthalte im Osten, teils lagen ihnen östliche Schriften vor, die sie genau kannten und mit denen sie arbeiteten, und viele davon wurden dann auch im Rahmen der großen Übersetzungswelle ins Lateinische übertragen und so an das Abendland vermittelt.

Unter den Vergleichsmotiven der arabischen Dichter nehmen Himmelsphänomene eine bevorzugte Stellung ein. Da in H manche einschlägigen Anspielungen und Namen nicht richtig erfaßt sind und da die Bespre-

[1] Hg. v. IḤSĀN ʿABBĀS. Beirut 1966 (im folgenden zitiert als A¹); hg. v. ʿA. M. I. ḤASANEIN. Kiel 1969 (zitiert als A²).

[2] W. HOENERBACH: *Dichterische Vergleiche der Andalus-Araber*. I—II. Bonn 1973 (zitiert als H). Längere Besprechungen dazu liegen u.a. vor von M. ULLMANN in: WO 9 (1977), 104—124 (zitiert als U), und von F. VIRÉ in: REI 43 (1975), 183—192 (zitiert als V). Einzelne Textstellen zitiere ich nach folgendem Schema: 1,6,1 [10], wobei die erste Ziffer das „Kapitel" (*bāb*), die zweite die laufende Nummer des Zitats innerhalb des Kapitels und die dritte den Vers innerhalb des jeweiligen Zitats bezeichnet; in eckigen Klammern folgt die Seitenzahl bei H.

chungen (U, V) naturgemäß nur Einzelnes herausgreifen konnten, soll dieser Themenkreis hier im Zusammenhang besprochen werden.

Allgemeine Vergleiche der Dichter mit „Stern(en)" sind so zahlreich und leicht verständlich, daß sie im Rahmen einer spezielleren Analyse nicht im einzelnen behandelt zu werden brauchen.

Am auffälligsten sind die Stellen, an denen spezielle (Fix-)Stern- und Planetennamen für Vergleiche gebraucht werden, wobei oft auf die „Höhe" (Polnähe, meist: *al-farqad*[*ān*], je einmal auch *nuǧūm al-quṭb* und *as-suhā*), gelegentlich auf die Farbe (weiß: Jupiter, rot: Mars, *al-farqad* = β und/oder γ Ursae Minoris), oder auf astrologische Eigenschaften („gute" und „böse", bei den Planeten, usw.) angespielt wird. Gebäude, oder Gekreuzigte, ragen hoch hinauf bis zu den Sternen, (funkelnde) Lanzenspitzen gleichen Sternen, Jagdtiere (Hunde, Vögel) schießen wie Sterne (oder Sternschnuppen) auf ihre Beute.

In den insgesamt 670 Zitaten der Anthologie, die auf 66 thematisch geordnete Kapitel verteilt sind, habe ich an 50 Stellen die Namen von 15 Fixsternen bzw. Sternbildern und 3 Planeten ermittelt. 21 verschiedene Dichter sind es, die derartige Vergleiche anstellen, die meisten davon mehrfach (in verschiedenen Gedichten). In vielen Fällen werden auch zwei (oder mehrere) verglichene Sterne innerhalb desselben Stückes zueinander in Beziehung gesetzt.

Im einzelnen treten folgende Gestirnnamen auf[3]:

al-asad, das Tierkreiszeichen Löwe. 9,7,5 [61]: in altarabischer Tradition dehnt sich der 'Löwe', als Himmelsbild, weit über das klassische Sternbild hinaus und umfaßt acht — oder, unter Hinzunahme von *al-ġafr*, gar neun — Mondstationen statt einer „Norm" von $2\frac{1}{3}$ (cf. K, p. 23ff.); der *naw'* („[Morgen-]Aufgang") des Löwen ist also nicht exakt zu erfassen (cf. die Vergleichstabelle bei Ch. Pellat in: *Arabica* 2 [1955], 32f., Stationen VII—XIV bzw. XV. Ihre Morgenaufgänge beanspruchen einen Zeitraum von mehr als drei Monaten). Die Erklärung von U, p. 116,

[3] Wichtigstes Hilfsmittel zur Identifizierung arabischer (Fix-)Sternnamen ist die Monographie von P. Kunitzsch: *Untersuchungen zur Sternnomenklatur der Araber*. Wiesbaden 1961 (im folgenden zitiert als K + lfd. Nr.). H hat sich leider nur auf die seit langem überholte Darstellung von L. Ideler (von 1809!) und auf eine Spezialarbeit von O. J. Tallgren (sic, nicht Tellgren!) gestützt (cf. H, p. 7); auch V beruft sich für die Sternnamenidentifizierung auf eine oft spekulative, völlig unzulängliche Arbeit (A. Benhamouda, 1951; von mir in meinem Buch *Arabische Sternnamen in Europa*. Wiesbaden 1959, häufig stark kritisiert! — Unter diesem quellengeschichtlichen Mangel leidet übrigens ebenfalls der mißglückte sternkundliche Teil in Virés Artikel "Kalb" in *EI*². Die von Viré auch angezogene Arbeit von C. Pellat über die *anwā'* reicht sachlich nicht für sämtliche Sternnamen aus). Für die Planetennamen vgl. unten Anm. 15 und 17.

240

ist demnach nur bedingt stichhaltig; es überwiegt das poetische Bild des wasserspeienden Löwen am Bassin, das hier nur locker und unbestimmt mit dem *naw'* des Löwen am Himmel in Vergleich gestellt wird. In H wäre besser zu übersetzen: „[Das Naß bei] sein[em] (Morgen-)Aufgang enthebt dich ...; kennst du einen reichlicheres Naß spendenden *naw'* ([Morgen-]Aufgang) als den des Löwen?" — 14,4,2 [94]: Warum hier gerade 'Löwe' und 'Widder' aus dem Tierkreis der die Pokale weiterreichenden Hände ausgenommen bleiben, erkenne ich nicht.

banāt naʿš 1,16,2 [15]: hier speziell *banāt naʿš al-kubrā*, die sieben Sterne des 'Großen Wagens', eines Teilstücks des 'Großen Bären' (K 55). Der kleine, nur schwer sichtbare Stern *as-suhā* (s.u.) steht neben der mittleren der drei *banāt* = dem mittleren der drei Deichselsterne des Wagens bzw. dem mittleren der drei Schwanzsterne des Bären, ζ Ursae Maioris.

ḏāt al-ḫiḍāb 1,19,2 [16f.]: Wie bereits von U, p. 115 richtig angemerkt, ist dieser Ausdruck sehr wahrscheinlich eine poetische Variante zu dem arabischen Gestirnnamen *al-kaff al-ḫaḍīb*, womit die fünf hellen Kassiopeia-Sterne βαγδε Cassiopeiae bezeichnet werden (K 136a/b). Statt H „die hennafingrigen (Plejaden)" wäre also zu übersetzen: „die fünf Kassiopeia-Sterne [βαγδε] sind ...". Der Anblick dieser Sterne am Himmel (bei uns häufig volkstümlich als großes W gedeutet) deckt sich sehr gut mit der weiteren Ausmalung des Bildes als „(gezacktes) Myrtenblatt", sofern derartigen Vergleichen überhaupt ein realer Wirklichkeitsgehalt innewohnt. Im übrigen gilt den Arabern diese „hennagefärbte Hand" als eine der beiden ausgestreckten Hände der *Ṯurayyā*, der Plejaden (K 136–137; 306); aber es handelt sich dabei um ein eigenes Himmelsobjekt, nicht etwa um die Plejaden selbst[4]. — Auf denselben Zusammenhang zwischen *aṯ-ṯurayyā* und ihrer hennagefärbten Hand, *al-kaff al-ḫaḍīb*, spielt offenbar auch 41,6,4 [170] an, wo es um die gelbliche Färbung der Beine eines Jagdfalken geht. — Das Bild der beiden Hände der *Ṯurayyā* (*al-kaff al-ḫaḍīb* „die hennagefärbte Hand", K 306, 136a, und *al-kaff al-ǧaḏmā'* „die verstümmelte Hand", K 306, 137) liegt ferner noch zugrunde in 1,10,1 [12], wo dem Betrachter auch die gesunde Hand (von den beiden) wie eine kranke (offensichtlich Anspielung auf *al-ǧaḏmā'*) erscheint (im Gedicht: *aydī ṯ-ṯurayyā*, poetischer Plural statt

[4] Daneben gibt es in der Dichtung auch viele Stellen, welche die Plejaden selbst mit einer "Hand" (*kaff, rāḥa, yad*) vergleichen, cf. al-ʿAskarī: *Dīwān al-maʿānī*. Kairo 1352 H. = 1933, I, 336,12—13; ʿAlī b. Ẓāfir al-Azdī al-Miṣrī: *Ġarā'ib at-tanbīhāt*. Kairo 1971, 20,6; 25,13; 36,8; 36,11; 37,5—6; 37,8—9; 37,11—12; 37, ult.; 40,9; 40,12; 40, ult.; an-Nuwayrī: *Nihāyat al-arab*. Kairo: Dār al-Kutub 1347 H. = 1929ff., I, 67,19—20; 68,2; aš-Šarīf ar-Raḍī: *Dīwān*. Beirut 1961, I 612,10. Auch hierselbst: 14,7,6 [94].

des sachlich vorgegebenen Duals); zur Gesamtschau des Bildes s.u. bei *al-ǧawzā'* 1,10,1.

al-farqadān (Dual, poetisch häufig auch im Plural *al-farāqid*), die beiden Sterne βγ Ursae Minoris (K 96): 1,12,7 [13] (cf. U, p. 114, wo statt 1,12,17 richtig 1,12,7 zu lesen ist). — 10,13,8 [70] (im Text ... *wa-farqadi*, der Singular ebenfalls lediglich aus poetischem Zwang); H's Übersetzung „Bär“ ist zwar im Hinblick auf die Sache zu frei, erfaßt indes mit geradezu RÜCKERT'scher Ausdruckskraft für deutsche Leser das Wesentliche der Aussage: der „Bär“ zur Umschreibung eines Gestirns ganz hoch (im Norden)[5]. — 41,9,4 [171] (Vergleich von hoch in den Lüften kreisenden Jagdfalken). Der poetische Plural ist von H nicht angemessen erfaßt; es wäre etwa zu übersetzen: „... sagten wir: die [beiden] Sterne *al-farqadān* [βγ Ursae Minoris; ein deutscher Name existiert nicht]! Und wie sie stürzten, sagten wir: die [beiden] Sterne *al-farqadān* im Untergang!“[6] — 43,1,2 [173]; besser zu übersetzen: „es [d.h. ein Pferd] wird

[5] Ein Gedicht im selben Metrum (*ṭawīl*), mit demselben Reim und mit einem fast völlig gleichlautenden *'aǧz* steht im Diwan von as-Sarī ar-Raffā'. Kairo 1355 (= 1937), S. 98,14: *wa-'ahdī bi-hā miṯla l-farāqidi tantaḍī * ḏawā'ibahā mā bayna nasrin wa-farqadī* (dagegen al-Ḥasan b. Ḥassān bei Ibn al-Kattānī, A[1] S. 78,11—12 = A[2] S. 42,11: *wa-'ulliyatun tud'ā l-munīfa ka-annamā * ḏawā'ibuhā nīṭat bi-nasrin wa-farqadī*). Beide Dichter lebten etwa gleichzeitig: ar-Raffā', gest. kurz nach 360 h = 970/971, im Osten (Aleppo und Bagdad); al-Ḥasan b. Ḥassān am Hofe 'Abdarraḥmāns III. (reg. 912—961) in Córdoba. Gegenseitige Bekanntschaft und Entlehnung sind also ausgeschlossen. So wird am Beispiel dieser beiden so weit auseinanderliegenden, voneinander völlig unabhängigen Dichter einmal schlagartig deutlich, wie schematisch-abstrakt zum Beispiel die Sternnamen in der Poesie verwendet werden: "konkret" ist nur die angestrebte Assoziation zu den Sternen. Die Wahl der betreffenden Namen ist von den poetischen Gegebenheiten bestimmt: Versmaß und *topos* (Gebäude, die zu den Sternen emporreichen). Ein unmittelbarer Wirklichkeitsbezug besteht nicht. Die Dichter verfügen, je nach Bildungsgrad, über einen Vorrat von Wörtern und Namen, der nach poetisch-technischem Bedarf eingesetzt wird. (Vgl. einen ähnlichen Fall bei G. VON GRUNEBAUM: *Kritik und Dichtkunst*. Wiesbaden 1955, S. 121, Fn. 117, Ibn Mayyāda und Ḥuṭay'a.) Noch befremdlicher ist das Vorgehen eines unbekannten (persischen) Dichters, der in einem Memorialvers die sieben Planeten aufzählt und dabei für "Saturn" aus metrischem Zwang mangels eines passenden dazugehörigen Namens ersatzweise *farqad* einsetzt (s. W. EILERS, in: *Der Orient in der Forschung. Festschrift Otto Spies*. Wiesbaden 1967, S. 133).

[6] Hier wäre der Dichter für schlechte Anwendung des Vergleichs zu tadeln. Im Gegensatz zu anderen Stellen, wo einfach z.B. Gebäude empor bis an die *farqadān* = βγ Ursae Minoris reichen (d.h. also, wo der Vergleich sich auf die Stellung, die Höhe bezieht und wo ohne Schaden auch ein poetischer Plural oder Singular statt des vorgegebenen Duals eingesetzt werden können), werden hier die Jagdfalken, *aš-šawāhīn* (Plural), die in größerer Anzahl als

geleitet durch etwas von der Art der [beiden] Sterne *al-farqadān*", in Anspielung auf das bekannte Motiv von der Wüstenorientierung nach den Sternen (*ihtidā'*), wobei hier die Sterne βγ Ursae Minoris, in der Nähe des Himmelspols, gewissermaßen den Polarstern vertreten, der die Nordrichtung anzeigt[7]. — 50,4,3 [189]; wieder poetischer Singular, etwa: „... aufgehängt am hängenden [d.h. in der Höhe schwebenden] Gestirn *al-farqadān*". — 61,4,1 [219], Farbvergleich: (silberner) Dirham — der weiße Jupiter, (goldener) Dinar — die (rötlichen) Sterne *al-farqadān*, ebenfalls wieder aus poetischen Gründen im Singular. Der Farbvergleich ist sachlich zutreffend: β und γ Ursae Minoris gehören — zufällig beide — der Spektralklasse K an und zeigen orangerötliche Färbung. In diesem Fall scheint dem Vergleich tatsächliche Naturbeobachtung zugrunde zu liegen, und nicht bloße literarische Tradition[8].

al-ǧawzā' nimmt in der Reihe der Tierkreiszeichen die Stelle der Zwillinge ein, ist aber als eigentliches arabisches Sternbild im Orion lokalisiert (cf. K p. 22). In Ibn al-Kattānīs meisten Zitaten ist, durch zusätzliche Namenelemente gesichert, eindeutig „Orion" zu verstehen. 1,10, Überschrift und v.1 [12]: *arǧul al-ǧawzā'* „die Füße des Orion",

zwei bei der Jagd eingesetzt sind, mit den *farāqid* verglichen, d.h. der Vergleich schließt die Zahl ein. Dabei läßt der Dichter außer Betracht, daß *al-farqadān* ein Begriff ist, der nur ein paarweise existierendes Objekt bezeichnet, und daß der Plural *al-farāqid* lediglich poetische Lizenz ist. In seiner Vorstellung wechselt er zu einem echten Plural über, als gäbe es viele *farāqid*. Damit fällt er aus dem System und zerstört gewissermaßen sein Vergleichsobjekt, verdirbt die Pointe. — Die Zweisamkeit von *al-farqadān* ist geradezu sprichwörtlich und hat ihre modellhafte Ausprägung in einem viel zitierten Vers gefunden: *wa-kullu aḫin mufāriquhū aḫūhū * la-ʿamru abīka illā l-farqadānī* (einschließlich einer partiellen Paronomasie: *fāraqa — farqadān*), cf. Belege bei A. Fischer/W. Bräunlich: *Schawāhid-Indices*. Leipzig 1934, S. 264a s.v. *al-farqadāni*; Herr Prof. A. Spitaler, München, zeigte mir aus seinen Sammlungen zwei Blätter (!) mit weiteren Fundstellen. Anspielungen auf dasselbe Motiv, die enge Verbundenheit der zwei *farqadān*, ebenfalls noch in weiteren Versen: aš-Šarīf ar-Raḍī: *Dīwān*. Beirut 1961, I 612,11—12 (Reim *sāmiʿū, ṭawīl*; u.a. auch in *Nihāya* [Dār al-Kutub] I 68, pu.-ult.; Ibn Manẓūr: *Niṯār al-azhār*. Konstantinopel 1298 H. = 1880, 118,1); anonym (Reim *al-firāqī, ḫafīf. Nihāya* [Dār al-Kutub] I 69,2—3); Ibn Harma: *Dīwān*. Ed. Naǧaf 1969, S. 114 (Nr. 89,2) = ed. Damaskus 1969, S. 119 (Nr. 47,2; auch in al-Marzūqī: *Azmina*. Ḥaidarābād 1332 H. = 1913, II 233,—2).

[7] Cf. auch Ibn Qutayba: *Anwā'*. Ḥaidarābād 1375 H. = 1956, 147,4—10 (teilweise besserer Wortlaut [Z. 8: *li-annahā lā taǧību*] bei al-Marzūqī: *Azmina* II 372).

[8] Andere Dichter haben diese Beobachtung nicht gemacht und reden stattdessen stereotyp vom "weißen Glanz" [*bayāḍ*] der beiden Sterne *al-farqadān*, s. ar-Rāʿī bei Ibn Qutayba: *Anwā'* 147,10 = al-Marzūqī: *Azmina* II 372,13.

Anspielung auf den bekannten Sternnamen *riǧl al-ǧawzā'*, β Orionis (K 251a), bzw. *riǧlā al-ǧawzā'*, κβ Orionis (K 251b). Die Deutung als „Orion“ wird durch den astronomischen Befund noch erhärtet, daß der Mond zwar „über“ dem Orion erscheint (vgl. die fünfte Mondstation, *al-haqʿa*, $\lambda\varphi^{1,2}$ Orionis, im „Kopf“ des Orion), aber höchstens „in“ oder „unter“ den Zwillingen; soweit, um „über“ den Zwillingen zu stehen, weicht er nicht nördlich von der Ekliptik ab. Die tragenden Elemente des Sternvergleichs sind also traditionell vorgegeben: Orion und seine Füße, die Plejaden und ihre Hände (cf. o. zu *ḏāt al-ḫiḍāb*). Eine Wirklichkeitsbeobachtung am Himmel kann dabei kaum vorliegen. Zwar vermag Orion den Eindruck zu erwecken, er laufe den Plejaden nach, zumal wenn diese sich dem Untergang zuneigen (cf. einen entsprechenden Vers von Ibn al-Muʿtazz, Reim *saqaṭā, basīṭ*: *Dīwān*. Ed. Lewin. III. Istanbul 1945, 165,4 (Nr. 265); auch bei Ibn Abī ʿAwn: *Tašbīhāt*. London 1950, 9 ult.; al-ʿAskarī: *Dīwān al-maʿānī* I 337,10 [zitiert H 12, Fn. 25, wo wiederum statt „die Zwillinge“ zu lesen ist „Orion“]). Doch die beiden „Hände“ der Plejaden, um die es hier geht, stehen so weit nordwestlich (βαγδε Cassiopeiae) und südwestlich (λαγδνμ Ceti) ab, daß sie kaum mit Orion unmittelbar in einem Bild zusammengesehen werden können. Es scheint sich auch hier wieder um eine rein literarische Konzeption aufgrund der bekannten vorgegebenen Nomenklatur zu handeln, das ganze vermutlich zum topos „quälende, lange Nacht, die nicht vergehen will“ gehörig: sogar die gesunde Hand der Plejaden wirkt wie eine (oder: wie die) kranke, die Füße des Orion kommen nicht vom Fleck. H's Kommentar in Fn. 25 gewinnt damit andere Aspekte. — 1,11,2 [12]: *al-ǧawzā'* = Orion hat ein Schwert, *sayf*, umgegürtet; cf. den Sternnamen *sayf al-ǧabbār*, $c\vartheta^{1,2}\iota$ Orionis, K 266. — 1,12,6 [13], *kawākibu l-ǧawzā'i*, ohne nähere Ausmalung; aber auch so, für sich, als das charakteristische altarabische Sternbild Orion aufzufassen. — 10,3,6 [64]: als „Orion“ gesichert durch die zusätzliche Nennung von *nuǧūm an-naẓm*, s.d.; auch in Fn. 226, im Sarī-Zitat aus *Nihāya*. Nachdruck Kairo 1383/1963, I, 407 (= ed. Dār al-Kutub, I 392,5), ist „Orion“ zu verstehen, dort zusätzlich abgesichert durch die Anspielung auf den Sternnamen *minṭaqat* bzw. *niṭāq al-ǧawzā'*, K 162 und 204. In derselben Fußnote ist übrigens auch das Zitat vom „allerhöchsten Glück“ aus *Nihāya* I, 407 (= ed. Dār al-Kutub I 392,16) verkannt, lies: „Sein Ort ist [so hoch] wie [das Gestirn] *saʿd as-suʿūd* [= βξ Aquarii und c^1 Capricorni, die 24. Mondstation, siehe K 257,10]“, wobei für Dichter wie Leser allerdings eine etymologische Assoziation zu *saʿd* = „Glück“ sicherlich mitschwingt. — 50,6,4 [190], *al-ǧawzā'* = Orion in Verbindung mit *aṯ-ṯurayyā*, wie in 1,10,1. Desgleichen ist in Fn. 45 auf S. 15, Zitat aus *Maʿānī* I,337 (Z. 13), statt „sein Zwillingsgestirn“ richtig zu lesen: „Der Sirius [sic, nicht Syrius, und *aš-šiʿrā*, nicht *aš-šuʿrā*!] leuchtete, während [der mit ihm zusammengehörige] Orion ...“ (die Vorstellung von Sirius, dem Hundsstern, als Hund des Orion war auch den Arabern bekannt, cf. K 139—140. Der betreffende Dichter, Ibn al-Muʿtazz, lebte spät genug, um diese, vermutlich auf der griechischen Tradition beruhende, Vorstellung kennenlernen zu können).

al-ḥamal, das Tierkreiszeichen Widder. 10,10,12 [68]: Im Widder steht die Sonne zu Frühlingsbeginn, also etwa vom 20. März bis 20. April, wenn erfahrungsgemäß die Luft rein und klar ist, nicht von Sommerglast oder Winterdunst beeinträchtigt. — 14,4,2 [94]: s. oben *al-asad*.

an-nasr 10,13,8 [70]: Von den beiden arabischen „Adler“-Gestirnen ist, da hier auf „Höhe“ abgezielt wird, vermutlich der weiter nördlich, auf *al-farqadān* zu, stehende *an-nasr al-wāqiʿ* gemeint (K 195a); vgl. im übrigen oben, Fn. 5.

nuǧūm an-naẓm 10,3,6 [64]: Sternname für die drei Sterne des Oriongürtels, δεζ Orionis (K 202a). Der zweite Halbvers ist von H mißverstanden, da er *an-naẓm* nicht als Sternnamen erkannte; zu lesen ist: „..., wie die Gürtelsterne [*nuǧūm an-naẓm*, δεζ Orionis] in ihrem Orion passend eingeordnet sind“.

as-simāk, altarabischer Name für α Virginis, Spica (*as-simāk al-aʿzal*, K 269) und für α Bootis, Arcturus (*as-simāk ar-rāmiḥ*, K 270); ohne die unterscheidenden Beiwörter ist nicht in jedem Fall klar zu erkennen, welcher dieser beiden Sterne im Text gemeint ist. Eindeutig ist 13,20,2 [91]: α Virginis, Spica. — Auch in 11,2,2 [71] muß Spica gemeint sein, da es sich um den im Tierkreis stehenden Stern dieses Namens handelt, was nur auf Spica, nicht aber auf den weiter nördlichen Arctur (α Bootis) zuträfe[9]. — In 11,7,6 [73] hat die arabische Handschrift laut Apparat A^2 *as-simāki*, was die richtige Lesart sein dürfte (von H zu Recht beibehalten, nur leider nicht passend übertragen; von A^1 [S. 83, ohne Notiz im Apparat!] und A^2 unnötig in *as-samā'i* emendiert), denn in der Fachliteratur wird der *naw'*, „(Morgen-)Aufgang“ von *as-simāk* (in diesem Falle = *as-simāk al-aʿzal*, α Virginis, die 14. Mondstation) ausdrücklich als regenreich (*ġazīr*) bezeichnet[10]. — In 43,1,2 [173] bleibt die Entscheidung offen, welcher der beiden *simāk* gemeint sein könnte (ganz falsch jedoch H: „die Wacht der ‘Fische’“, wobei es für *raʿy* statt „Wacht“ ebenfalls besser „Beobachtung“ heißen sollte, denn man

[9] V, p. 186, Nr. 7 (“En fait, on ne peut deviner à laquelle des deux pense le poète”) hat in diesem Fall also nicht recht.

[10] Cf. u.a. Ibn Qutayba: *Anwā'* 64 ult. f.; al-Marzūqī: *Azmina* I 310 ult.; al-Qazwīnī: *Kosmographie*. Ed. F. Wüstenfeld. 1: *ʿAǧā'ib al-maḫlūqāt*. Göttingen 1849, 47,11 ff.

„beobachtet“ den Stern, nach dem man sich orientieren will, behält ihn im Auge). In Fortführung des *ihtidā'*-Bildes (cf. o. sub *al-farqadān*, z. St.) ist also zu lesen: „und ersetzt die Beobachtung (*raʿy*) der Spica bzw. des Arctur durch sein brennendes Herz“ (das ihm den Weg besser zu zeigen weiß als der Stern). — Der Vergleich der beiden Planeten Jupiter (weiß) und Mars (rot) (wobei der metaphorische Sinn von v. 2 nicht ganz klar ist; vielleicht Gleichnis für Gläser mit weißem und rotem Wein?) mit zwei Lanzen (*rumḥāni*, v. 3—4) in 1,15,3—4 [15] hat nichts mit dem (Fix-)Stern *as-simāk ar-rāmiḥ* zu tun; es ist ein Vergleich Lanzen(spitzen)/ Sterne, der auch sonst auftritt: cf. 45,2,1 [178] und 45,6,1 [179] (wo allerdings für H's „Stern(e)“ im Text *šuhubun* bzw. Sing. *šihābun* „Sternschnuppe(n)“ steht); 45,5,1 [178] (vielfacher Vergleich: eine größere Zahl Lanzen wie ein Tierkreis [*burūǧ*], mit Sternen [*kawākibun*] als Spitzen, denen die Herzen der Übeltäter als Ziele [*manāzilu*; als himmelskundlicher terminus: „Mondstationen“, H schlecht „Häuser“] dienen); 1,16,1 [15] (Saturn durchsticht, wie eine Lanze, den Morgen); 41,7,6 [171] (ein Jagdfalke stürzt sich auf die Beute, als „bewerfe, erlege“ er sie [*ramāhā*, ein Ausdruck, der auch auf eine Lanze, einen Speer anwendbar ist] mit dem Planeten Saturn); s. auch *WKAS* 444 b 35.

aš-šiʿrā al-ʿabūr, Sirius (K 289a). 1,19,3 [17] (H *aš-šuʿrā*, Syrius, statt *aš-šiʿrā*, Sirius; so übrigens auch p. 15, Fn. 45; p. 17, Fn. 57); H's Übertragung „der große Sirius“ ist unnötig aufgetrieben, es reicht völlig zu sagen: “und (der) Sirius ...“.

as-suhā, „das Reiterlein“, Fl. 80 (*g*) Ursae Maioris (K 271). 1,16,2 [15]: der kleine, nur schwach sichtbare Stern, der sich gleichsam dicht neben seinem größerem Nachbarn ζ Ursae Maioris zu verstecken scheint vor dem rasenden Saturn; auch hier wieder eine primär literarisch erdachte Pointe, denn in der Himmelswirklichkeit steht Saturn, in der Ekliptik, viel zu weit von *as-suhā* ab, um mit diesem noch ein zusammenhängendes Bild zu ergeben. — 50,3,1 [189]: als hoch im Norden, in Polnähe stehender Stern; soll hier die Höhe des senkrechten Kreuzbalkens umschreiben.

suhayl, α Carinae, Canopus (K 272a). 14,7,6 [94]: von H falsch als „Venus“ übertragen (cf. auch U, p. 116). Die Kombination *suhayl/aṯ-ṯurayyā* scheint deutlich östlichen Ursprungs zu sein und auf zwei viel zitierte Verse zurückzugehen, die ʿUmar ibn Abī Rabīʿa zugeschrieben werden[11].

[11] *Dīwān*. Ed. P. Schwarz. Leipzig 1901. Die Verse sind nicht in der Diwan-Handschrift selbst überliefert, sondern werden vom Herausgeber unter den zugeschriebenen Stücken aufgeführt: S. 247 (arab.), Nr. 439, mit Fundstellen vorn S. 126. Vgl. ferner noch Ibn Qutayba: *Anwā'* 152,4—7 mit Fn. 2 (dort weitere Stellen). Zahlreiche zusätzliche Stellen machte mir

aṯ-ṯurayyā, die Plejaden (K 306), ein bekanntes und für dichterische Vergleiche besonders beliebtes Objekt[12]; ich zähle 12 Stellen. Auch mit ihrem Alternativnamen *an-naǧm* (K 186) kommen sie vor: 1,12,5 [13]. An anderen Stellen scheint *an-naǧm* aber wohl nur das Appellativ „Stern" darzustellen, nicht den Nebennamen der Plejaden, cf. z.B. 50,5,1 [189]; 50,6,2 [189f.]; 57,6,2 [214. Für diese Stelle nimmt V, p. 190, Nr. 32 den Sternnamen an]. Die *maṭāliʿ* der Plejaden in 18,8,3 [104] gibt H schlecht als „ihre Horizonte" wieder; wörtlich wäre(n) es „ihre Aufgangsstelle(n)", hier aber vielleicht in etymologischer Figur „die Stelle(n) am Himmel, von wo sie herabscheinen" (*ṭālaʿatnī*), nämlich „die schwarzen, tiefdunklen Höhen [des Himmels]" (*sūda ḏ-ḏurā mudlahimmahā*, H weniger passend: „mit den schwarzen Gipfeln, den tiefdunklen").

Der Planet Jupiter, *al-muštarī*: 1,12,8 [13] (die Verwendung von *ḫanasū* „(die) zurückgingen" zeigt Assoziation zu *al-ḫunnas*, einer alten Bezeichnung der fünf Planeten, die vermutlich im Hinblick auf die zeitweilige „Rückläufigkeit" in deren Bahnen geprägt wurde); 1,15,2 [15]; 61,4,1 [219].

Der Planet Mars, *al-mirrīḫ*[13]: 1,15,2 [15]. — In 15,1,5 [96] ist er mit seinem persischen Namen *bahrām* genannt[14]. — Der Wortlaut von 1,12,9 [14]: *ka-anna l-aḥmara l-mirrīḫa* ... (H „als ... Mars, der Rote") stellt nicht einfach eine Qualifikation (*al-aḥmar*, der rote) neben den eigentlichen Eigennamen *al-mirrīḫ*; vielmehr ist *al-aḥmar* selbst ein Alternativname für Mars, zumal im spanisch-arabischen Raum[15].

Herr Prof. A. Spitaler aus seinen Sammlungen namhaft. Sogar noch um 1500 bei Aḥmad ibn Māǧid: *Kitāb al-fawāʾid*. Ed. Ibrahim Khoury. Damaskus 1971, S. 41,8ff. (in englischer Übersetzung bei G. R. Tibbetts: *Arab Navigation in the Indian Ocean*... London 1971, S. 84); auch in eigenen Versen von Aḥmad ibn Māǧid selbst: 164 pu.-ult. = Tibbetts S. 148. Der Überlieferung nach waren in diesem Fall Suhayl und Ṯurayyā die Namen historischer Personen, auf deren Situation dann erst im Gedicht der Vergleich mit den (fern auseinander stehenden) Gestirnen *suhayl* und *aṯ-ṯurayyā* angewendet wurde.

[12] Cf. auch E. Perpiñá: *Las pléyades y la poesía árabe*. In: Al-Andalus 18 (1953), 439—444.

[13] A² vokalisiert in 1,15,2 und 1,12,9 unklassisch *al-marrīḫ*.

[14] Die persischen Planetennamen sind teilweise auch den Arabern bekannt, cf. u.a. Ibn Qutayba: *Anwāʾ* 126 unten; al-Marzūqī: *Azmina* II 365; Ibn Sīda: *Muḫaṣṣaṣ*. Būlāq 1316—21 H. = 1898—1903, IX, 36; Ibn al-Aǧdābī: *Azmina*. Damaskus 1964, 91,4 (*bahrām*); Ibn Manẓūr: *Niṯār al-azhār* 182 pu. ff.

[15] Belegstellen für diese spanisch-arabischen Nebenformen aus der astronomischen Fachliteratur bei P. Kunitzsch. In: Zeitschr. f. deutsche Sprache 25 (1969), 171—173, und in: *Wolfram-Studien*. Hg. v. W. Schröder. II. Berlin 1974, S. 29 mit Anm. 74.

Der Planet Saturn: Mit seinem üblichen arabischen Namen *zuḥal* erscheint er in 10,10,2 [67][16]. — Ferner wird auch er, wie oben schon Mars, einmal — innerhalb desselben Zitats — mit seinem persisch-altorientalischen Namen *kaywān* genannt[17]: 15,1,2 [96][18]. — Zweimal endlich wird er mit seinem spanisch-arabischen Alternativnamen *al-muqātil* „der Kämpfer" eingeführt (also nicht „Mars"! Von H an beiden Stellen verkannt)[19]: 1,16,1 [15; desgleichen in Fnn. 45 a.E. und 46; in A¹ ebenfalls unbekannt und daher falsch gedeutet: S. 23, Fn. 3][20] und 41,7,6 [171], beidemal übrigens von demselben Dichter, Ibn Hu̱dayl.

Die Planeten Merkur und Venus kommen in Ibn al-Kattānīs Verszitaten nicht vor, soweit ich sehe. Und auf die häufigen Beispiele mit Sonne und Mond braucht hier nicht eigens eingegangen zu werden.

Was die allgemeineren himmelskundlichen und astrologischen Vergleiche angeht, so sind sie meist leicht verständlich und in H passend wiedergegeben. Einige Besonderheiten mögen aber hier noch Erwähnung finden.

Neben den „Sternen" schlechthin (*nağm, kawkab*) treten oft auch „Sternschnuppen" (*šihāb*, pl. *šuhub*)[21] in Erscheinung, von H meist zutreffend erfaßt. In 41,2,3 [169] (ein davonschießender Jagdhund, *kaannahu l-kawkabu fī nṣibābih*) hat H: „als wäre er ein Komet in seiner Bahn". Kometen stehen für den Beobachter längere Zeit, zuweilen über Monate, nahezu unverändert am Himmel und wandern nur ganz langsam

[16] In Übereinstimmung mit dem Wissensstand der Astronomie hebt der Dichter die "Höhe" des Saturn hervor: er ist von den "sieben" dem Altertum und Mittelalter bekannten "Planeten" der höchste, d.h. der am weitesten von der im Zentrum des Systems angenommenen Erde entfernte.

[17] Cf. *WKAS* 518 b 9; W. Eilers: *Sinn und Herkunft der Planetennamen*. München 1976. (SB Bayer. Akad. d. Wiss. Philos.-Hist. Kl. Jahrg. 1975, Heft 5), S. 90f. (zu *bahrām*, ib. 75f.).

[18] A² vokalisiert abweichend vom „klassischen" Gebrauch *kīwān*. — Der Dichter, ʿAbdallāh ibn aš-Šimr, war Astrologe ʿAbdarraḥmāns II. (cf. A² p. 31 und p. 267 [arab.], Nr. 42); er kennt sich also im Fache aus und weiß daher auch richtig die *nuḥūs* (H „Verderblichkeiten") des Saturn zu nennen. Das muß nicht immer so sein: nach eigener Intention kann ein Dichter auch einmal einen Glücksplaneten poetisch als „Unglücksstern" bezeichnen, wie es al-Buḥturī mit Jupiter (*al-muštarī*) tut, cf. G. von Grunebaum: *Kritik und Dichtkunst*. Wiesbaden 1955, S. 59 (v. 5) = *Dīwān*. Ed. Ḥ. K. aṣ-Ṣayrafī. II. Kairo: Dār al-Maʿārif 1963, S. 1159, v. 38.

[19] Für Belegstellen cf. oben Anm. 15.

[20] Der „Zorn" (*iġtāẓa*) des Saturn dürfte aus seinem astrologischen Grundtypus als alter, grämlicher, freudloser Charakter hergeleitet sein; cf. R. Klibansky/E. Panofsky/F. Saxl: *Saturn and Melancholy*. London 1964, S. 127—195 (Saturn in the Literary Tradition).

[21] Vgl. C. A. Nallino: *Raccolta di scritti*. V. Rom 1944, S. 376ff., speziell 382ff.

weiter. Der Vergleich des Davonschießens legt es daher nahe, den arabischen Ausdruck ebenfalls als „Sternschnuppe" zu deuten, die bekanntlich blitzartig über den Himmel schießt. In 41,3,2 [169] (wieder ein Jagdhund, *yanṣā'u ka-l-kawkabi*) hat H zutreffend „Er jagt wie die Sternschnuppe". Schwieriger erscheint die Stelle 50,3,1 [189] (ein Gekreuzigter, ... *yaġšāhu naǧmun qāḏifū*, H „den ein steinschleudernder Stern überfällt"); vielleicht ist auch hier eine Sternschnuppe zu verstehen, die unversehens, blitzartig herniederstürzt, wobei womöglich sogar Meteoriten mit niedergehen. Dagegen „... ein Stern, der fällt" in 13,2,1 [82] (*naǧmun hawā*) ist in diesem Fall keine Sternschnuppe, die aufglühend niederschießt, sondern wohl einfach „ein untergehender Stern"[22].

Interessant ist das zweimalige Vorkommen des Motivs „Sterne (Sternschnuppen), die einem (zu vertreibenden) *šayṭān* nachgeworfen werden", das aus der ostarabischen Dichtung, sogar schon aus sehr alter Zeit, bekannt war und das mehrfach ebenfalls im Koran erscheint; s. hier 13,10,3 [88] und 18,14,1 [105][23]. Dadurch ist auch der zunächst befremdende Ausdruck „Stern des Teufels" (*kawkab al-mirrīd*) sachlich aufgeklärt, der also nicht zu den wirklichen Sternnamen gehört.

In 10,14,4 [70] scheint überhaupt kein Sternvergleich vorzuliegen: halbmondförmige Arkaden von Bauten an schönen Gärten, *wāfaqat su'ūda l-maǧārī* (H „schienen mit der Glückskonstellation übereinzu-

[22] Cf. auch oben sub *al-farqadān* 41,9,4 [171], wo die genannten Fixsterne natürlich nicht herniederschießen, sondern ihrer Natur gemäß allmählich herabsinken und untergehen. Bei dem bekannten Koranvers *wa-n-naǧmi iḏā hawā* (53,1) dagegen ist die Sachlage unklar. Es fehlt jeder nähere Anhaltspunkt, und so schwanken dort die Kommentare zwischen einem Stern (oder speziell: den Plejaden), der untergeht, und einer Sternschnuppe, die herabschießt.

[23] Es handelt sich um den „Sternschnuppenmythos" im Koran: Geister (Dschinnen) bzw. *šayṭāne*, die am Himmel heimlich die Engel belauschen wollen, werden durch (nachgeworfene) Sternschnuppen (*šihāb/šuhub*) vertrieben; cf. Koran 15,16—18; 37,6—10; 67,5; 72,8—9; dazu P. A. Eichler: *Die Dschinn, Teufel und Engel im Koran*. Diss. Leipzig 1928 (Druck: Lucka in Thür.), 30—32. Entsprechende Dichterstellen sind: Umayya b. Abī ṣ-Ṣalt. Ed. F. Schulthess. Leipzig 1911, Nr. 25,27—28, aus al-Ǧāḥiẓ: *Ḥayawān*. Ed. Hārūn. VI. Kairo 1943, 275,5—6 und ar-Rāġib al-Iṣfahānī: *Muḥāḍarāt al-udabā'*. II. Kairo 1287, 369,—7f.; Ḏū r-Rumma und Ibn ar-Rūmī, bei Ibn Abī 'Awn: *Tašbīhāt* 41,1—4 = al-'Askarī: *Ṣinā'atayn*. Ed. Kairo 1952, 82; (der Vers von Ḏū r-Rumma steht in ed. Macartney. Cambridge 1919, S. 27 [No. 1, v. 105] = ed. 'A. Abū Ṣāliḥ. I. Damaskus 1972, S. 111, v. 100). Die Stellenangaben stammen aus den Sammlungen von Herrn Prof. A. Spitaler, München. — *kawkab al-mirrīd* paßt als Gleichnis sehr gut zu dem vom Dichter angenommenen Bild: ein Spottgedicht so scharf treffend wie das Schwert ... oder wie einer von jenen Sternen (Sternschnuppen), die dem Teufel nachgeschleudert werden.

stimmen“). Vielmehr dürfte *al-maǧārī* Wasserläufe, Bäche bedeuten, in denen sich die *ahilla* (Halbmonde, bildlich für die Arkaden) spiegeln, so daß sich die Halbkreise zu einem vollen Rund zu schließen scheinen und die Halbmonde ihre Vollständigkeit (quasi als Vollmonde) herstellen. Man könnte also etwa übersetzen: „Als wären ihre Arkaden Halbmonde, die so passend an den [daran vorüberfließenden] Nebenarmen (?) der Wasserbäche stehen, daß sie sich zu Vollmonden zu komplettieren scheinen“. — 61,7,1 [220], in einer Beschreibung von Dinaren: „Aufgehende Glück verheißende [Sterne] (*ṭawāliʿu bišrin*), als wären sie, da wir sie blank sehen, [richtige] Sterne, die [wirklich] aufgehen“. *ṭawāliʿu bišrin* (H „Glücksterne“) ist hier ganz allgemein aufzufassen und nicht auf bestimmte Gestirne bezogen[24].

„Helle“ Sterne werden in der Poesie, aber auch von den Lexikographen häufig als *azhar*, pl. *zuhr* bezeichnet; im Gegensatz dazu stehen die kleinen, schwachen Sterne, deren Qualifikation *ḫafīy* lautet, und zwar auch in der astronomischen Fachliteratur[25]. Aus der „scheinbaren Helligkeit“ der Sterne können keine unmittelbaren Schlüsse auf ihre Entfernung gezogen werden. Es ist also nicht ganz angemessen, wenn H dabei die Vorstellung von „nahen“ und „weiter entfernten“ Sternen ins Spiel bringt: p. 7, Fn. 4; 1,14,2 [14]; ebenfalls in 12,11,4 [76], wo aber der Text statt *ḫafīya* poetischer sagt *diqāqu kawākibin*.

Die „Sieben Schimmernden“ (H), *ad-darārī as-sabʿ*, in 44,7,1 [177], in einem Schwertvergleich, sind in diesem Zusammenhang zweifellos „die sieben Planeten“ (im Altertum und Mittelalter, unter Einschluß von Sonne und Mond)[26]; das Schwert ist mit ihnen *mumtaziǧu*, nicht nach H „verquickt“, sondern „mit ihren ‘Temperamenten’ ausgestattet“[27]. — Auch in 10,10,2 [67] sind mit *ad-darārī* sicherlich „die Planeten“ gemeint, und nicht nach H allgemein „die Himmelskörper“; Sonne und Saturn werden dabei speziell herausgehoben.

[24] Über eine falsche Interpretation von *saʿd as-suʿūd* (H 64, Fn. 226) s.o. sub *al-ǧawzāʾ*, 10,3,6 [64].

[25] Cf. P. Kunitzsch: *Der Almagest. Die Syntaxis Mathematica des Claudius Ptolemäus in arabisch-lateinischer Überlieferung*. Wiesbaden 1974, S. 214.

[26] Cf. Ibn Sīda: *Muḫaṣṣaṣ* IX, 36; Ibn al-Aǧdābī: *Azmina* 77,2; 94,9; Ibn Manẓūr: *Niṯār al-azhār* 153, 183. Sonst können *ad-darārī* auch die fünf (echten) Planeten (ohne Sonne und Mond) oder helle Sterne überhaupt sein, also auch die Fixsterne der ersten Größe (cf. al-Marzūqī: *Azmina* II 369,13—370,5 [in 370,4 ist statt *al-yadān* richtig zu lesen: *ad-dabarān*]).

[27] Nach dem astrologischen Fachausdruck *mizāǧ*, griech. κρᾶσις; cf. P. Kunitzsch in: *ZDMG* 118 (1968), 64f. mit Fn. 11. In der Astrologie ist den Fixsternen der *mizāǧ* (das “Temperament”) jeweils eines oder zweier Planeten zugeteilt. Der Dichter schreibt dem gerühmten Schwert nun gar die Temperamente aller sieben Planeten gleichzeitig zu.

Unter den besprochenen stern- und himmelskundlichen Vergleichen und Anspielungen scheinen nur zwei (die Sondernamen der Planeten Mars und Saturn) spezifisch westlicher, spanisch-arabischer Herkunft zu sein. Alle übrigen gehören zum gemeinarabischen, d.h. im Osten geprägten, literarischen bzw. fachwissenschaftlichen Traditionsgut. Damit wird zugleich deutlich, daß die Bilderwelt der sternkundlichen Vergleiche weitgehend literarischen Charakters ist und — bei den hier betroffenen spanisch-arabischen Dichtern — zumeist keinen unmittelbaren Realitätsbezug enthält. Ein solcher wird nur in zwei Fällen erkennbar[28], aber auch da ist natürlich nicht erwiesen, ob es der Dichter selbst war, der die entsprechende Naturbeobachtung angestellt hat, oder ob nicht auch in diesen Fällen bereits fertige literarische Vorlagen existierten. Bis auf den tief südlich stehenden *suhayl* wären alle hier besprochenen Gestirne sehr wohl in Spanien zu beobachten gewesen. Aber anders als im alten Beduinenmilieu, wo tatsächliche Naturbeobachtung nicht nur in die Dichtung einfloß, sondern darin breiten Raum einnahm und ein tragendes Element darstellte, wurden die Naturerscheinungen bei den Dichtern späterer seßhafter, „Bildung" pflegender Generationen, im Osten ebenso wie im Westen, zu literarischen Versatzstücken, die nur in assoziativ-motivischer oder gar technisch-poetischer Absicht Verwendung fanden[29].

Wohl zeigen auch solche „sekundären", nicht selbst gefundenen Naturvergleiche die Kunstfertigkeit eines Dichters und vermögen die Leser/Zuhörer durchaus zu entzücken, umso mehr, je treffender sie gewählt sind. Andererseits können bei dieser realitätsfernen Methode auch manche Mißgriffe unterlaufen[30], wie ja auch bei uns ein Dichter durchaus einmal die schmale Sichel des abnehmenden Mondes am Abendhimmel „sehen" kann.

Ähnliche Beobachtungen und Schlüsse wie hier bei den stern- und himmelskundlichen Vergleichen werden sich, mutatis mutandis, vermutlich auch bei den dichterischen Bildern aus den anderen Naturreichen ergeben.

Die Übertragung derartiger Poesie ins Deutsche stellt einen Übersetzer vor kaum lösbare Probleme. Wie soll er die vielen Bezeichnungen und Begriffe wiedergeben, für die es im Deutschen keine unmittelbaren

[28] Form des Kassiopeia-Bildes, bei Saʿīd b. ʿAmrūn (1,19,2 [16f.]); Farbnuance von *al-farqadān*, bei Aḥmad b. Farağ (61,4,1 [219]).

[29] Auch etymologisch-wortspielerische Bilder kommen vor, cf. H. RITTER: *Über die Bildersprache Niẓāmīs*. Berlin/Leipzig 1927, S. 53 (letzter Vers); 54 (vorletzter Vers); 58 Fn. (Ibn al-Muʿtazz); 69 (6. Vers) — alle zu den beiden „Adlern", *an-nasr aṭ-ṭāʾir* und/oder *al-wāqiʿ*.

[30] Vgl. ʿUbaydallāh b. Idrīs al-Wazīr und seinen verunglückten *farqadān*-Vergleich: 41,9,4 [171].

Äquivalente gibt, seien es nun Tiere, Pflanzen, Sterne und anderes Einschlägige. Gewiß, auch für arabische Leser/Hörer, ja, wie wir vielfach gesehen haben, für die Dichter selbst waren diese Dinge mit zunehmender Entfernung von der alten Beduinenwelt nur abstrakte Wortgebilde, mit denen sie immer weniger konkrete Vorstellungen verbinden konnten. Man kann versuchen, bei der Übertragung vor allem den Gang der Assoziationen nachzuzeichnen, wie es in höchster Vollendung wohl Friedrich Rückert gelungen ist, wobei aber eben viele philologisch-sachliche Details der eingängigen Lesbarkeit geopfert werden. Oder man paraphrasiert absolut wortgetreu — doch was könnte ein deutscher Leser mit einem Gedicht anfangen, in dem es von Fachausdrücken und wissenschaftlichen Bezeichnungen aus den Naturreichen wimmelt. (So gehäuft wie gerade in dieser Anthologie dichterischer Vergleiche, die immer nur einschlägige „Stellen“ aus Gedichten vorführt, treten diese Schwierigkeiten in einem kompletten längeren Gedicht natürlich nicht auf.) Man wird also wohl häufig zweigleisig verfahren und einer glatteren assoziationsfähigen Übertragung die fremdartigen sachlichen Details in einem kommentierenden Apparat beigeben müssen. Ein gutes Beispiel ist H's Übertragung des Verses 10,13,8 [70]: „... dessen Zinnen an 'Adler' und 'Bär' zu stoßen scheinen“, wobei zu „Bär“ dann aber angemerkt werden müßte, daß im Text dafür *al-farqadān* „die beiden Kälber“ = die Sterne βγ Ursae Minoris steht, Sterne, für die es in unserer einheimischen Tradition nur die fachwissenschaftliche Bezeichnung, jedoch keinen eigenen eingebürgerten Namen gibt.

Ganz fremd ist das Phänomen einer solchen sachlich und fachlich schwer beladenen Dichtung für uns nicht. Man denke an mittelalterliche Dichter wie Wolfram von Eschenbach, oder an unsere Barockdichtung und Klassik, die zu ihrem Verständnis einen hohen Bildungsgrad nach jeweils zeitgenössischem Standard voraussetzen. Hier wie dort darf man sich bei der Interpretation natürlich stets dessen bewußt bleiben, daß der Dichter wohl ein Gebildeter, aber kein Gelehrter ist und daß sein Werk der Kategorie der Kunst, nicht der der Wissenschaft angehört. Dem Interpreten freilich bleibt es nicht erspart, über das bloß Assoziative hinaus auch konsequent Philologie zu betreiben und über Wörter und Sachen genauestens Auskunft zu erteilen.

Wie unübersetzbar Gedichte im letzten Grunde überhaupt sind, das haben schon die Araber selbst im Zuge ihrer Übertragungen aus dem Griechischen am Beispiel griechischer Verse erkannt[31].

[31] Abū Sulaymān al-Manṭiqī, al-Ǧāḥiẓ, Moše b. ʿEzra, in deutscher Übersetzung bei F. Rosenthal: *Das Fortleben der Antike im Islam*. Zürich/Stuttgart 1965, S. 35f. (Nr. 3—5).

XXVII

Zeus in Bagdad. Zu einem Gedicht von Abū Nuwās

An der arabischen Poesie überrascht immer wieder die Vielfalt der Variationen und Vergleiche, mit denen die Dichter die begrenzte Zahl der ihnen zur Verfügung stehenden Grundmotive ausgestalten. Das gilt umso mehr, je weiter man sich von der klassischen Beduinenpoesie entfernt und in die Zeit verfeinerter Stadt- und Hofkultur, zu den *muḥdaṯūn*, vordringt. Ein weiteres Element der Ausweitung kann darin liegen, daß ein Dichter durch Herkunft oder Milieu nicht mehr allein dem arabischen Bildungskreis angehört, sondern etwa auch persischen Einflüssen ausgesetzt ist. Das alles trifft auf unseren Dichter Abū Nuwās zu, dessen Mutter Ǧullabān Perserin war, der in der Residenzstadt Bagdad lebte, am Kalifenhof verkehrte und engen Umgang mit bedeutenden persischen Familien wie den Barmak und den Nawbaḫt pflegte. So ist denn auch seine Dichtung, über ihren anerkannten literarischen Wert hinaus, überreich an Material von höchstem kulturgeschichtlichem Interesse. Das Gedicht, dem dieser Aufsatz gewidmet ist, kann so mindestens in dreifacher Hinsicht als bedeutungsvoll gelten: es werden hier in die Poesie nicht nur Himmelskunde und Astrologie ganz allgemein als motivische Elemente eingeführt, sondern es werden regelrecht fachtechnische Details in ihrem fachlichen Sinne funktional eingesetzt; zweitens enthält das Gedicht, als Kuriosum, einen griechischen Planetennamen; drittens muß das Gedicht, ganz allgemein, als interessantes Dokument der arabischen Bekanntschaft mit antikem Wissen in jener Frühzeit gewertet werden, aus der primäre arabische Quellentexte noch kaum vorliegen.

Das Gedicht, mit dem wir uns hier befassen wollen, gehört in das dritte Kapitel (Lobgedichte) von Abū Nuwās' Dīwān. Es umfaßt sieben Verse im Metrum *ḫafīf* und wird in beiden Dīwān-Rezensionen, von Ḥamza al-Iṣfahānī und von aṣ-Ṣūlī, überliefert[1]. Es ist an den Barmakidenwesir Yaḥyā ibn Ḫālid ibn Barmak gerichtet, der im Januar 803, bei dem spektakulären Sturz der Barmakiden unter Hārūn, nach ar-Raqqa ins Gefängnis gebracht wurde, wo er im November 805 im Alter von 70 Jahren starb[2]. Also ist die Entstehung des Gedichtes der Zeit

1 *Dīwān* I 151.

2 Cf. D. Sourdel in EI², s.v. *al-Barāmika*.

vor 803 zuzuschreiben, als der Adressat noch in Amt und Würden war und die Sonne seiner Gnade über seinen Anhang strahlen lassen konnte.

Wir lassen nun zunächst eine Übersetzung des Gedichtes folgen[3]:

1. Ich nehme [meinem Pferd] al-Maḥḏūf[4] den Zaum nicht freiwillig ab, bevor ich nicht bei dem Sohn Ḫālids, dem Freigebigen, [angelangt bin].
2. Denn wenn ich in Abū l-Faḍls Meer hineinsteige, so streife ich [damit] das Unglück von meinen Gewändern ab.
3. [Du bist] das Bild des Jupiter, wenn er im 'Haus' des 'Lichtes der Nacht' steht; die Sonne bist du bei [ihrer] Kulmination[5].
4. Nicht ist Jupiter, wenn er vor den 'Fischen' seine Bahn zieht, noch der Vollmond, wenn er sich dem Untergang zuneigt,
5. Freigebiger als du mit dem, womit die Seelen geizen, wenn die Milch beim Melken[6] knapp wird.
6. Und nicht ist Mars, wenn er zusammen mit dem 'Skorpion' kulminiert und dabei in seiner ihm zugeschriebenen Kraft noch zunimmt,
7. Schärfer treffend als du in den Kriegen und nicht schrecklicher dem Auge [als du] beim Abschlagen der Köpfe[7].

In Vers 2 erfolgt bereits eine Einstimmung auf die folgende breit angelegte astrologische Metaphorik: »das Unglück«, *an-nuḥūs* (pl.), erweckt eine Assoziation zu der astrologischen Klassifizierung der Planeten als »glücklich« (*as-sa'dān*: Jupiter und Venus) bzw. »unglücklich« (*an-naḥsān*: Saturn und Mars).

Vers 3 enthält sodann eine geballte Ladung an astrologischem Gut, und zwar nicht lediglich in assoziativer Anspielung, sondern in voller fachterminologischer Direktheit. Allgemein gehalten ist lediglich noch der Gebrauch von *ṣūra* »Bild«, das im spezifischen Sinne ein Sternbild bezeichnen würde und das hier allgemeiner als »Bild, Figur, Anblick« gemeint ist. Der Jupiter (= *as-sa'd al-akbar*, der stärkere der beiden Glücksplaneten) ist mit seinem bekannten arabischen Namen *al-muštarī* genannt[8]. *nūr al-layl*, »das Licht der Nacht«, ist der Mond; sein *bayt*, »Haus«,

3 Die Verse 3-7 in deutscher Übersetzung auch bei WAGNER: *Studie* 392. In den ausgewählten Übersetzungen bei A. VON KREMER: *Diwan des Abu Nuwâs*. Wien 1855, ist unser Gedicht nicht vertreten.

4 So nach der Erklärung von *al-maḥḏūf* in der *ḥāšiya* in ms P; cf. *Dīwān* a.a.O.

5 D.h. wenn sie ihren höchsten Stand am Himmel erreicht hat, gerade über einem steht. WAGNER: *Studie* 392 unzutreffend: »beim Aufgang«. *al-intiṣāb* (bzw. *an-niṣāb*) werden im *Dīwān* a.a.O. richtig glossiert als *al-irtifā' fī s-samā'* und ... *fī wasaṭ as-samā'*. Cf. TIBBETTS: *Navigation* 543 (und die Stellen): *intaṣaba 'alā* »[senkrecht] über einem anderen Stern stehen«.

6 Oder: die Milch des Milcheimers, *darr al-ḥilāb*.

7 Wörtlich: der Hälse, *ar-riqāb*.

8 Cf. W. HARTNER in EI[1], s.v. *al-Mushtarī*.

ist der Krebs. Jupiter aber hat im Krebs seine »Erhöhung« (ὕψωμα, *šaraf*, *exaltatio*), in dieser Stellung kann er seine (glückbringende) Kraft am vollsten entfalten. (Daß der Dichter dabei die Präposition *ladā* »bei« statt direkterem *fī* »in« gebraucht, dürfte auf metrischem Zwang beruhen.) Man darf die Stelle also nicht als bloßen Vergleich des *mamdūḥ* mit dem glückbringenden Planeten Jupiter verstehen, sondern hat darin zugleich die Steigerung des Vergleichs zur Höchstform zu erkennen, die kunstvoll durch die astrologische Determination »Jupiter in seiner Exaltation« erzeugt wird, welche aber ihrerseits nochmals umschrieben ist durch die Bezeichnung des Krebses als »Haus des Lichtes der Nacht« (wiederum = des Mondes). Erstaunlich daran ist nicht nur, wie glatt offenbar Abū Nuwās selbst diese astrologischen Kenntnisse besitzt und handhabt, sondern daß er sie so unverhohlen auch bei seinem Publikum voraussetzt. Gewiß ist bekannt, wie groß der Einfluß der Astrologie im Mittelalter in Ost und West in allen Bevölkerungsschichten war. Hier kommt natürlich noch erleichternd hinzu, daß am Kalifenhofe wie auch im Hause des Barmakiden genügend Astrologen versammelt waren, um dem Verständnis einer solchen Stelle notfalls nachhelfen zu können. Das Bild von der Sonne in ihrem höchsten Stande, in der zweiten Vershälfte, ist dann wieder allgemeiner gehalten und beruht auf der alltäglichen Erfahrung aller Menschen, daß die Sonne eben zu dieser Zeit, also am Mittag, ihre stärkste Strahlungskraft entfaltet. Dies braucht daher nicht notwendig ein spezifisch astrologischer Topos zu sein[9].

In Vers 4, bei der zweiten Nennung, wird Jupiter auffälligerweise mit dem griechischen Götter- und Planetennamen Zeus (*Zāwīš*) angeführt (s. weiter unten). Auch hier aber ist er wieder nicht eben bloß so erwähnt, sondern in einer ganz bestimmten Position dargestellt: er zieht seine Bahn (*sāra*) vor den Fischen (*amāma l-ḥūt*[10]). Im Tierkreiszeichen der Fische hat Jupiter sein 'Nachthaus', ebenfalls also eine besonders wirkungsmächtige Stelle ähnlich der Exaltation (im Krebs), oben in Vers 3. Neben dieser astrologisch-funktionalen Angabe steht im zweiten Halbvers, offensichtlich bewußt parallel zum Aufbau von Vers 3, wieder eine volkstümlich-allgemeinere Beobachtung: der Vollmond im Untergangsstadium[11]. Jedermann weiß, daß der Mond über dem Horizont scheinbar viel

9 Cf. aber auch Ptolemäus: *Tetrabiblos* I,4 (S. 35): Sonne im Zenit.

10 *amāma* »vor« wohl ebenfalls wieder metrisch bedingt statt des direkten *fī* »in«.

11 Das Verb *hawā* bezieht sich auf das normale Untergehen von Gestirnen. *inṣibāb* dagegen bezeichnet anderenorts das schnelle Dahinschießen einer Sternschnuppe (im Vergleich eines davonschießenden Jagdhundes); cf. KUNITZSCH: *Dicht. Vergl.* 247f. An unserer Stelle wird also der Reimzwang für den Gebrauch des Wortes ausschlaggebend gewesen sein, das an sich dem allmählichen Untergehen eines Gestirns nicht ganz angemessen ist. Für Jagdhund- und weitere Jagdtiervergleiche mit dahinschießenden Sternen gebraucht Abū Nuwās *inhirāṭ* (*Dīwān* II 181,9), *inṣā'a ka-l-kawkabi fī nkidārihī* (II 192,17),

größer, mächtiger, eindrucksvoller auf den Beschauer wirkt, als wenn er klein und fern hoch am Himmel steht[12]. Eine gewisse Verbindung aber ergibt sich immerhin noch daraus, daß der Mond der Herr der Nachthälfte des Tierkreises ist, welcher auch das Zeichen Fische angehört.

An die Wiedergabe einer exakten Gestirnstellung zu einem bestimmten Zeitpunkt, den der Dichter im Sinne gehabt haben könnte, ist hierbei freilich nicht zu denken, da er den Jupiter einmal in seiner Exaltation, im Krebs, und das andere Mal in seinem Nachthaus, in den Fischen erscheinen läßt. Bei seiner langen siderischen Umlaufszeit von rd. 12 Jahren würde Jupiter für die Strecke vom Ende des Krebses bis zum Anfang der Fische rd. 7 Jahre brauchen, oder vom Ende der Fische bis zum Anfang des Krebses rd. 3 Jahre. Wenn wir als Periode, innerhalb deren unser Gedicht entstanden sein könnte, einmal die Jahre 780-800 annehmen, so befand sich in diesem Zeitraum Jupiter zweimal im Krebs (Juni 780 bis Juli 781 und Mai 792 bis Juni 793, danach erst wieder ab Mai 804) und zweimal in den Fischen (April 788 bis April 789 und März 800 bis März 801). Aus der großen zeitlichen Differenz zwischen Jupiters Aufenthalt im Krebs und in den Fischen ergibt sich bereits eindeutig, daß der Dichter mit seinen astronomischen Positionsangaben nicht auf ein ganz bestimmtes historisches Ereignis abgezielt haben kann. Diese Feststellung wird weiter erhärtet, wenn wir hinzunehmen, daß er weiter unten ebenfalls noch den Mars im Skorpion erwähnt. Dann würde, streng genommen, diese Konstellation genau zu den entsprechenden Jupiterstellungen passen müssen. Das ist theoretisch nicht unmöglich, da Mars nur eine siderische Umlaufszeit von knapp zwei Jahren hat, also etwa alle zwei Jahre einmal an Jupiter vorüberzieht. Auf die oben genannten vier Daten bezogen, ergäbe sich dann: 780/1: Mars im Skorpion Sept./Okt. 780; 792/3: Mars im Skorpion erst Okt./Nov. 793; 788/9: Mars im Skorpion nur Jan. 788 und dann erst wieder Nov./Dez. 789; und 800/1: Mars im Skorpion Jan./Febr. 801. Das heißt, mit Jupiters Stellung im Krebs würde Mars im Skorpion nur einmal genau übereinstimmen: Sept./Okt. 780; und mit Jupiter in den Fischen ebenfalls nur einmal: Jan./Febr. 801. Auch das sind aber zwei zu weit auseinander liegende Daten, als daß sie in dem Gedicht angesprochen sein könnten. Es bleibt also bei der zuvor gemachten Feststellung: Abū Nuwās verwendet die astronomischen Positionsangaben der beiden Planeten

inqiḍāḍ al-kawkab al-munṣāḥ (II 262,3), *kawkab yanqaḍḍu* (II 288,17), *al-kawkab al-munqaḍḍ* (II 297,7), und einmal auch *hawā*: *kawkabu ʿifrītin hawā li-ʿaddihī* (II 290,1).

12 Im Gegensatz zur Kulmination der Sonne hat Ptolemäus a.a.O. für den Mond keine Bemerkung über eine spezielle Stellung im Hinblick auf seine Wirkung. Generell sind dem Mond feuchte, befeuchtende Eigenschaften zugeschrieben, was wenigstens assoziativ zu der hier intendierten »Freigebigkeit« in Beziehung stehen könnte (cf. arab. *nadā*, *nadī al-kaff* bzw. *al-yadayn*, etc.).

als isolierte Aussagen zum Zwecke poetischer Hyperbeln in Form einer astrologischen Metaphorik, nicht aber als tatsächliche Sternkonstellation bei einem Ereignis im Leben des Adressaten seines Gedichtes[13].

Eine weitere astronomisch-astrologische Spezialität findet sich dann noch in Vers 6: der Mars (mit seinem persischen Namen *bahrām* genannt[14]), im Zeichen Skorpion stehend, welches sein 'Taghaus' ist — also ebenfalls eine Stelle von besonders starker Wirksamkeit —, in der Nacht kulminierend[15]. Obendrein ist der Skorpion dasjenige Zeichen, in dem der Mars sein *gaudium* hat (χαίρει, *yusarru*), was sich ebenfalls steigernd auf seine Mächtigkeit auswirkt. Wiederum eine Hyperbel: die Härte des *mamdūḥ* im Kampf und bei der Durchsetzung von Gesetz und Ordnung übertreffen noch die Macht des blutigen Kriegsplaneten Mars, sogar wenn dieser in seinem Taghaus steht, in welchem er zugleich sein *gaudium* hat, was beides ihn besonders wirkungskräftig macht. Die mitternächtliche Kulmination des Skorpions fällt etwa in den Juni. Die Beziehung der letzten Worte des Verses: (... *bi-l-layli*) *zā'idan fī l-ḥisābi* auf die länger werdenden Nächte des Herbstes[16] dürfte also kaum zutreffen. Der *ḥāl*-Satz *zā'idan* ... muß demnach auf das Subjekt bezogen werden: Mars, wenn er, in seinem Taghaus stehend, kulminiert und dadurch in der ihm zugeschriebenen Kraft noch zunimmt.

Das Gedicht ist also ganz auf astrologischer Metaphorik aufgebaut. Die guten Eigenschaften des *mamdūḥ* werden durch astrologisch-funktionale Aussagen dargestellt: seine reichliche Gebefreudigkeit durch zweimalige Einführung des Jupiter, ergänzt durch die weniger fachspezifische Hereinnahme von Sonne und Mond, und seine männlichen Tugenden der Tapferkeit und Härte durch Heranziehung des Mars. In allen drei Fällen wird der *mamdūḥ* aber nicht einfach mit dem Wesen des betreffenden Planeten verglichen, sondern mit ganz besonderen, gesteigerten Wirkungsformen dieser Planeten, wie sie in den Lehren der Astrologie niedergelegt sind. Die Steigerung der Begriffe wird also nicht auf sprachlich-poetischem Wege erzielt, sondern durch die funktionale Anwendung dieser astrologisch-fachlichen Aussagen. Der ganze Gedichtkomplex ist überdies auch dadurch im astrologischen Sinne zu einer Einheit zusammengehalten, daß die drei

13 Die obigen Planetenstellungen sind entnommen aus B. TUCKERMAN: *Planetary, Lunar, and Solar Positions A.D. 2 to A.D. 1650*. Philadelphia 1964 (Memoirs of the American Philosophical Society, vol. 59).

14 Die persischen Planetennamen sind auch den Arabern bekannt und werden, abgesehen von der astronomisch-astrologischen Fachliteratur, auch in der Dichtung benutzt und von den arabischen Philologen aufgeführt. Cf. KUNITZSCH: *Dicht. Vergl.* 246 mit Anm. 14.

15 Zu *istaqalla* »kulminieren« cf. auch TIBBETTS: *Navigation* 539.

16 Bei WAGNER: *Studie* 392, offenbar im Anschluß an die *ḥāšiya* in ms P: *wa-l-laylu zā'idun ya'nī layāliya l-ḫarīf*.

darin vorkommenden Tierkreiszeichen, Krebs, Fische und Skorpion, alle demselben 'Trigon', dem vierten des Tierkreises, angehören.

*

Bei einem Dichter, der derartiges Spezialwissen so glatt und nahtlos in einem Gedicht verarbeiten kann, sind fraglos weitere einschlägige Stellen zu erwarten. Einige seien hier noch in Auswahl erwähnt[17]. Dabei können wir drei Stadien der Durchdringung unterscheiden.

Zunächst einmal gibt es bei Abū Nuwās, wie bei vielen — wenn nicht den meisten — anderen arabischen Dichtern auch einfache Vergleiche mit Himmelsobjekten, ohne spezifische funktionale Aussagen: *Dīwān* I 80,15 (= Wagner: *Studie* 164 und 400); *Dīwān* I 183,4-8 (= *Studie* 255); *Dīwān* II 254,6-7 (= *Studie* 286 und 400); *Dīwān* ed. Ġazzālī 238,3; *ib.* 723,3; *Dīwān* II 246,4; *Dīwān* II 305,7; und *Dīwān* II 290,1 (Vergleich eines Jagdgeparden, *fahd*: *kawkabu ʿifrītin hawā li-ʿaddihī*[18]).

Zweitens finden wir in seinen Gedichten speziellere Fachkenntnisse, die aber nicht funktional verwertet werden: z. B. Wagner: *Studie* 194 oben[19], eine Aufzählung von Planeten mit ihren persischen Namen, die in der Religion der Zoroastrier angerufen werden[20]. Die Namen werden nur aufgezählt, den einzelnen

17 Wagner: *Studie* hat einiges Astronomisch-Astrologische mitübersetzt bzw. gelegentlich erwähnt, dies aber leider nicht unter einem zugehörigen Stichwort im Index zusammengefaßt. Im folgenden werden die betreffenden Stellen mit genannt.

18 Vgl. mit diesem *kawkabu ʿifrītin* den *kawkab al-mirrīd* des Andalus-Dichters Muḥammad ibn ʿAbd al-ʿAzīz al-ʿUtbī bei Kunitzsch: *Dicht. Vergl.* 248 mit Anm. 23 (den dort zitierten Stellen wäre noch hinzuzufügen *Tausendundeine Nacht*, ed. Macnaghten, II [Calcutta 1839] 526,11-12 = Übersetzung E. Littmann, Wiesbaden 1953, III 681 unten, in der Erzählung von der Sklavin Tawaddud). Dieser »Sternschnuppenmythus« klingt bei Abū Nuwās noch ein zweites Mal an: *Dīwān* II 297,7 (Vergleich einer »schwarzflügligen Gabelweihe« [so nach Wagner: *Studie* 282], *zurraq*: ... *ka-l-kawkabi l-munqaḍḍi fī raġmatihī*.

19 Den arabischen Text finde ich, in sehr verderbter Form, in Ibn Manẓūr: *al-Fukāha wa-l-ītinās fī muǧūn Abī Nuwās*. Kairo 1316, 79,1.

20 Ein Übergang zu funktionalem Gebrauch wird erkennbar bei der Nennung von vier Planeten in einem Gedicht des persischen Dichters Daqīqī (siehe Eilers: *Planetennamen* 81), wo immerhin die beiden »bösen« und die beiden »glückbringenden« Planeten absichtsvoll zugeordnet erscheinen. Eine weitere ähnliche Stelle ist bekannt aus dem *Parzival* Wolframs von Eschenbach (Buch XV; 782,6-12). Hier soll eine astrologisch fundierte Verkündung mitgeteilt werden; aus diesem Anlaß werden alle sieben an der astrologischen Voraussage beteiligten Planeten (mit ihren arabischen Namen!) aufgezählt, ohne daß jedoch einzelnen Planeten individuelle Wirkungen innerhalb der Handlung zugewiesen werden, also nur eine nicht funktional ausgewertete Erwähnung.

Gestirnen wird aber keine unmittelbare Funktion im Rahmen des Gedichtes zugewiesen.

Drittens endlich gibt es bei ihm eine Reihe von Fällen, bei denen astronomische Elemente im Sinne ihrer fachlichen Bedeutung funktional in die Darstellung eingearbeitet sind, z.B. WAGNER: *Studie* 38 (= *Dīwān* ed. ĠAZZĀLĪ 63,2[21]): Eintritt der Sonne in das Tierkreiszeichen Widder und Augenblick der (Frühlings-) Tagundnachtgleiche, zur Umschreibung des Frühlingsanbruchs[22]. Ebenso wird einmal auch der Herbst astronomisch umschrieben: der Sirius ist am Himmel unsichtbar geworden (*Dīwān* I 182,10 = WAGNER: *Studie* 255 und 400). Desgleichen wird die Frühlingszeit mit ihrer klaren Luft astronomisch umschrieben in dem Gedicht WAGNER: *Studie* 416 (= *Dīwān* ed. ĠAZZĀLĪ 74,9)[23]. Jahreszeitliche Bestimmungen sind astronomisch fixiert in dem Gedicht *Dīwān* ed. ĠAZZĀLĪ 209,-1 ([Morgen-]Aufgang der Mondstation *aš-šaraṭān*, d.h. der Sterne βγ Arietis[24], etwa Mitte April[25], als Zeit des Beginns der Dattelblüte) und 210,5 (Morgenaufgang des *suhayl* = Kanopus, α Carinae[26], etwa Mitte August[25], als Zeit der Dattelreife)[27]. Auch in dem Gedicht *Dīwān* II 82,15 (= WAGNER: *Studie* 371) wird eine Jahreszeit astronomisch umschrieben, und zwar eine besonders heiße: *anḍaǧatnā kawākibu l-ǧawzā'i*, »die Sterne der Zwillinge [oder: des Orion[28]] kochen uns«; die Angabe bezieht sich auf den Morgenaufgang von Sternen aus dem Orion, der auf Anfang Juni fällt, oder auf Morgenaufgänge von Sternen aus den Zwillingen, die auf Ende Juni/Anfang Juli fallen[29]. Das Ende einer durchzechten

21 Auch angeführt bei G. SCHOELER: *Arabische Naturdichtung*. Beirut 1974 (Beiruter Texte und Studien, Bd. 15) 63. Ebenfalls vorhanden in A. VON KREMERS Übersetzung (s. oben Anm. 3), als Nr. IX der Weingedichte.

22 Das *tarā* der Verses widerspricht der Realität insofern, als man dasjenige Sternzeichen, in dem sich die Sonne befindet, gerade nicht »sehen« kann. Aber dem Gedicht darf man nicht so pedantisch gegenübertreten, und es wäre einfach an eine Übersetzung wie »Hast du nicht bemerkt« o.ä. zu denken.

23 Cf. ähnlich KUNITZSCH: *Dicht. Vergl.* 244. Bei WAGNER a.a.O. wäre nach unserem Sprachgebrauch *al-ḥūt* richtiger mit dem Plural »(in) den Fischen« wiederzugeben.

24 Cf. P. KUNITZSCH: *Untersuchungen zur Sternnomenklatur der Araber*. Wiesbaden 1961, 110f., Nr. 286.

25 Cf. Ch. PELLAT: *Dictions rimés, anwā' et mansions lunaires chez les arabes*. In Arabica 2 (1955) 17ff.; s. die Tabelle S. 32f.

26 Cf. KUNITZSCH: *Untersuchungen* (wie Anm. 24), 106, Nr. 272a.

27 Die Kenntnis der Stelle verdanke ich einem brieflichen Hinweis von Herrn Prof. E. Wagner, Giessen.

28 Zu dieser Ambivalenz vgl. KUNITZSCH: *Dicht. Vergl.* 242f.

29 Cf. die Tabelle von PELLAT (wie Anm. 25), Nr. V, VI und VII. WAGNER a.a.O. faßt hier unpassend »die Zwillinge« als das Tierkreiszeichen auf, in dem die Sonne vom

Maiennacht ist astronomisch dargestellt in dem Weingedicht WAGNER: *Studie* 300,9-10 (= *Dīwān* ed. ĠAZZĀLĪ 111,14): Sirius ist schon lange untergegangen, und jetzt am Morgen haben die Plejaden ihren Aufgang (der Morgenaufgang der Plejaden fällt etwa auf Anfang bis Mitte Mai; um diese Zeit geht Sirius am frühen Abend unter; zwischen diesen beiden Phasen liegen, in unseren Breiten, über 8 Stunden, sie sind also nicht gleichzeitig am Himmel zu beobachten. In der Übersetzung sollte es darum auch besser heißen: »Sirius war schon [lange] untergegangen, und die Plejaden kamen am Morgen herauf, um sich wieder [aufs neue] zu zeigen«). Ohne Vorliegen des arabischen Textes nicht erklärbar bleibt mir die Stelle WAGNER: *Studie* 300,24-26, deren deutsche Übersetzung die arabische Terminologie nicht klar genug durchscheinen läßt. Farbvergleiche mit dem rötlichen Mars sind in der arabischen Dichtung nicht selten[30], so auch bei Abū Nuwās in dem Gedicht WAGNER: *Studie* 401,8-9.

Es fällt auf, daß unter den genannten Beispielen die astronomischen Datierungen von Jahreszeiten stark dominieren. Hier scheint sich die Tradition der orientalischen, schon von den Babyloniern gepflegten[31] und von den Arabern noch bis heute gebrauchten astronomischen Bauern- und Hirtenkalender[32] auszuwirken (vgl. in diesem Sinne besonders auch die Heranziehung der Morgenaufgänge der Plejaden, der *ǧawzā'*-Sterne, des *suhayl* und der Mondstation *aš-šaraṭān*).

*

Bei den detaillierten astrologischen Kenntnissen, die Abū Nuwās in unserem Gedicht entfaltet, erhebt sich die Frage, woher der Dichter diese Kenntnisse zu seiner Zeit in Bagdad haben konnte. Immerhin gehört die Astrologie zum Komplex jener antik-hellenistischen Wissenschaften, die den Arabern durch systematische Übersetzungen erst im Laufe des 9. Jahrhunderts in größerem Umfang bekannt wurden. Insofern kann man das Gedicht noch mit zu den interessanten Dokumenten

21.5. bis 21.6. steht. Diese Phase ist — im Gegensatz zu den Morgenaufgängen — optisch nicht beobachtbar und wird im allgemeinen auch nicht für Zeitbestimmungen herangezogen; cf. die entsprechenden Angaben in den *anwā'*-Büchern. Bei Ibn al-Aǧdābī: *al-Azmina wa-l-anwā'*. Damaskus 1964 (Iḥyā' at-Turāṯ al-Qadīm 9) 135,7-8 heißt es ausdrücklich: *fa-iḏā sami'tahum yansibūna l-ḥarr ilā ... al-ǧawzā' ... fa-innamā ḏālika 'inda ṭulū'ihā*; und Ibn Qutayba: *Kitāb al-anwā'*. Hyderabad 1956, 43,2-3 sagt: *wa-innamā ya'nūna bi-ṭulū'i »l-ǧawzā'i« l-haq'ata wa-l-han'ata* (d.i. Nr. V und VI in PELLATS Tabelle).

30 Cf. KUNITZSCH: *Dicht. Vergl.* 246.

31 Vgl. die kalendarischen *mul*APIN-Texte; ähnlich auch Ptolemäus' *Phainomena*.

32 Vgl. die kalendarisch geordneten *anwā'*-Bücher; dazu auch P. KUNITZSCH in Sudhoffs Archiv 61 (1977) 110 mit Anm. 13.

der Frühzeit der Übernahme antiken Gutes rechnen, über die bisher noch verhältnismäßig wenig und Unzuverlässiges bekannt ist.

Immerhin wurde die Astrologie unter den ʿAbbāsiden schon seit sehr früher Zeit aktiv praktiziert. Bei Gründung der Stadt Bagdad 762 wurde der günstige Zeitpunkt für die Grundsteinlegung durch ein Konsortium von Astrologen errechnet[33]. Die Kalifen hatten ihre Hofastrologen. Wir kennen eine Reihe individueller Namen, die zum Teil auch als Autoren einschlägiger Werke oder Übersetzungen erwähnt werden[34]. An Zeitgenossen des Abū Nuwās auf diesem Fachgebiet wären zu nennen al-Faḍl ibn Nawbaḫt, aus jener bekannten persischen Familie, zu der Abū Nuwās in engem persönlichem Verhältnis gestanden haben soll, Māšā'allāh, ʿUmar ibn al-Farruḫān aṭ-Ṭabarī, der ebenfalls Perser war, Angehörige der Familie Banū l-Munaǧǧim, und last not least Yaḥyā ibn Ḫālid selbst, der Adressat unseres Gedichtes, der als Anreger von Übersetzungen mehrfach hervorgetreten ist[35] und in dessen Umkreis Abū Nuwās ebenfalls reichlich Gelegenheit gehabt haben dürfte, mit entsprechenden Fachleuten zusammenzutreffen. Es gibt also gar keinen Zweifel, daß zu Abū Nuwās' Zeit in Bagdad die Astrologie bereits fest verankert war, so daß er Gelegenheit hatte, sowohl im persönlichen Umgang als auch aus allerlei Fachtexten sich entsprechende Kenntnisse anzueignen. Im übrigen ist es interessant, daß in diesem frühen Stadium das persische Element bei der arabischen Übernahme antiken Gutes merklich dominierte.

An einzelnen Texten aus jener Zeit, die dazu in Frage kämen und die inzwischen durch Editionen zugänglich sind, will ich hier zwei anführen: das astrologische Lehrwerk des Dorotheus von Sidon, das durch den bereits erwähnten ʿUmar ibn al-Farruḫān aus einer persischen Fassung ins Arabische übersetzt wurde[36], und das grundlegende Lehrwerk der Astrologie, die *Tetrabiblos* des

33 Cf. P. Kunitzsch: *Über das Frühstadium der arabischen Aneignung antiken Gutes.* In Saeculum 26 (1975) 268 ff.; hier 275 mit Anm. 41; ders., in Der Islam 52 (1975) 212 mit Anm. 15.

34 Siehe M. Ullmann: *Die Natur- und Geheimwissenschaften im Islam.* Leiden 1972 (Hdb. d. Orientalistik, 1. Abt., Ergänzungsbd. 6, 2. Abschnitt) 302-307. Von F. Sezgins *Geschichte des arabischen Schrifttums* lag bei Abfassung dieser Arbeit Band 7 noch nicht vor, der die Astrologie behandeln soll; auch dort sind hierüber weitere Informationen zu erwarten.

35 Cf. P. Kunitzsch: *Der Almagest. Die Syntaxis Mathematica des Claudius Ptolemäus in arabisch-lateinischer Überlieferung.* Wiesbaden 1974, 18 und 19 mit Anm. 22.

36 Arabisch, mit englischer Übersetzung, ediert von D. Pingree, s. das Abkürzungsverzeichnis.

Ptolemäus[37], von der im Lauf der Zeit mehrere arabische Übersetzungen angefertigt wurden und mit der sich auch bereits derselbe ʿUmar ibn al-Farruḫān beschäftigt hat, sei es daß er die *Tetrabiblos* selbst übersetzt oder eine vorliegende Übersetzung kommentiert hat[38].

Die oben erwähnten von Abū Nuwās verwerteten astrologischen Details sind in diesen beiden Werken auch nachzuweisen. Über die 'Häuser' der Planeten handeln Dorotheus S. 4,21-25 und *Tetrab.* I,17 (S. 79-83); die 'Exaltation' der Planeten ist behandelt bei Dorotheus S. 4,31-5,6 und in *Tetrab.* I,19 (S. 89-91); über die *gaudia* der Planeten handeln Dorotheus S. 4,25-27 und *Tetrab.* I,23 (S. 111-113); und die vier Trigone der Tierkreiszeichen sind besprochen bei Dorotheus S. 3,24-4,6 und in *Tetrab.* I,18 (S. 83-87). Über die Einteilung der Planeten in »gute« und »böse« siehe *Tetrab.* I,5 (S. 39).

Damit ist nicht gesagt, daß Abū Nuwās aus diesen beiden Werken, oder aus einem von ihnen, geschöpft hat. Es soll nur gezeigt werden, daß es sich bei Abū Nuwās' Kenntnissen um Dinge handelt, die zu seiner Zeit in Bagdad durchaus bekannt und auch schriftlich greifbar waren.

Neben diesen Kenntnissen fremden, nichtarabischen Ursprungs waren aber Abū Nuwās auch die einheimischen himmelskundlichen Traditionen der alten Araber bekannt, wie man aus seiner passenden funktionalen Verwendung der Angaben über *aš-šiʿrā*, *aṯ-ṯurayyā*, *suhayl*, *aš-šaraṭān* und die *ǧawzāʾ*-Sterne ersehen kann.

Als letztes haben wir uns nun noch dem auffälligsten Stück an Abū Nuwās' Gedicht zuzuwenden: dem griechischen Götter- und Planetennamen Zeus, *Zāwīš*[39].

Von der Sache her handelt es sich hierbei um etwas Bekanntes, nicht Ungewöhnliches. Das Besondere des Falles liegt aber darin, daß der griechische Name in einem sehr frühen arabischen Text, und noch dazu in einem Gedicht auftritt, wo man am allermeisten eine reine Arabizität erwarten würde, und dazu in einer Form, die von den »gängigen« arabischen Wiedergaben des griechischen Götternamens sichtlich abweicht.

Die geschichtliche Entwicklung der Planetennamen war, kurz gesagt, so verlaufen, daß die Griechen, im Anschluß an ihre babylonischen Vorgänger, die Planeten nach bestimmten Göttern benannten, wobei zunächst der Typus »Stern

37 Bisher ist noch keine der arabischen Fassungen ediert. Es liegt nur der griechische Text vor, s. das Abkürzungsverzeichnis.

38 Je nachdem, wie man in diesem Zusammenhang den Terminus *fassara* im *Fihrist* (ed. FLÜGEL 268,6-7; 273,15) auffassen will. Zu *fassara* im Sinne von »übersetzen« cf. KUNITZSCH: *Almagest* (wie Anm. 35), 17 f., Anm. 11.

39 Wegen dieses Namens war das Gedicht bereits SULAYMĀN AL-BUSTĀNĪ aufgefallen, so daß er es als Belegstück in seiner arabischen Ilias-Übersetzung mit abdruckte: *Ilyāḏat Hūmīrūs*. Kairo 1904, 205, Anm. 1. AL-BUSTĀNĪ selbst verwendet in seiner Übersetzung die der neugriechischen Aussprache nachgebildete Form *Zafs* = Ζεύς.

des [Gottes] N.N.« (hier also ὁ τοῦ Διὸς ἀστήρ, »Stern des Zeus«) vorherrschte, der allmählich, in einer langen Übergangsperiode durch die verkürzte direkte Benennung mit dem Götternamen selbst ersetzt wurde (Jupiter dann also: Zeus, Ζεύς). Dem griechischen Vorbild schlossen sich die Römer an, und so kamen im Westen die bekannten Planetennamen in allgemeinen Gebrauch[40].

Die arabische Sprache besitzt demgegenüber aus sehr alter Zeit eigene Namen für die Planeten, deren Herkunft im einzelnen nicht sicher geklärt ist[41]. Der Planet Jupiter heißt hier *al-muštarī*[42]. Daneben wurden gelegentlich auch die persischen Planetennamen benutzt[43] (cf. hier bei Abū Nuwās: *bahrām* für Mars).

Es steht außer Frage, daß den Arabern durch Übersetzungen antiker Texte unter anderem auch der Name des griechischen Göttervaters Zeus bekannt geworden ist. Dabei ist jedoch zu beobachten, daß in den bisher bekannten astronomischen und astrologischen Texten die griechischen Planetennamen normalerweise mit den entsprechenden einheimischen arabischen (und gelegentlich mit den persischen) Namen wiedergegeben werden; die griechischen Götternamen werden dabei nicht beibehalten. Wohl begegnen uns diese auch in der arabischen Literatur, aber an anderen Stellen.

Erstens einmal natürlich in unmittelbaren Übersetzungen aus dem Griechischen bzw. in Texten, die direkt auf solche zurückgehen. Als Beispiele nenne ich: al-Fārābī (*Zāwuš*)[44]; Artemidor, *Traumdeutung* (*Zaws*)[45]; die arabische Übersetzung der *Placita Philosophorum* (*Zāwus*)[46]; Arat, *Phainomena*, bei al-Bīrūnī, *India* (*Zaws*)[47]. An allen diesen Stellen ist Zeus aber nur Götter-, nicht Planetenname.

Als Planetenname in astronomischem Kontext erscheint der Name ganz selten, und auch da, wie nochmals betont sei, nicht im aktiven Textgebrauch, sondern lediglich in einer tabellarischen Übersicht der Planetennamen bei verschiedenen

40 Cf. W. und H. GUNDEL: Artikel »Planeten«, in PAULY-WISSOWA-KROLL: *Real-Encyclopädie der classischen Altertumswissenschaft*. 40. Halbbd. (1950), col. 2029-2032.

41 Siehe die Deutung der einzelnen Namen bei EILERS: *Planetennamen*.

42 Cf. oben Anm. 8 sowie EILERS: *Planetennamen* 84f.

43 Vgl. oben Anm. 14.

44 Alfarabius: *Compendium Legum Platonis*, ed. F. GABRIELI. London 1952, 5,3 (arab. Text). (Für die Besorgung von Kopien der einschlägigen Seiten des in Deutschland seltenen Buches danke ich Herrn Prof. R. Sellheim, Frankfurt/M.).

45 Vgl. am besten E. SCHMITT: *Lexikalische Untersuchungen zur arabischen Übersetzung von Artemidors Traumbuch*. Wiesbaden 1970, 228; dort die übrige Literatur.

46 H. DAIBER: *Die arabische Übersetzung der Placita Philosophorum*. Saarbrücken 1968 (Diss., Masch.-schr.) 9,13.15 (arab. Text).

47 *India*, ed. SACHAU 46ff. = ed. Hyderabad 1958, 72ff. = Übers. SACHAU 95ff.; dazu J. KRAEMER in ZDMG 106 (1956) 270ff.

Völkern bzw. in verschiedenen Sprachen: al-Bīrūnī, *Chronology* (*Zāwus*)[48]; eine ähnliche Übersicht findet sich in einer Dubliner alchimistischen Handschrift (*Zāwuš*)[49]. Die arabisch-islamische Astronomie und Astrologie scheinen also ganz offensichtlich die griechischen Namen für die Planeten nicht anzuwenden.

Ein Gebiet, auf dem »Zeus« dagegen recht häufig anzutreffen ist, ist die Alchimie. Wie bekannt, werden in der Geheimsprache der Alchimisten die Elemente gern mit »Decknamen« bezeichnet, die den verschiedensten Bereichen entnommen sind. Beliebt sind darunter auch Planeten- und Sternnamen. Für bestimmte Stoffe stehen dabei bestimmte Planetennamen als Decknamen fest. Zinn (*qalaʿī*) bzw. Weißblei (*raṣāṣ*) werden so, einer astrologischen Zuordnung der Metalle an die Planeten folgend, durch »Jupiter« umschrieben, wobei teils der arabische Name *al-muštarī* gebraucht wird, häufig aber auch der griechische Planetenname Zeus, cf. z. B. Stellen bei Ǧābir ibn Ḥayyān (*Zāwuš*, *Zāwus*)[50]; WIEDEMANN (*Zāwuš*, *Rāwus* [mit Verlust des Punktes über dem ersten Buchstaben, also *R* statt *Z*], *Zawāš*)[51]; SIGGEL, *Decknamen* (*Zāwuš*, *Zāʾūš*, *Zawš*)[52]; und eine Dubliner Handschrift (*Zāwuš*)[53]. Auch in einem syrischen alchimistischen Text erscheint der Name als *Zaws* (*Zws*[54]).

Für uns ist es nach all diesem interessant zu sehen, daß Abū Nuwās Kenntnis von dem griechischen Götternamen im Sinne eines Planetennamens hatte und ihn so, als astronomische Benennung, in sein Gedicht einfügte. Die Benennung muß also im zeitgenössischen Bagdad doch bekannter gewesen sein, als wir es aus den spärlichen textlichen Zeugnissen, die zum Teil auch erst späteren Jahren entstammen, entnehmen können.

Dabei stoßen wir auf ein weiteres Problem. Wie oben ersichtlich, erscheint der Name Zeus in arabischer Form in zwei Gundtypen, deren einer den griechischen Diphthong ευ durch den einfachen arabischen Diphthong *aw*

48 *Chronology*, ed. SACHAU 192 = Übers. SACHAU 172.

49 Siehe M. ULLMANN: *Katalog der alchemistischen Handschriften der Chester Beatty Library*. I. Wiesbaden 1974, 149 ult. (hier auch mit astrologischer Klassifizierung als *saʿd as-suʿūd* entsprechend dem sonstigen *as-saʿd al-akbar* bzw. *al-aʿẓam*). Eine gleichartige mehrsprachige Planetenliste enthält ferner ebenfalls der *Picatrix*; hier wird für Jupiter zitiert «*Zāūsh*», in der lateinischen Übersetzung *Raus* (aus dem unpunktierten arabischen Schriftzug); siehe D. PINGREE in Viator 7 (1976) 178.

50 P. KRAUS: *Jābir ibn Ḥayyān*. II. Kairo 1942, 33 Anm. 1, 232, 260. (Einen Hinweis auf diese Stellen verdanke ich Herrn Prof. M. Ullmann, Tübingen.)

51 E. WIEDEMANN: *Aufsätze zur arabischen Wissenschaftsgeschichte*. 2 Bände. Hildesheim/New York 1970, II 604 f. (Nr. V, 15. 17; und Dresden, 23).

52 A. SIGGEL: *Decknamen in der arabischen alchemistischen Literatur*. Berlin 1951, 15, 18, 41.

53 Bei ULLMANN: *Katalog* (wie in Anm. 49), 64,15.

54 Nach C. BROCKELMANN: *Lexicon Syriacum*. Halle ²1928, 192 b.

wiedergibt (*Zaws*; so auch die syrische Form *Zws*), während der andere mehr buchstabengetreu das ε des Diphthongs separat mit *alif* umschreibt (Typus *Zāwus*[55])[56]. Dieselbe Ambivalenz ist auch an den arabischen Transkriptionen anderer griechischer Eigennamen auf -εύς zu beobachten: vgl. die Sternbildnamen Kepheus, Κηφεύς (arab. *qyfws* und *qyf'ws*) und Perseus, Περσεύς (arab. *brsws* und *brs'ws*) aus dem *Almagest* des Ptolemäus, in denen die Form mit *alif* offenbar den jüngeren, gegenüber dem Griechischen »exakteren« Typus repräsentiert[57,58].

Abū Nuwās' Form *Zāwīš*[59] scheint von dem Typus *Zāwus*, mit *alif*, auszugehen. Für uns unerkennbar bleibt, ob die Einfügung des *y* vor dem *š* absichtlich erfolgte, um eine metrisch erforderliche zweite lange Silbe *-wī-* zu gewinnen, oder ob die Form lediglich aus einer schlecht lesbaren schriftlichen Vorlage hervorgegangen ist, in welcher vor den drei Zähnen des *sīn* (*šīn*) noch ein weiterer Zahn erkennbar war, der dann als *y* aufgefaßt wurde, oder ob gar die Form überhaupt nicht

55 Hiervon leiten sich auch die neupersischen Formen ab, die W. EILERS anführt: *Zāvuš, Zāvūš, Zāvaš, Zavāš* (*Planetennamen* 80f.) und die gelegentlich sogar für andere Planeten als Jupiter stehen sollen (*ib.* 53f., 55: *Zāvuš, Zūš* = Merkur; 63, 66 [mit Nachtrag S. 102]: *Zāvars, Zāvas* = Venus).

56 Auf die schwankende Schreibung des *s*-Lautes als *sīn* oder *šīn* braucht hier nicht näher eingegangen zu werden, da es sich dabei um eine rein graphische Variante handeln kann, die den eigentlichen Transkriptionstypus nicht berührt. (Vgl. dazu auch KRAUS: *Jābir* [wie in Anm. 50], 260 Anm. 6.)

57 Cf. KUNITZSCH: *Almagest* (wie oben Anm. 35), 173 und 180f.

58 In der aus einer syrischen Vorstufe entstandenen, leider sehr korrupt überlieferten arabischen Übersetzung von Aristoteles' *Poetik* ist der Name des Zeus an der einzigen Stelle, wo er im griechischen Text vorkommt, leider ausgelassen (siehe J. TKATSCH: *Die arabische Übersetzung der Poetik des Aristoteles*. 2 Bände. Wien 1928-1932; hier I 278, ult.). Der ähnlich anlautende Name des Malers Zeuxis (Ζεῦξις, gen. Ζεύξιδος) erscheint hier einmal mit *alif*: I 280,12 (*z'wkss*), und zweimal korrupt, anscheinend ohne ursprüngliches *alif*: I 232, ult. (*zwksyds*) und I 234,2 (*wksyds*). Namen auf -εύς sind dort z.B. Peleus, Πηλεύς (I 258, ult.: *fylws*) und Odysseus, Ὀδυσσεύς (I 238,8.12: *'wdwsws*; I 266,13: *'dwsws*; aber I 280,8: *'wdws'ws*). Offenbar sind hier beide Transkriptionstypen, mit und ohne *alif*, durcheinander gebraucht.

59 *Dīwān* ed. ĠAZZĀLĪ 476,5 enthält den Namen ohne jeden Punkt: *Rāwis* (so auch die *Dīwān*-Ausgabe Dar Sader/Dar Beyrouth. Beirut o. J., 84,6). WAGNER hat in *Dīwān* I 151,10 *Zāwiš*, ohne Variante im Apparat; diese Form auch in der Wiedergabe von S. AL-BUSTĀNĪ (cf. oben Anm. 39) und im *Dīwān*, ed. M. AS-SAMĀLŪṬĪ. Kairo 1277, 17. (In der *Dīwān*-Ausgabe von I. M. I. AN-NABAHĀNĪ. Kairo 1322, ist unser Gedicht, soweit ich sehe, nicht enthalten.) Bei AL-ĠAZZĀLĪ a.a.O. Anm. 2 gibt es dazu übrigens die kurze, oberflächliche Erklärung *Rāwis wa-l-ḥūt: naġmān*, die sich auch in der genannten Beiruter Ausgabe wiederfindet, anscheinend direkt aus ĠAZZĀLĪs Ausgabe übernommen.

unmittelbar aus einer arabischen Quelle stammt, sondern aus einer persischen Vorform, bei der sich irgendwie der *i*-Laut vor dem *s/š* eingeschoben haben könnte. Den arabischen Philologen waren offenbar sowohl der Name an sich ungewöhnlich als auch die entsprechende Quelle nicht mehr greifbar, wie man der Glosse in Ḥamzas Rezension entnehmen muß: *Zāwīš ism al-muštarī bi-r-rūmīya fa-innahā tusammīhi Zayyūš*[60] (*Dīwān* I 151,10). Immerhin hatte Ḥamza bei Zusammenstellung seiner Dīwān-Rezension im 10. Jahrhundert schon die gesamte Palette der oben aufgezählten Texte zur Verfügung, in denen »Zeus« vorkommt. Aber die »arabischen« Philologen werden diese unklassischen Texte der Übersetzungs- und Wissenschaftsliteratur nicht sehr intensiv ausgewertet haben.

Wir scheiden von diesem interessanten Gedicht, das uns so viele Einblicke in die kulturgeschichtlichen Zusammenhänge des literarischen und wissenschaftlichen Lebens im Bagdad des ausgehenden 8. Jahrhunderts beschert, indem wir noch eine erwähnenswerte Beobachtung anschließen. Bei aller Reichhaltigkeit der fernliegenden, zum Teil »artfremden« Elemente, die Abū Nuwās in sein Gedicht eingeschlossen hat, hat er darin im Grunde doch das Wesentliche arabischer Poesie bewahrt. Wie wir gesehen haben, können sich seine astronomischen Angaben nicht auf ein wirkliches Ereignis beziehen. Damit wäre also jeder »epische« Einschlag ausgeschlossen, und das Gedicht ist in seinem Grundcharakter, in Übereinstimmung mit dem der echten arabischen Poesie, eine rein beschreibende Darstellung, ein *waṣf*.

60 Auch hier könnte das *y* sekundär eingeschoben sein, so daß diese Form dem Typus *Zws* (*Zaws*) zuzuordnen wäre. (Wie weit Schreiberirrtümer hierbei für das zusätzliche *y* verantwortlich sind, läßt sich schwer entscheiden. An sich erscheint die Überlieferung des Dīwāns, wie ein Blick in WAGNERS Apparat zeigt, relativ stabil.) Eine Form des gleichen *rasm*, *Zyws*, erscheint auch in der arabischen Version von Aristoteles' *De caelo* (ed. A. BADAWI. Kairo 1961, 278,7). Diese Stelle ist aber für eine genauere Auswertung nicht ausreichend, da BADAWI nur eine — nicht sehr wichtige — von mehreren Handschriften abgedruckt hat (cf. G. ENDRESS: *Die arabischen Übersetzungen von Aristoteles' Schrift De caelo*. Diss. Frankfurt/M. 1966, 21), da die arabische Fassung mehrfach überarbeitet wurde und obendrein ursprünglich aus einer syrischen Version übersetzt ist, und da hierbei im Griechischen der Genitiv Διός zugrunde lag, der möglicherweise in dieser Form transkribiert worden ist und so von vornherein eine arabische Form *Ḏyws* ergab (die später im Hinblick auf den bekannteren Nominativ Zeus mit *Z* am Anfang geschrieben worden sein mag). Hierzu bleibt eine künftige kritische Edition nach allen Versionen und Handschriften abzuwarten.

Bibliographische Abkürzungen:

Abū Nuwās: *Dīwān* = Abū Nuwās: *Dīwān*, ed. E. Wagner. I. Kairo/Wiesbaden 1958; II. Wiesbaden 1972.

Dīwān ed. Ġazzālī = *Dīwān Abī Nuwās*. Ed. A. 'A.M. al-Ġazzālī. Kairo 1953.

Dorotheus = *Dorothei Sidonii Carmen Astrologicum*. Ed. D. Pingree. Leipzig 1976 (Bibliotheca Teubneriana).

Eilers: *Planetennamen* = W. Eilers: *Sinn und Herkunft der Planetennamen*. München 1976 (SB Bayer. Akademie der Wissenschaften, Philos.-Hist. Kl., Jahrg. 1975, Heft 5).

Kunitzsch: *Dicht. Vergl.* = P. Kunitzsch: *Der Sternhimmel in den »Dichterischen Vergleichen der Andalus-Araber«*. In ZDMG 128 (1978) 238-251.

Ptolemäus: *Tetrabiblos* = Ptolemy: *Tetrabiblos*. Ed. u. übers. F.E. Robbins. Cambridge, Mass./London 1971 (The Loeb Classical Library).

Tibbetts: *Navigation* = G.R. Tibbetts: *Arab Navigation in the Indian Ocean Before the Coming of the Portuguese*. London 1971.

Wagner: *Studie* = E. Wagner: *Abū Nuwās. Eine Studie zur arabischen Literatur der frühen 'Abbāsidenzeit*. Wiesbaden 1965.

Nachtrag zu S.103, 14ff. (zā'idan fī l-ḥisābi)

Nach Abschluss des Druckes stiess ich noch auf eine Stelle bei al-Bīrūnī, *Tafhīm* ed. Wright p. 111f. (§ 203), wo mehrere Stellen von *ziyāda* und *nuqṣān* der Planeten angeführt werden. Davon bezieht sich die erste Gruppe auf die Örter der Planeten auf ihren Deferenten und Epizykeln. Die vierte Unterart dieser Gruppe (a.a.O., p. 112, Nr. 4) heisst *az-ziyāda fī l-ḥisāb* ("increase by calculation"). "This occurs when the equation of the anomaly has to be added to the true longitude of the centre to obtain the true position of the planet in the ecliptic." Im Falle der Addition heisst der Planet dann *zā'id fī l-ḥisāb* ("mit hinzugefügtem [Wert zur Bestimmung des wahren Ortes auf der Ekliptik, gemäss den hier angegebenen Regeln]"), im Falle der Subtraktion *nāqiṣ fī l-ḥisāb* ("mit abgezogenem [Wert etc.]"). Ich glaube indes nicht, dass Abū Nuwās mit seinem gleichlautenden Ausdruck diesen Spezialfall einer genauen Positionsbestimmung im rein theoretisch-astronomischen Zusammenhang im Sinne hatte. Das gesamte Gedicht bewegt sich in astrologischen Gedankengängen und verwendet termini der astrologischen Fachsprache. Es wäre daher im doppelten Sinne unpassend, wenn *zā'idan fī l-ḥisābi* hier in dem von al-Bīrūnī beschriebenen theoretisch-astronomischen Sinne verwendet wäre: einmal wäre damit die an sich einheitliche rein astrologische Assoziationsreihe gesprengt, und zweitens ist die Position des Mars ja bereits hinreichend charakteristisch bezeichnet (im Skorpion, kulminierend), um einen astrologischen Aussagewert zu erzielen, während dagegen die rein astronomische verfeinerte Positionsbeschreibung innerhalb des Systems der zusammengesetzten Planetenkreisbahnen per se zu keiner astrologischen Schlussfolgerung führen würde.

Note to article XXVIII

In an article in the subsequent issue of the same journal (*Der Islam* 60 [1983], pp. 297-301), Prof. O. Neugebauer (Princeton, N.J.), from an improved interpretation of the position of Venus, has derived the date of A.D. 968 for the horoscope.

XXVIII

The 'Description of the Night' in Gurgānī's *Vīs u Rāmīn*

The Persian verse epic of *Vīs u Rāmīn,* composed by Faḫr ad-Dīn As'ad Gurgānī in Isfahan around A.D. 1050, does not occupy a very prominent rank in Persian literary history.[1]) It has gained, however, a certain importance in comparative literary studies during the last hundred years, on the grounds that its plot – the love of both Šāh Mawbad and his younger brother Rāmīn for Vīs – and some of its details have been assumed to have served as a model for the composition of the mediaeval European story of King Mark, Tristan, and Isolt, an assumption which, however, cannot be supported by any realistic historical evidence.[2]) One highly interesting feature of this epic is that according to the author's own words its story dates back to older Persian times, Sasanian, or even Arsacid.[3]) The task remains, therefore, of looking for special details that could prove – or disprove – older datings of the story. With regard to this problem, Dr A. de Mandach (Université de Neuchâtel), at the Arthurian Congress of Regensburg,[4]) drew my attention to King Mawbad's 'horoscope' raising the question whether this 'horoscope' might contain astronomical elements which could contribute to the dating of the poem. Following this suggestion, the present article was composed with the two aims of commenting upon this section of Gurgānī's text, and examining the possibility of deriving a date from the astronomical elements in the 'horoscope'.

King Mawbad of Marv was promised by princess Šahrū of Hamadan to give him a daughter who was still to be born, in marriage. When the daughter, Vīs, was born

---

[1]) See P. Horn, *Geschichte der persischen Litteratur,* Leipzig, 1901, pp. 43 and 179ff.; E. G. Browne, *A Literary History of Persia,* ii, Cambridge, 1951, p. 274f.; J. Rypka, *Iranische Literaturgeschichte,* Leipzig, 1959, pp. 176–178; A. Pagliaro and A. Bausani, *Storia della letteratura persiana,* Milano, 1960, pp. 621–626; H. Massé, article "Gurgānī, Fakhr al-Dīn As'ad," in *EI*², ii, Leiden/London, 1965. For text editions and translations, see the *References* at the end of this article.

[2]) See the latest analysis and conclusions by P. Kunitzsch, presented in a paper at the Twelfth International Arthurian Congress, held at Regensburg, August 1979: "Are There Oriental Elements in the Tristan Story?", published in *Vox Romanica* 39 (1980), 73–85.

[3]) These problems have recently been studied by Minorsky, Minovi, and others without, however, yet attaining uncontested results.

[4]) Cf. above, footnote 2.

and grown up, the princess forgot her promise and married Vīs to her son Vīrū, Vīs's brother. Eventually King Mawbad, with the help of Šahrū, abducted Vīs to make her his wife. The night when Mawbad came to the castle of Gūrāb to abduct Vīs – the night that was so decisive for the King's fate, his troubles with Vīs, and his end – is described at great length in the poem under astrological aspects. It is interesting to see that the poet, here, does not content himself with giving a normal horoscope – i.e., the positions of the planets for the respective moment – but he extends the astronomical description by adding a portrayal of the whole sky with all the constellations of the fixed stars, which is unparalleled in mediaeval oriental poetry.

This important section, the 'Description of the Night', consists of 57 verses which, in some MSS. and printed editions, are counted as a separate chapter (ch. 29: *andar ṣifat-i šab,* ed. MINOVI pp. 80–83; ed. MAḤǦŪB pp. 60–63).[5]) In P,[6]) the 'Description of the Night' is incorporated in the chapter '*Nāma nivištan-i Mawbad nazd-i Šahrū . . .*' ("Mawbad's writing a letter to Šahrū . . ."), of which it comprises the verses 53 to 108 (pp. 87–91; 56 verses only).

The fatal night is described as being dark and evil, indicating imminent dangers and bad luck. The section begins with a general account of the bad aspects of the night (P 53–64). Then follow specialized descriptions of the twelve zodiacal constellations (P 65–78), of nineteen constellations of the northern celestial hemisphere (P 82–90), and of the fifteen constellations of the southern celestial hemisphere (P 91–101). Finally comes what is to be called a horoscope proper, giving the positions of some of the planets, and some remarks on the bad aspects to be derived from these positions (P 102–108).

Thus, out of a total of 57 (or 56) verses, only seven verses really form a horoscope, while all the rest is just a descriptive passage serving to depict the sinister aspects of the moment and to emphasize the prominent rôle the sky played in foreboding the coming troubles.

In an analysis of the 'Description of the Night' we may thus say that it falls into three sequences of increasing importance: First, general topics which require no special knowledge from the author, and which might have been composed in any epoch, early Persian as well as Islamic. Second, descriptions of the constellations. This is an astronomical speciality going, for the most part, beyond the level of a mediaeval Islamic poet's general literary education. It requires the knowledge, and

---

[5]) Trans. MASSÉ, pp. 72-75; trans. MORRISON, pp. 56–58. The Georgian version of *ca.* A.D. 1200 omitted the major part of the celestial description and ends with the zodiacal sign of Pisces (corresponding to P 78b); cf. trans. WARDROP, p. 43. K. H. GRAF, in his account of *Vīs u Rāmīn* in *Zeitschrift der Deutschen Morgenländischen Gesellschaft* 23 (1869), pp. 375–433, gives only a superficial résumé of the 'Description of the Night', in a few lines, without entering upon details (p. 389, at the bottom).

[6]) I.e., in the edition of TODUA/GWAKKHARIA/AINI, 1970. See the *References* at the end of this article. In the following, references to P are always made by verse numbers of this same section, the letters a and b indicating the first and second half-verses.

use, of special sources by the poet which, however, might have been available in pre-Islamic Persia as well as in Islamic Persia. For the particular purpose of the section – the astrological prediction of Mawbad's fate with Vīs – this item is of no direct significance, the horoscope being constituted by the positions of the planets only. But the lengthy description of those strange, and perhaps intimidating figures was devised by the poet in order to impress his readers/listeners and to convey to the scene more weight and a higher degree of astronomical authenticity. Third, the horoscope proper. This, finally, is the device which is used to establish predictions for the future. But these astrological techniques were also known in pre-Islamic times as well as in the Islamic period, so that their use in the poem is of no help with regard to its dating. The only chance of finding a real hint to the date of the composition of this passage, or the story itself, would be, then, the examination of the planets' positions, about which we shall speak later in this article.

Let us, now, examine more closely item two, the description of the constellations. Even a superficial inspection of the section (P 65–101) shows that the poem here offers a complete 'catalogue' of the classical constellations. These constellations had been formed in classical antiquity, most of them after Babylonian models. Around the year A.D. 138, Ptolemy, the famous Alexandrian astronomer, following his predecessors, established a catalogue of 1025 individual fixed stars, arranged in 48 constellations. This catalogue, part of his great astronomical handbook Μαθηματικὴ Σύνταξις (*"Almagest"*), served as a basis for stellar astronomy until the seventeenth century, and was adopted, together with the *Almagest* itself, by the Mediterranean and neighbouring civilizations. Thus, it is not improbable that the *Almagest*, or selected parts of it, became known in Persia under Šāhpūr I (3rd century A.D.), and again under Ḫusraw Anūširvān (6th century A.D.).[7]) In the late eighth and the ninth centuries A.D. several Arabic translations of the *Almagest* were made at Baghdad. The best known and most widely used book on the fixed stars and their constellations, in Islamic times, was written (in Arabic) around A.D. 964 by Abū l-Ḥusayn 'Abd ar-Raḥmān aṣ-Ṣūfī, in close relation to the *Almagest*; its detailed text was accompanied by star tables and by drawings of every single constellation, one as it is seen in the sky, and the other one as it is seen on a celestial globe.

Now that we have found that the poem contains the complete list of the 48 classical constellations (with just two names missing), the important questions arise, when and whence this item was added to the story of *Vīs u Rāmīn* – in pre-Islamic times, or in the Islamic period, by Gurgānī himself? Taking into account a number of details in the descriptions of those figures, and certain linguistic details in the nomenclature used in the Persian poem, we arrive at the conclusion that the description of the 48 constellations in *Vīs u Rāmīn* was added to the poem in later times only, most probably by Gurgānī himself. This part of the 'Description of the Night', therefore, cannot contribute to the problem of the dating of *Vīs u Rāmīn*, not astronomically because the constellations do not show remarkable changes in so short a period as a thousand years or less, and not historically because the descrip-

---

[7]) For this and the following remarks on the transmission of the *Almagest*, cf. P. Kunitzsch, *Der Almagest. Die Syntaxis Mathematica des Claudius Ptolemäus in arabisch-lateinischer Überlieferung*, Wiesbaden, 1974.

tions obviously are borrowed from the classical astronomical material that was translated into Arabic and had spread throughout the Islamic world from the late eighth century onwards.

When looking for sources on which Gurgānī might have drawn in establishing the descriptions of the constellations, there are two sets of things that come into close consideration: astronomical books and celestial globes, both of which were widely spread in his time in the Islamic world. As for the books, we should look first to works in the Persian language, because Gurgānī only exceptionally uses Arabic words in his descriptions which generally are given in Persian. There are three major works, in Persian, containing complete surveys, or catalogues of the 48 constellations two of which came into existence much too late for Gurgānī to use them,[8]) so that only the third one – composed in A.D. 1029, in Ghazna – is of relevance here: al-Bīrūnī's *Kitāb at-tafhīm li-awā'il ṣināʿat at-tanǧīm.*[9]) This work exists both in an Arabic and in a Persian edition, and until now the problem is still unsolved whether one of the two versions is a translation from the other, or whether both versions of the book were put down by al-Bīrūnī himself.[10]) Some similarities between Gurgānī's descriptions and the *Kitāb at-tafhīm* point to the fact that Gurgānī obviously had seen the Persian version, some twenty years after the composition of the work.

Celestial globes showing the figures of the constellations, the positions of the individual fixed stars, and the basic great circles of the celestial sphere, were a well known and much used device since the times of classical antiquity.[11]) Together with all the other classical scientific material they also became known in the Islamic world, and a good number of celestial globes of Islamic fabrication, of varying provenience, are still preserved today. It is, therefore, quite probable that Gurgānī could have seen and used such a celestial globe for the composition of his celestial descriptions.[12]) As for the drawings of the constellations in aṣ-Ṣūfī's book, it cannot be

---

[8]) I.e., Persian translations of aṣ-Ṣūfī's (Arabic) book on the 48 constellations of the fixed stars (composed around A.D. 964), the first one of which was completed in A.D. 1250 by Naṣīr ad-Dīn aṭ-Ṭūsī, and which was followed by two or three more translations in A.D. 1630/31, 1681, etc. (see C. A. Storey, *Persian Literature,* vol. ii, part i, (reprint), London, 1972, p. 41 f.); and the *Zīǧ* of Uluġ Bēg (epoch of the star catalogue: A.D. 1437) which contained the complete Ptolemaic star catalogue in a Persian version (see Storey, *op. cit.,* p. 67 ff., no. 2).

[9]) See Storey, *op. cit.,* p. 44 f.; C. Brockelmann, *Geschichte der arabischen Litteratur,* i (2nd ed.), Leiden, 1943, p. 627 (no. 3); *id., Supplement,* i, Leiden, 1937, p. 872 f. (work no. 3); M. Ullmann, *Die Natur- und Geheimwissenschaften im Islam,* Leiden, 1972, p. 335; F. Sezgin, *Geschichte des arabischen Schrifttums,* vol. 7, Leiden, 1979, p. 189 f. It can hardly be assumed that Gurgānī, who composed *Vīs u Rāmīn* in Isfahan around A.D. 1050, had met al-Bīrūnī, who came to Ghazna in A.D. 1017/18, where he continued to live until his death in A.D. 1048.

[10]) For the text editions, see the *References* at the end of this article.

[11]) Cf., e.g., P. Kunitzsch, *Der Almagest* (*op. cit.* in footnote 7), p. 76 f.

[12]) It may be noteworthy, here, that according to certain reports there existed in A.D. 1043/44 in Cairo a celestial globe of aṣ-Ṣūfī's fabrication (together with a

assumed that they were a source for Gurgānī's descriptions. For aṣ-Ṣūfī only gave isolated drawings of each of the 48 constellations, thus not showing the total view of the stellar sky and thereby the relative positions of the constellations, whereas Gurgānī, who describes some of the constellations in relation to others, seems to have relied on a representation that showed the whole of all the constellations, as was the case with the globes.

When examining Gurgānī's text we are, however, not able to decide safely whether the poet has directly drawn upon the *Kitāb at-tafhīm*, or upon a globe, or both, for his descriptions show a number of discrepancies with respect to both of them. Thus, in the distribution of the 48 constellations,[13]) and in a few linguistic details (which we shall see soon), he shows relations to the *Kitāb at-tafhīm*. On the other hand, he does not assign to many of the constellations the names or designations given by al-Bīrūnī, but describes them as if just collecting their figures – which were absolutely unknown to him as a non-astronomer – from the images on a globe;[14]) also he does not follow the succession of the constellations as given by al-Bīrūnī,[15]) which again suggests that he collected the figures from a globe.

A remark should be added here as to the scientific character of Gurgānī's description of the sky. As we have stated before, Gurgānī's astronomical section is nothing but the mere reproduction of a complex of classical tradition which had, ultimately, been formed around A.D. 138 by Ptolemy and which was transmitted, unchanged and revered like a canon, in the great stream of scientific transmission from antiquity to the Islamic world. He does not describe the stellar sky as he, or some of his contemporaries, or somebody else in other periods or regions, actually observed it. All the astronomical elements involved, as the northern or southern

---

second one said to have been made by Ptolemy himself); cf. the quotations in Kunitzsch, *Der Almagest* (*op. cit.* in footnote 7), p. 76, footnote 196. In our present study we have included, for comparison, a number of Islamic celestial globes of which nice and easily accessible reproductions exist in the literature (globes A, C, E, F, and L; see the *References* at the end of this article).

[13]) Both al-Bīrūnī in his survey and Gurgānī present first the twelve zodiacal signs, then the constellations of the northern celestial hemisphere, and finally the constellations of the southern celestial hemisphere.

[14]) Also, he omits two of the northern constellations which could scarcely have happened if he had followed an edited text, but which is easily comprehensible if he took the figures from a globe where – especially in the northern hemisphere – in some regions the constellations are rather densely packed. But cf. footnote 21a, below.

[15]) In accordance with Ptolemy's astrological *Tetrabiblos* (i,9). In the star catalogue of the *Almagest* Ptolemy has the same sequence of the constellations, but the zodiacal signs are inserted, according to latitude, between the northern and the southern constellations. This system was followed by Abū Ma'šar in his *Kitāb al-mudḫal al-kabīr* (ii,1). Another succession was adopted by Muḥammad ibn Aḥmad al-Ḫwārizmī in his *Kitāb mafātīḥ al-'ulūm*, ed. van Vloten, Leiden, 1895, pp. 210–213; but this one is also different from Gurgānī's.

delimitation of the constellations, and their visibility, etc., singularly are referred to Ptolemy's place and time, and to nothing else.[16])

Now we shall proceed to a detailed examination of the constellations described in *Vīs u Rāmīn.*[17])

A. The twelve zodiacal signs

Here it has to be especially stressed that Gurgānī does not describe these signs in their traditional forms. Moreover, his poetical genius has induced him to turn many details of their descriptions into corrupt, or somehow spoilt forms. Thus, as a poetical device, he has created a travesty of the classical figures to underline the pernicious influence of that night: not only were the planets standing in bad positions, but even the constellations of the fixed stars – renowned for never changing their places and forms – had changed their forms, had turned into broken and damaged figures.[18]) The sky itself broken, the planets in bad positions: what omen could be worse for Mawbad's disastrous undertaking in that night? But only the twelve zodiacal signs and one of the southern constellations are thus travestied. The reason for this may be that the knowledge of the zodiac – as also of the planets – might have formed part of the poet's general education, so that he felt able freely to dispose of this material. The other constellations, however, were less well known (or unknown) to him, so that he had difficulty enough in recognizing and describing their figures, and did not feel able to introduce essential modifications into their descriptions. The same holds good for his readers, who might have been able to follow his intention with the zodiacal signs but certainly could not have done so with the other constellations, which were far too recondite for the common public.[19])

1 and 2: The Ram, and the Bull (P 65 calls them by their Arabic names, *ḥamal* and *ṯawr*). Gurgānī describes them as standing "face to face" (*rūy dar rūy*) which

---

[16]) For the benefit of correctness we shall add that the only change that was introduced to this classical complex by the astronomers of later times was the addition of the value of precession to the longitudes of the stars, according to their respective epochs.

[17]) I usually quote the English translation of Morrison. The corresponding verses of P are always added. al-Bīrūnī and the globes are cited according to the system given in the *References* at the end of this article. Maḥǧūb, in a commentary appended to his edition of the poem, compares Gurgānī's text to that of al-Bīrūnī's *Kitāb at-tafhīm* (pp. 451–455). Reference to this commentary is made in the present article only in case of errors and false interpretations. The horoscope itself is not discussed in Maḥǧūb's commentary.

[18]) Travesties of single constellations, based on etymological speculations, have been observed in Arabic poetry and in Niẓāmī's works; cf. P. Kunitzsch, in *Zeitschrift der Deutschen Morgenländischen Gesellschaft* 128 (1978), pp. 240f. and 242f. (*aydī ṯ-ṯurayyā*, and *arǧul al-ǧawzā'*), and 250 footnote 29.

[19]) It is significant in this context that the Georgian version of *Vīs u Rāmīn,* of *ca.* A.D. 1200, broke off the celestial description at the end of the zodiacal signs leaving out the description of the other constellations and the horoscope.

departs from the pictorial situation on the globes where the Ram's tail is on the eastern side, and its head on the western side (but turned backwards, eastward), while the Bull, which follows east of the Ram, has its head towards the east.[20])

3: The Twins (*do paykar,* P 67. The common Persian name, but not in BP, p. 90. Wright, however, in B, p. 70, quotes *do paykar* from some Persian MSS. of the *Kitāb at-tafhīm*).

4: The Crab (P 68). In reality, the Crab lies near the Twins' heads, not their feet.

5: The Lion (called by its Arabic name, *asad.* P 69). On the globes, the Lion's tail is never bent back towards, or over its head. It is always in its hind, eastern, part, near the bright star β Leonis (Ptolemy: "on the end of the tail").

6: The Virgin (P 71: *zan-i dōšīza*; not in BP, p. 90, but quoted by Wright in B, p. 70, from Persian MSS). To the description given in BP and B Wright adds, from some Persian MSS.: "in her hand an ear or two of corn . . .". This is the traditional figure of the Virgin, and thus, in P 71a, *do ḫūša* has to be translated "two ears of corn" (in her hand), not "two bunches of vines" as Morrison has it (cf. the dictionaries);[21]) Massé has correctly "deux épis" (p. 73). The Georgian version seems to have taken *ḫūša* in the wrong sense, cf. Wardrop, p. 43, "vine sprouts."

7: The Scales (P 72). Its lines, or ropes, are broken, and so is the beam; the two pans are "dismissed, or abandoned" (I take *mānda* in this sense, registered by the dictionaries, instead of "left," as in Morrison, p. 56).

8: The Scorpion (P 73). "Head and tail brought together," sounds as if the animal was going to kill itself. In reality, the tail is bent away from the Scorpion's body in a large curve thus giving the constellation its characteristic view.

9: The Archer (P 74).

10: The Capricorn (P 75–76). Several MSS. and the editions of Minovi and Maḥǧūb have the reading *barra* for this sign (followed by Massé, p. 74: "le Bélier," and by Morrison, p. 57: "the Lamb," and not corrected in Maḥǧūb's commentary, p. 451) which, however, is the Persian name for Aries, no. 1. The better reading, instead, is *buza* (P 75–76), a derivation from *buz* ("a she-goat; the goat species"), and a fully appropriate Persian designation for Capricorn (since Ptolemy: Αἰγόκερως, "the goat-horned"), the common Persian name being *buzġāla.* The reading *buza* is also confirmed by the poem's metre (*hazaǧ* [*maḥḏūf*]) which requires, in the beginning of the verse, the quantities ⏑ – which

---

20) Morrison's translation "*strength* drained . . ." very suitably follows the variant reading *rūzišān* (P 66b, where the text itself has, less good, *rawšanān.* This reading was followed by Massé in his translation, p. 73: "tandis que s'enfuyaient promptement *les étoiles*"). "Hind legs," here, can apply to the Ram alone, because the Bull is depicted in the sky as a προτομή (first half) only.

21) In the translation of verse P 71b, Morrison has omitted *za-sustī,* "of debility, or languor" (. . . rooted to the spot).

are exactly furnished by the word *buza* (*barra,* instead, would be – –). The Georgian version has correctly "the Goat" (WARDROP, p. 43). Contrary to MORRISON, p. 57, the sign of Capricorn begins already at P 75 ("the Lamb sleeping . . ."), and not at P 76 ("then the Lamb, . . ."). The "grass" and the "tulips" are of course not seen in the sky, the poet added these details from his mind.

11: The Water Carrier (P 77). It may be noted that Gurgānī calls the "bucket" by the Arabic word *dalw* which, in the indigenous Arabic tradition, was the designation for the sign of Aquarius itself (referring to the vessel only and not to the person handling it). The word does not appear in B, p. 70 (here: *kūz*), and in BP, p. 91 (here: *kūza*). Gurgānī might have taken it from a globe, or it may have formed part of his general education (just as the Arabic names of the signs of Aries, Taurus, and Leo, nos. 1, 2, and 5 above). The Water Carrier, originally raising a vessel in his right hand and pouring out water that runs down beneath his feet, has lost the vessel; it fell down the well (which, properly, does not exist in the constellation), and he remains confounded (*ḫīra*) like one who has lost his way (*gumrāh*).

12: The Fish (P 78). Properly this sign is made up of two fishes; cf. B, p. 70, and BP, p. 91.

B. The constellations of the northern celestial hemisphere

As told above, Gurgānī in enumerating the northern constellations does not follow the Ptolemaic sequence. It seems, also from his several indications of constellations as being placed "in front of," or "behind," or "opposite" others, that the poet has used a celestial globe in the elaboration of this section. For some reason he omits two constellations so that, of Ptolemy's 21 northern constellations, he only names 19 (the missing two constellations are the Lyre, Lyra, = Ptolemy, no. 8, and the Triangle, Triangulum, = Ptolemy, no. 21).[21a)]

1: "The Dragon," Draco[22)] (= Ptolemy, no. 3), P 82. B, p. 71 = BP, p. 91 states that the Dragon is twisted around the north pole of the ecliptic. This more specifically applies to the front half of the Dragon's body, while the back half is winding, at a greater distance, around the equatorial north pole, that is, the pole generally called "The North Pole" by us. Gurgānī's reference is to this latter pole, whether intentionally, or by just copying it from a globe, or by confounding al-Bīrūnī's statement, it is not possible to decide. The Dragon is called, in the poem, by its Arabic name, *tinnīn,* taken either from B/P, or from a globe.

[21a)] It is not at all improbable that Gurgānī had given the descriptions of these two constellations, too, but the respective verse, containing in its two halves the descriptions of the two constellations, was omitted by some redactor, or scribe, of the poem. Cf. the apparatus of P which indicates that, in some of the Persian MSS., the verses 67, 72, 83, and 91–93 were omitted.

[22)] I add, in all cases, the Latin designations of the constellations being used in modern astronomy.

2–3: The Greater and the Lesser Bear, Ursa Maior and Ursa Minor (= Ptolemy, nos. 2 and 1), P 83.

4: "A woman bound with a chain," Andromeda (= Ptolemy, no. 20), P 84a. It is a greater leap in the sky from the two Bears over to Andromeda, and several other constellations fall between them. The "chain" of Andromeda is shown on some of the globes (see L), and was also mentioned in B, p. 72 = BP, p. 93; the formula *ba-zanǧīr basta* in P is identical with BP.

5: "A man kneeling," Hercules (= Ptolemy, no. 7), P 84b. There is again a greater leap from Andromeda to Hercules, and one wonders how Gurgānī places Hercules "in front of" Andromeda (*ba-pīšaš,* Morrison: "before her"). BP has *zānū zada,* but *zānū nišasta* is quoted by WRIGHT in B from some Persian MSS., and this is quite close to Gurgānī's formula.

6: "An Eagle," Aquila (= Ptolemy, no. 16), P 85a. The Eagle is not so far distant from Hercules, so that P's *barābar* means "just opposite, or by," and not "over against," i.e., in the opposite half of the sky.

7: "An arrow," Sagitta (= Ptolemy, no. 15), P 85b. Both on the globes and with B the Eagle is placed right over the Arrow.

8: "A Youth standing like a guard," Bootes (= Ptolemy, no. 5), P 86a. Persian *pāsbān* can be translated as "a guard" and as "a shepherd," both of which are appropriate for Bootes who in the classical tradition is regarded as a cowherd (hence, Arabic *al-baqqār*) or as Arctophylax, Guardian of the North (Arabic *ḥāris aš-šamāl*).[23]) It is difficult to explain how Gurgānī arrived at this definition of Bootes because in B neither the shepherd nor the guard are mentioned,[24]) and from the figures on the globes such qualifications could hardly be derived.

9: "A goblet and tray of gold," Corona Borealis (= Ptolemy, no. 6), P 86b. The double description of this constellation as a "goblet" (*ṭašt*) and "tray" (*ḫwānī*) probably depends on B/BP where it is stated that the constellation, besides its Ptolemaic name, had the popular designation "the cup, or plate of the poor," Arabic *qaṣʿat al-masākīn,* Persian *kāsa-i miskīnān.* Originally it was not situated in the hands of Bootes, but it might appear to be so on a globe, if these two neighbouring constellations were drawn too close to each other.

10–11: P 87a. This half-verse is the most difficult to analyse in the whole section. There are no variants of greater importance in the Persian text, and both translations, MASSÉ and MORRISON, are correct: "deux poissons tout pareils à deux outres gonflées," "two Fish for all the world like two bags full of wind." According to the classical tradition, there is no constellation of "two Fish" in the northern celestial hemisphere. The explanation must be sought in another direction: Near the "Duck" (see no. 12) which is described in the second half-verse

---

[23]) Cf. P. KUNITZSCH, *Der Almagest* (*op. cit.* in footnote 7), pp. 174–176.

[24]) At least here, in the survey of the Ptolemaic constellations. However in B, p. 78, in a list of star names of indigenous Arabic origin, *ḥāris aš-šamāl* ("the Guardian of the North") is mentioned as an occasional name for the star α Bootis (Persian, in BP, p 101: *nigahbān-i šamāl*).

(P 87b), there is in the sky (and on the globes) the constellation Delphinus, the Dolphin (= Ptolemy, no. 17); this was recognized by Gurgānī[25]) on the globe he used, as a "Fish." The comparison of this Fish's shape to an inflated water-skin was first made by al-Bīrūnī (B, p. 72: *šibh az-ziqq al-manfūẖ*; BP, p. 92: *hamčūn ẖīk*); the use of the same comparison, even using the same word *ẖīk* as in BP, makes it almost sure that Gurgānī had seen al-Bīrūnī's *Kitāb at-tafhīm,* in the Persian version.[26]) So much for one "Fish." But what of the second one included by Gurgānī under the name "two Fish?" There seem three of the 21 Ptolemaic northern constellations to be missing in Gurgānī's catalogue: Lyra, Equuleus, and Triangulum. Of these three, Triangulum is situated near Aries, far away from Delphinus. Lyra, sometimes drawn by the Islamic globemakers as a tortoise[27]) (following one of the Arabic translations of the *Almagest*), and sometimes as something resembling a vessel,[28]) is in the farther neighbourhood of Delphinus, separated from it by Aquila and Cygnus. The third one, Equuleus, the Foal (= Ptolemy, no. 18), consisting of the Foal's head only (Ptolemy: Ἵππου προτομή), is situated immediately by the side of Delphinus. On some globes its figure is drawn very carelessly, so that it looks like the inflated Dolphin (cf. especially globe A). It can be assumed, therefore, that Gurgānī had before his eyes a globe of this type and took Equuleus to be a second individual of the same kind as the neighbouring Dolphin. Consequently he combined the two similar figures under one name as "two Fish." If we identify the two constellations in this half-verse, Gurgānī's numbers 10 and 11, as Ptolemy's numbers 17 and 18 (Delphinus and Equuleus), the remaining two Ptolemaic constellations which were omitted from Gurgānī's catalogue will be Ptolemy's numbers 8 and 21, the constellations of Lyra and Triangulum. (The problem of these "two Fish" was not solved by Maḥǧūb in his commentary, p. 453.)

12: "A Duck . . .," Cygnus (= Ptolemy, no. 9), P 87b. Here again, Gurgānī obviously follows al-Bīrūnī who explains that the constellation of Cygnus (*ad-daǧāǧa*) resembles the figure of a duck (B, p. 71: *baṭṭa*; BP, p. 92, equally: *baṭṭī*). The globes – as far as can be judged from the surviving examples – carried only the

---

[25]) Perhaps aided by al-Bīrūnī who explains the foreign name Dolphin (*ad-dulfīn*) as "a marine animal" (B, p. 71 – Arabic: *ḥayawān baḥrī*; BP, p. 92: *ǧānvarī-yi daryā'ī*).

[26]) It must be added that the shape of the Dolphin in some pictorial reproductions of the stellar sky really looks like an inflated water-skin as stated by al-Bīrūnī (cf. the globes C, F, and especially L. See also Quṣayr ʿAmra, article Beer, 1932, pp. 300 and 301, figs. 355 and 356, and the isolated drawing in article Saxl, p. 291, fig. 341, at "Q"). But the chance that Gurgānī could have found, from the appearance of the figure on a globe, the same term *ẖīk* that had been introduced by al-Bīrūnī some twenty years earlier cannot be regarded as very great. Much more plausible is it to assume that Gurgānī here followed al-Bīrūnī.

[27]) Cf. globe A.

[28]) Cf. globes C, E, F. Globe L, instead, has another indeterminable figure. In the cupola of Quṣayr ʿAmra, Lyra is partially destroyed (cf. figs. 355 and 356 in Beer's article of 1932).

official (Arabic) names of the constellations which, in this case, was *ad-daǧāǧa* ("the Hen," translated from Ptolemy's Ὄρνις, "a bird.") Gurgānī, however, calls the figure *yakī baṭṭ,* "a Duck," using the Arabic word *baṭṭ* which he most probably picked up from al-Bīrūnī's *Kitāb at-tafhīm.* Numbers 10–12 are all situated in the near neighbourhood, whereas 13, 14/15, and 16 are separated from them and from each other by greater distances.

13: "A man without horse but still holding reins," Auriga (= Ptolemy, no. 12), P 88a. Also in this case, Gurgānī shows signs of having used al-Bīrūnī's *Kitāb at-tafhīm,* retaining, for the "reins" in the hand of the figure, the Arabic word *ʿinān* which occurs in al-Bīrūnī's description of the figure (B, p. 71; BP, p. 92). Massé translates this half-verse: "l'un sans cesse tenant . . ." (p. 74), which should read: "un sans cesse . . .".

14–15: "A snake charmer" with "a snake," Ophiuchus and Serpens (= Ptolemy, nos. 13 and 14), P 88b. In Massé's translation, instead of "l'autre comme un charmeur . . ." (p. 74), it should read: "un autre . . .".

16: "A *man* [!] sitting on a silver throne," Cassiopeia (= Ptolemy, no. 10), P 89a. The constellation described is a female figure, the mythical ancient queen Cassiopeia. The Persian,[29]) therefore, has to be translated in the feminine form (or in a neutral way), and not – as done by both Massé and Morrison – in the masculine form: "A figure sitting . . .", or "A female sitting . . ." (for Morrison, p. 57); "une autre, étant assise . . ." (for Massé, p. 74). For the "throne," Gurgānī has retained the Arabic designation *kursīy,* either from the official Arabic name on the globe, or from al-Bīrūnī's explanation of the figure.

17: "A horse . . .," Pegasus (= Ptolemy, no. 19), P 89b. Pegasus is not immediately near Cassiopeia, but not too far distant, so that Gurgānī's *pīš-i ū,* "before *her*" [sic, not *him* – it refers to Cassiopeia!] might be appropriate here. The horse's description: *az band rasta,* "escaped, liberated from the bridle, or reins" (Massé, p. 74: "échappé de ses liens"), is less well translated by Morrison, p. 57: "broken from its halter."[30]) Perhaps this description goes back to the horse's pictorial figure on the globes: with far outstretched front legs, as if gallopping on, escaped from the reins. (Pegasus, in the sky, is another of the animals of which the front half only is shown.)[31])

---

29) *yakī . . . nišasta,* "one (male or female) . . . seated."

30) Obviously following the variant reading *ḫasta* (in the apparatus of P, in one MS.). Another variant, there, is *ǧasta* (in two MSS.) which comes near to the sense of *rasta,* above.

31) Maḥǧūb, in his commentary, p. 454, wrongly understands this "horse" as a combined representation of the two Ptolemaic constellations Pegasus and Equuleus. Wright has misunderstood al-Bīrūnī's explanation: "the fore part of a horse with two wings; *there is no man with it,* for . . ." (*wa-huwa ka-niṣf faras muǧannaḥ lā riǧl^a* [*sic legendum,* not *lā raǧul^a*] *lahū*), which, of course, means: ". . . with two wings, and without *hind legs*; it is cut . . ." (B, p. 72). Thus, BP, p. 93, has correctly: *bā do par va-pāy na-dārad.*

18: "A man with a demon's head in his palm," Perseus (= Ptolemy, no. 11), P 90a.[32])

19: "A crown-bearer . . .," Cepheus (= Ptolemy, no. 4), P 90b. Between Perseus and Cepheus, there is still the figure of Cassiopeia. Gurgānī, therefore, is not quite right placing Cepheus "before him" [= Perseus], *ba-pīšaš*. With Ptolemy, it is a special feature of Cepheus to wear, on his head, a τιάρα, an oriental king's head-dress. This was translated by the Arabic translators of the *Almagest* as *qalansuwa* (whence, Latin *pileus*),[33]) which is rather a cap (of a high conical shape) than a crown.[34]) This cap was called in Persian *kulāh*[35]) (for which *kulah* also occurs, as here in *Vīs u Rāmīn*). Gurgānī most probably has taken this word from BP. Another problem is mentioned by Maḥǧūb in his commentary, p. 454, that is that Gurgānī describes the figure as *īstāda,* "standing" (as if standing upright on both his feet), while in the classical tradition one of his two legs is bent as if he were kneeling.[36]) Maybe pure reasons of metre urged Gurgānī to depart, in this detail, both from the literary and the pictographic traditions.

C. The constellations of the southern celestial hemisphere

Ptolemy has catalogued, in the southern celestial hemisphere, fifteen constellations. These are described by Gurgānī completely, without omission. The sequence of the constellations in *Vīs u Rāmīn* is almost regular, close to that on the globe, starting from Eridanus and ending at Cetus, proceeding clockwise (= eastward, on the globe), with the sole exception of Centaurus and Lupus (verses 99–100), which are deplaced and described between Piscis Austrinus and Cetus (verses 98 and 101) instead of between Crater and Ara (verses 96 and 97) – a defect which perhaps was due to an interchange of verses by some redactor rather than to Gurgānī's interpretation of the celestial globe.

1: "A dark stream," Eridanus (= Ptolemy, no. 3), P 91.

2: "A Hare," Lepus (= Ptolemy, no. 4), P 92a.

3–4: "Two hounds," Canis Maior and Canis Minor (= Ptolemy, nos. 5 and 6), P 92b.

---

[32]) Massé's translation (p. 74), "un autre, ayant placé sur sa tête un démon," is wrong. The reason for the error is not quite clear. The correct translation would be: "un autre, ayant placé dans sa main la tête d'un démon."

[33]) See Kunitzsch, *Der Almagest* (*op. cit.* in footnote 7), p. 226, no. 51.

[34]) See, e.g., globes C, E, F, L; Saxl's article on Quṣayr ʿAmra, p. 295, fig. 348; E. Wellesz, *An Islamic Book of Constellations,* Oxford, 1965, plate 2 (drawings of the constellations from one of the oldest existing MSS. of aṣ-Ṣūfī's book on the constellations of the fixed stars).

[35]) BP, p. 91. Uluġ Bēg, in his star catalogue of A.D. 1437, also has *kulāh* (see Kunitzsch, as in footnote 33).

[36]) Thus explained by B, p. 71 = BP, p. 91. Cf. the globes and figures mentioned above in footnote 34. A similar case occurs with Orion, see below.

5: "A man with a belt,"[37]) Orion (= Ptolemy, no. 2), P 93. For the interpretation of Orion's figure as "standing like a *king*," Gurgānī must have relied on sources other than al-Bīrūnī,[38]) perhaps Arabic traditions that might have spread in the Orient and might have formed part of the poet's general education.[39]) In the pictographic tradition, Orion is not standing, but moreover running, or jumping, cf. the globes, Quṣayr 'Amra, and aṣ-Ṣūfī.[40])

6: "A Ship," Argo Navis[41]) (= Ptolemy, no. 7), P 94. "Its anchor made of rubies," refers to the bright first-magnitude star α Carinae, the fortyfourth star of the Ship in Ptolemy's star catalogue, known in the Islamic Orient by its classical Arabic name *suhayl.*[42]) It was described in Arabic traditions as *aḥmar,* "red,"[43]) though, astrophysically, it belongs to the spectral class F and, therefore, appears to the observing eye as yellowish-white. Locating this star on the ship's "anchor" (Pers. *langar*), is Gurgānī's free interpretation of what he saw on the globe. Originally the part of the ship on which this star was placed was the 'helm', or 'rudder' (Ptolemy: πηδάλιον; in the Arabic translations, *al-miǧdāf,* and *as-sukkān*).[44])

7: "A Snake thin . . .," Hydra (= Ptolemy, no. 8), P 95a. The description of this snake, *bārīk mārī,* seems to follow BP, p. 94 (*mār-i bārīk-i dirāz*).

8: "A Crow," Corvus (= Ptolemy, no. 10), P 95b. Of course, the globe shows no "meadow"; this is a poetical addition of Gurgānī.

9: "A Cup," Crater (= Ptolemy, no. 9), P 96. The rest of the Cup's description, again, is made up of Gurgānī's poetical invention.

10: "A Censer," Ara (=Ptolemy, no. 13), P 97a. Here, Gurgānī uses the constellation's Arabic name: *miǧmar* (in Arabic, properly, *al-miǧmara,* for Ptolemy's Θυμιατήριον); BP, p. 94, had *'ūd sūzī,* instead. That the Censer is made of silver, is again Gurgānī's free invention.

11: "A Crown," Corona Australis (= Ptolemy, no. 14), P 97b. The figure, on the globes, can hardly be recognized as a "royal crown" (Pers. *šāhāna afsar*);

---

[37]) Persian *kamardārī,* lit. "having a belt." According to the dictionaries, the word means "a servant, serving-man," and so it was translated by MASSÉ, p. 75: "un serviteur." Orion being famous for his "belt" both in classical ancient and in indigenous Arabic stellar lores, it is more appropriate, however, to understand *kamardārī,* here, literally as "a man with a belt" (MORRISON, p. 57).

[38]) B, p. 72, and BP, p. 93, do not mention the "king," *šāh.*

[39]) Cf. P. KUNITZSCH, *Untersuchungen zur Sternnomenklatur der Araber,* Wiesbaden, 1961, p. 23f., quoting Ibn Qutayba, *Kitāb al-anwā'*, Hyderabad, 1956, p. 45.

[40]) E. WELLESZ, (*op. cit.* in footnote 34), plate 19.

[41]) In modern astronomy it is cut into four separate constellations: Carina, Puppis, Pyxis, and Vela.

[42]) Cf. KUNITZSCH, *Untersuchungen* (*op. cit.* in footnote 39), p. 106, no. 272a.

[43]) See the preceding footnote, and KUNITZSCH, *Der Almagest* (*op. cit.* in footnote 7), p. 321f.

[44]) See KUNITZSCH, *Der Almagest* (*op. cit.* in footnote 7), p. 332, no. 572.

Gurgānī, here, rather seems to follow al-Bīrūnī's text. The "jewels" are Gurgānī's poetical addition.

12: "A fish," Piscis Austrinus (= Ptolemy, no. 15), P 98. Its further description, again, is to be attributed to Gurgānī's poetical invention,[45]) the "scales" probably being taken from the drawing on the globe (cf. globes C and F).

13: "A horse like a human," Centaurus (= Ptolemy, no. 11), P 99–100a. The "confusion," "like a drunken man," has been added by Gurgānī who, now, seems to resume his travesties of the constellations first introduced with the zodiacal signs. The "beautiful flowers blooming on its body" are perhaps inspired by certain decorations with which the body of the animal was covered on Gurgānī's globe (cf. the globe C made, according to von Klüber, p. 12, "in unverkennbar persischem Stil"; see also F).

14: "a lion" (Centaurus . . . had taken a *lion's* paw in its hands), Lupus (= Ptolemy, no. 12), P 100b. This constellation was just called τὸ Θηρίον ("the Animal," perhaps a beast of prey) by Ptolemy, and *as-sabu*ʿ ("the beast of prey") by the Arabic translators of the *Almagest* (hence, *lupus* in Gerard of Cremona's Latin translation of the Arabic *Almagest*). BP, p. 94, translates the Arabic word into Persian as *šīr,* "lion." Thus, here, Gurgānī seems to follow al-Bīrūnī's text, because the figure of the animal as drawn on the globes cannot be understood as a "lion," specifically.

15: "A sign like a bird . . .," Cetus (= Ptolemy, no. 1), P 101. Gurgānī may have derived his description of the legendary marine animal from the figure on a globe[46]) as well as from al-Bīrūnī's text who explicitly states that it has "a tail like a bird's" (B, p. 72: *ḏū . . . wa-ḏanab ka-ṭ-ṭayr*; BP, p. 93: *va-dunbālaš čūn ān-i murġ*).[47]) The interpretation of the "bird" as a "peacock" (*ṭāvūsī*) was Gurgānī's poetical invention.

D. *The horoscope*

Having described, in the preceding lines, the sinister aspects of the fatal night in a general way, Gurgānī now proceeds to giving the horoscope proper, i.e., the positions of the planets which are supposed to indicate King Mawbad's fate with Vīs, whom he is going to abduct that night.

P 102a: From the east had risen (*bar kašīda*) a sinister ascendant (*ṭāliʿ-i bad*). The word *ṭāliʿ* may either designate, more generally, a 'horoscope', or, as a special term

---

[45]) Massé, p. 75: "semblable à un poisson d'argent," as if the Persian text had *ba-sān-ī māhi-yī sīm.* All the MSS. and editions, however, read *māhi-yī šīm,* "a spotted fish" (cf. the dictionaries: *šīm,* "a small spotted scaly fish").

[46]) See the globes, and Quṣayr ʿAmra (especially Saxl's article of 1932, p. 291, fig. 340). Cf. also a recent paper by A. Bausani, "Κῆτος fra le stazione lunari?", in *Oriente Moderno* 58 (1978), 275–283, with plates I–VI (see especially plate V!).

[47]) Morrison, p. 57: "without wings," should better read: "without feathers and wings," corresponding to the Persian, verse 101a: *bī par-ū bāl.*

of astrology, the 'ascendant', that is the 'house'[48]) that is just rising above the eastern horizon at a given moment. The poem does, however, not indicate which portion of the zodiac, or which planet, was actually located in the ascendant.

P 103a: The sun was in conjunction with the moon. This configuration, under certain conditions, can produce a solar eclipse (see also below). But at night, of course, both the luminaries are under the horizon, invisible, as is explicitly stated by Gurgānī (P 56, 58, 62).

P 104a: They were accompanied by "Old Saturn".[49]) Saturn, astrologically, is the worst of the planets (*an-naḥs al-akbar*). Its being in conjunction with the sun and the moon confers upon them, and the horoscope generally, the most inauspicious aspects. According to the critical apparatus in P, only the Paris MS. has this reading, whilst the other MSS., and this edition, have "Mercury and Saturn," instead.[50]) This reading is also possible, metrically, graphically, and astrologically: Mercury, as an astrological potency, is neutral, it adopts and intensifies the qualities of other planets if in conjunction with them. Thus, here, Mercury would intensify Saturn's bad influence.

P 104b: The conjunction of these three, or four, planets (sun, moon, and Saturn, and perhaps Mercury also) took place in the fourth house. To this house are assigned predictions, or interrogations, regarding parents, descendants, real estate, etc.[51]) 'House' in this half-verse is called *čarḫ*; according to the apparatus, three MSS. have *burǧ* instead, which is less good because this (Arabic) word generally designates the twelve signs, or constellations, of the zodiac, which are not to be confused with the twelve astrological 'houses' (Arabic: *bayt,* plural *buyūt*).

P 105: Just opposite the ascendant,[52]) in the seventh house,[53]) there are the Descendant Node (of the moon)[54]) and "aggressive Mars." Mars is the lesser

---

[48]) The twelve 'houses' are not identical with the twelve zodiacal signs, or constellations. There were, since antiquity, several methods of calculating the houses, which, in most of these systems, do not occupy straight twelfths of the zodiac but portions of different size. (Cf. Ptolemy, *Tetrabiblos,* book iii; B, pp. 275–279.)

[49]) Text: *pīr-i kayvān,* and so translated by MASSÉ (p. 75: "le vieux Saturne"), and by MORRISON, p. 58.

[50]) Pers. *tīr u kayvān.*

[51]) Cf. B, p. 275.

[52]) MASSÉ, p. 75, wrongly: "et, face à la station septième du même astre, . . .".

[53]) 'House' here is called by its appropriate Persian designation, *ḫāna* (corresponding to Arabic *bayt,* and distinct from *burǧ*).

[54]) The Persian text has: *ḏanab,* an Arabic word designating, in the horoscope, the "tail" (of the Dragon), i.e. the descending node of the moon's orbit. MORRISON (p. 58) wrongly took the word to be the name of the fixed star α Cygni and thus translated it as "Deneb," the Arabic star name in modern astronomical use (cf. P. KUNITZSCH, *Arabische Sternnamen in Europa,* Wiesbaden, 1959, p. 155, no. 81). A fixed star, however, usually has no function in a horoscope where only the planets (and the two nodes of the moon's orbit) are of influence. MASSÉ, p. 75, seems to have omitted *ḏanab* from his translation.

one of the two maleficent planets (*an-naḥs al-aṣġar*), the stronger one being Saturn (see above). The two nodes of the moon's orbit, the ascending and the descending node, are often counted, in Islamic astrology (following Indian models), among the planets, and are assigned similar virtues in the calculation of a horoscope.[55]) Thus, the Descending Node is of a maleficent influence.[56]) So, here, we have another conjunction of two 'maleficent' planets, viz., the Descending Node and Mars, this time in the seventh house which is dedicated to women, concubines, giving in marriage, and similar matters,[57]) which are of special interest for King Mawbad and Vīs in this particular moment.

P 106a: Venus was placed between the two (*miyān-i har duvān*), that is, obviously, between the two last named planets, the Descending Node and Mars, in the seventh house.[58]) Thus, the planet of love was enclosed and threatened by two extremely maleficent elements.

P 107: This verse was translated by Morrison, p. 58: "None of the influences militating for justice, whose conjunction might have been propitious for that enterprise, were favorable." A more precise rendering of the Persian text would be: "None of the fortune-telling individuals [= the planets] was friendly, [friendly, that is, in such a way] that his [= King Mawbad's] familiar relationship [viz., to Vīs, by marrying her], through that deed [= Vīs's abduction], might have turned out happily."

Summing up we find that the horoscope indicates the positions of six (or seven) out of nine 'planets' used in mediaeval Islamic astrology: sun, moon, Saturn, (Mercury) – in conjunction, in the fourth house; the Descending Node, Mars, Venus – also in conjunction, in the seventh house. (Not mentioned are Jupiter, the greater of the two beneficent planets, and the Ascending Node. The case of Mercury is uncertain and depends on what reading we adopt in P 104a.)

Further, we see that the positions of these planets are only given by 'houses', and not by zodiacal signs. This is a conclusive observation which means that the horoscope contains no astronomical elements which would allow a

---

[55]) See W. Hartner, "The Pseudoplanetary Nodes of the Moon's Orbit . . .," in *Ars Islamica* 5 (1938), 113 ff., esp. pp. 131–134; idem, *Le Problème de la Planète Kaïd,* Les Conférences du Palais de la Découverte, Série D, N° 36, Paris, 1955; idem, "The Mercury Horoscope of Marcantonio Michiel of Venice," in *Vistas in Astronomy* 1 (1955), pp. 84–138, esp. pp. 86, 87, 96 f. (here, in the horoscope the Ascending Node is employed). All these three articles are also reprinted in W. Hartner, *Oriens – Occidens,* Hildesheim, 1968, pp. 349 ff. 268 ff., and 440 ff., respectively. – Astronomically, the two Nodes are important for the occurrence of eclipses; according to the moon's and the sun's location in the same node or different nodes, there occur solar or lunar eclipses.

[56]) Cf. L. P. Elwell-Sutton, *The Horoscope of Asadullāh Mīrzā,* Leiden, 1977, p. 61 (s.v. *dhanab*).

[57]) See B, as in footnote 51.

[58]) It is not at all impossible that this phrase means: between the two said houses, the fourth and the seventh, or the two groups of planets located there (but, then, without definite indication of Venus' own house).

dating of this section, or of the whole poem. The 'houses' are a stationary division of the zodiac, depending on the geographical latitude of the place involved.[59]) Dates could only be obtained from the planets' positions in the zodiacal signs (*burūǧ*), that is their ecliptical longitudes.[60])

Thus, we may conclude that the author – Gurgānī, or his anonymous historical predecessors – inserted this horoscope, which in itself is correct as to astrological principles but which is kept void of astronomical elements pointing to fixed dates, as a literary element only,[61]) obviously with the intention of not giving the reader a hint towards a historical identification of the story told in the poem.

References

A. Text editions and translations of Gurgānī's Vīs u Rāmīn

Persian text: ed. W. Nassau Lees, Calcutta, 1864–65 (from the Calcutta MS.); ed. M. Minovi, Teheran, 1314/1935; ed. M. Ǧ. Maḥǧūb, Teheran, 1337/1959 (these two mainly based on the Paris MS.); ed. M. A. Todua, A. A. Gwakharia, and K. S. Aini, Teheran, 1970 (based on five complete and five fragmentary MSS., and using also the preceding printed editions and the Georgian translation) [quoted as P in this article]. French translation: H. Massé, *Gorgâni. Le Roman de Wîs et Râmîn,* Paris, 1959 (from Minovi's printed edition, using also the Calcutta edition. Inferior to the English translation, in this section); English translation: G. Morrison, *Vis and Ramin, Translated from the Persian . . .*, New York and London, 1972 (using all the printed editions, and some MSS. directly, and the Georgian and Tadzhikistan versions). Georgian version of *ca.* A.D. 1200: most recent editions by A. Gwakharia and M. Todua, Tiflis, 1962, and by G. Tsereteli, Tiflis, 1964; English translation: O. Wardrop, *Visramiani. The Story of the Loves of Vis and Ramin . . .*, London, 1914 (repr. 1966). German translation: R. Neukomm and K. Tschenkéli, *Wisramiani oder die Geschichte der Liebe von Wis und Ramin,* Zurich, 1957 (much abbreviated and, therefore, not

59) See, e.g., L. P. Elwell-Sutton, *The Horoscope* (*op. cit.* in footnote 56), pp. 66–78, s.v. *khāna*.

60) For examples of 'true' horoscopes, giving the planets' astronomical positions in the signs, see Elwell-Sutton, *op. cit.*, pp. 1–4, from Niẓāmī's *Ḫusraw va-Šīrīn,* and al-Bayhaqī's biographical work *Tatimmat ṣiwān al-ḥikma.* Two horoscopes from Sasanian times are given in the story of the foundation of the Sasanian dynasty (*Kārnāmag ī Artaḫšēr*), cf. Th. Nöldeke, "Geschichte des Artachšîr i Pâpakân, aus dem Pehlewi übersetzt . . .," in *Beiträge zur Kunde der indogermanischen Sprachen* 4 (Göttingen, 1878), 22–69, see pp. 41 f. and 44; the translation (and/or the text) are, however, rather vague and inconsistent and need a more detailed investigation (cf., on this work, B. Reinert, "Die Anfänge des spanisch-arabischen Epos," in *Saeculum* 29 [1978], 231–268, esp. p. 242 f.).

61) The 'Description of the Night' is shortly mentioned by H. Ringgren, *Fatalism in Persian Epics,* Uppsala and Wiesbaden, 1952, p. 69. More fatalistic and astrological lore in *Vīs u Rāmīn* is discussed here, pp. 63–65, and 69–71.

quoted here; for the 'Description of the Night' see p. 24). Tadzhikistan edition: ed. by A. DEWANAKULOV, Dushanbeh, 1966 (used by MORRISON, see above).

B. Islamic celestial globes etc.

(in assigning the abbreviation letters, I follow the system introduced by H. VON KLÜBER, see below)

A Ibrāhīm ibn Saʿīd, Valencia, Ṣafar 478 h = A.D. 1085. See F. MEUCCI, *Il globo celeste Arabico del secolo XI . . .*, Florence, 1878, chart at the end of the book. (Cf. MAYER, p. 51 f., no. iv.)

C Muḥammad ibn Hilāl al-Mawṣilī, 674 h = A.D. 1275/76. See B. DORN, "Description of the Celestial Globe belonging to Major-General Sir John Malcolm . . .," in *Transactions of the Royal Asiatic Society . . .*, ii (1830), pp. 371–392, with plates A and B. (Cf. MAYER, p. 68.)

E Muḥammed ibn Maḥmūd ibn ʿAlī aṭ-Ṭabarī, 684 h = A.D. 1285/86. See P. CASANOVA, "Une sphère céleste de l'an 684 de l'Hégire," in *Mémoires publiés par les Membres de la Mission Archéologique Française du* [fasc. 2: *au*] *Caire*, t. 6 (fasc. 2), Paris, 1897 [fasc. 2: 1892], pp. 313–330, with one plate. (Cf. Mayer, p. 72.)

F Muḥammad ibn Mu'ayyid al-ʿUrḍī, Maragha, *ca.* A.D. 1279 (or 1289). See A. DRECHSLER, *Der arabische Himmels-Globus . . . zu Dresden*, Dresden, 1873 (2nd ed., Dresden, 1922), plates i–iv at the end of the book. (Cf. MAYER, p. 72 f.)

L Ḍiyā' ad-Dīn Muḥammad ibn ʿĪsā Allāh-dād, Lahore, 1070 h = A.D. 1660/61. See H. VON KLÜBER, "Über einen arabischen Himmelsglobus aus Indien," in *Baessler-Archiv. Beiträge zur Völkerkunde* 18 (1935), 1–21, and figures 5–10. (Not registered by Mayer.)

For a comprehensive bibliography on Islamic (astrolabes and) celestial globes, see L. A. MAYER, *Islamic Astrolabists and their Works*, Genève, 1956.

The celestial cupola of Quṣayr ʿAmra (*ca.* 60 kilometres east of Amman, Jordan; datable perhaps *ca.* A.D. 711-715, under the Omayyad caliph al-Walīd I): see F. SAXL, "The zodiac of Quṣayr ʿAmra," in K. A. C. CRESWELL, *Early Muslim Architecture*, vol. i, Oxford, 1932, pp. 289–295; A. BEER, "The Astronomical Significance of the Zodiac of Quṣayr ʿAmra," *ib.*, pp. 296-303; A. BEER, "Astronomical Dating of Works of Art," in *Vistas in Astronomy* 9 (1967), 177–187 (with an addition by W. HARTNER, *ib.*, p. 225); M. ALMAGRO etc., *Qusayr ʿAmra. Residencia y baños omeyas en el desierto de Jordania*, Madrid, 1975.

C. Astronomical sources

B al-Bīrūnī (*d.* A.D. 1048), *Kitāb at-tafhīm li-awā'il ṣināʿat at-tanǧīm* (The Book of Instruction in the Elements of the Art of Astrology), facsimile of the Arabic text and English trans. by R. R. WRIGHT, London, 1934.

BP Persian version of the same: *Kitāb at-tafhīm li-awā'il ṣināʿat at-tanǧīm*, ed. ǦALĀL AD-DĪN HUMĀ'Ī, 2nd ed., Teheran, Bahman 1353/Jan.–Febr. 1976 [*Silsila-i Intišārāt-i Anǧuman-i Āṯār-i Millī*, Vol. 109]. Containing corrections and additions to the first ed., by the same editor, of 1318/1939–40.

XXIX

ANNOTATED INDEX
of Nautical Star Names

mentioned and discussed in some writings

of Paul Kunitzsch

<u>Abbreviations</u> (chronologically):

AS = Arabische Sternnamen in Europa. Wiesbaden 1959 (quoted by pages).

Unt = Untersuchungen zur Sternnomenklatur der Araber. Wiesbaden 1961 (quoted by current numbers).

N = "Zur Stellung der Nautikertexte innerhalb der Sternnomenklatur der Araber", in: Der Islam 43 (1967), p.53-74 (quoted by pages).

S^{i} = "Die arabischen Sternbilder des Südhimmels" (part 1), in: Der Islam 51 (1974), p.37-54 (quoted by pages).

R = Review of G.R.Tibbetts: Arab Navigation in the Indian Ocean... (Transl. of Aḥmad b. Mā\u01e7id, Kitāb al-fawā\u02beid), in: Der Islam 51 (1974), p.344-353 (quoted by pages).

S^{ii} = "Die arabischen Sternbilder des Südhimmels" (part II), in: Der Islam 52 (1975), p.263-277 (quoted by pages).

- -

al-aḥmarān الأحمران N 65.

al-aḥyam الأحيم N 58 -- S^{i} 49.

al-uḥaymir الأحيمر Unt 310 (here, in line 2 instead of "fol.16r ff." read "fol.16v, 8-9"; omit the quotation of fol.82v, 4) -- S^{i} 49.

āḫir an-nahr آخر النهر , α Piscis Austrini. N 61. 67.

al-aḍlāᶜ الأضلاع , the stars of Perseus. N 69f.

al-aᶜraǧān الأعرجان Unt 13.

al-aᶜlām الأعلام N 55f. -- R 350.

al-iklīl al-ǧanūbī الإكليل الجنوبي N 60 (A.39).

al-iklīl aš-šamālī الإكليل الشمالي N 60 (A.39).

al-awwalān الأولان Unt 22.

al-bār البار Unt 60 -- N 66.

al-barkān البركان N 59.

al-baᶜīr البعير N 68 (A.92).

taḥta l-farš تحت الفَرْش N 64 -- S^{i} 50 -- R 350 (in the Almagest = γ Velorum, but here for ζ Puppis).

at-tinnīn التنين N 55. 62.

tawābiᶜ al-ᶜayyūq توابع العيوق N 55f.

at-tīr التير Unt 302 -- N 62. 65f.

aṯ-ṯānī الثاني N 59.

(Index of nautical stars -- p.2)

(Index of nautical stars - p.3)

(Index of nautical stars -- p.4)

al-qaws القوس N 61. 63 -- R 351 -- S^{ii} 274f. (here, not the zodiacal sign of Sagittarius, but a recently invented constellation of the navigators, in Phoenix, Eridanus and/or Aquarius, being completed by the "arrow" sahm al-qaws and its preceding star sahm al-awwal, qq.v.).

al-kāsir الكاسر Unt 146.

al-kafīt الكفيت Unt 138.

Luqmān لقمان N 59.

al-maḥnat̲, al-muḥannit̲ المحنث N 55. 70.

marāzim al-ǧawzāʾ مرازم الجوزاء N 70.

al-murabbaᶜ (and its constituents) المربع N 58 -- S^{i} 49.

al-misḥalān المسحلان AS 216 (n.) -- Unt 168.

al-mᶜql (al-maᶜqil ?) المعقل Unt 178 -- S^{i} 44. (the first one of the two al-ḫimārān, and sowith β Centauri).

al-muqaddamān المقدّمان Unt 176.

miqd̲āf as-safīna مقذاف السفينة N 55.

munīr an-nāqa منير الناقة N 56.

al-mīḫ الميخ N 65 -- S^{ii} 274.

al-mīzān الميزان , $\delta \epsilon \zeta$ Orionis. Unt 171.

an-nāḥis (al-bāǧis ?) الباجس / الناحس Unt 188 -- N 57 (A.22).

naǧm ad-daǧāǧa نجم الدجاجة N 64 (cf. ridf al-wāqiᶜ al-kabīr).

nuǧūm al-aḫd̲ نجوم الأخذ N 66 (A.76).

an-nasr aš-šāmī النسر الشامى N 60.

an-nasr aṣ-ṣaǧīr الصغير Unt 194b.

an-nasr aṭ-ṭalīq الطليق N 59 -- R 351f.

an-nasr al-kabīr الكبير Unt 195b.

an-nasr al-yamānī اليمانى N 60.

an-naṭḥ النطح , α Arietis. N 66 (A.78) -- R 351.

an-nahr النهر (not Eridanus, but part of Aquarius !) N 61. 67 -- R 352.

hādī an-naǧm هادى النجم N 71 (A.116).

al-hulba الهلبة R 348.

al-hīrāb الهيراب N 65.

al-hīrān الهيران N 65. 72.

al-wāqiᶜ الواقع N 64 -- S^{i} 50.

INDEX

(Arabic names are transliterated here according to the system current in English, even for the articles in German; the article *al-* is ignored; *ibn*, in genealogical applications, is abbreviated as "b.")

I—General Index

II—List of star tables discussed or edited